Nonlinear Systems and Complexity

Volume 32

Nonlinear Systems and Complexity provides a place to systematically summarize recent developments, applications, and overall advance in all aspects of nonlinearity, chaos, and complexity as part of the established research literature, beyond the novel and recent findings published in primary journals. The aims of the book series are to publish theories and techniques in nonlinear systems and complexity; stimulate more research interest on nonlinearity, synchronization, and complexity in nonlinear science; and fast-scatter the new knowledge to scientists, engineers, and students in the corresponding fields. Books in this series will focus on the recent developments, findings and progress on theories, principles, methodology, computational techniques in nonlinear systems and mathematics with engineering applications. The Series establishes highly relevant monographs on wide ranging topics covering fundamental advances and new applications in the field. Topical areas include, but are not limited to: Nonlinear dynamics Complexity, nonlinearity, and chaos Computational methods for nonlinear systems Stability, bifurcation, chaos and fractals in engineering Nonlinear chemical and biological phenomena Fractional dynamics and applications Discontinuity, synchronization and control.

More information about this series at http://www.springer.com/series/11433

Panayotis G. Kevrekidis · Jesús Cuevas-Maraver ·
Avadh Saxena
Editors

Emerging Frontiers in Nonlinear Science

Springer

Editors
Panayotis G. Kevrekidis
Department of Mathematics and Statistics
University of Massachusetts
Amherst, MA, USA

Avadh Saxena
Theoretical Division
Los Alamos National Laboratory
Los Alamos, NM, USA

Jesús Cuevas-Maraver
Grupo de Fisica No Lineal, Departamento de Fisica Aplicada I, Escuela Politécnica Superior
Universidad de Sevilla
Seville, Spain

Instituto de Matemáticas de la Universidad de Sevilla (IMUS)
Seville, Spain

ISSN 2195-9994 ISSN 2196-0003 (electronic)
Nonlinear Systems and Complexity
ISBN 978-3-030-44991-9 ISBN 978-3-030-44992-6 (eBook)
https://doi.org/10.1007/978-3-030-44992-6

This Springer imprint is published by the registered company Springer Nature Switzerland AG
The registered company address is: Gewerbestrasse 11, 6330 Cham, Switzerland

The authors dedicate this lovingly to the memory of George I. Kevrekidis (1929–2018) whose endless prompting for and immeasurable delight about such books kicked off this project a few days after his passing.

Preface

The concept of Nonlinearity is a universal feature in Mathematics, Physics, Chemistry, and Biology as a characterization of systems whose behavior does not amount to a superposition of simple building blocks, but rather features quite complex and often chaotic patterns and phenomena. The systematic study of nonlinear systems, arguably, has only been initiated in the last slightly over half a century. Initially, the watershed moment was the study of lack of equipartition of coupled nonlinear oscillators in the famous work of Fermi, Pasta, Ulam, and Tsingou in one of the first (if not the first) reported examples of a computational experiment in Los Alamos in 1955. Then, it was the seminal study of solitons and solitary waves in the mid-1960s by Zabusky and Kruskal, and also the famous chaotic three-dimensional dynamical system obtained by Ed Lorenz (that now bears his name, together with the associated attractor) in 1968. Subsequently, there was the explosive growth that ensued in systems both with few and many degrees of freedom, in theory (e.g., inverse scattering) and in applications (from mechanical and electrical systems to chaotic signal transmission, and from superconducting Josephson junctions, to spiral waves, e.g., in the heart, and chemical reactions such as the Belousov-Zhabotinsky one), to mention only a select few out of countless examples in Physics, Chemistry, and Biology.

As we enter 2020, we are at a crucial juncture in this still young, but now far more mature than in its first half-century, field of Nonlinear Science. As such, it is a particularly apropos time to reflect on some of the major successes of the field, on the key lessons we have learned and the techniques that have been developed in this time frame. It is also an excellent opportunity to consider the major challenges lying ahead and the perspectives opening up through the emergence of major new tools including large scale computational resources, the explosion of data, and the techniques of quantum computing, machine learning, and artificial intelligence, among many others. It is thus within the scope of this book to summarize some of the principal impact areas and themes of the past and also to ponder on the enticing vistas opening up towards the future. Given this very timing, we wanted to term this volume: "Nonlinear Science: a 20/20 Vision", but in their marketing wisdom, the Springer Nature Editors thought that the associated intended pun might render the book obsolete after 2020. That is how we converged on terming this journey *Emerging Frontiers in Nonlinear Science*.

The resulting volume discusses a broad range of themes including time-honored ones such as biology, geometry, topology, but also modern ones encompassing networks, metamaterials, and artificial intelligence, to name a few. Each chapter recaps key elements of recent progress, but importantly also charts the territory of challenges lying ahead. The volume consists of an interdisciplinary collection of the views of a diverse group of experts and active scholars across different fields, in an unprecedented effort to provide an overarching picture of the *field*: past, present, and future. This book is expected to be of wide interest to both beginners and seasoned researchers in the field of Nonlinear Science, in numerous areas of Physics (Optics, Quantum Physics, Waves, Materials Physics, Biophysics), as well as Applied Mathematics (ODEs, PDEs, Dynamical Systems, Machine Learning and beyond), and Engineering.

Here is a brief summary of the different Chapters and what they, respectively, offer:

Mark Ablowitz provides a systematic perspective on the history and recent developments of one of the most celebrated directions in the study of nonlinear waves, namely the notion of integrability and integrable systems. After going through the basics of the methodology of the inverse scattering transform in some of the prototypical relevant PDEs, he goes on to present some of the recent extensions thereof in the corresponding nonlocal problems such as the Korteweg-de Vries and the nonlinear Schrödinger equation.

Kirill Kalinin and Natalia Berloff present one of the major challenges and opportunities in the near future, namely the use of nonlinear systems as tools for unconventional computing. They identify numerous physical systems including nonlinear oscillator networks, lasers, and polariton condensates that have been proposed and realized with the aim of addressing hard computational problems stemming from a diverse range of areas. Aside from reviewing the relevant physical systems, they discuss the physically inspired algorithms used to solve a variety of problems.

Li Ge and Wenjie Wan describe one of the most versatile platforms used recently in optical physics and quantum mechanics, namely the non-Hermitian systems. These feature a different class of symmetries (e.g. parity-time reversal or PT) than the standard Hermitian systems that we are accustomed to, and the authors examine the manifestation of the nonlinear features in this novel class of systems.

Guofei Pang and George Em Karniadakis develop a comparison of a diverse palette of very modern techniques aimed at studying linear and nonlinear PDEs. This is accomplished through Gaussian processes and also via neural networks. Both continuous-time and discrete-time models are developed for performing the comparison. A number of case examples in both 1D (Burgers) and 2D (Navier-Stokes) are provided for concreteness.

Edgar Knobloch studies pattern formation in dissipative systems, both in the un-driven and in the driven (by an external force) case. He focuses on the prototypical snakes and ladders scenario of snaking bifurcation diagrams and offers a wide variety of examples to illustrate the generic emergence of such a scenario in dissipative nonlinear systems.

Stefano Iubini, Stefano Lepri, Roberto Livi, Antonio Politi, and Paolo Politi explore yet another aspect of important implications of nonlinear interactions. In particular, they focus on their impact on non-equilibrium statistical mechanics. By reviewing a number of classical anharmonic oscillator models, they present the connection of such anharmonicity with slow relaxation processes and the potential violations of the Fourier law for heat transport.

Alexandre Rosas, Daniel Escaff, and Katja Lindenberg focus on an insightful example of emergence of collective behavior: they explore a system of discrete units with all-to-all interaction. Their emphasis is on the widely important concept of synchronization and how it emerges in such systems for finite and infinite populations.

Boris Malomed offers an overview of solitary waves as they appear in nonlinear dynamical lattices. Among the prototypical examples used are the Toda and Frenkel-Kontorova (discrete sine-Gordon), as well as the discrete nonlinear Schrödinger lattice and its integrable analog, the so-called Ablowitz-Ladik model. The emphasis is on one-dimensional and two-dimensional theoretical and computational findings, however, these are complemented with some of the main experimental observations (e.g., in nonlinear optics) of such waves as well.

Nicholas Manton presents the geometric effects of a wide variety of intriguing higher dimensional waveforms such as vortices, lumps, and monopoles. He shows how these waves and their interactions can be well approximated via a finite-dimensional dynamical system of collective coordinates. This also offers the possibility of quantization and associated quantum models, such as the quantized Skyrmion model, are shown to hold promise for understanding aspects of nuclear physics and topics such as nuclear fusion more concretely.

Lei Xu, Moshen Rahmani, David A. Powell, Dragomir Neshev, and Andrey Miroshnichenko contributed a chapter in an emerging and important area of nonlinear science, namely nonlinear metamaterials. These are engineered structures designed to exhibit exotic electromagnetic properties (although more recently they have also been extended to other areas such as acoustics). This was an effort originally geared towards linear properties but has been substantially extended to the nonlinear realm more recently. Along with a historical overview of the field, a number of concrete examples such as strong local field enhancement or exotic phase-matching conditions are presented.

Network theory is another recently emerging major area interfacing with ideas from nonlinear systems. Mason Porter in the relevant chapter reviews topics from temporal networks, stochastic and deterministic dynamical processes, adaptive networks (where the structure is coupled to the dynamics), and even higher-order (rather than just pairwise) processes. Numerous important applications of the fields, e.g., in contagion dynamics, opinion models, and coupled oscillator networks are given.

Undoubtedly, Biology is an extremely wide and often largely unexplored field where nonlinearity has been shown to play an enormous role. Zoi Rapti presents an overview of select themes where the role of nonlinearity has been established often in close connection with biological experiments, notably in ecology and epidemiology. She also touches upon a number of additional fascinating areas of studies including the regulation of glucose-insulin, models of cancer progression, pattern formation in limbs, and DNA dynamics.

Naturally, for all of the above settings, numerical computation is of paramount importance for understanding the nonlinear dynamical systems that model the physical (or chemical or biological) reality. Thus carefully crafted and computationally efficient numerical methods are critical. Laurette Tuckerman reviews a number of state-of-the-art methods for identifying steady states, periodic orbits and bifurcation diagrams, coupling relevant aspects to the stability, and spectral characteristics of the solutions (i.e., eigenvalues and Floquet multipliers). A series of highly demanding classical examples are used to illustrate the methods drawn from the theory of fluids (such as Rayleigh-Bénard convection or flow past a cylinder), as well as that of superfluids and Bose-Einstein condensates.

Kazuya Fujimoto and Masahito Ueda explore one of the timely and important research directions in quantum systems, through their potential universal relaxation dynamics. They focus on emerging theories such as the Kibble-Zurek mechanism of pattern formation (that is also relevant to cosmology), the dynamics of coarsening, and the important recent notion of non-thermal fixed points. They showcase these types of relaxation dynamics by various mathematical examples of physical models, including quantum spin models, the famous Gross-Pitaevskii PDE, and kinetic equations.

Finally, Avadh Saxena, Panayotis G. Kevrekidis, and Jesús Cuevas-Maraver explore the interplay of nonlinearity and topology in a wide variety of physical systems such as chiral magnets, Bose-Einstein condensates, liquid crystals, topological materials, and photonic resonator arrays. They focus on a slew of topological defects, e.g., skyrmions, merons, hopfions, vortex rings, and monopoles in these settings and discuss the effect of nonlinearity on various emergent properties in the presence of such defects. They also illustrate the powerful technique of Bogomolnyi decomposition in terms of studying soliton-like defects on curved manifolds.

When this Preface was written, the world around us was a different place, i.e. it was prior to the pandemic of COVID-19 and the associated virus SARS-CoV-2. As we put these proofs together, it did not elude us that the epidemiological study of disease spreading or mitigation (or, for that matter, the fluid mechanical studies of the viral shedding, deposition, inhalation, sedimentation, etc.) are based on fundamentally nonlinear models. This has turned out to be yet another—and perhaps even more urgent than emergent—frontier for the nonlinear science of the next few years. In this context we emphasize that the first chapter by Zoi Rapti covers certain aspects of epidemiological modeling and the sixth chapter by Mason A. Porter addresses contagion models.

Amherst, MA, USA — Panayotis G. Kevrekidis
Seville, Spain — Jesús Cuevas-Maraver
Los Alamos, NM, USA — Avadh Saxena

Contents

Contributors

Mark J. Ablowitz Department of Applied Mathematics, University of Colorado, Boulder, CO, USA

Natalia G. Berloff University of Cambridge, Cambridge, United Kingdom; Skolkovo Institute of Science and Technology Russian Federation, Moscow, Russia

Jesús Cuevas-Maraver Departamento de Fisica Aplicada I, Grupo de Fisica No Lineal, Universidad de Sevilla, Sevilla, Spain;
Instituto de Matemáticas de la Universidad de Sevilla (IMUS), Edificio Celestino Mutis, Sevilla, Spain

Daniel Escaff Complex Systems Group, Facultad de Ingeniería y Ciencias Aplicadas, Universidad de los Andes, Las Condes, Santiago, Chile

Kazuya Fujimoto Department of Applied Physics, Institute for Advanced Research, Nagoya University, Nagoya, Japan

Li Ge Department of Physics and Astronomy, College of Staten Island, CUNY, Staten Island, NY, USA;
The Graduate Center, CUNY, New York City, NY, USA

Stefano Iubini Dipartimento di Fisica e Astronomia, Università di Padova, Padova, Italy;
Consiglio Nazionale delle Ricerche, Istituto dei Sistemi Complessi via Madonna del Piano 10, Sesto Fiorentino, Italy

Kirill P. Kalinin University of Cambridge, Cambridge, United Kingdom

George Em Karniadakis Division of Applied Mathematics, Brown University, Providence, RI, USA

Panayotis G. Kevrekidis Department of Mathematics and Statistics, University of Massachusetts, Amherst, MA, USA

Edgar Knobloch Department of Physics, University of California at Berkeley, Berkeley, CA, USA

Stefano Lepri Consiglio Nazionale delle Ricerche, Istituto dei Sistemi Complessi via Madonna del Piano 10, Sesto Fiorentino, Italy;
Istituto Nazionale di Fisica Nucleare, Sezione di Firenze, Sesto Fiorentino, Italy

Katja Lindenberg Department of Chemistry and Biochemistry and BioCircuits Institute, University of California San Diego, La Jolla, CA, USA

Roberto Livi Consiglio Nazionale delle Ricerche, Istituto dei Sistemi Complessi via Madonna del Piano 10, Sesto Fiorentino, Italy;
Istituto Nazionale di Fisica Nucleare, Sezione di Firenze, Sesto Fiorentino, Italy;
Dipartimento di Fisica e Astronomia, Università di Firenze, Sesto Fiorentino, Italy

Boris A. Malomed Department of Physical Electronics, Faculty of Engineering, School of Electrical Engineering, and Center for Light-Matter Interaction, Tel Aviv University, Tel Aviv, Israel

Nicholas S. Manton Department of Applied Mathematics and Theoretical Physics, University of Cambridge, Cambridge, UK

Andrey E. Miroshnichenko School of Engineering and Information Technology, University of New South Wales, Canberra, ACT, Australia

Dragomir Neshev Research School of Physics, ARC Centre of Excellence for Transformative Meta-Optical Systems (TMOS), The Australian National University, Canberra, ACT, Australia

Guofei Pang Division of Applied Mathematics, Brown University, Providence, RI, USA

Antonio Politi Institute for Pure and Applied Mathematics & SUPA University of Aberdeen, Scotland, United Kingdom

Paolo Politi Consiglio Nazionale delle Ricerche, Istituto dei Sistemi Complessi via Madonna del Piano 10, Sesto Fiorentino, Italy;
Istituto Nazionale di Fisica Nucleare, Sezione di Firenze, Sesto Fiorentino, Italy

Mason A. Porter Department of Mathematics, University of California, Los Angeles, CA, USA

David A. Powell School of Engineering and Information Technology, University of New South Wales, Canberra, ACT, Australia

Mohsen Rahmani Research School of Physics, Nonlinear Physics Centre, The Australian National University, Canberra, ACT, Australia;
Advanced Optics and Photonics Laboratory, Department of Engineering, Nottingham Trent University, Nottingham, UK

Zoi Rapti Department of Mathematics, University of Illinois at Urbana-Champaign, Urbana, IL, USA;
Carl R. Woese Institute for Genomic Biology, University of Illinois at Urbana-Champaign, Urbana, IL, USA

Alexandre Rosas Departamento de Física, CCEN, Universidade Federal da Paraíba, João Pessoa, Brazil

Avadh Saxena Theoretical Division, Los Alamos National Laboratory, Los Alamos, NM, USA

Laurette S. Tuckerman Laboratoire de Physique et Mécanique des Milieux Hétérogènes (PMMH), CNRS, ESPCI Paris, PSL Research University, Sorbonne Université, Université Paris Diderot, Paris, France

Masahito Ueda Department of Physics, University of Tokyo, Bunkyo-ku, Tokyo, Japan;
RIKEN Center for Emergent Matter Science (CEMS), Wako, Saitama, Japan

Wenjie Wan The State Key Laboratory of Advanced Optical Communication Systems and Networks, University of Michigan-Shanghai Jiao Tong University Joint Institute, Shanghai Jiao Tong University, Shanghai, China;
MOE Key Laboratory for Laser Plasmas and Collaborative Innovation Center of IFSA, Department of Physics and Astronomy, Shanghai Jiao Tong University, Shanghai, China

Lei Xu School of Engineering and Information Technology, University of New South Wales, Canberra, ACT, Australia

Acronyms

AD	Activity driven (model)
AKNS	Ablowitz, Kaup, Newell and Segur (formalism; system)
AL	Ablowitz-Ladik (equation)
AMO	Atomic, molecular and optical (systems)
BEC	Bose-Einstein condensate
CH	Cahn-Hilliard (equation)
CIM	Coherent Ising machine
CMOS	Complementary metal–oxide–semiconductor
CPA	Coherent perfect absorption
cSH	conserved Swift-Hohenberg (equation)
DB	Discrete breather
DNLS	Discrete Nonlinear Schrödinger (equation)
DOPO	Degenerate optical parametric oscillator
DS	Davey-Stewartson (equation)
ED	Electric dipole
FK	Frenkel-Kontorova (model)
FPGA	Field-programmable gate array
FPU;FPUT	Fermi-Pasta-Ulam (model); also known as the Fermi-Pasta-Ulam-Tsingou (model)
FR	Fano resonance
FWM	Four-wave mixing
GL	Ginzburg-Landau (equation)
GP	Gaussian process
GPE	Gross-Pitaevskii equation
HOBO	Higher order binary optimisation
IC	Intersite-centered (discrete soliton or vortex)
IST	Inverse scattering transform
JJ	Josephson junction
KdV	Korteweg-de Vries (equation)

KP	Kadomtsev-Petviashvili (equation)
KPZ	Kardar-Parisi-Zhang (equation)
KZ	Kibble-Zurek (mechanism)
LS	Localized structure
LTEM	Lorentz Transmission Electron Microscopy
MD	Magnetic dipole
mKdV	modified Korteweg-de Vries (equation)
NHPH	Non-Hermitian particle-hole (symmetry)
NLD	Nonlinear Dirac (equation)
NLS	Nonlinear Schrödinger (equation)
NLS	Nonlinear Weyl (equation)
NN	Neural network
NTFP	Non-thermal fixed point
OC	Onsite-centered (discrete soliton or vortex)
ODE	Ordinary differential equation
OL	Optical lattice
OPA	Optical parametric amplification
OPO	Optical parametric oscillator
PDE	Partial differential equation
PFC	Phase field crystal
PIGP	Physics-informed Gaussian process
PINN	Physics-informed neural network
PN	Peierls-Nabarro (potential)
PT	Parity-time (symmetry)
QCO	Quadratic continuous optimisation
QoI	Quantity of interest
QUBO	Quadratic unconstrained binary optimisation
RC	Reservoir computing
RNN	Recurrent neural networks
RST	Reverse space-time (symmetry)
RT	Reverse time (symmetry)
SAD	Simplicial Activity Driven (model)
SEM	Scanning electron microscopy
sG	sine-Gordon (equation)
SH	Swift-Hohenberg (equation)
SHG	Second harmonic generation
SI	Susceptible–infected (model)
SIR	Susceptible–infected–recovered (model)
SIS	Susceptible–infected–susceptible (model)
SM	Salerno model
SRR	Split-ring resonator
SSB	Spontaneous symmetry breaking
SSH	Su-Schrieffer-Heeger (model)
TD	Toroidal dipole

TFI	Transverse field Ising (model)
TH	Third harmonic
THG	Third harmonic generation
TL	Toda lattice (equation)
VA	Variational approximation
VL	Vortex line
VR	Vortex ring
WTM	Watts threshold model

Chapter 1
Nonlinearity and Biology

Zoi Rapti

Abstract In this chapter we discuss the origin and role of nonlinearities in some classes of biological models. We describe underlying biological mechanisms that generate nonlinearities and how they have been modeled in subfields, such as ecology and epidemiology. We present examples of recent models to highlight the importance of indirect effects and the emergence of alternative stable states, and trade offs. At the same time, we emphasize recent developments and unresolved challenges in biological modeling, such as data-theory coupling, parameter estimation and the generalization of results from low- to high-dimensional systems. We finish with recent examples of mathematical models of the glucose-insulin regulatory system, cancer treatment, limb development and pattern formation, and DNA.

1.1 The Origin of Nonlinearities in Biology

Living organisms and their interactions are, in general, complicated and variable. Foraging ants leave pheromone trails aiming to recruit more foragers to a particular food source [1]. The positive feedback generated by their infochemicals has been found to be nonlinear: the strength of the trail is not proportional to the reaction it elicits from the ants [1]. This nonlinearity generates a bias towards food sources with the largest number of initial visitors, despite the fact that qualitatively it might be equivalent to other nearby sources.

Female túngara frogs select their mate based primarily on acoustic cues. To attract females, male frogs emit simple calls (a whine only) or complex calls (a whine followed by one or more chucks) and females respond preferentially to complex calls [2]. Frog groups range in size from a few to a hundred, hence competition

Z. Rapti (✉)
Department of Mathematics, University of Illinois at Urbana-Champaign, 1409 W. Green Street, Urbana, IL 61801, USA
e-mail: zrapti@illinois.edu

Carl R. Woese Institute for Genomic Biology, University of Illinois at Urbana-Champaign, 1206 W. Gregory Drive, Urbana, IL 61801, USA

P. G. Kevrekidis et al. (eds.), *Emerging Frontiers in Nonlinear Science*, Nonlinear Systems and Complexity 32, https://doi.org/10.1007/978-3-030-44992-6_1

among males is fierce. It has been determined that túngara frogs follow Webber's law when comparing mating cues, namely they prefer one call over another not based on the difference of the number of chucks, but rather based on the ratio of number of chucks (a chuck is defined as a complex call consisting of two or more whines). Lack of privacy makes mating success even more uncertain due to predation risk: frog-eating bats also listen attentively to these mating calls in order to locate their next meal.

It has been generally noted that the outcome of interactions between two species often depends on which other species are present and interact with them [3, 4]. Moreover, these interactions may be direct or indirect. Competition for common resources, parasitism, and predation are examples of direct effects. The effect of species A on species B is indirect if it takes place by A altering the effect a third species, C, has on B. Indirect effects can, in turn, be density- or trait-mediated. A density-mediated indirect effect occurs when, for instance, a parasite affects only the superior of two competitors, thus reducing its density, which in turn aids the inferior competitor. An example of a trait-mediated indirect effect is when the presence of a predator evokes anti-predatory behavior in one of two competitors, thus affecting the outcome of their pairwise competition. Such indirect effects [4] were the focus of a recent study of a host-parasite-competitor nonlinear model [5], where it was shown that indirect effects are qualitatively and quantitatively as strong as direct effects. This importance of indirect effects has long been advocated for by biologists [4].

In experiments of the same host-parasite system, kairomones from different predators evoked different effects [6]: fish predators increased host susceptibility to the parasite, while an invertebrate predator induced no such effect. On the other hand, the fish predator caused a decrease in the infective propagule yield per host, compared to the invertebrate predator. Higher susceptibility aids the parasite, while lower propagule yield harms its chances of successive infections. Therefore, it is nontrivial to determine what the net effect of each predator on the disease dynamics and the parasite is.

These examples illustrate not only the emergence of nonlinear phenomena in biological systems, but also how nuances in each system may lead to lack of uniformity in modeling approaches, which in turn may partially explain the relative scarcity of general "laws" and a general theory in biological modeling, when compared with other disciplines, such as physics.

1.1.1 Interactions of the Lotka–Volterra Type

Few systems are probably more well-studied in mathematical biology than those of prey-predator interactions. The most easily recognizable prey-predator model is the Lotka–Volterra nonlinear system

$$\frac{dx}{dt} = (b - fy)x \tag{1.1}$$

$$\frac{dy}{dt} = (efx - d)y. \tag{1.2}$$

Here, x and y denote the prey and predator population density, respectively, and t denotes time. The parameter b denotes the net per capita growth rate of the prey in the absence of the predator, f the predator feeding rate, e the conversion efficiency of prey to predator biomass, and d the per capita predator death rate. The model was introduced almost 100 years ago independently by Lotka in 1925 [7] and Volterra in 1926 [8]. It has the trivial equilibrium $E_0 = (0, 0)$ where both prey and predator go extinct, which is a saddle for all realistic parameter values, and the co-existence equilibrium $E_c = \left(\frac{d}{ef}, \frac{b}{f}\right)$. Its dynamics is fully understood and it has the following constant of motion:

$$L(x, y) = b \ln y - fy - efx + d \ln x. \tag{1.3}$$

Plotting the contours of L yields periodic solutions around the coexistence equilibrium.

The terms $b - fy$, $efx - d$ in (1.1)–(1.2) are referred to as the per capita growth rates and are linear functions of the dependent variables x, y. This linearity has several underlying biological assumptions.

- First, it is assumed that predators have an insatiable appetite. The more prey exists, the more they feed on it.
- Second, it is assumed that the prey has no defensive mechanisms and the more predators exist, the greater its predation risk is.

Both of these assumptions have been contested, giving rise to different nonlinear prey-predator models.

1.1.2 Holling-Type II Functional Response and the Rosenzweig-MacArthur Model

The first assumption was amended by Holling in 1959 [9] and later analyzed by Rosenzweig and MacArthur in 1963 [10]. Holling replaced the linear function fx in the predator feeding rate by

$$F(x) = \frac{f_m x}{h + x}, \tag{1.4}$$

where f_m is the maximum feeding rate and h is the half-saturation constant, namely the value of prey density where the feeding rate achieves half its maximum value $\frac{f_m}{2}$. This is referred to as *Holling Type-II functional response*, whereas the linear feeding rate is referred to as *Holling Type-I functional response*. The saturating functional response is used in the case of predators that require time for both searching for and handling of the prey, while the linear functional response is used in the case of predators that require no time for handling their prey, so all time can be used for searching. In [11] the authors have recently used the framework of stochastic processes (microscopic description at the individual level) to derive ODE models (macroscopic description at the population level) of predator-prey interactions, while carefully considering the entire life-cycle of the species involved. What is interesting is the fact that the macroscopic equations contain the original Holling functional responses, which were derived for individual predators. Rosenzweig and MacArthur [10] also included the more realistic assumption that in the absence of the predator, the prey does not grow exponentially according to bx in (1.1), but rather logistically, according to:

$$rx\left(1-\frac{x}{K}\right), \tag{1.5}$$

where r is the maximum intrinsic per capita growth and K the carrying capacity.

1.1.3 Higher-Order Interactions

The assumption that prey is defenseless and predation risk increases with predator density was challenged, among others, by Abrams in 1983 [12]. Specifically, he argued that some prey may actively engage in predator avoidance which would lead to a decrease, rather than an increase, of predation rate with predator density. He subsequently advocated for *higher-order interactions* in predator-prey models. Higher order-interactions have indeed been incorporated in both prey (bacterium)-predator (phage) models [13] and in other interactions, such as competitive ones [14]. These interactions were found to have a stabilizing effect on the dynamics. Specifically, it was shown in [13] that despite the system having a conserved quantity like (1.3), the dynamics are not necessarily oscillatory: for certain initial conditions, the trajectories converge to an attracting equilibrium point. Using a simple Lyapunov function, it was shown in [14] that while competitive interaction of Lotka–Volterra type (quadratic nonlinearities) lead to cyclic dynamics, higher-order interactions yield a globally stable equilibrium point.

1.1.4 Lotka–Volterra in Disguise

Another equally well-known model is the classic epidemic model [15, 16]:

$$\frac{\mathrm{d}s}{\mathrm{d}t} = -\beta s i \tag{1.6}$$

$$\frac{\mathrm{d}i}{\mathrm{d}t} = \beta s i - \gamma i. \tag{1.7}$$

In this model, s and i denote, respectively, the fractions of susceptible and infected classes in the total population, which also includes the recovered class r, and t denotes time, as previously. The equation for the recovered class $r' = \gamma i$ is omitted for the following reason. The total population is assumed to be constant $s(t) + i(t) + r(t) \equiv 1$, hence $r(t)$ can be trivially obtained once $s(t), i(t)$ have been found. The parameter β denotes the contact rate between the two classes and $1/\gamma$ is the average infectious period. One readily notices that system (1.6)–(1.7), as far as nonlinearities are concerned, is identical to the Lotka–Volterra one (1.1)–(1.2). However, the focus of mathematical analysis of this model, is different than of that of the Lotka–Volterra system. Specifically, the question of interest is whether the infected population i will first rise to a maximum and then decrease (existence of an epidemic outbreak) or whether it will just decrease to zero monotonically (no epidemic outbreak taking place). The answer depends on the threshold parameter $\frac{\beta}{\gamma}s_0$, which is the product of the contact number $\frac{\beta}{\gamma}$ and the initial condition of the susceptible class $s(0) = s_0$.

1.1.5 Holling-Type II Functional Response in Disguise

While the Holling-type II nonlinearity has its origins in ecology, the same nonlinearity has two more aliases reflecting its alternative origins. The same type of saturating nonlinearity was also introduced by Monod [17] in the context of bacterial growth in an environment of limiting resources. Before that, Michaelis and Menten in 1913 were actually the first to use it in the context of enzyme-catalyzed reactions [18]. Specifically, in their seminal paper they derived a model for the rate of an enzymatic reaction, where the binding of the reaction products competes with the binding of the substrate, thus leading to saturation. More recently saturating functions have been used to model the infection of bacteria by phages in order to account for biofilm formation due to quorum-sensing [19].

1.1.6 Holling-Type III Functional Response and the Hill Equation

Sigmoidal functional responses of the form

$$G(x) = \frac{g x^n}{h^n + x^n}$$

are used in prey-predator models [20], chemical reactions [21] and gene regulatory networks [22, 23]. Here, g is the maximum of function which is achieved asymptotically and h can be seen as a half-saturation constant, since $G(h) = \frac{g}{2}$. The parameter n is referred to as the Hill coefficient, after Archibald Hill who was the first to introduce a sigmoidal function when studying the binding of oxygen to haemoglobin [21]. The value of n controls how abrupt the switch is, with high n yielding steeper transition. In some prey-predator systems, the predator exhibits a learning behavior, where in low prey densities it will not feed, but as the prey densities increase above a threshold, the predator feeding rate increases until it saturates. As outlined in [20], such a switch in predator feeding rate may occur for other reasons too, such as the discovery of alternative areas or search paths and the learning of hunting techniques or increased prey handling efficiency. In gene regulatory networks, Hill's functions are used to describe the control of one variable by another. In that context, g is interpreted as the maximum of the activation/induction rate and h as the half-maximal induction concentration.

1.1.7 Hunting for Universal Biological Models

This phenomenon of using essentially the same model and reinterpreting it in different contexts permeates mathematical modeling in biology. It also demonstrates the lack of general rules in biological modeling. However, the need for generalization has been recently recognized [24]. In [24] the authors advocate the use of universal consumer-resource models, which encompass prey-predator and susceptible-infected interactions, that can then be modified to describe particular systems while still accounting for their intricacies. Specifically, many well known models, such as Lotka–Volterra, susceptible-infected-recovered, and chemostat, have been recast as a general consumer-resource model.

1.2 Dynamics in Nonlinear Biological Systems

Having identified some of the origins and modeling approaches of nonlinearities in biological applications, the next step is to explore the types of dynamics they generate. In particular, we will focus on indirect effects, the role of nonlinearities in the creation of alternative steady states, and the emergence of trade-offs in biological models.

1.2.1 Indirect Effects

Cross-feeding in polymicrobial systems [25] is a prime mechanism demonstrating the complicated relationships that arise due to indirect effects. The relevant biological system consists of two competitors that are also indirectly linked due to the fact that competitor A produces a toxic (to A) by-product, E, that is however beneficial to competitor B, since B utilizes it as a resource. Thus, although the direct effect of A on B, and vice versa, is competitive, the indirect effect of one competitor onto the other is mutualistic. Hence, the net ecological interaction between A and B could be of either kind and this was showcased in [25] using a simple mathematical model. In other words, there is a trade-off between the negative effect (competition for primary resources between A and B) and the positive effect (A offering an additional resource to B without any cost to itself, and B removing the toxic for A by-product) each competitor has on the other. Despite the fact that biologists acknowledge that interactions are not permanent, neither in time, nor in space, mathematical models are, with few exceptions, lagging behind. Namely, most theoretical studies focus on interactions of fixed type (such as competitive: both species are harmed, mutualistic: both species have a net benefit, antagonistic: one species benefits, while the other is harmed) and assume that they retain their type throughout.

Indirect effects and trade-offs have been observed and demonstrated mathematically in a plethora of other systems. For instance, in invertebrate disease dynamics they seem to be a prevailing theme. In [26], the role of the inferior competitor *Daphnia pulicaria* as diluter in fungal epidemics of the focal host *Daphnia dentifera* was investigated. The two species compete for a common food resource (algae), but *D. pulicaria* is not a competent host. Hence, when it consumes infectious fungal propagules from the water column while filter-feeding, it decreases the disease risk in the focal host. However, in that study it was found that the positive effect of dilution (decrease in infection risk) was not enough to counterbalance the negative direct effect of competition. Actually, depending on the competitive ability of the two species, the density of the focal host was decreased, and in some cases, the focal host was completely eliminated from the system. Only at low carrying capacity for the algal resources was the net role of the inferior competitor positive, namely it increased the density of the focal host through the dilution effect.

In the following example we use the simple 3-dimensional ODE model of [25] to show the changing type of interactions due to indirect effects.

Example

The simple ODE model of [25] reads

$$\frac{\mathrm{d}A}{\mathrm{d}t} = (r(1 - \alpha A - \beta B) - fE)A \tag{1.8}$$

$$\frac{dB}{dt} = ((1 - \gamma A - B) + ghE)B \tag{1.9}$$

$$\frac{dE}{dt} = yA - hEB - uE, \tag{1.10}$$

where A, B and E denote the producer, cross-feeder, and by-product density, respectively. In this model, r stands for the relative intrinsic growth rate of A, α, β, γ are competition coefficients, f is the toxicity rate of E on A, h represents the consumption rate of E by B, and g the conversion efficiency. Finally, y is the production rate of E by A and u the decay rate of E. All equilibrium points were found in [25] and their stability was analyzed. The model is in non-dimensional form.

To showcase the variation in the interaction between the two competitors, we plot the stable steady states as a function of byproduct toxicity f at two different values for consumption rate h of E by B. Specifically, we pick a low consumption rate $h = 0.01$ and a high consumption rate $h = 0.8$ and compare the density of the two competitors when the other is absent and when they coexist. As it is shown in Fig. 1.1, when the consumption rate of the toxic byproduct E by B is small, both competitors have higher density when the other is absent (panels a and b). On the other hand, when the consumption rate is high and byproduct toxicity is above a threshold, both competitors are benefited when they coexist, since their density is higher than the one they would reach in the absence of the other (panels c and d). Hence, for $h = 0.01$ their net interaction is competitive, while at $h = 0.8$, their net interaction is mutualistic, for high enough toxicity. However, at very low toxicity f and high consumption rate h, one can see from panel c that competitor A has higher density in the absence of B. Thus, for low toxicity rate, the interaction is antagonistic: positive for B, but negative for A.

1.2.2 Alternative Stable States

It is well-established that gene regulatory networks have feedback loops which induce the bistability that underlies their binary (switch-like) behavior [27]. Actually, twenty years ago, a mathematical model was used to experimentally construct a bistable gene network [28].

What has been shown in the past two decades is that the same nonlinear mechanisms that explain the switch-like behavior in regulatory networks, e.g. hysteresis, are also at play in self-organization and collective behavior of certain species, such as Pharaoh's ants [29]. Specifically, it was shown experimentally that there exists a behaviorally imposed hysteresis phenomenon which explains the transition of forager ants from a disorganized state to an organized one (trail formation towards a

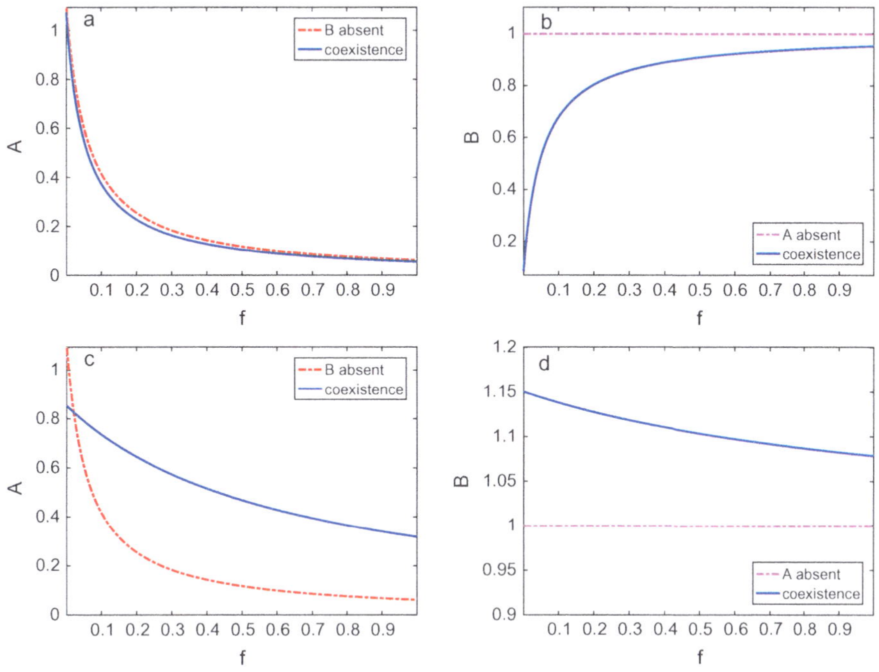

Fig. 1.1 Equilibrium density of A as a function of toxicity rate f in the absence (red dashed-dotted curve) and presence (blue solid curve) of B is shown in panel a for $h = 0.01$. Equilibrium density of B in the absence (magenta dashed-dotted line) and presence (blue solid curve) of A is shown in panel b for $h = 0.01$. Panels c and d are analogous to panels a and b, respectively, but correspond to $h = 0.8$. The other parameter values are set at $r = 1$, $\alpha = 0.9$, $\beta = 0.2$, $\gamma = 1$, $u = 0.1$, $g = 1$, and $y = 1.5$

food source). The authors also wrote a simple nonlinear ODE model which exhibits bistability and reproduces the experimental results.

Later on, they expanded the one-dimensional system to a coupled system of ODEs in order to capture the ant dynamics when two food sources are available, as outlined in Sect. 1.1. Since the results in [1] were mostly numerical, in this section we present some analytical results on the two-dimensional model, as an illustrative example of bistability in biological systems.

Example

In the ant-trail model with two food sources [1], X_A and X_B denote the number of forager ants walking to food source A and B, respectively, and N is the total number of ants able to forage. Each individual ant may find the food source through a random search at a rate α or it may follow the pheromone trail of other ants at a per-capita rate $\beta_i X_i$, $i = A, B$. These per capita rates are multiplied with the number of ants that have not selected a food source yet, namely $N - X_A - X_B$. Since pheromones are quite volatile, their trail diminishes over time. However, it is assumed that trail strength increases with the number of ants using it and that the stronger the trail, the less likely it is for ants to lose it. Hence, the loss rate of ants forming a trail was assumed to depend nonlinearly on their number as $\frac{sX_i}{K+X_i}$, $i = A, B$, where s is the maximum rate at which ants lose the trail when the trail is saturated and K is the number of ants that result in a rate loss of $s/2$. Finally, it is assumed that time t is measured in minutes. The coupled nonlinear ODE model then reads [1]

$$\frac{\mathrm{d}X_A}{\mathrm{d}t} = (\alpha + \beta_A X_A)(N - X_A - X_B) - \frac{sX_A}{K + X_A} \tag{1.11}$$

$$\frac{\mathrm{d}X_B}{\mathrm{d}t} = (\alpha + \beta_B X_B)(N - X_A - X_B) - \frac{sX_B}{K + X_B}. \tag{1.12}$$

An interesting bimodality was observed in experiments [1]. Namely, when the two food sources A, B are of identical quality, the one with the largest number of ants after 60 min (almost at steady-state), is the one that has the largest number of ants initially. Bistability in the corresponding mathematical model reflects this bimodality in the biological data.

It is straightforward to verify that $X_i = 0$ implies $\frac{\mathrm{d}X_i}{\mathrm{d}t} = \alpha(N - X_j) \geq 0$, $j \neq i$, $j = A, B$ and $X_A + X_B = N$ implies $\frac{\mathrm{d}X_i}{\mathrm{d}t} = -\frac{sX_i}{K+X_i} \leq 0$, $i = A, B$. These, in turn, imply that the triangular region bounded by $X_A = 0$, $X_B = 0$, and $X_A + X_B = N$ is positively invariant for system (1.11)–(1.12).

If, in addition, the food sources are of equal quality, it is assumed that $\beta_A = \beta_B = \beta$ and the following are also true.

- The set $X_A = X_B$ is invariant under the dynamics of (1.11)–(1.12).
- The equilibrium points satisfy $X_A = X_B$ or $X_A X_B = \frac{\alpha K}{\beta}$, which yield a saddle point on the invariant set $X_A = X_B$ and two positive linearly stable equilibrium points.

This information is summarized graphically in Fig. 1.2. The saddle point is depicted by a red square. Any trajectory originating on $X_A = X_B$ (the stable manifold of the saddle point) ends us at the saddle point. On the other hand, any trajectory originating in either of the smaller triangular regions is trapped there and ends up in the respective green dot depicting the stable equilibrium point.

How important were the nonlinearities in this model? It is noted in [1] that, any change that would remove the cubic terms in (1.11)–(1.12) would result in model predictions of equal final trails to both food sources, regardless of the initial conditions. This is an undesirable outcome, as it violates the experimental findings.

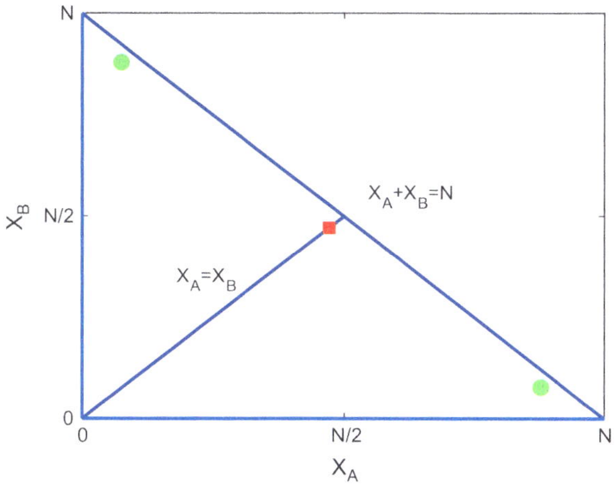

Fig. 1.2 Equilibrium points and positively invariant regions for the ant trail model with two equal-quality food resources. The red square corresponds to the saddle point and the green circles correspond to the linearly stable equilibrium points. The triangles bounded by the blue line segments are positively invariant regions for the system. Parameter values are set to $N = 100$ ants, $\alpha = 0.0052$ 1/min, $\beta_A = \beta_B = 0.0015$ 1/(min × ant), $s = 1$ 1/min and $K = 10$ ants

1.2.3 Trade Offs

One of the central questions in host-parasite interactions is related to the optimum virulence from the pathogen's perspective. Since most pathogens depend on their host for survival and transmission, common sense would suggest that it is to their benefit not to cause harm to their host [30]. This is the so-called "conventional wisdom" of avirulence. It has, however, been suggested that due to feedbacks and adaptations between pathogens and their hosts, pathogens will select for the virulence that maximizes their fitness, which is referred to as the "enlightened theory" [31]. In most studies, the basic reproduction number, R_0, namely the number of secondary infections generated by a single infected host when introduced in a completely susceptible population [16] is used as a proxy for fitness [31–33]. Recently, both empirical [30] and mathematical studies [32, 33] suggest that an intermediate level of virulence maximizes the pathogen's fitness.

There are various trade-offs that cause this intermediate optimum. First, there is the trade-off between transmission and disease-induced mortality [31], where disease-induced mortality increases with transmission rate. R_0 increases with disease transmission, but decreases when the period over which hosts remain infectious decreases, thus generating an intermediate level of optimum disease-induced mortality. Second, there might be a trade-off between contact rate and transmissibility [33]. This happens when the severity of the infection, which increases with pathogen load in the host, causes a decrease in contacts due to the host experiencing too severe symptoms to function normally. Increased pathogen load, on the other hand, increases the probability of disease transmission given contact. Third, in infectious diseases with recovery, a trade-off between transmission and recovery is possible [34]. This occurs when the activation of a host's immune defenses depends on the pathogen's growth rate. Hence, the pathogen may increase its transmission rate, while simultaneously decreasing the infectious period. In this setting, the definition of virulence has been expanded to also include non-lethal effects caused by the pathogen.

In the following example, we highlight how biologically imposed trade-offs result in intermediate levels of optimum virulence.

Example

We consider a PDE model [32], where susceptible *Daphnia* hosts, S, become infected, I, by accidentally ingesting fungal propagules, Z, while filter-feeding on algae, A. The model is a combination of a prey-predator model (S and A), similar to (1.1)–(1.2), and a susceptible-infected model (S and I), similar to (1.6)–(1.7). Host feeding rate $f_S(A)$ is given by a Holling type-II functional response (1.4), and algal resources grow logistically, according to (1.5). The infection rate $\mu \frac{f_S(A)}{A} Z$ per susceptible host S consists of the host susceptibility μ and the contact rate with spores, which is given by the product of the ingesting rate $\frac{f_S(A)}{A}$ and the spore density Z. This arises because if we set $f_S(A) = w(A)A$ for the functional response, then since spores Z are ingested in the same way as algae A, the ingesting rate of spores is $w(A)Z = \frac{f_S(A)}{A}Z$. Since most species are part of interaction networks involving pathogens, competitors, and predators, such eco-epidemiological models are quite common. In the following model though, we combine eco-epidemiological interactions with the process of within-host pathogen growth. A second time variable a, having the same units as time t, is introduced to keep track of the within host-dynamics. Here, $I(t, a)$ denotes the density of infected hosts at time t that have been infected for a days. Due to the growth rates of the *Daphnia* host and its fungal pathogen, it is assumed that $\frac{\mathrm{d}a}{\mathrm{d}t} = 1$. The within-host pathogen load is denoted by $W(a)$. Moreover, the term virulence encompasses all harm inflicted unto the host by the pathogen, such as reduced fecundity modeled by $0 \leq \rho(a) \leq 1$, higher predation rate for the infected hosts by visual predators modeled by $\theta(a) \geq 1$, as well as disease-induced mortality $v(a) > 0$. The full nonlinear model is:

$$\frac{\mathrm{d}S}{\mathrm{d}t} = e_S f_S(A)\left(S + \int_0^{a_0} \rho(a) I(t,a)\mathrm{d}a\right) - (d + p_S)S - \mu \frac{f_S(A)}{A} SZ \tag{1.13}$$

$$\frac{\partial I}{\partial t} + \frac{\partial I}{\partial a} = -(d + v(a) + \theta(a)p_S)I, \quad I(t,0) = \mu \frac{f_S(A)}{A} SZ \tag{1.14}$$

$$\frac{\mathrm{d}Z}{\mathrm{d}t} = \sigma e_S f_S(A) \int_0^{a_0} (d + v(a)) I(t,a) W(a) \mathrm{d}a - \lambda Z -$$

$$f_S(A)\left(S+\int_0^{a_0} I(t,a)\mathrm{d}a\right)\frac{Z}{A} \tag{1.15}$$

$$\frac{\mathrm{d}A}{\mathrm{d}t}=r\left(1-\frac{A}{K}\right)A-f_S(A)\left(S+\int_0^{a_0} I(t,a)\mathrm{d}a\right). \tag{1.16}$$

Here, e_S denotes the conversion efficiency of algal to host biomass, d denotes the background mortality rate, p_S the predation rate, σ is the propagule release parameter, and λ is the fungal propagule loss rate. The integral $\int_0^{a_0} \rho(a)I(t,a)\mathrm{d}a$ in (1.13) gives the total host biomass generated by the infected hosts $I(t,a)$ that reproduce at a rate modified by $\rho(a)$ (compared to the susceptible hosts) during their life-span a_0. The integral $\int_0^{a_0}(d+v(a))I(t,a)W(a)\mathrm{d}a$ in (1.15) denotes all the spores that are produced by the infected hosts $I(t,a)$ that either die due to natural causes at rate d or due to the infection at rate $v(a)$ and which after being infected for a days have a within-host spore load of $W(a)$. The integral $\int_0^{a_0} I(t,a)\mathrm{d}a$ in (1.15)–(1.16) denotes the total biomass of all infected hosts, regardless of their age of infection a.

A trade-off exists due to the fact that as the pathogen proliferates within the host, the normally transparent host becomes opaque, and hence is easier located and preyed-on by visual predators (fish). Unfortunately for the pathogen, most of the propagules contained in infected hosts consumed by fish are lost from the water-column. Therefore, for given predator behavior $\theta(a)$, if the pathogen kills the host too soon, the dead infected host will not release enough propagules, whereas, if the pathogen kills the host too late, it might be consumed by fish first, effectively removing all propagules from the water column. The disease induced mortality $v(a)$ is given by a sigmoidal function as seen in panel a of Fig. 1.3. In panel b of Fig. 1.3 the basic reproductive number R_0 is shown as a function of the time of the switch h_v for various values of the maximum disease induced mortality v_0. One can see that for fixed v_0, the maximum value of R_0 in attained at intermediate values of h_v.

1.3 Experiments, Observations, and the Pursuit of Realistic Biological Models

Oftentimes, nonlinearities are ignored in experimental and field studies. Due to time and cost limitations, when experiments and observational studies are designed, restrictive choices are made regarding the range of varied parameters and observed conditions. However, as highlighted in a 2013 note [35], not accounting for non-linear relationships may yield inconsistent conclusions and contradictory findings. Consider for instance the case where variable Q depends on variable p according to

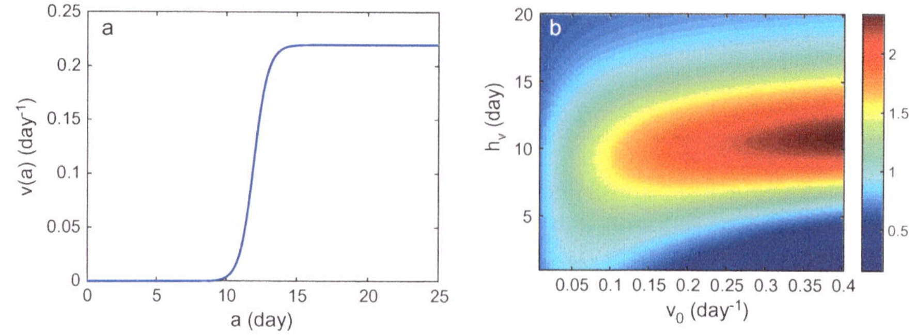

Fig. 1.3 The sigmoidal function $v(a) = \frac{v_0}{2}(1 + \tanh(a - h_v))$ is shown in panel a for $v_0 = 0.22$ day^{-1} and $h_v = 12$ days. The basic reproductive number R_0 is shown in panel b, as a function of v_0 and h_v. The other parameter values follow [32] and are well known in the literature

$$Q(p) = pe^{-p},$$

which is the Ricker nonlinearity used in [35]. Then, depending on whether the region, before, close to, or after the maximum at $p = 1$ is explored, the outcome of the study could be that Q increases, does not depend on, or decreases with p, respectively.

Another way in which data-theory coupling may become nontrivial is due to the timescales considered and the existence of intrinsic oscillations in the data. This was the case in a statistical analysis of global jellyfish blooms [36]. The motivation behind the study was the widespread awareness of increases in jellyfish numbers. Nevertheless, the major finding of the study, which was possible thanks to observational studies spanning several decades, was that the perceived increase was part of decadal oscillations in jellyfish abundance. This echoes the warning issued by Robert May 35 years ago [37], when he demonstrated how transient damped oscillations operating on a long timescale might, when not promptly identified as such, erroneously suggest the failure of immunization campaigns.

There are currently two routes to obtain realistic models given biological data. One is the so-called sparsity-promoting optimization that has been recently developed to select nonlinear models from a library of existing ones [38]. This is a data-driven method requiring no prior knowledge of the underlying biology, thus falling into the category of an unsupervised learning algorithm. The main idea is to synthesize the time series data into a nonlinear function library, relate the elements of the library to the time derivatives using regression and enforce sparsity to ensure only a few nonzero coefficients. Limitations of this method include the requirement of large data sets and the treatment of noisy data when noise is too high.

An alternative method is to first use biological insight to create an appropriate nonlinear model and then fine-tune its parameters, after having examined whether they are indeed identifiable [39]. Identifiability may fail in two ways: due to the structure of the model and the nature of the measurements (structural identifiability), or due to the specific dataset used (practical identifiability), for reasons such as noise or having the wrong sampling frequency. The modeling process may first involve rescaling the model to eliminate parameters. Afterward, the rescaled model is fitted to data using least squares. One may also generate simulated data to assess estimation errors and investigate alternative measurement scenarios. This method also detects combinations of parameters that are unidentifiable based on the given data, and hence informs of additional required measurements. The limitations of the method are mostly computational in nature.

1.4 Recent Advances and Open Problems

Even before considering what types of models and nonlinearities should be examined, there is a more fundamental question to be asked by modelers. Namely, one should ponder which biological systems, processes, or populations are important to be modeled and studied in the first place. Models and theoretical studies should be grounded on original biological questions. Within the last decade, such key questions are related to the comparative study of direct and indirect species interactions, the role of hysteresis in ecological systems, and the study of life-history trade-offs in relation to evolution, as well as the resilience of populations and ecosystems to disturbances and perturbations, the quantification of extinction risk, the prediction of tipping points and the reconciliation of multiple timescales [40].

Recent technological progress has resulted in a deluge of biological data, some of which are time series. A key question regarding time series is the existence of regular patterns (such as periodicity) or randomness. In some applications, such as microbiota network analysis, periodicity, alternative states and the ability to construct predictive models are of major importance [41].

Time Series Analysis

Using permutation entropy as a measure is an efficient way to detect and analyze unusual patterns in biological time series data [42, 43]. It is simple to implement, computationally fast, and robust [42]. It has been used to study qualitative and quantitative changes in simulated dynamical systems (such as the Lorenz system and the logistic map [43]) and real data (such as those from epileptic seizures [44]). Permutation entropy is defined as follows. At each time instance s of a given time series $\mathcal{X} = \{x_t : t = 1, \ldots, T\}$, a vector composed of m subsequent values is constructed $s \mapsto (x_s, x_{s+1}, \ldots, x_{s+(m-1)})$. Here, m is called the embedding dimension and it has been shown empirically that it should be in the range $5 \leq m \leq 7$. To each vector, an ordinal pattern is associated, defined as the permutation $\pi = (r_0 r_1 \ldots r_{m-1})$ of $(01 \ldots m-1)$ with the property $x_{s+r_0} \leq x_{s+r_1} \leq \cdots x_{s+r_{m-2}} \leq x_{s+r_{m-1}}$. An embedding delay L can be introduced in a straightforward way $s \mapsto (x_s, x_{s+L}, \ldots, x_{s+(m-2)L}, x_{s+(m-1)L})$ [43, 44] to consider different temporal resolutions. To each time series, it is possible to associate a probability distribution Π, whose elements π_i are the frequencies associated with the i-th possible permutation pattern, where $i = 1, 2, \ldots, m!$. Then, the permutation entropy PE is defined as the Shannon entropy of this distribution:

$$PE = -\sum_{i=1}^{m!} \pi_i \log \pi_i .$$

Unlike other complexity measures, the permutation entropy is able to distinguish noise from chaos and can be used to reveal relationships between two time series [44]. Computer memory limitations and the length requirements are two of the restrictions of the method.

Alternative Stable States
While the initial choice of a food source might not be catastrophic for a colony of ants, alternative stable states may make all the difference to the state of coral reefs (healthy vs. diseased) [45] or to the state of lakes (clear vs. turbid) [46]. Techniques from bifurcation theory [47] have been used to study tipping points in the context of local bifurcations. More recently, progress has been made to both define and quantify resilience in the case when disturbances are repeated and discrete [48]. Specifically, the authors introduced the novel kick-flow method to study the effect of disturbances on the state of a dynamical system. The main idea of the method is to define a map $G_{\tau,\kappa}$ that describes one cycle of the flow-kick process, where τ is the flow time (the time interval between disturbances) and κ is the kick size (the size of the disturbance). Then, if $\phi_\tau(x)$ is the flow of the ODE describing the unperturbed system, it holds

$$G_{\tau,\kappa}(x) = \phi_\tau(x) + \kappa.$$

Iteration by $G_{\tau,\kappa}$ amounts to repeated disturbances. Flow-kick fixed points and their stability can then be used to determine the resilience boundary in $\tau\kappa$-space (disturbance space). Unlike methods used to predict tipping-points [47], the flow kick method does not require closeness to the equilibrium of the unperturbed system. This method, however, has only been used in low-dimensional ODE systems, and, to the best of our knowledge it has not been examined either for spatially explicit, or for age-structured systems.

Models and Predictions
In [32] it was demonstrated how the age-of-infection structured model (1.13)–(1.16) yields the same R_0 and steady-state densities as a corresponding ODE model, under an appropriate averaging of those parameters that depend on the age-of-infection. In [5], it was shown that using different values in several parameter sets yields very similar steady states. Finally, in [49], in comparisons of five cholera epidemic models, it was found that R_0 was poorly estimated by all models, due to an identifiability issue between the parameter of pathogen transmission and the pathogen decay rate. At the same time, other parameters were able to be estimated in all models. Hence, any model predictions should be scrutinized and carefully interpreted.

In conclusion, due to noise, heterogeneity in the observed and measured variables, and interactions that are necessarily omitted in mathematical models, direct comparison between simple models and data might not always be feasible [37]. Nevertheless, mechanistic models can be used, in conjunction with biological insight and data, to understand the drivers of observed patterns.

1.5 Models Beyond Ecology and Epidemiology

In this final section we present several recent studies of nonlinear models in biological applications, other than those that are ecological or epidemiological in nature. In particular, we focus on two diseases that are not infectious, namely diabetes and cancer, on models of limb development and pattern formation, and finally on DNA models.

1.5.1 Diabetes

The sugar glucose is the primary energy source for the human body. Glucose concentration in the blood increases through ingestion or through hepatic production, namely when it is secreted in the liver. Insulin and glucagon are two hormones that are produced in the pancreas and regulate the concentration of glucose. Glucagon is secreted by α-cells when glucose in the bloodstream is low, while insulin is secreted by β-cells when glucose levels are high. Insulin promotes the absorption of glucose by cells in the body. High levels of insulin in the blood prohibit glucose secretion in the liver.

Diabetes is a chronic metabolic disorder, occurring when either a person's pancreas does not produce enough insulin (Type I diabetes), or when insulin is produced, but is not used efficiently (Type II diabetes). The resulting health problems of diabetes include major cardiovascular diseases, stroke, vision loss, and kidney disease. According to the Centers for Disease Control and Prevention [50], it is estimated that more than 30 million people in the U.S. had diabetes (of any type) in 2015. In the same report, it is noted that more than 14 million visits to emergency departments around the country listed diabetes as the diagnosis in 2014. The disease also carries great cost, since the total direct and indirect expenses were listed as 245 billion dollars in 2012.

It is of little surprise then that numerous mathematical studies exist, which model the glucose-insulin regulatory system and the effect of drugs and other factors, such as physical exercise, on the regulation of high levels of blood glucose. For a review of the various types of models ranging from ODE-based, delay differential equation-based, as well as stochastic differential equation-based models, see [51]. Delay models, in particular, have played a prominent role, since it is known that time delays exist in the glucose-insulin regulatory system due to the time lag τ_1 between the signal of elevated glucose concentration and the secretion of insulin from the β-cells, as well as the time lag τ_2 between the signal of increased insulin level and the suppression of hepatic production [52, 53]. Besides using delay differential equations, delays can also be modeled by increasing the number of compartments in systems of ordinary differential equations. Another feature of the dynamics of the glucose-insulin regulatory system is the existence of *ultradian oscillations*, namely oscillations with period less than one day. These oscillations occur even in healthy individuals and have a period of about two hours and are characterized by blood glucose fluctuations of about 10% [53].

At a minimum, models of the glucose-insulin system include two compartments, namely one of the glucose concentration $G(t)$ and one for the insulin concentration $I(t)$, where t denotes time and is measured in minutes [52, 53]. The delay differential equation model in [52] reads

$$\frac{\mathrm{d}G}{\mathrm{d}t} = G_{in} - f_2(G(t)) - f_3(G(t))f_4(I(t)) + f_5(I(t - \tau_2)) \tag{1.17}$$

$$\frac{\mathrm{d}I}{\mathrm{d}t} = f_1(G(t - \tau_1)) - d_i I(t), \tag{1.18}$$

where G_{in} denotes the glucose infusion rate, $f_1(G(t - \tau_1))$ denotes the increase of insulin production as a response to the glucose levels, $f_2(G(t))$ and $f_3(G(t))f_4(I(t))$ denote the insulin-independent and insulin-dependent glucose utilization by cells, respectively, $f_5(I(t - \tau_2))$ denotes the hepatic production, and finally d_i is the insulin degradation rate. More recently, a similar model was presented and analyzed [53], where instead of constant insulin degradation d_i, a Michaelis–Menten (1.4) degradation was included, a prefactor was included in front of $f_3(G(t))f_4(I(t))$ to model the difficulty in the uptake of glucose by cells in a diabetic person, as well as an additional factor accounting for the positive effect of exercise on insulin sensitivity. In the insulin compartment (1.18), a constant insulin infusion rate was added and also a prefactor in front of $f_5(G(t - \tau_1))$ modeling the decreased pancreatic efficiency in a diabetic person. This study focused on the role of physical exercise and a medication regime in controlling glucose levels. The study offered a promising glimpse of what mathematical models have to offer in personalized medicine and the design of the so-called artificial pancreas.

ODE models of the glucose-insulin system based on Lotka–Volterra type interactions (1.1)–(1.2) have also been considered [54]. In [54], insulin was the predator and glucose the prey, and the density of β-cells was also included. The dynamics of the system included interesting phenomena, such as a parameter regime of periodic behavior and another of chaotic dynamics. This further highlights the fact that basic ecological and epidemiological models are often used as the building blocks of models for diverse biological systems, as mentioned in Sect. 1.1.7.

1.5.2 Cancer

Another disease with high incidence and high cost of treatment is cancer. Not only is it among the leading causes of death in about 100 countries, but the number of new cases is continuously increasing [55]. There were 18.1 million new cases and 9.6 million cancer deaths in 2018 alone. While surgery, radiation, and chemotherapy remain the most common treatments, combination therapies also exist, including, for instance, immunotherapy and chemotherapy.

Mathematical models of cancer-related processes are as diverse as the cancers that affect the human body. In [56] a timeline is presented that summarizes 50 years of prominent mathematical models of cancer. Such models have evolved from phenomenological ones [57], aiming to discover the mechanism of tumor invasion in healthy tissue, to those studying the effect of combination therapies parameterized by actual patient data [58, 59].

Phenomenological models, namely those that are constructed using observations of general relationships between variables, may be similar in nature to ecological models. For instance, the interaction between healthy and neoplastic (tumor) tissue has been modeled by reaction-diffusion equations [57] that include Lotka–Volterra competitive interactions, since the two tissue types compete for space and resources. Lotka–Volterra interactions have also been used in an ODE model of immunotherapy [58], that eventually gives rise to a *hybrid model*, namely one where a deterministic framework is used as long as population values exceed a threshold. When they decrease below it, stochastic simulations are used for the populations. The growth and invasion of cancer cells is a complicated process, affected by environmental conditions, such as hypoxia [60], and including other processes, such as angiogenesis [56] from the regular tissue to the tumor.

When spatial dynamics is taken into account, healthy tissue is not diffusing, whereas neoplastic tissue diffuses at a rate depending on the healthy tissue concentration. In cancer and other biological processes, it is quite common to encounter interacting diffusing and non-diffusing variables, resulting in interesting stability problems of the model steady-states [61]. In the case of cancer invasion, stability analysis reveals correlations between the structure of the healthy-tumor tissue interface and the propagation speed of the tumor tissue [57]. More recently, a macroscopic mathematical model combined with diffusion tensor imaging data was used to study the anisotropic invasion of gliomas [62]. By neglecting tumor-host interactions beyond the directional spread of cancer cells, the model reads

$$c_t = \nabla\nabla\left(D_C(x)c\right) + f(c), \tag{1.19}$$

where c denotes the macroscopic cell density as a function of time t and position in space x, $D_C(x)$ the macroscopic glioma cell diffusion tensor, and $f(c)$ the cell proliferation rate, which is a standard logistic growth rate (1.5). The model was developed by considering the individual-level dynamics of cells and then obtaining the mascroscopic model (1.19) by scaling techniques, hence it differs from phenomenological models.

1.5.3 Limb Development and Animal Coat Patterns

Models of animal limb development and coat patterns, also heavily rely on reaction-diffusion equations. Alan Turing in his seminal work in 1952 [63], demonstrated how morphogens, namely chemical substances that react together and diffuse through tissue, form the basis of pattern formation. Since then, his idea has been used to reproduce the spot- and stripe-like patterns on the coat of many living organisms [64], and to model limb and fin development in animals ranging from *Brachypterygius* (a type of *Ichthyosaur*) [65], to catsharks [66], and in mice [67]. Patterns through the Turing mechanism are generated through an instability of the homogeneous state as two or more morphogens diffuse and react, giving rise to stripes, spots, rings, and spirals.

As an example of such a system we use the following model for vertebrate limb development [65]

$$\frac{\partial c_a}{\partial t} = D_a \nabla^2 c_a + U(c_a) - k_a c_a c_i \tag{1.20}$$

$$\frac{\partial c_i}{\partial t} = D_i \nabla^2 c_i + V(c_a) - k_a c_a c_i, \tag{1.21}$$

where c_a and c_i denote the concentrations of two morphogens, the activator and the inhibitor, respectively, D_a, D_i denote the diffusion coefficients, U and V are functions that control morphogen production through feedback relationships, and finally, k_a is a decay constant. The reaction-diffusion system (1.20)–(1.21) was simulated on domains resembling limb-bud contours using finite element algorithms. The simulation patterns showed a striking resemblance with actual limbs and the authors proposed several future directions, including an extension of the model to take into account the Hox gene regulatory network. Hox genes regulate segmentation, are considered conserved across animals, and their expression patterns have also been found to correspond to coloration patterns in bumblebees [68].

1.5.4 DNA

Besides mathematical biology, biophysics is another area that has contributed notable models of biological processes. We will mainly focus on one such model, namely the Peyrard–Bishop DNA model [69, 70], because its analysis has generated interesting mathematics [71] and because it has been used to address significant biological questions [72] (see also [73] for a review of other models).

For each pair of bases n (Adenine-Thymine or Guanine-Cytosine), a variable for the out-of phase motion y_n is attributed, since this is the motion that stretches the hydrogen bonds. The stretching of the bonds plays a key role in DNA thermal denaturation, which is a stepping stone for understanding DNA transcription. Like thermal denaturation, transcription requires the two strands of the DNA helix to be separated. The potential representing the hydrogen bonds is assumed to be a Morse potential

$$V(y_n) = D_n(e^{-a_n\sqrt{2}y_n} - 1)^2,$$

where D_n denotes the depth of the potential and a_n sets the range of the interaction. To represent the interaction between nearest-neighbor bases belonging to the same strand, a stacking interaction term is included:

$$W(y_n, y_{n-1}) = \frac{k_n}{2}\left(1 + \rho e^{-b(y_n+y_{n-1})}\right)(y_n - y_{n-1})^2,$$

where k_n is a spring constant. Originally $\rho = 0$ was taken [69], but a few years later $\rho > 0$ was introduced to effectively modify the spring constant, taking into account the cooperativity between and bases and capturing the sharp melting transition of the double helix [74].

The model has been used to study the formation of DNA bubbles, namely regions along the helix where the strands are separated due to thermal destabilization [75, 76], the unzipping of DNA by mechanical forces [72, 77], and the identification of promoter regions [78]. Ising type models, where the base pair is assumed to be in either the open or closed state [79, 80] have been introduced as a coarse-grained version of models such as the Peyrard–Bishop model, where the continuous variable y_n is used to characterize the state of the hydrogen bonds.

DNA denaturation based on the Peyrard–Bishop model is also related to multibreathers in discrete Klein–Gordon equations [81–83], in either the original model or a modified model accounting for base interactions beyond the nearest neighbors [84]. Multibreathers are time-periodic solutions that are spatially localized and are solutions to the equations of motion of the Hamiltonian

$$H(y) = \sum_n \left(\frac{m}{2}(y_n')^2 + W(y_n, y_{n-1}) + V(y_n)\right).$$

They have nontrivial existence and stability properties, which have been studied both numerically [85] and analytically with the use of Floquet theory [86]. They have been demonstrated to correlate with promoter strength and the location of transcription start sites [85], and it is believed that stable strand separations could alter gene expression in the presence of electromagnetic THz radiation [86].

1.5.5 Concluding Remarks

In conclusion, there is a wealth of open problems in both eco-epidemiology and in other types of biological models, such as those presented in this section. One of the common themes is the call for a strong collaboration between experimentalists and theoreticians, and also data-theory coupling. As the modeling of human diseases such as diabetes and cancer has demonstrated, one of the opportunities of mathematical modeling is the development of personalized treatments. Time series analysis and parameter identifiability methods are going to be as important for these kind of models as they have been for eco-epidemiological ones.

To end on a positive note, quite recently, persisting predator-prey oscillations, that have long been predicted by mathematical models, were realized experimentally [87]. Under constant experimental conditions, the population cycles persisted for approximately 300 generations of the rotifer predator, which is unprecedented. Moreover, a mathematical model was created that exhibits a Hopf bifurcation, mirroring the transition between a steady state and oscillations, that was observed in the experiments. This recent development highlights the power and usefulness of mathematical models when combined with experiments.

Acknowledgements We gratefully acknowledge support through grants NSF-DMS-1815764 and UIUC RB17060.

References

1. D.J.T. Sumpter, M. Beekman, Anim. Behav. **66**, 273 (2003)
2. M.J. Ryan, R.A. Page, K.L. Hunter, R.C. Taylor, Anim. Behav. **147**, 189 (2019)
3. M.J. Hatcher, J.T.A. Dick, A.M. Dunn, Ecol. Lett. **9**, 1253 (2006)
4. E.E. Werner, S.D. Peacor, Ecology **84**, 1083 (2003)
5. Z. Rapti, T.E. Stewart Merrill, B. Mueller-Brennan, J.H. Kavouras, C.E. Cáceres, Theor. Popul. Biol. **130**, 132 (2019)
6. C.R. Bertram, M. Pinkowski, S.R. Hall, M.A. Duffy, C.E. Cáceres, Oecologia **173**, 1023 (2013)
7. A. Lotka, *Elements of Physical Biology* (Williams and Wilkins, Baltimore, 1925)
8. V. Volterra, Mem. R. Accad. Naz. dei Lincei **2**, 31 (1926)
9. C.S. Holling, Can. Entomol. **91**, 385 (1959)
10. M.L. Rosenzweig, R.H. MacArthur, Amer. Nat. **97**, 209 (1963)
11. J.H.P. Dawes, M.O. Souza, J. Theor. Biol. **327**, 11 (2013)
12. P.A. Abrams, Amer. Nat. **121**, 887 (1983)
13. J.S. Weitz, J. Dushoff, Theor. Ecol. **1**, 13 (2008)
14. J. Grilli, G. Barabás. M.J. Michalska-Smith, S. Allesina, Nature **548**, 210 (2017)
15. W.O. Kermack, A.G. McKendrick, Proc. R. Soc. A **115**, 700 (1927)
16. H.W. Hethocote, SIAM Rev. **42**, 599 (2000)
17. J. Monod, Annu. Rev. Microbiol. **3**, 371 (1949)
18. K.A. Johnson, R.S. Goody, Biochemistry **50**, 8464 (2011)
19. S.M. Clifton, T. Kim, J.H. Chandrashekhar, G.A. O'Toole, Z. Rapti, R.J. Whitaker, mSystems **4**, e00221-19 (2019)
20. L.A. Real, Amer. Nat. **111**, 289 (1977)
21. A.V. Hill, J. Physiol. **40**(Suppl.), i–vii (1910)
22. J.-D. Bézanet, M. Bischofberger, E. Tiecke, A. Gonçalves, J.F. Martin, A. Zuniga, F. Naef, R. Zeller, Science **323**, 1050 (2009)
23. K.E. Sears, J.A. Maier, M. Rivas-Astroza, R. Poe, S. Zhong, K. Kosog, J.D. Marcot, R.R. Behringer, C.J. Cretekos, J.J. Rasweiler IV, Z. Rapti, PLoS Genet. **11**, e1005398 (2015)

24. K.D. Lafferty, G. DeLeo, C.J. Briggs, A.P. Dobson, T. Gross, A.M. Kuris, Science **349**, 854 (2015)
25. S. Estrela, C.H. Trisos, S.P. Brown, Amer. Nat. **180**, 566–576 (2012)
26. C.E. Cáceres, G. Davis, S. Duple, S.R. Hall, A. Koss, P. Lee, Z. Rapti, Math. Biosci. **258**, 148 (2014)
27. O.S. Venturelli, H. El-Samad, R.M. Murray, Proc. Natl. Acad. Sci. USA **109**, E3324 (2012)
28. T.S. Gardner, C.R. Cantor, J.J. Collins, Nature **403**, 339 (2000)
29. M. Beekman, D.J.T. Sumpter, F.L.W. Ratnieks, Proc. Natl. Acad. Sci. USA **98**, 9703 (2001)
30. J.C. de Roode, A.J. Yates, S. Altizer, Proc. Natl. Acad. Sci. USA **105**, 7489 (2008)
31. R.E. Lenski, R.M. May, J. Theor. Biol. **169**, 253 (1994)
32. Z. Rapti, C.E. Cáceres, B. Math, Biol. **78**, 235 (2016)
33. C.-J. Lin, K.A. Deger, J.H. Tien, Math. Biosci. **277**, 15 (2016)
34. S. Alizon, Amer. Nat. **172**, E113 (2008)
35. M.A. Gil, Ecology **94**, 1871 (2013)
36. R.H. Condon, C.M. Duarte, K.A. Pitt, K.L. Robinson, C.H. Lucas, K.R. Sutherland, H.W. Mianzan, M. Bogeberg, J.E. Purcell, M.B. Decker, S-i. Uye, L.P. Madin, R.D. Brodeur, S.H.D. Haddock, A. Malej, G.D. Parry, E. Eriksen, J. Quiñones, M. Acha, M. Harvey, J.M. Arthur, W.M. Graham, Proc. Natl. Acad. Sci. USA **110**, 1000 (2013)
37. R.M. May, Proc. Roy. Soc. Lond. B **228**, 241 (1985)
38. N.N. Mangan, S.L. Brunton, J.L. Proctor, J.N. Kutz, I.E.E.E. Trans, Mol. Biol. Multi-Scale Commun. **2**, 52 (2016)
39. Y.H. Kao, M. Eisenberg, Epidemics **25**, 89 (2018)
40. W.J. Sutherland, R.P. Freckleton, H.C.J. Godfray, S.R. Beissinger, T. Benton, D.D. Cameron, Y. Carmel, D.A. Coomes, T. Coulson, M.C. Emmerson, R.S. Hails, G.C. Hays, D.J. Hodgson, M.J. Hutchings, D. Johnson, J.P.G. Jones, M.J. Keeling, H. Kokko, W.E. Kunin, X. Lambin, O.T. Lewis, Y. Mahli, N. Mieszkowska, E.J. Milner-Gullard, K. Norris, A.B. Phillimore, D.W. Purves, J.M. Reid, D.C. Reuman, K. Thompson, J.M.J. Travis, L.A. Turnbull, D.A. Wardle, T. Wiegand, J. Ecol. **101**, 58 (2013)
41. K. Faust, L. Lahti, D. Gonze, W.M. de Vos, J. Raes, Curr. Opin. Microbiol. **25**, 56 (2015)
42. C. Bandt, B. Pompe, Phys. Rev. Lett. **88**, 174102 (2002)
43. Y. Cao, W-w. Tung, J.B. Gao, V.A. Protopopescu, L.M. Hively, Phys. Rev. E **70**, 046217 (2004)
44. M. Zanin, L. Zunino, O.A. Rosso, D. Papo, Entropy **14**, 1553 (2012)
45. P.J. Mumbly, A. Hastings, H.J. Edwards, Nature **450**, 98 (2007)
46. M. Scheffer, S.H. Hosper, M.-L. Meijer, B. Moss, E. Jeppesen, Trends Ecol. Evol. **8**, 275 (1993)
47. M. Scheffer, J. Bascompte, W.A. Brock, V. Brovkin, S.R. Carpenter, V. Dakos, H. Held, E.H. van Nes, M. Rietkerk, G. Sugihara, Nature **461**, 53 (2009)
48. K. Meyer, A. Hoyer-Leitzel, S. Iams, I. Klusky, V. Lee, S. Ligtenberg, E. Bussmann, M.L. Zeeman, Nat. Sustain. **1**, 671 (2018)
49. E.C. Lee, M.R. Kelly Jr., B.M. Ochocki, S.M. Akinwumi, K.E.S. Hamre, J.H. Tien, M.C. Eisenberg, J. Theor. Biol. **420**, 68 (2017)
50. National Diabetes Statistics Report, 2017 (Centers for Disease Control and Prevention, U.S. Department of Health and Human Services, Atlanta, USA, 2017)
51. P. Palumbo, S. Ditlevsen, A. Bertuzzi, A. De Gaetano, Math. Biosci. **244**, 69 (2013)
52. J. Li, Y. Kuang, SIAM J. Appl. Math. **67**, 757 (2007)
53. S.M. Kissler, C. Cichowitz, S. Sankaranarayanan, D.M. Bortz, J. Theor. Biol. **359**, 101 (2014)
54. P.S. Shabestari, S. Panahi, B. Hatef, S. Jafari, J.C. Sprott, Chaos. Soliton Fract. **112**, 44 (2018)
55. F. Bray, J. Ferlay, I. Soerjomataram, R.L. Siegel, L.A. Torre, A. Jemal, CA-Cancer J. Clin. **68**, 394 (2018)
56. H.M. Byrne, Nat. Rev. Cancer **10**, 221 (2010)
57. R.A. Gatenby, E.T. Gawlinksi, Cancer Res. **56**, 5745 (1996)
58. G.J. Kimmel, F.L. Locke, P.M. Altrock, Evolutionary dynamics of CAR T cell therapy (2019). https://doi.org/10.1101/717074
59. L.G. de Pillis, W. Gu, A.E. Radunskaya, J. Theor. Biol. **238**, 841 (2006)
60. T. Alarcón, H.M. Byrne, P.K. Maini, J. Theor. Biol. **229**, 395 (2004)

61. A.K. Barreiro, J.C. Bronski, Z. Rapti, SIAM J. Math. Anal. **51**, 256 (2019)
62. K.J. Painter, T. Hillen, J. Theor. Biol. **323**, 25 (2013)
63. A.M. Turing, Phil. Trans. Royal Soc. B **237**, 37 (1952)
64. J.D. Murray, Sci. Am. **258**, 80 (1988)
65. J. Zhu, Y.T. Yang, M.S. Alber, S.A. Newman, PLoS ONE **5**, e10892 (2010)
66. K. Onimaru, L. Marcon, M. Musy, M. Tanaka, J. Sharpe, Nat. Commun. **7**, 11582 (2016)
67. A. Badugu, C. Kraemer, P. Germann, D. Menshykau, D. Iber, Sci. Rep. **2**, 991 (2012)
68. Z. Rapti, M.A. Duennes, S.A. Cameron, Biol. J. Linn. Soc. **113**, 384 (2014)
69. M. Peyrard, A.R. Bishop, Phys. Rev. Lett. **62**, 2755 (1989)
70. M. Peyrard, Nonlinearity **17**, R1 (2004)
71. J. Cuevas-Maraver, P.G. Kevrekidis, in *A Dynamical Perspective on the ϕ^4 Model: Past, Present and Future*, ed. by P.G. Kevrekidis, J. Cuevas-Maraver (Springer, Cham, 2019), p. 137
72. N.K. Voulgarakis, A. Redondo, A.R. Bihop, K.Ø. Rasmussen, Phys. Rev. Lett. **96**, 248101 (2006)
73. L.V. Yakushevich, *Nonlinear Physics of the DNA*, 2nd edn. (Wiley, Hoboken, 2004)
74. T. Dauxois, M. Peyrard, A.R. Bishop, Phys. Rev. E **47**, R44 (1993)
75. Z. Rapti, A. Smerzi, K.Ø. Rasmussen, A.R. Bishop, C.H. Choi, A. Usheva, Europhys. Lett. **74**, 540 (2006)
76. Z. Rapti, A. Smerzi, K.Ø. Rasmussen, A.R. Bishop, C.H. Choi, A. Usheva, Phys. Rev. E **73**, 051902 (2006)
77. Z. Rapti, K.Ø. Rasmussen, A.R. Bishop, J. Nonlinear Math. Phys. **18**, 381 (2011)
78. C.H. Choi, Z. Rapti, V. Gelev, M.R. Hacker, B. Alexandrov, E.J. Park, J.S. Park, N. Horikoshi, A. Smerzi, K.Ø. Rasmussen, A.R. Bishop, A. Usheva, Biophys. J. **95**, 597 (2008)
79. A. Krueger, E. Protozanova, M.D. Frank-Kamenetskii, Biophys. J. **90**, 3091 (2006)
80. M.R. Kantorovitz, Z. Rapti, V. Gelev, A. Usheva, BMC Bioinform. **11**, 604 (2010)
81. J. Cuevas-Maraver, P.G. Kevrekidis, D.E. Pelinovsky, Stud. Appl. Math. **137**, 214 (2016)
82. Z. Rapti, Phys. Lett. A **377**, 1543 (2013)
83. V. Koukouloyannis, P.G. Kevrekidis, J. Cuevas, V. Rothos, Phys. D **241**, 16 (2013)
84. Z. Rapti, Eur. Phys. J. E **32**, 209 (2010)
85. B.S. Alexandrov, V. Gelev, S.W. Yoo, L.B. Alexandrov, Y. Fukuyo, A.R. Bishop, K.Ø. Rasmussen, A. Usheva, Nucleic Acids Res. **38**, 1790 (2010)
86. P. Maniadis, B.S. Alexandrov, A.R. Bishop, K.Ø. Rasmussen, Phys. Rev. E **83**, 011904 (2011)
87. B. Blasius, L. Rudolf, G. Weithoff, U. Gaedke, G.F. Fussmann, Nature **577**, 226 (2020)

Chapter 2
Nonlinearity and Topology

Avadh Saxena, Panayotis G. Kevrekidis, and Jesús Cuevas-Maraver

Abstract The interplay of nonlinearity and topology results in many novel and emergent properties across a number of physical systems such as chiral magnets, nematic liquid crystals, Bose–Einstein condensates, photonics, high energy physics, etc. It also results in a wide variety of topological defects such as solitons, vortices, skyrmions, merons, hopfions, monopoles to name just a few. The interaction among and collision of these nontrivial defects itself constitute topics of significant interest. Curvature and underlying geometry also affect the shape, interaction and behavior of these defects. Such properties can be studied using techniques such as, e.g. the Bogomolnyi decomposition. Some applications of this interplay, e.g. in nonreciprocal photonics as well as topological materials such as Dirac and Weyl semimetals, are also elucidated.

A. Saxena (✉)
Theoretical Division, Los Alamos National Laboratory,
Los Alamos, NM 87545, USA
e-mail: avadh@lanl.gov

P. G. Kevrekidis
Department of Mathematics and Statistics, University of Massachusetts, Amherst, MA 01003-4515, USA
e-mail: kevrekid@umass.edu

J. Cuevas-Maraver
Departamento de Fisica Aplicada I, Grupo de Fisica No Lineal, Universidad de Sevilla, Escuela Politécnica Superior, C/ Virgen de África 7, 41011 Sevilla, Spain
e-mail: jcuevas@us.es

Instituto de Matemáticas de la Universidad de Sevilla (IMUS), Edificio Celestino Mutis, Avda. Reina Mercedes s/n, 41012 Sevilla, Spain

P. G. Kevrekidis et al. (eds.), *Emerging Frontiers in Nonlinear Science*, Nonlinear Systems and Complexity 32, https://doi.org/10.1007/978-3-030-44992-6_2

2.1 Introduction

The main context of this chapter is how topological effects in nonlinear systems give rise to a rich playground of excitations and properties. Topology, whether in real space or momentum space or more generally in a parameter space, is associated with certain system properties remaining unaltered under continuous deformation. It follows that during deformation neighboring points remain close to each other. Topology could be local, e.g. change in the lattice or network due to a defect, or global. The latter means attributes such as the genus (g) or Euler characteristic (χ) are overall or global features of a system.

Boundary conditions play an important role (through χ), for example in finite carbon nanotubes or *edge states* in topological materials: quantum Hall systems, topological insulators [1], topological superconductors [2], Dirac and Weyl semimetals [3], etc. The latter are three dimensional analogs of graphene featuring gapless electronic excitations that are protected by topology and (time reversal, space inversion or other crystalline) symmetry [3]. Very recent experiments indicate that these materials, specifically Weyl semimetals, may also provide a realization of axions, very weakly interacting neutral particles in quantum field theory and potential candidates for dark matter, in condensed matter [4, 5].

In many physical systems and materials [6] there are point defects as well as extended or topological defects. The topological defects can significantly alter the physical properties and dynamics of the system. Apart from the celebrated soliton-like defects there is a whole slew of more elaborate ones that include skyrmions, merons, hopfions, monopoles, dislocations, and disclinations among others. In this chapter we discuss how such defects arise in chiral magnets, nematic liquid crystals, Bose–Einstein condensates (BECs), etc. We also discuss the role of topology in the momentum space, particularly in the context of topological materials.

The combination of nonlinearity and topology also provides a highly desirable functionality in photonics, namely nonreciprocity, which is quite important for a variety of photonic devices including optical isolators [7]. We thus provide examples of the interplay between nonlinearity and topology in photonics as well as condensed matter analogs. In addition, we illustrate the role of geometry and topology in determining spin textures via the so-called Bogomolnyi decomposition [8]. Finally, we discuss several open problems and future directions with regard to the role of topology in the presence of nonlinearity.

2.2 Topological Defects in Nonlinear Field Theories

We consider a variety of topological defects that arise in a number of field theories including the nonlinear σ-model [9, 10]. The Hamiltonian for the latter is given by

$$H = \int (\nabla \mathbf{n})^2 \mathrm{d}^2 x \,, \quad \mathbf{n}^2 = 1 \,, \tag{2.1}$$

where the unit vector $\mathbf{n}$ lives on a unit sphere. This model can support scalar soliton configurations under special conditions. However, under a scaling transformation $x \rightarrow \lambda x$ and $y \rightarrow \lambda y$ the Hamiltonian H remains invariant and thus a soliton solution can be trivially scaled to a point because there is no length scale in the plane. We will return to this point later when we consider the nonlinear σ-model on curved manifolds.

We note here that to describe disordered Weyl semimetals an anisotropic topological term can be analytically derived from the action of the nonlinear σ model [11]. In the next section we consider a variety of topological defects such as skyrmions, merons and hopfions. Given their extensive interest and applicability, subsequently in Sect. 2.4, we consider vortices and vortex loops/rings. We then turn to different prototypical applications such as liquid crystals (and the emergence of skyrmions in them) in Sect. 2.5, as well as Bose–Einstein condensates in Sect. 2.6. After providing an example of a theoretical tool for the study of topology in curved manifolds via the Bogomolnyi decomposition (Sect. 2.7), we present a broader perspective of the impact of topological ideas in Materials (Sect. 2.8), Optics (Sect. 2.9) and Acoustics and beyond (Sect. 2.10). Then in Sect. 2.11, we summarize our findings and present our Conclusions, as well as some directions for future work.

2.3 Skyrmions, Merons and Hopfions

Beyond the well known solitons, there are more exotic topological defects such as vector field or spin textures called skyrmions [10], which can have topological charge of one (or two or even more). As shown in Fig. 2.1, in a skyrmion at the outer boundary all spins point up (red arrows) whereas at the center there is a spin pointing down (blue arrow). Therefore somewhere in the middle the spins have to lie in the plane (green arrows). Half skyrmions are also referred to as merons and have a topological charge of one half: the outer spins point up whereas the spin in the center lies in the plane. There are many other related topological defects such as sphalerons and bags (or lumps) known in high energy physics [9]. Similarly, three dimensional defects, e.g. vortex lines, vortex loops (rings) and knots appear in many physical systems, including, e.g. polymeric knots [12].

2.3.1 *Skyrmions in Chiral Magnets*

Before discussing magnets we note that beyond magnetic materials (e.g. ferroelectrics) skyrmions have been observed at interfaces [13]. In this case, the texture

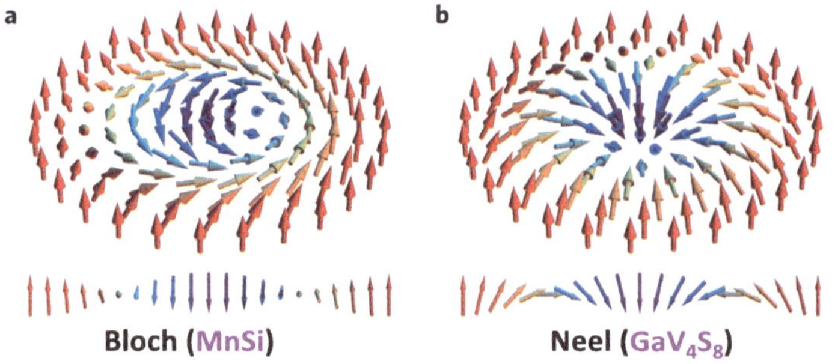

Fig. 2.1 **a** Bloch and **b** Néel type skyrmions in a chiral magnet MnSi and a lacunar spinel GaV_4S_8, respectively. Below the skyrmion textures their radial cross sections show Bloch and Néel wall, respectively. Reproduced from [16]. ©2015 by the Nature Publishing Group

is given in terms of polar vectors or electric dipoles. Nanodots and nanocomposites can also stabilize skyrmions (due to boundary conditions) in polar materials [14].

The Hamiltonian for a chiral magnet (i.e. lacking spatial inversion symmetry in its crystal structure, e.g. MnSi) consists of the nonlinear σ-model plus the Zeeman term in addition to the Dzyaloshinskii–Moriya interaction [15]

$$H = \int \left[J(\nabla \mathbf{n})^2 + D\mathbf{n} \cdot (\nabla \times \mathbf{n}) - \mathbf{n} \cdot \mathbf{B} \right] \mathrm{d}^2 x \,, \tag{2.2}$$

where J denotes the (magnetic) exchange constant, $\mathbf{B}$ is the external magnetic field in the last term in the Hamiltonian representing the Zeeman interaction and D represents the strength of the Dzyaloshinskii–Moriya interaction. The latter arises from the spin-orbit interaction at the microscopic level. Here $\mathbf{n}$ is a unit vector describing the direction of the magnetic moment.

The topological charge associated with a skyrmion spin configuration is given by [15]

$$Q = \frac{1}{4\pi} \int \mathrm{d}r^2 \left[\mathbf{n} \cdot (\partial_x \mathbf{n} \times \partial_y \mathbf{n}) \right] = \pm 1 \,. \tag{2.3}$$

For a metallic material when a conduction electron traverses across a skyrmion it gets spin polarized as a result of its interaction with the spin configuration of the skyrmion. In addition, the conduction electron is subjected to an effective electric and magnetic field [15], respectively given by

$$E = \frac{\hbar}{2e} [\mathbf{n} \cdot (\nabla \mathbf{n} \times \partial_t \mathbf{n})], \tag{2.4}$$

$$B = \frac{\hbar c}{2e} [\mathbf{n} \cdot (\partial_x \mathbf{n} \times \partial_y \mathbf{n})]. \tag{2.5}$$

These emergent fields give rise to the topological and skyrmion Hall effects.

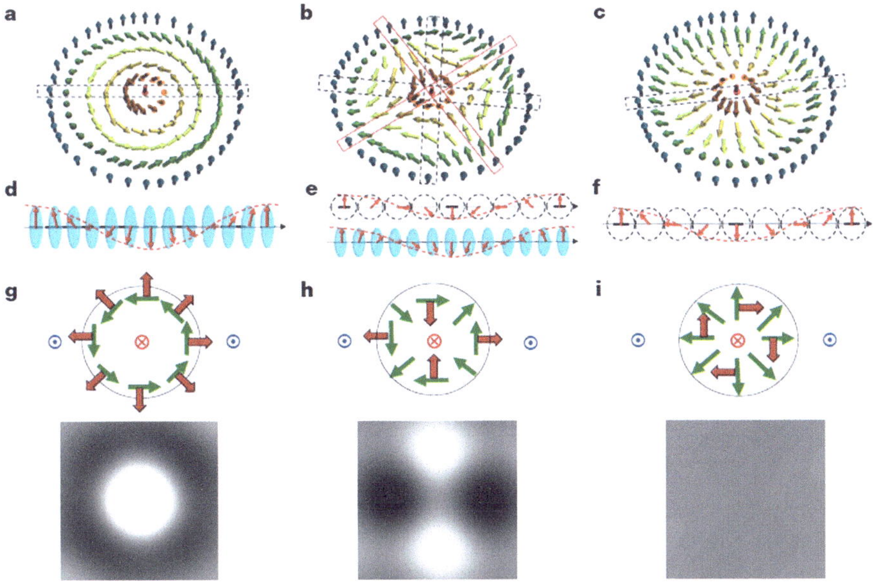

Fig. 2.2 **a** Anti-skyrmion (center panels) in $Mn_{1.4}Pt_{0.9}Pd_{0.1}Sn$ and its comparison with the Bloch (left panels) and Néel skyrmions (right panels). The radial cross sections are shown below the spin textures. Schematics of the magnetic moments (green arrows) and the associated Lorentz deflections of transmitted electrons (red arrows) are shown in the middle row. The bottom row shows the corresponding simulated Lorentz Transmission Electron Microscopy (LTEM) patterns with dark and bright lobes. Reproduced from [17]. ©2017 by the Nature Publishing Group

Depending on the material or system, skyrmions can have chirality (i.e. Bloch skyrmion as in MnSi) or no chirality (i.e. Néel skyrmion as in GaV_4S_8) as depicted in Fig. 2.1. They are so called because the radial cut of a Bloch skyrmion provides a magnetic Bloch domain wall (which is a strictly 3D structure) whereas a radial cut of the Néel skyrmion leads to a magnetic Néel domain wall (which is a 2D structure). Note that the spin configurations of the two types of skyrmions are topologically equivalent. Anti-skyrmions, which have structural characteristics of both the Bloch and Néel skyrmions, have also been observed using Lorentz Transmission Electron Microscopy (LTEM) [17] in tetragonal Heusler materials even above the room temperature, as shown in Fig. 2.2. We note here that spin-1 photonic skyrmions have also been described in the literature [18].

2.3.2 *Merons*

Half-skyrmions whose field covers only a hemisphere (i.e. topological charge 1/2) are known as merons. They have been observed (and modeled) in liquid crystals [19] and magnetic multi-layers [20]. If we add a magnetic anisotropy energy term

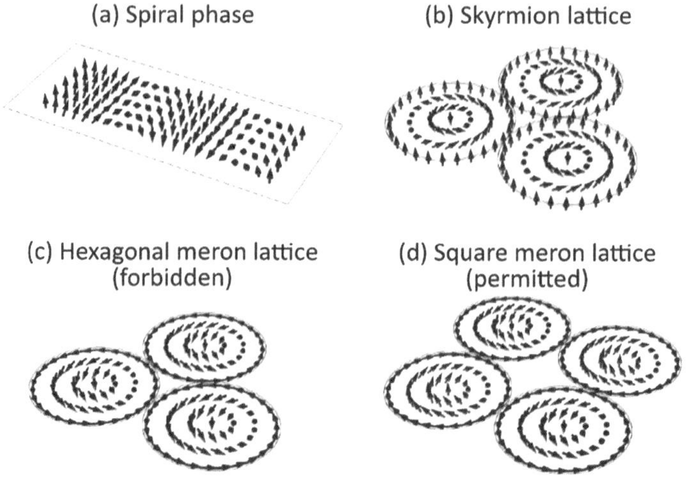

Fig. 2.3 Comparison of skyrmion and meron latices in chiral magnets. A triangular skyrmion lattice arises from a spiral phase (top row) but a triangular meron lattice is at best metastable (bottom left). In contrast, a square meron lattice is stable (bottom right) but a square skyrmion lattice is at best metastable. On the other hand, in nematic liquid crystals a triangular meron lattice is stable but a triangular skyrmion lattice is unstable (not shown). Reproduced from [19]

An_z^2 to the skyrmion Hamiltonian in (2.2), in the large anisotropy limit a skyrmion breaks into merons [21]. When $A > 0$ it is called the easy-plane anisotropy which is what we will consider here. The case of $A < 0$ is called the easy-axis anisotropy. As A is increased the skyrmion size increases, particularly due to the expansion of the equatorial region of the skyrmion. At a certain large value of A, skyrmions become unstable and merons emerge.

In chiral magnets a stable triangular lattice of skyrmions emerges from the spiral (or helical) phase as depicted in Fig. 2.3. In the spiral phase all spins are parallel and point in the same direction at one end. After a helical twist of 2π they come back to the original parallel state, see Fig. 2.3a. The square skyrmion lattice is at best metastable; however, a square meron lattice is allowed. On the contrary, in nematic liquid crystals the triangular skyrmion lattice is at best metastable but a triangular meron lattice is stable [19].

Merons have been experimentally observed at room temperature in a chiral-lattice magnet $Co_8Zn_9Mn_3$, a material exhibiting in-plane magnetic anisotropy [22]. In this material a meron-antimeron square lattice emerges from the helical state of spins and then transforms into a triangular lattice of skyrmions when a magnetic field is applied, see Fig. 2.4. Interestingly, in analogy with the baryon model in high energy physics, skyrmion bags have been observed both in chiral magnets and nematic liquid crystals [23]. These bags are multi-skyrmion configurations where a large skyrmion contains a variable number of anti-skyrmions inside it, as depicted in Fig. 2.5. Merons can be compared and contrasted with magnetic vortices; there is a difference in their spin configurations.

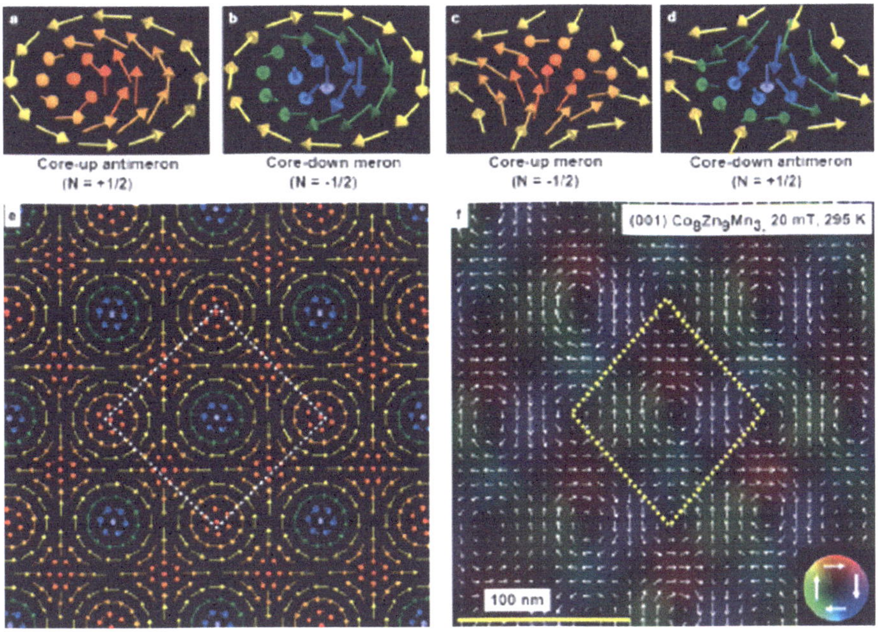

Fig. 2.4 a–d Various experimentally observed meron and anti-meron spin structures in the chiral magnet $Co_8Zn_9Mn_3$ at room temperature. **e** Theoretically predicted and **f** LTEM observed meron/anti-meron square lattice. Reproduced from [22]

2.3.3 *Hopfions and Torons*

Three dimensional topological solitons (appearing in $3 + 1$ dimensional scalar field theories) that can be characterized by the integer-valued Hopf invariant are known as hopfions [10]. Ludwig Faddeev proposed their existence in the 1970s. They represent one of the best known examples of knot solitons in field theory. In 1931 Heinz Hopf considered a link of two loops, thus paving the way for the linking number of circles as a topological invariant, i.e. the Hopf number.

The topological charge or linking number of a hopfion is the homotopy group of the Hopf map $\pi_3(S^2) = \mathbf{Z}$, where $\mathbf{Z}$ is the group of relative integers. The Hopf fibration is a topologically stable texture of a smooth, global configuration of a field. In effect, it is an interwoven structure of preimages. A preimage is defined as the set of all points where a field orientation takes a specific value. The Hopf fibration has been observed in liquid crystals [24] and so are hopfions [25], see Figs. 2.6 and 2.7. Similarly, there are light controlled torons in liquid crystals [26, 27]. The toron is essentially a tube of double twist which is wrapped upon itself such that its boundary forms a torus. It contains two point defects, which can be manipulated to create a defect free structure topologically equivalent to a Hopf fibration. Finally, hopfions

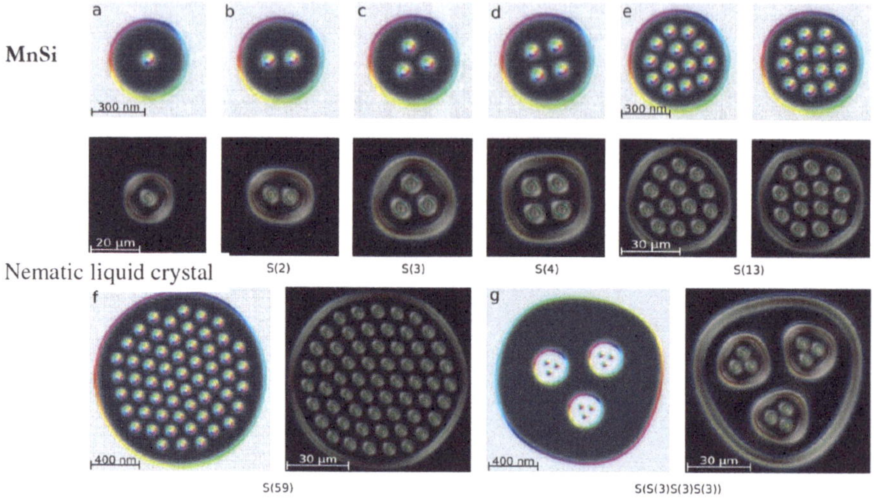

Fig. 2.5 Skyrmion bags in chiral magnets (MnSi) and nematic liquid crystals. Top panels correspond to experimental (optical micrographs) observations in MnSi whereas middle panels are simulations of the above states. Second and fourth panels in the bottom panel are the result of simulations for the state of the optical micrograph at their left. This figure has been adapted from the preprint version of [23] (arXiv:1806.02576v1). Only some of its panels were finally published in [23]. ©2019 by the Nature Publishing Group

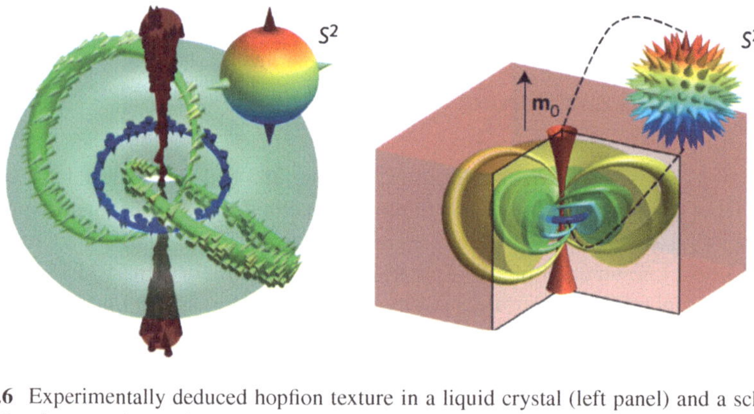

Fig. 2.6 Experimentally deduced hopfion texture in a liquid crystal (left panel) and a schematic of the hopfion as a knotted soliton (right panel). The hopfion is obtained by linking the circle-like preimages residing on nested tori in the material's 3D space. The preimages correspond to color-coded points on S^2. Reproduced from [25]. ©2017 by the Nature Publishing Group

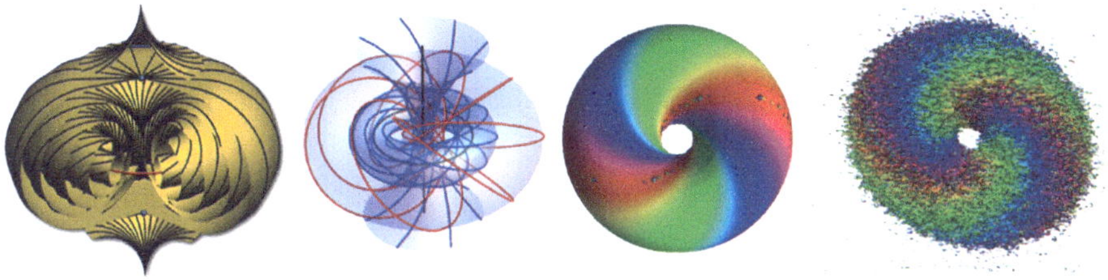

Fig. 2.7 Left panel: The texture of a toron (reproduced from [26]). ©2010 by the Nature Publishing Group). Second panel: Flow lines of the Hopf fibration. Third panel: The preimage surface of the Hopf fibration. Right panel: The experimental preimage of a Hopf fibration. Reproduced from [24]. ©2010 by the American Physical Society

have also been considered in chiral magnets [28] and torus knots have been described as hopfions [29].

2.3.4 Monopoles

As such, free magnetic monopoles do not exist in nature but recent advances in condensed matter and atomic physics have demonstrated the existence of effective magnetic monopoles in artificial spin ice [30, 31], chiral magnets [32] and BECs [33, 34]. The latter are created in ^{87}Rb atom condensates in a synthetic magnetic field. Note that a monopole-antimonopole pair is necessarily connected by a Dirac string. When two skyrmion tubes touch at a point it creates an effective magnetic monopole because the emergent magnetic field (see (2.5)) at that point is radially outward [32], see Fig. 2.8. Similarly, a moving hedgehog at the end of a skyrmion line (in a ferromagnetic nanowire) constitutes an emergent magnetic monopole [35], see Fig. 2.9. An experimentally observed and simulated monopole in a Bose–Einstein condensate is depicted in Fig. 2.10.

2.4 Vortices and Vortex Loops

In many physical systems such as fluids, superconductors and magnets (and those modeled by the 3D Heisenberg model) vortices, vortex lines and vortex loops (rings) [37] are observed. Creation and dynamics of trefoil-like (and other) knotted vortices have been studied in water using specially shaped hydrofoils [38]. Since magnetic, superconducting and other vortices as well as their dynamics have been studied extensively, this is a fully developed area of research and thus we will not dwell on this general theme, but rather limit ourselves to a number of recent developments.

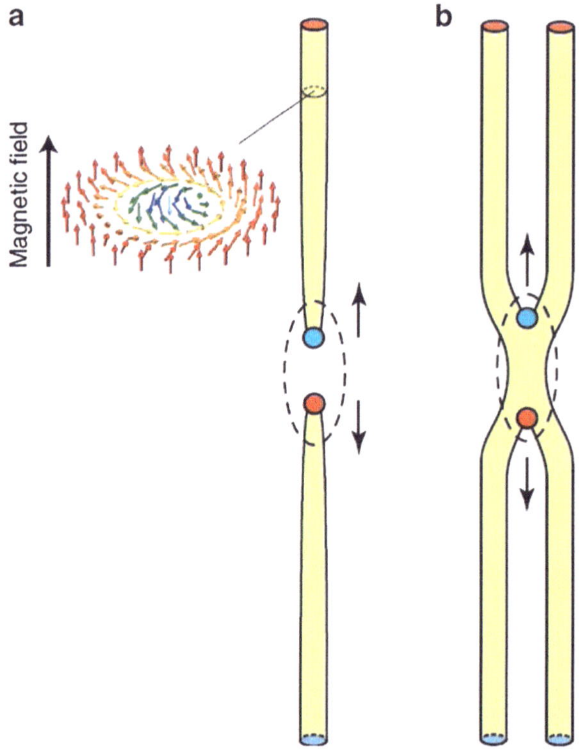

Fig. 2.8 Emergence of magnetic monopoles and anti-monopoles (colored circles). They result either **a** from pinching off of a skyrmion string or **b** a partial merging of two neighboring skyrmion strings. Reproduced from [36]. Creative Commons Attribution License (CC BY) https://creativecommons.org/licenses/by/4.0/

Vortices are persistent circulating flow patterns that occur in diverse scientific contexts [39], ranging from hydrodynamics, superfluids, and nonlinear optics [40, 41] to specific instantiations in sunspots [42], dust devils [43], and plant propulsion [44]. The study of the associated 2D effective particle dynamics that results from the logarithmic interaction potential is a theme of broad interest in physics. Not only it is relevant for the prototypical fluid/superfluid applications (see e.g. the review of Aref et al. [45] and the book of Newton [46]), but also for a variety of other settings. As such, we mention electron columns in Malmberg–Penning traps [47] and magnetized, millimeter sized disks rotating at a liquid-air interface [48, 49], among others.

The realm of atomic BECs [50–52] has produced a novel and pristine setting where numerous features of the exciting nonlinear dynamics of single- and multi-charge vortices, as well as of vortex lattices, can be not only theoretically studied, but also experimentally observed. Although BEC is known to be a fundamental phenomenon connected, e.g. to superfluidity and superconductivity [53], BECs were only experimentally realized 70 years later: this major achievement took place in 1995

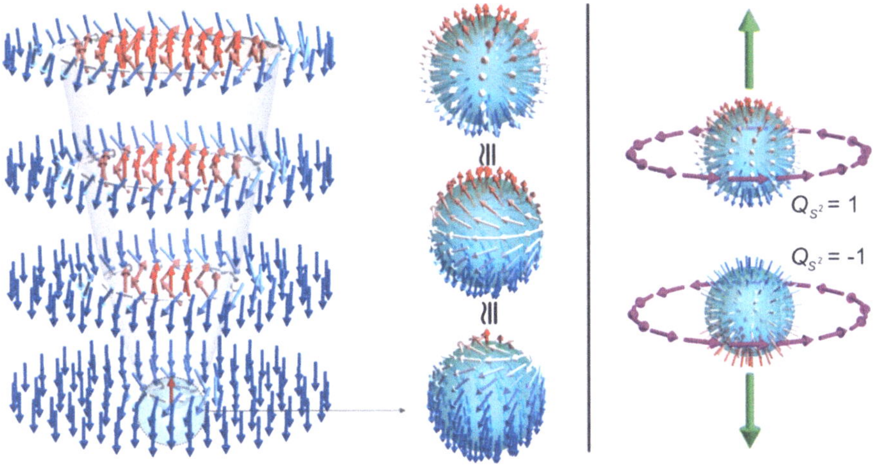

Fig. 2.9 Left panel: Spin texture at the end of a skyrmion line. Middle panel: The magnetization configuration is topologically equivalent to a hedgehog with radially outward magnetization, i.e. a magnetic monopole. Right panel: Two separating hedgehogs (or monopoles) with opposite topological charge with an emergent solenoidal electric field (purple arrows). Reproduced from [35]. ©2018 by the American Physical Society

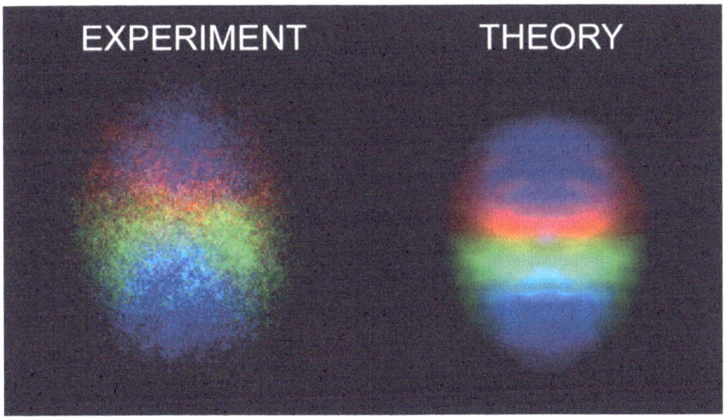

Fig. 2.10 Experimental observation and theoretical prediction of a monopole in BEC. Reproduced from [34]. Creative Commons Attribution License (CC BY) https://creativecommons.org/licenses/by/4.0/

[54–56] and has already been recognized through the 2001 Nobel prize in Physics [57, 58]. The role of vortices and the remarkable manifestation of highly ordered, triangular vortex lattices were, in turn, cited in the 2003 Nobel Prize in Physics [59]. Importantly, vortex dipoles (pairs of oppositely charged vortices) that will be relevant in what follows have played a quintessential role in the Kosterlitz–Thouless (KT) transition [60] from a gas of dipoles to configurations of unbound vortices, earning its discoverers the 2016 Nobel Prize in Physics. This transition has, moreover, found one of its most canonical realizations in the context of atomic BECs [61].

In addition to being at the epicenter of some of the most important physical notions of the past few decades, the coherent structures considered herein have been recognized as having both practical, as well as more exotic applications. For instance, solitary waves have been argued to provide the potential for 100-fold improved sensitivity for interferometers to phase shifts [62], while their lifetime of a few seconds enables precise force sensing applications [63]. Moreover, vortices present their own potential for applications. An intriguing example is the so-called "analogue gravity", whereby they may play a role similar to spinning black holes. This allows to observe in terrestrial, experimentally controllable environments associated phenomena such as the celebrated Hawking radiation or simpler ones such as super-radiant amplification of sonic waves scattered from black holes [64]. It has also been recently argued that vortices of a rotating BEC can collapse towards the generation of supermassive black holes [65] and that supersonically expanding BECs can emulate properties of an expanding universe in the lab [66].

The first experimental observation of BEC vortices [67] paved the way for a systematic investigation of the dynamical properties of such entities. Stirring the BECs [68, 69] above a certain critical angular speed [70–72] led to the production of few vortices [72] and vortex lattices [73, 74]. Other vortex-generation techniques were also used in experiments, including the breakup of the BEC superfluidity by dragging obstacles through the condensate [75], as well as nonlinear interference between condensate fragments [76]. In addition, apart from unit-charged vortices, higher-charged vortex structures were produced [77, 78] and their dynamical (in)stability was examined.

The majority of these early experiments focused on creating individual vortices and large vortex arrays. However, in 2008, the work of [79] enabled the use of the so-called Kibble–Zurek (KZ) mechanism to quench a gas of atoms rapidly across the BEC transition. The result of this is that phase gradients do not have sufficient time to "heal" but rather freeze, resulting in the formation of vortices. Then, in 2010 another technique was devised that enabled for the first time the dynamical visualization of vortices [80] during an experiment. This, in turn, spearheaded the work of [81, 82] where particle models were developed that predicted the dipole dynamics (equilibria, near-equilibrium epicyclic precessions and far from equilibrium quasi-periodic motions) observed in these experiments. A nearly concurrent development concerned the production in the lab of such vortex dipoles (one or multiple such), by the superfluid analogue of dragging a cylinder through a fluid [83]. More recently, the KZ mechanism together with rotation (i.e. injection of angular momentum) have been used to "dial in" and observe the dynamics of vortex clusters of, controllably, any number of vortices between 1 and 11. This is because rotation favors the formation of vortices of the same charge and in this way, depending on the angular momentum provided, different charge configurations (vortex clusters) arise. The resulting configurations may suffer symmetry breaking events [84]. As a result, instead of the commonly expected anti-diametric pair, equilateral triangle, or square configurations that one may expect, it is possible to observe symmetry broken

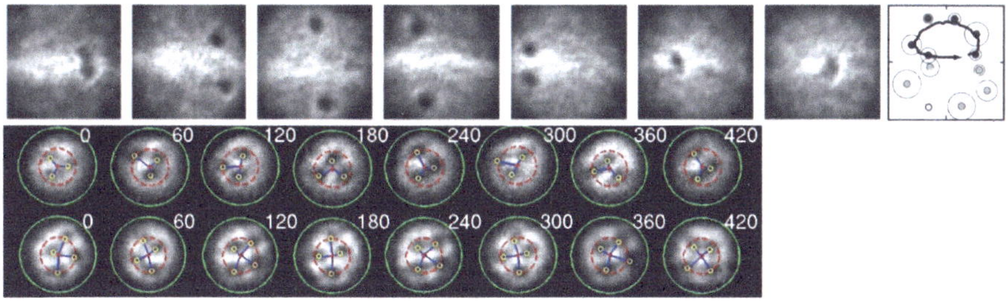

Fig. 2.11 Sampler of small vortex clusters in recent BEC experiments. Top row: vortex dipole dynamics; reproduced from [83] (©2010 by the American Physical Society). Bottom row: dynamics for three (top row) and four (bottom row) same-charge vortices reported in [84]

configurations featuring asymmetric pairs, isosceles triangles, and rhombi or general/asymmetric quadrilaterals [85]. To further add to these developments, 3-vortex configurations of (same but also of) alternating charge in the form of a tripole (i.e. a positive-negative-positive or its opposite) have been experimentally produced [86, 87]. This turns out to be one of the simplest setups where chaotic dynamics can ensue [88, 89]. A sampler of experimental images from these different experimental efforts is depicted in Fig. 2.11.

It is evident from the above recent developments that there is a tremendous momentum toward the study of vortex dynamics in atomic physics. Moreover, this theme presents nontrivial twists in comparison to the classical fluid or superfluid playground [45, 46]. To mention a canonical difference between the two, atomic BECs are typically confined by parabolic traps [50–52, 90] constraining the density and hence the region within which the vortices evolve. This trapping induces vortices to rotate around the center of the trap [90–92]. The frequency of this precessional motion can be well approximated by a constant close to the center of the trap, yet as the edges of the BEC are approached and the density decreases, the relevant frequency increases drastically [84, 85, 93]. These modifications of the "standard" picture of the vortices interacting through a logarithmic potential are critically responsible for some of the phenomena observed recently. For instance, the competition between the rotation and the interaction in the case of the vortex dipole [81, 82] is responsible for the existence of stationary states or epicyclic/quasi-periodic trajectories; the deviation from a constant precession frequency is, in turn, responsible [85, 93] for the symmetry-breaking bifurcations enabling asymmetric vortex configurations [84].

2.5 Skyrmions in Liquid Crystals

One can describe a liquid crystal by a tensor order parameter $\mathbf{Q}(\mathbf{r})$, which is related to the director field $\mathbf{n}(\mathbf{r})$ and the scalar order parameter $S(\mathbf{r})$ by $Q_{\alpha\beta} = S(\frac{3}{2}n_\alpha n_\beta - \frac{1}{2}\delta_{\alpha\beta})$. According to the Landau–de Gennes theory, one can express the free energy density in terms of $\mathbf{Q}$ as [19]

$$F = \frac{1}{2} a Tr\mathbf{Q}^2 + \frac{1}{3} b Tr\mathbf{Q}^3 + \frac{1}{4} c \left(Tr\mathbf{Q}^2\right)^2 \qquad (2.6)$$
$$+\frac{1}{2} L(\partial_\gamma Q_{\alpha\beta})(\partial_\gamma Q_{\alpha\beta}) - 2Lq_0\epsilon_{\alpha\beta\gamma} Q_{\alpha\delta}\partial_\gamma Q_{\beta\delta}.$$

The first line represents the free energy of a uniform system, when we expand it in powers of the tensor order parameter. This part favors those eigenvalues of **Q**, which are associated with a specific magnitude of uniaxial nematic order. One assumes the coefficient a varies linearly with temperature, whereas b and c are constant with regard to temperature. The last two terms are the elastic free energy corresponding to variations in **Q** as a function of position. The first of these terms is the (equal) energy cost of splay, twist, and bend deformations, where L is an elastic coefficient. The last term allows a chiral twist of the nematic order, where q_0 is a characteristic inverse length that arises from the molecular chirality. Other possible elastic terms giving different energy costs for splay, twist, and bend, e.g. $\frac{1}{2}L_2(\partial_\alpha Q_{\alpha\gamma})(\partial_\beta Q_{\beta\gamma})$, are neglected here for simplicity.

The chiral twist term q_0 is analogous to the Dzyaloshinskii–Moriya interaction in the magnetic case (see (2.2)). Comparative analysis and simulations based on (2.2) for the magnetic case and (2.6) for the liquid crystals [19] provide the results depicted in Fig. 2.3. For chiral magnets the triangular (or hexagonal) skyrmion lattice is stable and it arises from the spiral (or helical) phase. However, a triangular meron lattice is not stable but a square meron lattice is allowed. In contrast, based on energetic grounds, in liquid crystals the skyrmion lattice is disfavored but a triangular lattice of merons is allowed. This difference arises from the nature of the order parameter: vector for chiral magnets versus tensor for nematic liquid crystals.

In addition to skyrmions and merons, stable skyrmion bags have been observed in liquid crystals [23]. An observed and simulated example is illustrated in Fig. 2.5. Corresponding skyrmions bags in chiral magnets are also possible as shown in Fig. 2.5 as a result of micromagnetic simulations. Interestingly, these bags are similar to the models of atomic nuclei containing different number of baryons, as originally surmised by T. H. R. Skyrme.

2.6 Bose–Einstein Condensates: From Vortex Lines to Rings, From Hopfions to Skyrmions and Knots

We already discussed in Sect. 2.4 the relevance of Bose–Einstein condensates as a prototypical playground where two-dimensional topological excitations in the form of vortices naturally arise. We now turn to a 3D extension of such structures, starting with the natural generalization of the vortex, namely the vortex line (VL). A VL, also referred to as a solitonic vortex, is the 3D extension of a 2D vortex by (infinitely and homogeneously) extending the solution into the axis perpendicular to the vortex plane. VLs might be rendered finite in length if their background is made bounded by an external potential. In that case, VLs are called vorticity "tubes" that are straight or bent in U and S shapes depending on the aspect ratio of the background [94, 95]. If a VL is bent enough to close on to itself or if two VLs are close enough to each other then they can produce a vortex ring [96]. Vortex rings (VRs) are 3D structures whose core is a closed loop with vorticity around it [97] (i.e. a vortex that is looped back into itself). VRs can also be produced by an impurity traveling faster than the speed of sound of the background [98], by nonlinear interference between colliding blobs of atomic matter [99, 100], by phase and density engineering techniques [101–103], or even by introducing "bubbles" of one component in the other component in two-component nonlinear Schrödinger (NLS) systems [104].

It should be noted that VRs inherently possess a velocity perpendicular to the ring plane due to Helmholtz's law [105] (unless they are stopped by the presence of an external trap [106, 107]). Also, another special feature of VLs and VRs is that they support intrinsic dynamics along the vortex line/ring. For example, it is possible to transversally excite the vorticity line to produce oscillations called Kelvin modes (or Kelvons) [108–111]. Kelvin modes not only have their own dynamics and interactions across vortex lines [112], but they can also self interact within a single VR and slowdown or even reverse the velocity of the VR [113, 114]. We note here that Kelvin modes have also been studied in the context of skyrmion tubes [115]. Another possibility for exciting the vorticity line of the VR is by creating varicose or capillary waves (periodic compressions of the vortex tube along its length) [109, 116]. Lastly, VRs interact in intriguing ways involving, e.g. leapfrogging motions when they are co-axial (see Fig. 2.12), but also more complex interactions when they are not [117]. Recently, an effective particle description has been utilized not only in order to understand the stability and dynamics of a single VR [118, 119], but also that of multiple or interacting VRs [120].

It is important to highlight here that the relevance of VLs and VRs goes beyond atomic BECs. They emerge ubiquitously in fluid mechanics [117] and in Helium and other related superfluid systems [97]. Rather, what is the case here is that atomic condensates present a pristine, well-controlled setting for the creation and exploration of these structures.

In addition to VRs and VLs, it was also realized that BECs offer also the potential for the formation of more complex topological structures. This is to a considerable extent due to the potential of creating atomic condensates either of different species (e.g. ^{87}Rb and ^{23}Na, i.e. hetero-nuclear mixtures) or of the same species (e.g. confining and condensing two different hyperfine states of the same gas, such as spin-1 and spin-2 states of ^{87}Rb) [90]. Among the early suggestions along this vein, is the multi-component skyrmion state creation in BECs. The topological properties of such a state enable its structural realization in a multi-component BEC. Here, as originally proposed in [122, 123], the skyrmion consists of a VR in one of the components, "trapping" a VL in a second component. Interestingly, more complex skyrmion states involving three-component (so-called spinor) BECs have been recently realized experimentally in both 2D [124] and 3D [125], involving, respectively, coupled states of topological charge $S = -1, 0, 1$ and $S = 0, 1, 2$; see also the recent theoretical work of [126].

Of increased interest recently has been not only this structure, but also its one-component counterpart, which in the relevant recent BEC literature is referred to

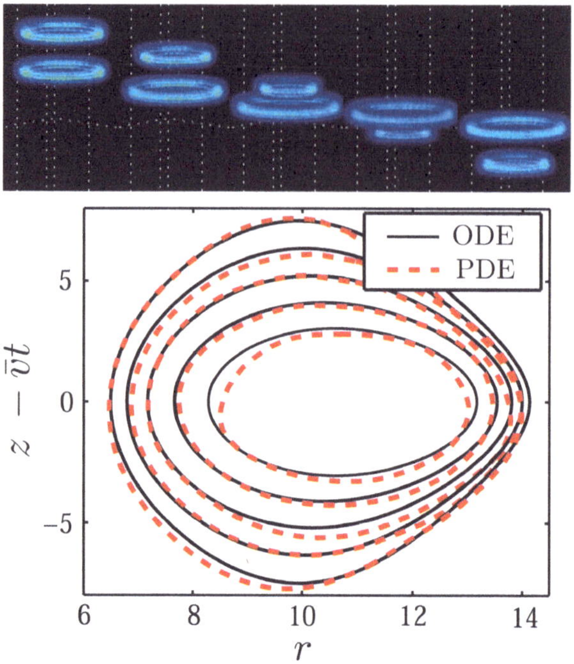

Fig. 2.12 The top panel shows the snapshots of the evolution of two vortex rings leapfrogging through each other. The bottom panel illustrates this type of motion in the radial-polar (r, z) plane on a co-moving reference frame. The solid lines correspond to the orbits predicted by ordinary differential equations for the radius and vertical position of the rings as a function of time. The dashed lines are the corresponding numerics of the full 3-dimensional NLS equation. The bottom panel is adapted from [121]. ©2016 by the American Institute of Physics

as a hopfion state [127–129]. The latter consists of a VR and a VL in the same component with the axis of the VR constituting the line of vorticity of the VL. A stable hopfion state was found to exist both in the setup of [127] which, however, involved the rather elaborate realization of radially increasing nonlinear interactions and the purely dynamical exploration of [128] for condensates that are being rotated (and was suggested to be stable only for some intermediate rotation rate).

Lastly, the study of quantum (vortex) knots is one that has only a relatively short history in the context of atomic BECs. To the best of our knowledge, the possibility of such complex topological structures was introduced in the work of [130] (see also [131]), illustrating how different torus knots $\mathcal{T}_{K,q}$, with co-prime K and q, can be generated in the wavefunction of an atomic species. Subsequently the work of [132] seemed to put a full stop on the subject through the extensive simulation of 1458 vortex knots from the so-called "knot atlas" [133], and finding that the *trapless* Gross–Pitaevskii equation (GPE) could not support stable knots: all of the simulated knots would eventually untie into simpler patterns. Nevertheless, the recent work of [134, 135], both at the level of the Biot–Savart dynamical law (for the vortex knot motion), as well as at that of the full 3D GPE has, perhaps counter-intuitively, indicated that the *trefoil knot can be a very long lived* structure in the context of a trapped atomic BEC [136].

2.7 Topology and Curved Manifolds: Bogomolnyi Decomposition

We now provide an example of how the types of configurations considered herein may be used to minimize the energy of a system of classical spins. The continuum limit of (classical) Heisenberg spins on a two-dimensional (planar or) curved manifold corresponds to the nonlinear σ model. That is, the corresponding Hamiltonian H is given by (2.1). If we impose homogeneous boundary conditions on the vector field $\mathbf{n}$ in the plane R^2, i.e. $\lim_{r\to\infty} \mathbf{n} \to \mathbf{n}_0$, then we can compactify the plane into the surface of a sphere S^2. This allows us to classify different configurations according to the homotopy class $\pi_2(S^2) = \mathbf{Z}$, where $\mathbf{Z}$ is the group of relative integers [137, 138].

Topology does not directly help us to state anything about the energy of the field configuration but indirectly, by invoking the so-called Bogomolnyi inequalities [8], it enables us to establish energy bounds for configurations belonging to equivalent homotopy classes labeled by $n \in \mathbf{Z}$. The inequality in the present case can be expressed as

$$(\partial_i \mathbf{n} - \epsilon_{ij}\partial_j \mathbf{n})^2 \geq 0\,, \tag{2.7}$$

whereby it follows that

$$H \geq \int \mathbf{n}\cdot(\partial_x \mathbf{n} \times \partial_y \mathbf{n})\mathrm{d}x\mathrm{d}y\,. \tag{2.8}$$

Thus, the minimum energy in each homotopy class is attained when

$$\partial_i \mathbf{n} = \pm\epsilon_{ij}\partial_j \mathbf{n}\,, \tag{2.9}$$

i.e. when these self-dual equations are satisfied by the field configurations.

If we consider this model on a plane (R^2), there is no characteristic length. As a result the nonlinear σ model Hamiltonian can be scaled and thus all the nontrivial field configurations (satisfying homogeneous boundary conditions) can be scaled as well. This situation is drastically changed if there is a characteristic length scale, e.g. if the underlying manifold is curved. We will introduce a length scale in two ways: (i) first we will consider the nonlinear σ model on a rigid cylinder and then (ii) we will also apply an axial magnetic field through the cylinder. In the first case the radius ρ_0 of the cylinder is the characteristic length whereas in the second case there is an *additional* length scale introduced by the magnetic field $\mathbf{B}$. There are other ways of introducing a length scale, e.g. through magnetic anisotropy, ellipticity of the cylinder cross section, etc. but we will not consider these different cases here.

We set the unit vector $\mathbf{n} = (\sin\theta\cos\Phi, \sin\theta\sin\Phi, \cos\theta)$ in terms of the colatitude θ and the azimuthal angle Φ. Next, we write the Hamiltonian [137] in terms of cylindrical coordinates (ρ, z, ϕ)

$$H = J \iint_{\text{cyl}} \left[(\partial_z\theta)^2 + \sin^2\theta(\partial_z\Phi)^2 + (\partial_\phi\theta)^2/\rho_0^2 + \sin^2\theta(\partial_\phi\Phi)^2/\rho_0^2\right] \rho_0 \mathrm{d}z\mathrm{d}\phi\,, \tag{2.10}$$

where J denotes the spin exchange interaction energy. In order to invoke topological considerations let us impose homogeneous boundary conditions, i.e. $\lim_{z\to\mp\infty} \equiv 0[\pi]$ and $\lim_{z\to\mp\infty} \mathrm{d}\theta/\mathrm{d}z = 0$. If we seek cylindrically symmetric solutions then $\Phi = \phi$ and $\partial\theta/\partial\phi = 0$. Thus the Hamiltonian simplfies to

$$H = 2\pi\rho_0 J \int_{-\infty}^{\infty} \left[(\partial_z\theta)^2 + \sin^2\theta/\rho_0^2\right] \mathrm{d}z\,. \tag{2.11}$$

The variation of this Hamiltonian ($\delta H = 0$), i.e. the Euler–Lagrange equation turns out to be the celebrated sine-Gordon equation

$$\mathrm{d}^2\theta(z)/\mathrm{d}z^2 = (1/2\rho_0^2)\sin 2\theta\,, \tag{2.12}$$

with the well known kink solution $\theta(z) = \arctan[\exp(z/\rho_0)]$. It is depicted in Fig. 2.13a. The energy for this configuration is $H = 8\pi J$, which is the minimum energy belonging to the first homotopy class.

By invoking the technique used by Belavin and Polyakov [138], or equivalently the Bogomolnyi decomposition [8], we note that the solutions that correspond to the minimum energy in each homotopy class satisfy the first order self-dual equations

$$\rho_0\partial_z\theta = \pm\sin\theta\,\partial_\phi\Phi\,, \quad \partial_\phi\theta = \mp\rho_0\sin\theta\,\partial_z\Phi\,. \tag{2.13}$$

If we apply an external magnetic field (**B**) along the axis of the cylinder then the Hamiltonian is modified as [140]

$$H_{\text{mag}} = J\iint_{\text{cyl}} (\nabla n)^2 \mathrm{d}S - g\mu \iint_{\text{cyl}} \mathbf{n}\cdot\mathbf{B}\,\mathrm{d}S\,, \tag{2.14}$$

where g is the gyromagnetic ratio and μ denotes the Bohr magneton. With homogeneous boundary conditions ($\theta = 0$ as $z \to \pm\infty$) the Hamiltonian simplifies as

$$H_{\text{mag}} = 2J(2\pi\rho_0)\int_{-\infty}^{\infty} [\theta_z^2 + \{\sin^2\theta/2\rho_0^2 + (1/\rho_B^2)(1-\cos\theta)\}]\mathrm{d}z\,, \tag{2.15}$$

where $\rho_B^2 = 2J/g\mu B$ is the magnetic length scale. Variation of this Hamiltonian leads to

$$\theta_{zz} = (1/2\rho_0^2)\sin 2\theta + (1/\rho_B^2)\sin\theta\,, \tag{2.16}$$

which is the double sine-Gordon equation. The corresponding 2π-kink solution that is consistent with the boundary conditions is given by

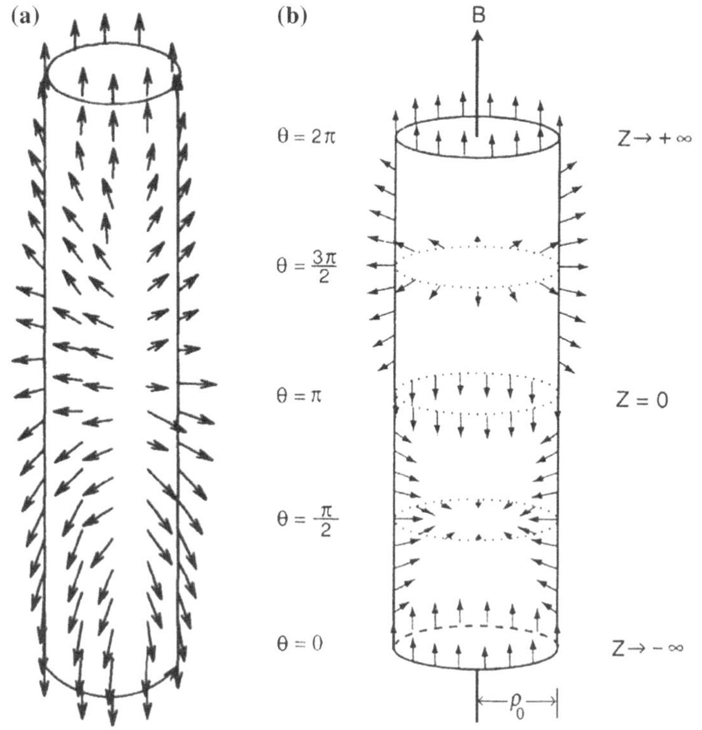

Fig. 2.13 **a** Heisenberg spins on a cylinder as a sine-Gordon π-soliton (reproduced from [139], ©1994 by Elsevier Science B.V.) and **b** in the presence of an axial magnetic field as a double sine-Gordon 2π soliton (reproduced from [140])

$$\theta(z) = 2 \arcsin \frac{1}{\sqrt{\cosh^2(z/\xi) - (\xi^2/\rho_0^2)\sinh^2(z/\xi)}}, \tag{2.17}$$

where the kink width $\xi = \rho_0 \rho_B / (\rho^2 + \rho_B^2)^{1/2}$ is another characteristic length in the problem. This solution is depicted in Fig. 2.13b.

Because of the homogeneous boundary conditions at the cylinder boundaries all the spins point in the same direction, see Fig. 2.13. Thus they can be compactified to a single spin and the spin configuration in Fig. 2.13a covers the unit sphere once, i.e. it is a skyrmion of topological charge 1. Similarly, the spin configuration in Fig. 2.13b covers the unit sphere twice, thus it is a skyrmion of topological charge 2. If the cylinder were semi-infinite, it will be topologically equivalent to a plane with a hole of radius ρ_0. The spin configuration in this case will be a half-skyrmion (or a meron) [141].

If the cylinder is elastic (i.e. deformable) then the geometric frustration caused by the mismatch of the cylinder radius and kink width can be relieved by a pulse-like deformation in the region of the magnetic kink [137, 140]. The Bogomolnyi technique is quite general and can be used in a broader context. Another application of the Bogomolnyi decomposition is in the calculation of elastic deformation energy of vesicles as a function of genus [142]. The latter has significance in the context of the Willmore conjecture [143].

2.8 Topological Materials

In conventional materials such as metals, insulators and semiconductors the non-relativistic Schrödinger equation describes the energy dispersion of low-lying electronic excitations, $E_S = p^2/2m^*$, which is quadratic in the electron momentum p with effective mass m^*. However, over the past decade there is a growing class of materials, which are known as Dirac materials [144] or more generally topological materials, exhibiting linear electronic dispersion in their band structure. Examples include topological insulators [1, 145], topological superconductors [2], topological crystalline insulators [146] as well as Dirac semimetals and Weyl semimetals [3, 147].

One of the distinguishing features of these materials are Dirac points, where the (conduction and valence) bands touch each other at an isolated set of points. The corresponding band features are called Dirac cones. These points are topologically protected due to specific (time reversal, spatial inversion or crystalline) symmetries in that they are robust under perturbations. An example is graphene where the protection comes from the sublattice symmetry (of the underlying honeycomb lattice) and the energy dispersion of its electrons is linear in the momentum. Specifically, it is given by the relativistic Dirac equation: $E_D = c\sigma \cdot p + mc^2\sigma$. Here $\sigma = (\sigma_x, \sigma_y)$ denotes Pauli matrices and the speed of light c is replaced by the Fermi velocity v_F. In d spatial dimensions (with $c = 1$) the Dirac equation is written as $(i\gamma^\mu \partial_\mu - m)\psi = 0$ [3], where $\mu = 0, 1, \ldots, d$ with $\mu = 0$ denoting time and the Dirac gamma matrices γ^μ anticommute. In odd dimensions ($d = 1, 3, \ldots$) it can be simplified. In particular, for $d = 1$ one gets $i\partial_t \psi = (\gamma^0\gamma^1 p + m\gamma^0)\psi$ with momentum $p = -i\partial_x$. If we further consider massless ($m = 0$) particles, we get the one-dimensional Weyl equation $i\partial_t \psi_\pm = \pm p\psi_\pm$. Thus we get simple linear dispersion $E_\pm = \pm p$ representing the right and left moving chiral particles or Weyl fermions.

Topology seems to enhance the nonlinear response of topological materials. A recent important experimental technique is TFISH (Terahertz Field-Induced Second Harmonic Generation) which allows to understand the nonlinear response of topological Dirac and Weyl semimetals, e.g. TaAs [148]. In particular, second harmonic generation (SHG) is found to be enhanced in Weyl semimetals. The latter contain Weyl points (or nodes) in their electronic structure at which linearly dispersing, non-degenerate bands cross. They also exhibit Fermi arc surface states that are attached to the Weyl nodes in the bulk material (Fig. 2.14). In these materials there is a transition between the topological and non-topological (or trivial) phases which proceeds via a gapless state.

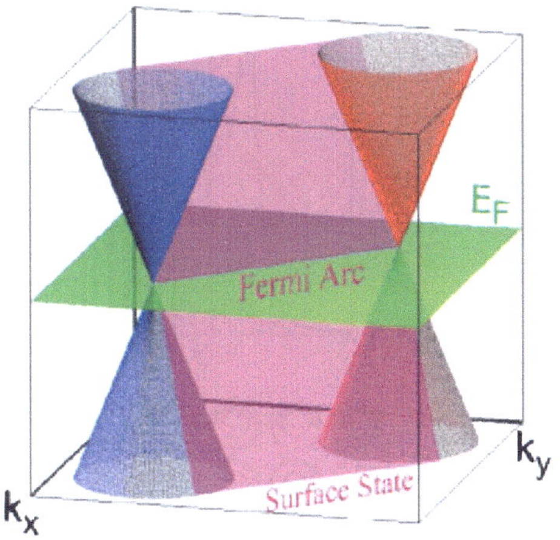

Fig. 2.14 The double Dirac cone structure of a Weyl semimetal with two Weyl nodes connected by a Fermi arc. Here E_F denotes the Fermi energy. These materials exhibit characteristic surface states which can be observed experimentally. Reproduced from [156]. ©2011 by the American Physical Society

When an appropriate nonlinearity is added to the usual (linear) Dirac equation and Weyl equation, a nonlinear Dirac (NLD) equation [149] and a nonlinear Weyl (NLW) equation [150] results, respectively. Their properties and attendant nonlinear (topological) excitations such as solitons and vortices [151] are quite different as compared to the nonlinear Schrödinger (NLS) equation. Currently, a significant amount of analytical and numerical effort is being devoted to understanding the stability and collision dynamics of solitons in NLD and NLW equations [152].

In closing this section we note that in addition to chiral magnets, liquid crystals and BECs, there are many other materials including topological materials in which a variety of topological defects can form under right conditions. In particular, ferroelectrics [13], multiferroic materials, e.g. Cu_2OSeO_3 [153] and magnetic shape memory alloys such as Ni_2MnGa [154] can also support skyrmion-like topological excitations. There has been a recent observation of skyrmions in the heterostructures of a ferromagnet (Cr_2Te_3) and a topological insulator (Bi_2Te_3) [155]. It would be highly desirable to observe hopfions in such materials as well.

Although we have not discussed dislocations separately here, they play an important role in determining the properties of topological materials [157].

2.9 Nonreciprocal Topological Photonics

Many of the modern photonic devices such as optical isolators and optical circulators are based on the principle of Lorentz reciprocity. It entails that in a (i) linear, (ii) time-independent material or medium with (iii) symmetric property (or constitutive optical) tensors, the received and transmitted fields are identical for both forward and time-reversed propagation directions [158]. However, in some cases reciprocity is deleterious, as, e.g. in self-echo in antennas. In addition, for many

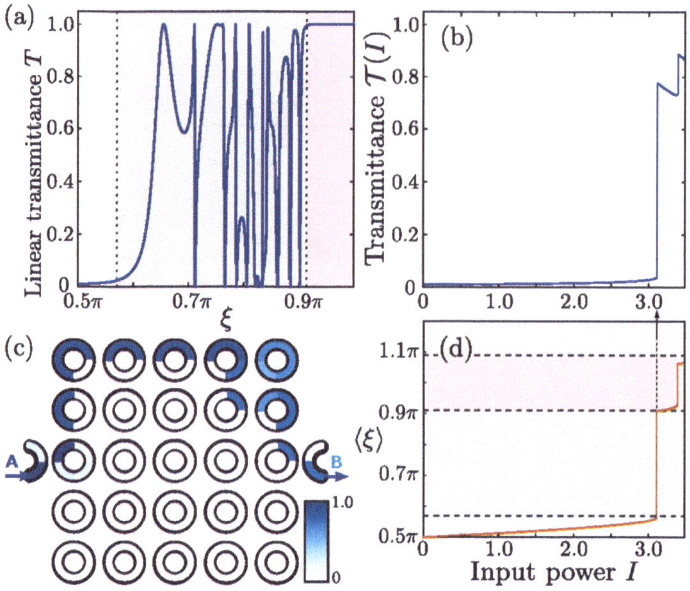

Fig. 2.15 Nonlinear coupled ring lattice or resonator array for transmittance studies related to nonreciprocal topological photonics. **a** Transmittance T through a linear lattice as a function of phase shift ξ. **b** Transmittance as a function of input power I through the nonlinear lattice. **c** Field intensity distribution (normalized) above the discontinuity ($I > 3$). **d** Average phase shift as a function of I. Reproduced from [159]. Creative Commons Attribution License (CC BY) https://creativecommons.org/licenses/by/3.0/

desired and emerging optical functionalities (e.g. optical circulators) it is important to break Lorentz reciprocity by relaxing any of the three conditions. Clearly, one way to obtain nonreciprocity is by way of introducing optical nonlinearity. The latter in conjunction with non-Hermitian photonics and *topological photonics* can significantly enhance nonreciprocity [7]. Examples of optical nonlinearities include the Kerr effect, two-photon absorption and the thermo-optic effect. An example of nonreciprocal topological photonic setup employing a nonlinear coupled resonator array is depicted in Fig. 2.15.

One can create topologically nontrivial photonic band structures in analogy with the electronic band structure of topological materials discussed above. In particular, one can create topological edge states that are robust against perturbations or defects. One way to create such states is by forming an interface between a topologically nontrivial and a trivial optical material. Just like in electronic topological materials, photonic nontrivial topological bands cannot be deformed to trivial bands in an adiabatic way. A photonic realization of a two-dimensional electronic Chern insulator uses a lattice of magnetized ferrite rods at microwave scale [160] in which the lattice edge state acts as an isolating waveguide. This structure leads to almost perfect forward transmission and exponentially suppressed backward transmission for frequencies in the photonic bandgap.

There are two ways of realizing nonlinear topological photonic structures. One can consider nonlinear propagation dynamics in an otherwise linear topological photonic system. Here the nonlinearity locally alters the system properties. An example is that of waveguide arrays with evanescent coupling between neighboring waveguides. Another way is to use a probe beam to induce a phase transition in the dynamics of a linearized probe beam. Coupled optical resonator lattices provide a realization of this type where one gets quite strong nonlinear effects due to the resonant light confinement as compared to waveguide lattices. Finally, we note that a one dimensional nontrivial topological lattice can be modeled by the photonic Su–Schrieffer–Heeger (SSH) model [161, 162].

2.10 Topological Modes in Acoustics and Beyond

Admittedly, the study of topological insulators has drawn considerable interest in a variety of fields, among other reasons because of the ability of such media to feature transport that is immune to the presence of defects [1, 145]. One of the most recent venues for such studies has been in the area of mechanical and acoustic systems, where the topological properties can inspire the design of unconventional mechanical structures with unique elastic and vibrational properties [163–167]. This, in turn, can lead to significant new paradigms in the realm of energy harvesting, as well as in that of vibration isolation [168].

One of the canonical examples that is possible to realize in this mechanical setting is a direct analogue of a dimer in the form of an SSH model as has been suggested e.g. in [169]. This enables through its corresponding phononic band-gap structure the emergence of a zero-frequency topological mode. Finite (non-zero) frequency topological modes can also be achieved [170]. There have been numerous recent efforts in this direction of harnessing topological properties of suitable mechanical media to improve the propagation or storage of energy. These include, among others, the examination of edge solitons [171] and their ability for nonlinear conduction in topological mechanical insulators [172, 173], the study of nonlinear edge states that arise in phononic lattices [174], as well as the examination of topological band transitions in tunable phononic systems, under the variation of suitable (e.g. stiffness) parameters [175, 176].

Lastly, we touch upon the theme of topology optimization which is an important method used for many industrial and technological applications. For structural robustness and additive manufacturing, nonlinear topology optimization [177] and topology optimization for geometrically nonlinear structures [178] have been recently studied. In the former case a nonlinear elastic model of the materials is considered along with plasticity aspects in conjunction with invoking the von Mises (yield) criterion. In the latter case it is assumed that the structures under consideration experience large displacement but small strain. In particular, high-resolution topology optimized solutions are obtained for structures that are geometrically nonlinear. Invariably, the results are quite important for many engineering applications.

2.11 Conclusions and Future Work

In this chapter, we have delineated the importance of topology in a variety of physical systems and discussed the ubiquity of topological defects such as skyrmions, merons, vortices, hopfions and monopoles in a number of distinct nonlinear systems, e.g. chiral magnets, liquid crystals, BEC, etc. We have also elucidated the pervasive role of topology in nonlinear condensed matter and photonic, as well as phononic systems. Although we did not discuss it here, topology in soft matter is also quite important e.g. in topological colloids [179]. In addition, we illustrated the interplay of nonlinearity, topology and geometry by considering the nonlinear σ model on simple curved manifolds. The resulting spin configurations are sine-Gordon or double sine-Gordon solitons.

Despite the above significant progress over the last few years, there are several important open problems related to the interplay of nonlinearity and topology. Nonlinearity with "fragile topology" in a quantum system is a topic for future research. Fragile topology (as opposed to strong topology) refers to a set of quantum phenomena that endow materials or systems with unusual properties [180–182]. Examples include the misaligned layers of graphene [183] and "knotty" electronic quantum states in some topological materials. In the latter case electrons are restricted to move along certain directions. Understanding these states properly may require considerations other than K-theory [184], as discussed in e.g. [180].

Akin to the study of soliton collisions and vortex interactions, it would be desirable to study the interaction between and the collision of different hopfions. This is certainly a challenging numerical problem.

A study of one-dimensional NLD and NLW equations on (planar and space) curves and higher (two and three) dimensional such equations on curved manifolds will provide important insights into the interplay of topology, geometry and nonlinearity. In particular, the shape and dynamics of soliton and vortex solutions will be modified by the curved geometry.

There are three fundamental (relativistic) fermions in nature: Dirac, Weyl (with zero mass) and Majorana (which are their own antiparticles and thus neutral). Similar to NLD and NLW equations there could exist a nonlinear Majorana equation; it would be intriguing to explore its soliton solutions and their dynamics. Majorana fermions also have potential applications in the growing field of (braiding-based) topological quantum computing [185], especially with fault tolerance. We note here that the nonlinear dynamics of Majorana modes has been studied using topological Josephson junctions [186]. Similarly, it was proposed that Majorana-like modes of light can also be realized in a one-dimensional array of nonlinear cavities [187].

During the last few years there have been studies of a class of nonlinear models that harbor kink solitons with non-exponential tails. Collision of such kinks with non-exponential tails, e.g. power-law [188] or super-exponential, in $1 + 1$ dimensional field theories is an important open question.

Recently the field of quantum time crystals [189] has emerged with many insights including topological considerations [190]. The role of nonlinearity in quantum time crystals with topological aspects remains an open field.

Finally, we note that we could not be exhaustive in our coverage of topological defects. For instance, many aspects related to hopfions [191, 192] and the Bogomolnyi decomposition [193, 194] among many other topics could not be included simply for the lack of space.

Acknowledgements A.S. and P.G.K. acknowledge the support of the U.S. Department of Energy. Specifically, LANL is operated by Triad National Security, L.L.C. for the National Nuclear Security Administration of the U.S. Department of Energy under Contract No. 892333218NCA000001. J.C.M. thanks financial support from MAT2016-79866-R project (AEI/FEDER, UE). This material is based upon work supported by the National Science Foundation under Grant No. DMS-1809074 (P.G.K.). Finally, P.G.K. gratefully acknowledges the support of the Leverhulme Trust towards a visiting fellowship at the University of Oxford and the kind hospitality of the Mathematical Institute of the University of Oxford.

References

1. M.Z. Hasan, C.L. Kane, Rev. Mod. Phys. **82**, 3045 (2010)
2. M. Sato, Y. Ando, Rep. Prog. Phys. **80**, 076501 (2017)
3. N.P. Armitage, E.J. Mele, A. Vishwanath, Rev. Mod. Phys. **90**, 015001 (2018)
4. J. Gooth, B. Bradlyn, S. Honnali, C. Schindler, N. Kumar, J. Noky, Y. Qi, C. Shekhar, Y. Sun, Z. Wang, B.A. Bernevig, C. Felser, Nature **575**, 315 (2019)
5. D.J.E. Marsh, K.C. Fong, E.W. Lentz, L. Smejkal, M.N. Ali, Phys. Rev. Lett. **123**, 121601 (2019)
6. S. Gupta, A. Saxena (eds.), *The Role of Topology in Materials*. Springer Series in Solid-State Sciences, vol. 189 (2018)
7. W. Chen, D. Leykam, Y.D. Chong, L. Yang, MRS Bull. **43**, 443 (2018)
8. E.B. Bogomol'nyi, Sov. J. Nucl. Phys. **24**, 449 (1976)
9. N. Manton, P. Sutcliffe, *Topological Solitons* (Cambridge University Press, Cambridge, 2004)
10. Y.M. Shnir, *Topological and Non-Topological Solitons in Scalar Field Theories* (Cambridge University Press, Cambridge, 2018)
11. Y.X. Zhao, Z.D. Wang, Phys. Rev. Lett. **114**, 206602 (2015)
12. K. Koniaris, M. Muthukumar, Phys. Rev. Lett. **66**, 2211 (1991)
13. S. Das, Y.L. Tang, Z. Hong, M.A.P. Gonçalves, M.R. McCarter, C. Klewe, K.X. Nguyen, F. Gómez-Ortiz, P. Shafer, E. Arenholz, V.A. Stoica, S.-L. Hsu, B. Wang, C. Ophus, J.F. Liu, C.T. Nelson, S. Saremi, B. Prasad, A.B. Mei, D.G. Schlom, J. Íñiguez, P. García-Fernández, D.A. Muller, L.Q. Chen, J. Junquera, L.W. Martin, R. Ramesh, Nature **568**, 368 (2019)
14. Y. Nahas, S. Prokhorenko, L. Louis, Z. Gui, I. Kornev, L. Bellaiche, Nat. Commun. **6**, 8542 (2015)
15. N. Nagaosa, Y. Tokura, Nat. Nanotechnol. **8**, 899 (2013)
16. I. Kézsmŕki, S. Bordács, P. Milde, E. Neuber, L.M. Eng, J.S. White, H.M. Rønnow, C.D. Dewhurst, M. Mochizuki, K. Yanai, H. Nakamura, D. Ehlers, V. Tsurkan, A. Loidl, Nat. Mater. **14**, 1116 (2015)
17. A.K. Nayak, V. Kumar, T. Ma, P. Werner, E. Pippel, R. Sahoo, F. Damay, U.K. Rößler, C. Felser, S.S.P. Parkinet, Nature **548**, 561 (2017)
18. T. Van Mechelen, Z. Jacob, Opt. Mater. Express **9**, 95 (2019)
19. A. Duzgun, J.V. Selinger, A. Saxena, Phys. Rev. E **97**, 062706 (2018)

20. S. Wintz, C. Bunce, A. Neudert, M. Körner, T. Strache, M. Buhl, A. Erbe, S. Gemming, J. Raabe, C. Quitmann, J. Fassbender, Phys. Rev. Lett. **110**, 177201 (2013)
21. S.-Z. Lin, A. Saxena, C.D. Batista, Phys. Rev. B **91**, 224407 (2015)
22. X.Z. Yu, W. Koshibae, Y. Tokunaga, K. Shibata, Y. Taguchi, N. Nagaosa, Y. Tokura, Nature **564**, 95 (2018)
23. D. Foster, C. Kind, P.J. Ackerman, J.-S.B. Tai, M.R. Dennis, I.I. Smalyukh, Nat. Phys. **15**, 655 (2019)
24. B.G.-G. Chen, P.J. Ackerman, G.P. Alexander, R.D. Kamien, I.I. Smalyukh, Phys. Rev. Lett. **110**, 237801 (2013)
25. P.J. Ackerman, I.I. Smalyukh, Nat. Mater. **16**, 426 (2017)
26. I.I. Smalyukh, Y. Lansac, N.A. Clark, R.P. Trivedi, Nat. Mater. **9**, 139 (2010)
27. H.R.O. Sohn, C.D. Liu, Y. Wang, I.I. Smalyukh, Opt. Express **27**, 29055 (2019)
28. P. Sutcliffe, J. Phys. A: Math. Theor. **51**, 375401 (2018)
29. M. Kobayashi, M. Nitta, Phys. Lett. B **728**, 314 (2014)
30. C. Phatak, A.K. Petford-Long, O. Heinonen, M. Tanase, M. De Graef, Phys. Rev. B **83**, 174431 (2011)
31. C. Nisoli, R. Moessner, P. Schiffer, Rev. Mod. Phys. **85**, 1473 (2013)
32. P. Milde, D. Köhler, J. Seidel, L.M. Eng, A. Bauer, A. Chacon, J. Kindervater, S. Mühlbauer, C. Pfleiderer, S. Buhrandt, C. Schütte, A. Rosch, Science **340**, 1076 (2013)
33. M.W. Ray, E. Ruokokoski, K. Tiurev, M. Möttönen, D.S. Hall, Science **348**, 544 (2015)
34. T. Ollikainen, K. Tiurev, A. Blinova, W. Lee, D.S. Hall, M. Möttönen, Phys. Rev. X **7**, 021023 (2017)
35. M. Charilaou, H.B. Braun, J.F. Löffler, Phys. Rev. Lett. **121**, 097202 (2018)
36. F. Kagawa, H. Oike, W. Koshibae, A. Kikkawa, Y. Okamura, Y. Taguchi, N. Nagaosa, Y. Tokura, Nat. Commun. **8**, 1332 (2017)
37. E.H. Fyhn, J. Linder, Phys. Rev. B **100**, 214503 (2019)
38. D. Kleckner, W.T.M. Irvine, Nat. Phys. **9**, 253 (2013)
39. L.M. Pismen, *Vortices in Nonlinear Fields* (Clarendon Press, Oxford, 1999)
40. Yu.S. Kivshar, B. Luther-Davies, Phys. Rep. **298**, 81 (1998)
41. Yu.S. Kivshar, J. Christou, V. Tikhonenko, B. Luther-Davies, L. Pismen, Opt. Commun. **152**, 198 (1998)
42. A. López Ariste, R. Centeno, E. Khomenko, Astron. Astrophys. **591**, A63 (2016)
43. H.J. Lugt, *Vortex Flow in Nature and Technology* (John Wiley and Sons, New York, 1983)
44. D.L. Whitaker, J. Edwards, Science **329**, 406 (2010)
45. H. Aref, J. Math. Phys. **48**, 065401 (2007)
46. P.K. Newton, *The N-Vortex Problem: Analytical Techniques*. Applied Mathematical Sciences, vol. 145 (Springer, Berlin, 2001)
47. D. Durkin, J. Fajans, Phys. Fluids **12**, 289 (2000)
48. B. Grzybowski, H. Stone, G. Whitesides, Nature **405**, 1033 (2000)
49. B. Grzybowski, H. Stone, G. Whitesides, Proc. Natl, Acad. Sci. USA **99**, 4147 (2002)
50. F. Dalfovo, S. Giorgini, L.P. Pitaevskii, S. Stringari, Rev. Mod. Phys. **71**, 463 (1999)
51. C.J. Pethick, H. Smith, *Bose-Einstein Condensation in Dilute Gases* (Cambridge University Press, Cambridge, 2002)
52. L.P. Pitaevskii, S. Stringari, *Bose-Einstein Condensation* (Oxford University Press, Oxford, 2003)
53. A. Griffin, D.W. Snoke, S. Stringari (eds.), *Bose-Einstein Condensation* (Cambridge University Press, Cambridge, 1995)
54. M.H.J. Anderson, J.R. Ensher, M.R. Matthews, C.E. Wieman, Science **269**, 198 (1995)
55. K.B. Davis, M.-O. Mewes, M.R. Andrews, N.J. van Druten, D.S. Durfee, D.M. Kurn, W. Ketterle, Phys. Rev. Lett. **75**, 3969 (1995)
56. C.C. Bradley, C.A. Sackett, J.J. Tollett, R.G. Hulet, Phys. Rev. Lett. **75**, 1687 (1995)
57. E.A. Cornell, C.E. Wieman, Rev. Mod. Phys. **74**, 875 (2002)
58. W. Ketterle, Rev. Mod. Phys. **74**, 1131 (2002)
59. A.A. Abrikosov, Sov. Phys. JETP **5**, 1174 (1957)

60. J.M. Kosterlitz, D.J. Thouless, J. Phys. C: Solid State Phys. **6**, 1181 (1973)
61. Z. Hadzibabic, P. Krüger, M. Cheneau, B. Battelier, J. Dalibard, Nature **441**, 1118 (2006)
62. A. Negretti, C. Henkel, J. Phys. B: At. Mol. Opt. Phys. **37**, L385 (2004)
63. M. Kasevich, Atom systems and Bose Einstein Condensates for metrology and navigation. Presented at the NASA Quantum Future Technologies Conference, 17–21 Jan 2012. https://web.archive.org/web/20161118024357/http://quantum.nasa.gov/materials/2012-01-18-B1-Kasevich.pdf. Accessed 27 Jan 2020
64. T.R. Slatyer, C.M. Savage, Classical Quant. Grav. **22**, 3833 (2005)
65. P. Das Gupta, E. Thareja, Classical Quant. Grav. **34**, 035006 (2017)
66. S. Eckel, A. Kumar, T. Jacobson, I.B. Spielman, G.K. Campbell, Phys. Rev. X **8**, 021021 (2018)
67. M.R. Matthews, B.P. Anderson, P.C. Haljan, D.S. Hall, C.E. Wieman, E.A. Cornell, Phys. Rev. Lett. **83**, 2498 (1999)
68. K.W. Madison, F. Chevy, W. Wohlleben, J. Dalibard, Phys. Rev. Lett. **84**, 806 (2000)
69. J.E. Williams, M.J. Holland, Nature **401**, 568 (1999)
70. A. Recati, F. Zambelli, S. Stringari, Phys. Rev. Lett. **86**, 377 (2001)
71. S. Sinha, Y. Castin, Phys. Rev. Lett. **87**, 190402 (2001)
72. K.W. Madison, F. Chevy, V. Bretin, J. Dalibard, Phys. Rev. Lett. **86**, 4443 (2001)
73. C. Raman, J.R. Abo-Shaeer, J.M. Vogels, K. Xu, W. Ketterle, Phys. Rev. Lett. **87**, 210402 (2001)
74. J.R. Abo-Shaeer, C. Raman, J.M. Vogels, W. Ketterle, Science **292**, 476 (2001)
75. R. Onofrio, C. Raman, J.M. Vogels, J.R. Abo-Shaeer, A.P. Chikkatur, W. Ketterle, Phys. Rev. Lett. **85**, 2228 (2000)
76. D.R. Scherer, C.N. Weiler, T.W. Neely, B.P. Anderson, Phys. Rev. Lett. **98**, 110402 (2007)
77. A.E. Leanhardt, A. Görlitz, A.P. Chikkatur, D. Kielpinski, Y. Shin, D.E. Pritchard, W. Ketterle, Phys. Rev. Lett. **89**, 190403 (2002)
78. Y. Shin, M. Saba, M. Vengalattore, T.A. Pasquini, C. Sanner, A.E. Leanhardt, M. Prentiss, D.E. Pritchard, W. Ketterle, Phys. Rev. Lett. **93**, 160406 (2004)
79. C.N. Weiler, T.W. Neely, D.R. Scherer, A.S. Bradley, M.J. Davis, B.P. Anderson, Nature **455**, 948 (2008)
80. D.V. Freilich, D.M. Bianchi, A.M. Kaufman, T.K. Langin, D.S. Hall, Science **329**, 1182 (2010)
81. S. Middelkamp, P.J. Torres, P.G. Kevrekidis, D.J. Frantzeskakis, R. Carretero-González, P. Schmelcher, D.V. Freilich, D.S. Hall, Phys. Rev. A **84**, 011605(R) (2011)
82. P.J. Torres, P.G. Kevrekidis, D.J. Frantzeskakis, R. Carretero-González, P. Schmelcher, D.S. Hall, Phys. Lett. A **375**, 3044 (2011)
83. T.W. Neely, E.C. Samson, A.S. Bradley, M.J. Davis, B.P. Anderson, Phys. Rev. Lett. **104**, 160401 (2010)
84. R. Navarro, R. Carretero-González, P.J. Torres, P.G. Kevrekidis, D.J. Frantzeskakis, M.W. Ray, E. Altuntaş, D.S. Hall, Phys. Rev. Lett. **110**, 225301 (2013)
85. A.V. Zampetaki, R. Carretero-González, P.G. Kevrekidis, F.K. Diakonos, D.J. Frantzeskakis, Phys. Rev. E **88**, 042914 (2014)
86. J.A. Seman, E.A.L. Henn, M. Haque, R.F. Shiozaki, E.R.F. Ramos, M. Caracanhas, P. Castilho, C. Castelo Branco, P.E.S. Tavares, F.J. Poveda-Cuevas, G. Roati, K.M.F. Magalhaes, V.S. Bagnato, Phys. Rev. A **82**, 033616 (2010)
87. S. Middelkamp, P.G. Kevrekidis, D.J. Frantzeskakis, R. Carretero-González, P. Schmelcher, Phys. Rev. A **82**, 013646 (2010)
88. V. Koukouloyannis, G. Voyatzis, P.G. Kevrekidis, Phys. Rev. E **89**, 042905 (2014)
89. N. Kyriakopoulos, V. Koukouloyannis, C. Skokos, P.G. Kevrekidis, Chaos **24**, 024410 (2014)
90. P.G. Kevrekidis, D.J. Frantzeskakis, R. Carretero-González, *Emergent Nonlinear Phenomena in Bose-Einstein Condensates* (Springer, Berlin, 2008)
91. A.L. Fetter, A.A. Svidzinsky J. Phys. Condens. Mat. **13**, R135 (2001)
92. A.L. Fetter, Rev. Mod. Phys. **81**, 647 (2009)

93. T. Kolokolnikov, P.G. Kevrekidis, R. Carretero-González, Proc. Roy. Soc. Lond. A **470**, 20140048 (2014)
94. A. Aftalion, I. Danaila, Phys. Rev. A **68**, 023603 (2003)
95. S. Komineas, N.R. Cooper, N. Papanicolaou, Phys. Rev. A **72**, 053624 (2005)
96. T.P. Simula, Phys. Rev. A **84**, 021603(R) (2011)
97. R.J. Donnelly, *Quantized Vortices in Helium II* (Cambridge University Press, Cambridge, 1991)
98. A.S. Rodrigues, P.G. Kevrekidis, R. Carretero-González, D.J. Frantzeskakis, P. Schmelcher, T.J. Alexander, Yu.S. Kivshar, Phys. Rev. A **79**, 043603 (2009)
99. R. Carretero-González, N. Whitaker, P.G. Kevrekidis, D.J. Frantzeskakis, Phys. Rev. A **77**, 023605 (2008)
100. R. Carretero-González, B.P. Anderson, P.G. Kevrekidis, D.J. Frantzeskakis, C.N. Weiler, Phys. Rev. A **77**, 033625 (2008)
101. J. Ruostekoski, Z. Dutton, Phys. Rev. A **72**, 063626 (2005)
102. B.P. Anderson, P.C. Haljan, C.A. Regal, D.L. Feder, L.A. Collins, C.W. Clark, E.A. Cornell, Phys. Rev. Lett. **86**, 2926 (2001)
103. I. Shomroni, E. Lahoud, S. Levy, J. Steinhauer, Nat. Phys. **5**, 193 (2009)
104. K. Sasaki, N. Suzuki, H. Saito, Phys. Rev. A **83**, 033602 (2011)
105. P.H. Roberts, J. Grant, J. Phys. A: Gen. Phys. **4**, 55 (1971)
106. M. Abad, M. Guilleumas, R. Mayol, M. Pi, Laser Phys. **18**, 648 (2008)
107. M. Guilleumas, D.M. Jezek, R. Mayol, M. Pi, M. Barranco, Phys. Rev. A **65**, 053609 (2002)
108. A.L. Fetter, Phys. Rev. A **69**, 043617 (2004)
109. T.P. Simula, T. Mizushima, K. Machida, Phys. Rev. A **78**, 053604 (2008)
110. T.-L. Horng, S.-C. Gou, T.-C. Lin, Phys. Rev. A **74**, 041603(R) (2006)
111. F. Chevy, S. Stringari, Phys. Rev. A **68**, 053601 (2003)
112. J.J. García-Ripoll, V.M. Pérez-García, Phys. Rev. A **64**, 053611 (2001)
113. F. Maggioni, S. Alamri, C.F. Barenghi, R.L. Ricca, Phys. Rev. E **82**, 026309 (2010)
114. J.L. Helm, C.F. Barenghi, A.J. Youd, Phys. Rev. A **83**, 045601 (2011)
115. S. Lin, J.-X. Zhu, A. Saxena, Phys. Rev. B **99**, 140408(R) (2019)
116. K. Sasaki, N. Suzuki, H. Saito, Phys. Rev. A **83**, 053606 (2011)
117. P.G. Saffman, *Vortex Dynamics* (Cambridge University Press, Cambridge, 1992)
118. R.N. Bisset, W. Wang, C. Ticknor, R. Carretero-González, D.J. Frantzeskakis, L.A. Collins, P.G. Kevrekidis, Phys. Rev. A **92**, 043601 (2015)
119. C. Ticknor, W. Wang, P.G. Kevrekidis, Phys. Rev. A **98**, 033609 (2018)
120. W. Wang, R.N. Bisset, C. Ticknor, R. Carretero-González, D.J. Frantzeskakis, L.A. Collins, P.G. Kevrekidis, Phys. Rev. A **95**, 043638 (2017)
121. R.M. Caplan, J.D. Talley, R. Carretero-González, P.G. Kevrekidis, Phys. Fluids **26**, 097101 (2014)
122. J. Ruostekoski, J.R. Anglin, Phys. Rev. Lett. **86**, 3934 (2001)
123. C.M. Savage, J. Ruostekoski, Phys. Rev. Lett. **91**, 010403 (2003)
124. J. Choi, W.J. Kwon, Y. Shin, Phys. Rev. Lett. **108**, 035301 (2012)
125. L.S. Leslie, A. Hansen, K.C. Wright, B.M. Deutsch, N.P. Bigelow, Phys. Rev. Lett. **103**, 250401 (2009)
126. J. Lovegrove, M.O. Borgh, J. Ruostekoski, Phys. Rev. Lett. **112**, 075301 (2014)
127. Y.V. Kartashov, B.A. Malomed, Y. Shnir, L. Torner, Phys. Rev. Lett. **113**, 264101 (2014)
128. Y.M. Bidasyuk, A.M. Chumachenko, O.O. Prikhodko, S.I. Vilchinskii, M. Weyrauch, A.I. Yakimenko, Phys. Rev. A **92**, 053603 (2015)
129. R.N. Bisset, W. Wang, C. Ticknor, R. Carretero-González, D.J. Frantzeskakis, L.A. Collins, P.G. Kevrekidis, Phys. Rev. A **92**, 063611 (2015)
130. D. Proment, M. Onorato, C.F. Barenghi, Phys. Rev. E **85**, 036306 (2012)
131. D. Proment, M. Onorato, C.F. Barenghi, J. Phys.: Conf. Ser. **544**, 012022 (2014)
132. D. Kleckner, L.H. Kauffman, W.T.M. Irvine, Nat. Phys. **12**, 650 (2016)
133. The Knot Atlas, http://katlas.org. Accessed 27 Jan 2020
134. V.P. Ruban, JETP Lett. **107**, 307 (2018)

135. V.P. Ruban, JETP Lett. **108**, 605 (2018)
136. C. Ticknor, V.P. Ruban, P.G. Kevrekidis, Phys. Rev. A **99**, 063604 (2019)
137. R. Dandoloff, S. Villain-Guillot, A. Saxena, A.R. Bishop, Phys. Rev. Lett. **74**, 813 (1995)
138. A.A. Belavin, A.M. Polyakov, JETP Lett. **22**, 245 (1975)
139. S. Villain-Guillot, R. Dandoloff, A. Saxena, Phys. Lett. A **188**, 343 (1994)
140. A. Saxena, R. Dandoloff, Phys. Rev. B **58**, R563 (1998)
141. A. Saxena, R. Dandoloff, Phys. Rev. B **66**, 104414 (2002)
142. J. Benoit, A. Saxena, T. Lookman, J. Phys. A: Math. Gen. **34**, 9417 (2001)
143. F.C. Marques, A. Neves, Annals Math. **179**, 683 (2014)
144. T.O. Wehling, A.M. Black-Schaffer, A.V. Balatsky, Adv. Phys. **63**, 1 (2014)
145. X.L. Qi, S.-C. Zhang, Rev. Mod. Phys. **83**, 1057 (2011)
146. Y. Ando, L. Fu, Annu. Rev. Condens. Matter Phys. **6**, 361 (2015)
147. O. Vafek, A. Vishwanath, Annu. Rev. Condens. Matter Phys. **5**, 83 (2014)
148. N. Sirica, R.I. Tobey, L.X. Zhao, G.F. Chen, B. Xu, R. Yang, B. Shen, D.A. Yarotski, P. Bowlan, S.A. Trugman, J.-X. Zhu, Y.M. Dai, A.K. Azad, N. Ni, X.G. Qiu, A.J. Taylor, R.P. Prasankumar, Phys. Rev. Lett. **122**, 197401 (2019)
149. F. Cooper, A. Khare, B. Mihaila, A. Saxena, Phys. Rev. E **82**, 036604 (2010)
150. J. Cuevas-Maraver, P.G. Kevrekidis, F.G. Mertens, A. Saxena, Phys. Rev. E **100**, 022210 (2019)
151. J. Cuevas-Maraver, P.G. Kevrekidis, A. Saxena, A. Comech, R.M. Lan, Phys. Rev. Lett. **116**, 214101 (2016)
152. J. Cuevas-Maraver, N. Boussaïd, A. Comech, R. Lan, P.G. Kevrekidis, A. Saxena, in *Nonlinear Systems*, vol. 1, ed. by V. Carmona, J. Cuevas-Maraver, F. Fernández-Sánchez, E. García-Medina (Springer, Cham, 2018), p. 89
153. S. Seki, X.Z. Yu, S. Ishiwata, Y. Tokura, Science **336**, 198 (2012)
154. C. Phatak, O. Heinonen, M. De Graef, A. Petford-Long, Nano Lett. **16**, 4141 (2016)
155. J. Chen, L. Wang, M. Zhang, L. Zhou, L. Jin, X. Wang, H. Qin, Y. Qiu, J. Mei, F. Ye, B. Xi, H. He, B. Li, G. Wang, Nano Lett. **19**, 6144 (2019)
156. X. Wan, A.M. Turner, A. Vishwanath, S.Y. Savrasov, Phys. Rev. B **83**, 205101 (2011)
157. R.-J. Slager, J. Phys. Chem. Solids **128**, 24 (2019)
158. L. Bi, MRS Bull. **43**, 408 (2018)
159. X. Zhou, Y. Wang, D. Leykam, Y.D. Chong, New J. Phys. **19**, 095002 (2017)
160. Z. Wang, Y. Chong, J.D. Joannopoulos, M. Soljacic, Nature **461**, 772 (2009)
161. W.P. Su, J.R. Schrieffer, A.J. Heeger, Phys. Rev. B **22**, 2099 (1980)
162. Y. Hadad, V. Vitelli, A. Alu, ACS Photonics **4**, 1974 (2017)
163. R.K. Pal, M. Schaeffer, M. Ruzzene, J. Appl. Phys. **199**, 084305 (2016)
164. R. Süsstrunk, S.D. Huber, Science **349**, 47 (2015)
165. L.M. Nash, D. Kleckner, A. Read, V. Vitelli, A.M. Turner, W.T.M. Irvine, Proc. Natl, Acad. Sci. USA **112**, 14495 (2015)
166. S.H. Mousavi, A.B. Khanikaev, Z. Wang, Nat. Commun. **6**, 8682 (2015)
167. P. Wang, L. Lu, K. Bertoldi, Phys. Rev. Lett. **115**, 104302 (2015)
168. S.D. Huber, Nat. Phys. **12**, 621 (2016)
169. C.L. Kane, T.C. Lubensky, Nat. Phys. **10**, 39 (2013)
170. R. Süsstrunk, S.D. Huber, Proc. Natl, Acad. Sci. USA **113**, E4767 (2016)
171. D.D. Snee, Y.-P. Ma, Extreme Mech. Lett. **30**, 100487 (2019)
172. B.G. Chen, N. Upadhyaya, V. Vitelli, Proc. Natl. Acad. Sci. U. S. A. **111**, 13004 (2014)
173. B. Deng, P. Wang, Q. He, V. Tournat, K. Bertoldi, Nat. Commun. **9**, 3410 (2018)
174. R.K. Pal, J. Vila, M. Leamy, M. Ruzzene, Phys. Rev. E **97**, 032209 (2018)
175. R. Chaunsali, E. Kim, A. Thakkar, P.G. Kevrekidis, J. Yang, Phys. Rev. Lett. **119**, 024301 (2017)
176. R. Chaunsali, G. Theocharis, Phys. Rev. B **100**, 014302 (2019)
177. T. Zhao, A.S. Ramos Jr., G.H. Paulino, Int. J. Numer. Meth. Eng. **118**, 804 (2019)
178. M. Abdi, I. Ashcroft, R. Wildman, Eng. Optimiz. **50**, 1850 (2018)

179. B. Senyuk, Q. Liu, S. He, R.D. Kamien, R.B. Kusner, T.C. Lubensky, I.I. Smalyukh, Nature **493**, 200 (2013)
180. A. Biuhon, A.M. Black-Schaffer, R.-J. Slager, Phys. Rev. B **100**, 195135 (2019)
181. J.O. Island, X. Cui, C. Lewandowski, J.Y. Khoo, E.M. Spanton, H. Zhou, D. Rhodes, J.C. Hone, T. Taniguchi, K. Watanabe, L.S. Levitov, M.P. Zaletel, A.F. Young, Nature **571**, 85 (2019)
182. Z. Song, L. Elcoro, N. Regnault, B.A. Bernevig, Fragile phases as affine monoids: full classification and material examples. arXiv:1905.03262
183. D. Castelvecchi, Nature **571**, 17 (2019)
184. M. Karoubi, *K-theory. An introduction*. Classics in Mathematics (Springer, Berlin Heidelberg, 1978). arXiv:math/0602082
185. S. Das Sarma, M. Freedman, C. Nayak, NPJ Quantum Inform. **1**, 15001 (2015)
186. J.-J. Feng, Z. Huang, Z. Wang, Q. Niu, Phys. Rev. B **98**, 13415 (2018)
187. C.E. Bardyn, A. Imamoglu, Phys. Rev. Lett. **109**, 253606 (2012)
188. I.C. Christov, R.J. Decker, A. Demirkaya, V.A. Gani, P.G. Kevrekidis, A. Khare, A. Saxena, Phys. Rev. Lett. **122**, 171601 (2019)
189. F. Wilczek, Phys. Rev. Lett. **109**, 160401 (2012)
190. K. Nakatsugawa, T. Fujii, A. Saxena, S. Tanda, J. Phys. A: Math. Theor. **53**, 025301 (2020)
191. E. Babaev, L.D. Faddeev, A.J. Niemi, Phys. Rev. B **65**, 100512(R) (2001)
192. F.N. Rybakov, J. Garaud, E. Babaev, Phys. Rev. B **100**, 094515 (2019)
193. K. Sokalski, L.T. Stepien, D. Sokalska, J. Phys. A: Math. Gen. **35**, 6157 (2002)
194. L.T. Stepien, J. Phys. A: Math. Theor. **49**, 175202 (2016)

Chapter 3
Nonlinear Metamaterials

Lei Xu, Mohsen Rahmani, David A. Powell, Dragomir Neshev, and Andrey E. Miroshnichenko

Abstract Metamaterials are engineered structures designed to exhibit exotic electromagnetic properties. Early on in the development of metamaterials, these properties were extended to exotic regimes of nonlinear behaviour, unknown in classical nonlinear optics. In this chapter, we give a historical overview of metamaterials, considering first their exotic linear properties, and show how these give rise to exotic nonlinear properties, at frequency ranges from RF to visible. We overview the main attractive features of metamaterials for nonlinear applications, namely their strong local field enhancement, their ability to achieve exotic phase matching conditions, and the possibility to create inclusions with the correct symmetry to enhance a chosen nonlinear process. We then summarise the two most important classes of nonlinear optical metamaterials, plasmonic and all-dielectric.

L. Xu · D. A. Powell · A. E. Miroshnichenko (✉)
School of Engineering and Information Technology, University of New South Wales, Canberra, ACT 2600, Australia
e-mail: Andrey.Miroshnichenko@unsw.edu.au

L. Xu
e-mail: Lei.Xu@unsw.edu.au

D. A. Powell
e-mail: David.Powell@adfa.edu.au

M. Rahmani
Research School of Physics, Nonlinear Physics Centre, The Australian National University, Canberra, ACT 2601, Australia
e-mail: Mohsen.Rahmani@anu.edu.au

Advanced Optics and Photonics Laboratory, Department of Engineering, Nottingham Trent University, Nottingham NG11 8NS, UK

D. Neshev
Research School of Physics, ARC Centre of Excellence for Transformative Meta-Optical Systems (TMOS), The Australian National University, Canberra, ACT 2601, Australia
e-mail: Dragomir.Neshev@anu.edu.au

P. G. Kevrekidis et al. (eds.), *Emerging Frontiers in Nonlinear Science*, Nonlinear Systems and Complexity 32, https://doi.org/10.1007/978-3-030-44992-6_3

3.1 Introduction: From Microwave to Optics

Metamaterials have opened a new and exciting chapter in electromagnetic theory and practice. The name originates from the Greek word "meta" meaning "beyond" or "more than", and it can refer to any artificial material with extraordinary electromagnetic properties not existing in Nature. Their design is typically based on a bottom-up approach, where the unusual electromagnetic properties arise from the combination of heterogeneous and hybrid materials with a carefully engineered response. The genesis of metamaterials can be traced back to Bose's artificial twisted structure in 1898, which rotated the plane of polarisation of a linearly-polarised wave [1]. More recently, metamaterials have come to be associated with materials having a negative refractive index, where permittivity and permeability are simultaneously negative. Such media was initially proposed by Veselago [2] as a Gedanken experiment. He showed several unexpected consequences such as backward electromagnetic waves (where the wave vector and energy flow point in opposite directions), inverse Doppler shift and Cherenkov radiation, and also negative refraction at the interface with conventional media. In particular, a slab with $\epsilon = \mu = -1$ should focus a divergent light beam, a finding which was later revisited by Pendry [3], leading to the concept of the perfect lens, which overcomes conventional diffraction limits on optical resolution. This generated considerable interest in the scientific community, trying to realise a negative index material at frequencies ranging from RF to the visible, in order to achieve imaging resolution of less than half a wavelength.

3.1.1 Negative Index and Backward Waves

The backward electromagnetic waves which occur in negative-index media have a much longer history, which can be traced back to Lamb [4], Pocklington [5], and Mandelshtam [6] in the early 20th century. They discussed media with negative dispersion, resulting in electromagnetic waves where the wavefronts move towards the source. Interesting enough, they analysed the properties of inhomogeneous materials with periodically varying permittivity, a concept which emerged again in the 1980's under the label of "photonic crystals" [7]. Another prominent example is microwave tubes known as backward wave oscillators, which exhibit opposite phase and group velocities. However, many of these structures have unit cells comparable in size to the wavelength, which prevents them being described by effective permittivity and permeability. To realise negative permittivity is relatively easy, since most metals exhibit this property. Metals typically exhibit a Drude type dispersion of their permittivity, having infinitely negative permittivity at DC, then increasing with frequency, up to the plasma frequency where it becomes positive. Most metals have their plasma frequency in the visible frequency range, and at frequencies just below their plasma frequency they possess relatively small negative values of permittivity, enabling electric fields to penetrate within, and for waves to be trapped at interfaces, giving rise to the field of plasmonics [8]. In the 1960s it was shown that an "artificial plasma" made of thin wire media could be described by a Drude-like permittivity with plasma frequency in the microwave range [9]. It was subsequently demonstrated that this model is only valid for waves propagating normally to the wires [10, 11], whereas for off-normal propagation, spatial dispersion is significant, resulting in new features and phenomena (see Sect. 3.1.3). Moreover, above their plasma frequency, such as wire and other media can be treated as an artificial dielectric, and this concept has been used successfully to design light-weight microwave lenses.

To achieve negative permeability with natural materials is much more difficult since the Larmor frequency of electrons limits the natural magnetic response. There have been several different attempts to achieve artificial magnetism without magnetic constituents. Schelkunoff and Friis [12] elaborated one of the pioneering ideas resulting in the development by split-ring resonators (SRRs), which later became a commonly used meta-atom, i.e. a fundamental building block of metamaterials. They were originally proposed for radio frequency antennas. Since their magnetic resonance frequency is inversely proportional to their size, advances in micro and nano-technology enabled SRRs to be subsequently extended from microwave to terahertz, and even near-infrared ranges. Importantly, an expression for the magnetic polarizability of SRRs was derived based on a Lorentzian model, exhibiting a negative response in a narrow band below the resonant frequency [13]. However, since most metals exhibit quite strong dissipative losses in the visible, the magnetic resonances can be very strongly damped, to the extent that negative magnetic polarizability may not occur. This motivated studies into alternative meta-atoms based on high-index dielectric nanoparticles, which can support strong magnetic resonance in the visible frequency range [14].

At microwave frequencies, a negative refractive index was demonstrated by combining split-ring resonators with the artificial plasma wire media. However, the complex shape of this structure makes it incompatible with fabrication techniques applicable to visible and near-IR frequencies. However, it was shown that two patterned metallic layers separated by a dielectric layer (known as the "fishnet" structure) could simultaneously exhibit the required electric and magnetic properties, while being compatible with planar fabrication techniques. Despite numerous theoretical studies, the realisation of full 3D metamaterials remains challenging due to fabrication limitations.

Although the concept of negative refractive index stimulated the modern development of metamaterials, the field has since broadened to cover a wide range of phenomena. One of the most promising areas of research is metasurfaces, the planar version of metamaterials, which avoid the requirement for sophisticated 3D manufacturing techniques. They can be viewed as a reincarnation of frequency selective surfaces or phased-array antennas, but in a much wider frequency range [15]. They allow efficient manipulation of the amplitude, phase, and polarisation of a transmitted or reflected electromagnetic wave [16].

3.1.2 Anisotropic Media

In general, the arrangement of a metamaterial's constituent elements may not be fully symmetric. Thus it can have a different material response for fields oriented in different directions. This can be modelled by permittivity and permeability tensors, which allow different responses to each field component. The ability of metamaterials to control multiple components of a material tensor led to the study of "Transformation Optics" [17], borrowing some ideas and approaches from General Relativity theory by relating the product of the effective permittivity and permeability tensors to a metric tensor. It even allows several interesting effects to be emulated, such as a gravitational lensing and black holes for light.

Given that metamaterials can exhibit both positive and negative permittivity and permeability, an interesting case arises when different components of the material tensors have different signs at the same frequency. Media with this property are commonly referred to as hyperbolic media, since their iso-frequency surfaces are hyperbolic [18], in contrast to regular media, where they are ellipsoidal. Hyperbolic media gained interest because they can be fabricated relatively easily at optical wavelengths using metal-dielectric multi-layers, and also have the potential to transfer sub-wavelength image details.

3.1.3 Bianisotropy, Chirality and Spatial Dispersion

The various inclusions (meta-atoms) and their spatial arrangements define the bulk electromagnetic properties of metamaterials. Only if their characteristic sizes and separations are smaller than the operating wavelength, is it possible to introduce effective bulk material parameters based on homogenization. Although homogenization has its origin in three-dimensional bulk materials, appropriate techniques have also been developed for metasurfaces [19]. Homogenization drastically simplifies modelling of a metamaterial or metasurfaces' scattering response. In the most general case, the formal homogenization of bulk media is described not only by permittivity and permeability tensors, but also by electromagnetic coupling or bi-anisotropy tensors [20]. These tensors then define the constitutive relations between electric and magnetic fields and the corresponding displacement fields. This requires the full machinery of the tensor or dyadic approach to rigorously cover all cases.

Coupling between electric and magnetic responses occurs in structures with low symmetry. There are two particular cases of common interest. Structures which have no mirror or inversion symmetry are referred to as chiral, based on the same considerations for chiral molecules, which are well known for their ability to rotate the plane of polarisation of light. Examples of chiral metamaterials include spirals [21], Swiss rolls [22] and multi-layered planar structures with twisted arrangement [23]. Another important type of coupled electric and magnetic response is Omega bianisotropy, which occurs in structures that have overall mirror symmetry, but which lack mirror symmetry along one of their principal axes [24]. Such media, which are named after the canonical Omega particle, consist of coupled electric and magnetic dipolar antennas and lead to more subtle effects, such as different characteristic impedance values for opposite propagation directions.

The picture of equivalent material parameters that we have considered so far is based on a local response, where the displacement fields at a point depend only on the electric and/or magnetic field at the same point. More general models are based on a non-local response, whereby displacement fields can be induced by electric or magnetic fields at a different location. This can occur when fields are conducted along long wires, or more generally when the unit cell is not sufficiently small compared to the wavelength. It is typical to describe the response of such media in the spectral domain, using a material tensor $\overline{\overline{\varepsilon}}(\omega, \mathbf{k})$, and we refer to the corresponding dependence on wavenumber k as *spatial dispersion* of the material properties. For such spatially dispersive media, only a single material tensor is required, and from it permittivity, permeability and bianisotropy tensors can be recovered as the lowest order terms [25]. This gives us an alternative view of exotic metamaterial properties, such as the artificial magnetism of split-ring resonators. From this point of view, the SRR is responding to a spatial gradient of the electric field, which we can treat as an effective magnetic field through the harmonic version of Maxwell's equations.

3.1.4 Overview of Nonlinear Metamaterials

In addition to their exotic linear properties, many exciting phenomena have been recently predicted and demonstrated using nonlinear metamaterials [26]. A key reason why metamaterials are interesting for nonlinear physics is their tendency to increase the effective field strength, through concentrating fields in small gaps, and resonant enhancement, both of which can lead to very large values of effective nonlinear susceptibility. They have the potential to realise novel effects based on magnetic type nonlinearity, which is otherwise impossible to obtain at optical frequencies. The complexity of linear properties of metamaterials leads to considerably rich but complex nonlinear phenomena, requiring the development of new tools and methods to analyse them. A wide variety of methods have been reported to realize nonlinear metamaterials, the most popular being nonlinear inclusions and immersion in a nonlinear host medium. Below, we focus primarily on the use of nonlinear elements in various frequency ranges. Since the dimensions of the elements are much smaller than the operating wavelength, their response is usually described by the induced polarizability. Alternatively, they can be treated as lumped elements with effective resistance, capacitance, and inductance. Their response can be described and analysed based on the transfer function approach.

To achieve the macroscopic nonlinearity, various nonlinear elements can be added to the resonant meta-atom. The nonlinear response of the element can be linked to the currents in the meta-atom, which may be induced by either electric or magnetic fields, leading to a corresponding nonlinear response. One of the earliest examples is

SRRs which were made nonlinear by loading them with varactor diodes [26]. More exotic nonlinear inclusions have also been investigated, including superconducting Josephson junctions [27]. Because of the complex nonlinearity of circuit elements, it is quite difficult to predict their nonlinear response analytically. In particular, it was suggested that an array of Superconducting Quantum Interference Devices (SQUIDs) could be considered as a magnetic metamaterial with extraordinary properties supporting counter-intuitive dynamic chimaera states [28]. Furthermore, since both electric and magnetic fields may be involved in any nonlinear process, the total number of nonlinear tensor elements is very large. Therefore, homogenization procedures based on numerical simulation have been proposed [29], with related methods also having been developed for metasurfaces [30].

Similar to nonlinear optics, the harmonic generation has been demonstrated in nonlinear metamaterials over a wider frequency range. An advantage of metamaterials for frequency conversion is that their response can be simultaneously engineered at multiple frequencies, greatly enhancing harmonic generation efficiency [31]. On the other hand, negative refraction and backward waves propagation lead to rather interesting peculiarities of wave interaction [32]. For example, second harmonic generation (SHG) can be achieved for the reflected wave of the pump excitation propagating in the backward regime [33, 34]. This allows subwavelength imaging to be achieved at the second harmonic. Phase matching conditions can also be altered, resulting in higher conversion efficiency, peculiar phase-locked harmonic generation, and trapped spatio-temporal pulse propagation. In transmission-line based microwave metamaterials nonlinear response can also be achieved by replacing capacitors with varactor diodes leading to exotic phase matching regimes of forward and backward propagating waves at fundamental and generated frequencies. They can be easily implemented and used for various applications such as, for example, tunable leaky-wave antennas [35]. In the microwave frequency range, high nonlinearity can be achieved at low power levels (fraction of a Watt). On the other hand, parametric amplification has been suggested and widely used as loss compensation mechanism mostly in microwave range for transmission lines, magneto-inductive arrays and SRRs [26].

Nonlinear self-action at higher powers results in the switching of metamaterials between different states, analogous to a Kerr-type nonlinear response. In addition to the continuous tuning of the linear properties with incident power, bistable and multistable behaviour can be observed [36]. The strong nonlinearities observable particularly in transmission-line type microwave metamaterials defy many of the common assumptions of stable steady-state behaviour. In such structures, a variety of autonomous frequencies can be generated, which can only be described by a more accurate nonlinear state-space model [37].

A very different type of nonlinear response can be observed when meta-atoms attract or repel each other due to electromagnetic forces. By adding an elastic restoring force to counter this electromagnetic force, the distance between a pair of meta-atoms will depend on the incident field strength [38]. The resonant nature of meta-atoms and their strong near-field leads to a very strong dependence of their frequency response on their separation. The combination of these effects leads to a response which depends on the strength of the incident field, which was experimentally demonstrated. To improve the strength of this nonlinear response, subsequent works used a torsional restoring force between a pair of rotating rings [39]. By using this softer elastic restoring force, a much stronger nonlinear response was observed. In experiments with a pair of chiral meta-atoms of opposite handedness, spontaneous chiral symmetry breaking was observed, where the structure was achiral to linear waves, and became chiral at a range of high incident powers. Most interestingly, for a range of incident powers, this structure was shown to exhibit self-oscillations, whereby a constant intensity electromagnetic wave led to approximately periodic mechanical oscillations of the rings.

While most studies have investigated the nonlinear response of microwave metamaterials to a continuous wave signal, there have also been studies where the time-domain response of a metamaterial or metasurfaces was considered. This allowed the development of a waveform-dependent absorbing metasurfaces, as seen in the review paper [40]. Note that nonlinear microwave metasurfaces have demonstrated many of the same effects of metamaterials, and they have the advantage of a compact structure which is readily fabricated with conventional circuit techniques. It is worth noting that nonlinear metamaterials and metasurfaces are closed related to their tunable counterparts, and often make use of similar inclusions. A varactor diode can be tuned by an external bias voltage, in order to shift the linear resonance, and also leads to a nonlinear response at high incident power [41]. Recently this idea has been developed further, by creating metasurfaces where the nonlinear coefficient is tunable, and can vary across the surface [42].

Strong nonlinear effects have also been observed at interfaces between conventional and nonlinear metamaterials [43]. In particular, nonlinear forward and backward propagating surface guided waves can be excited and controlled based on their nonlinear dispersion relation. Moreover, both positive and negative lateral Goos–Hänchen shift can also be observed at the interfaces depending on the incident power [44]. A particularly interesting class of edge states that have recently been observed in photonics are topologically protected states [45], which are immune to scattering by certain types of defects. Due to their ability to tailor the medium response, metamaterials have served as an important platform for the observations of such states. Recently it was shown that topological states can undergo nonlinear spectral shifts due to a pump beam, eventually being destroyed at very high pump powers [46].

Metamaterials have delivered an extended set of tools for the design of novel nonlinear electromagnetic media. The artificial magnetic response makes available many more nonlinear susceptibility components, offering greater control over the harmonic generation and wave mixing phenomena for novel nonlinear metadevices. One of the exceptional abilities of nonlinear metamaterials and metasurfaces is to control the direction of the nonlinear emission, resulting in significant improvement of the nonlinear signal generation and detection [26]. At microwave frequencies very strong nonlinearities are readily achieved, due to the availability of lumped components with a strong response. However, from an applications perspective, the greatest demand for new nonlinear functionality is expected to be at higher frequencies. In the terahertz (THz) frequency range nonlinear metamaterials can be realised by integrating subwavelength resonant inclusions with transition metal oxides or semiconductors such as GaAs and InAs [47–49]. The plasma frequency of semiconductors can be tuned by adjusting the doping level. Semiconductors exhibit strong nonlinear response at the THz frequency range, allowing highly efficient nonlinear metamaterials to be designed. At mid-infrared frequencies, achieving a nonlinear metamaterial response remains challenging, although related advances in externally tunable metamaterials have been made [50]. The ready availability of laser sources at near-infrared and visible wavelengths has enabled a wide variety of nonlinear phenomena to be exhibited within these ranges. In the following sections, we outline key achievements in nonlinear photonic metamaterials, considering both metallic and dielectric metamaterial structures.

3.2 Plasmonic Nonlinear Metamaterials

Metallic nanostructures have been among the initially explored platforms to achieve strong nonlinear responses at the nanoscale [51–61]. Such nanostructures interact with light through the oscillation of free electrons on the surface, known as surface plasmons, which cause strong near-field enhancements. These enhancements originate unusual light-matter interactions in both linear and nonlinear regimes [51–70]. Nanorod metamaterial slabs, so-called hyperbolic metamaterials, are good examples for demonstrating the plasmonic capabilities in nonlinear nanophotonics [71–73]. As compared to uniform dielectrics, Au nanorods with $\sim$10 nm diameter can increase the SHG intensity up to three orders of magnitude. Such a significant enhancement is caused by the generated near-field, as well as dark evanescent second harmonic components, which can be detected in the far-field.

3.2.1 *Effect of Symmetry*

Plasmonics can offer an additional nanoscopic degree of freedom, i.e. coupling, among the neighbouring nanoparticles. Such a characteristic can find tremendous applications in nonlinear nanophotonics. For example, a recent research work reported by Gennaro et al. shows that the scattering of SHG light can be manipulated by the symmetry of multi-resonant nanostructures [74]. This is a nonlinear process that converts two photons of frequency ω into one photon of frequency 2ω. Figure 3.1a illustrates two different symmetries with the same components. In the left case, two discs (2ω-particles), radiating in phase, lead to bright and directional SHG. However, in the right case, SHG is suppressed, where two discs radiate out of phase. Figure 3.1b shows the scanning electron microscopy (SEM) images of various studied symmetries. Subsequently, their corresponding experimentally measured SHG are plotted in Fig. 3.1c, demonstrating the significant effect of the symmetry.

More recently, Krauth et al. have studied the effect of coupling in a 3D-fashion via a dolmen type structure arrays (see Fig. 3.1d, e, f) [75]. They have studied the effect of

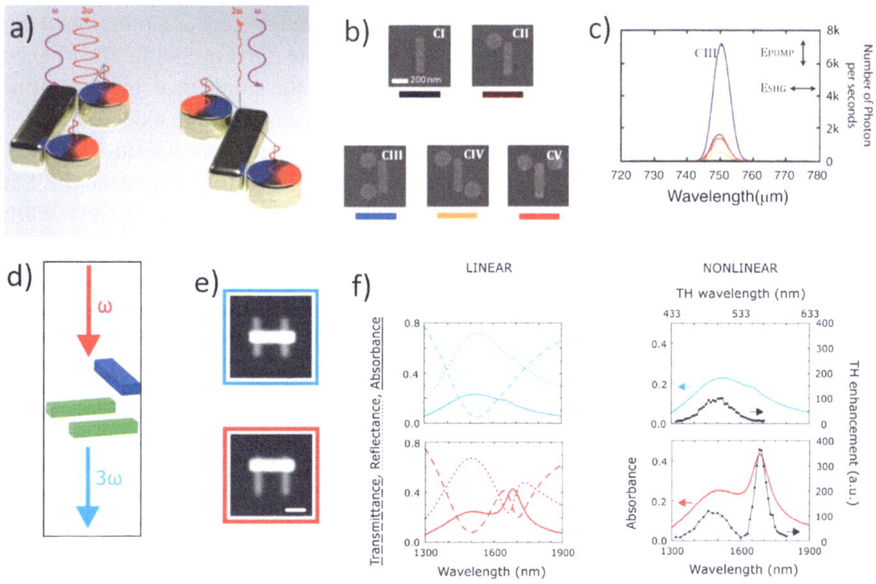

Fig. 3.1 **a** An illustration of the symmetry effect on SHG in complex plasmonic nanoantennas. **b** SEM images of various symmetrical configurations (CI-CV). In all cases, the dimension of the bar is $340 \times 80 \times 40$ nm in x, y and z, respectively, and the discs 160×40 nm in diameter and height, respectively. In all cases, the gaps between the discs and the bar are kept at 20 nm. **c** The experimentally measured SHG signals for different symmetrical configurations, shown in **b**. The pump and SHG signal are polarized along and perpendicular to the bar, respectively. **d** Illustration of the 3D plasmonic structure, with varied position of dipole rod (blue bar). **e** Two different SEM image representing two different coupling case: top case (blue image) exhibit no coupling; bottom case (red image) causes the maximal coupling. **f** Left: linear spectra for two different coupling strengths, where the dashed, dotted, and solid lines represent transmittance, reflectance, and absorbance spectra, respectively. Right: experimentally measured TH signals. Panels **a**, **b** and **c** are adapted from [60] (©2016 by the American Chemical Society) whereas panels **d**, **e** and **f** are adapted from [75] (©2019 by the American Chemical Society)

enhanced induced absorbance on third harmonic (TH) spectroscopy. The absorbance (A) is calculated using $A = 1 - T - R$ with T and R denoting the transmission and reflection, respectively. Figure 3.1f shows both linear and nonlinear spectra, corresponding to the investigated nanostructures. Left panels represent the measured spectra for two different coupling cases in the linear regime (see Fig. 3.1e), with increasing coupling from top to bottom. In the uncoupled regime (top panel), just an individual plasmon resonance exhibits at 1500 nm. This is originated from the single dipole rod. When the coupling is increased by shifting the dipole rod out of the centre position (bottom panel), the absorbance grows in amplitude, remarkably. As can be observed, with larger coupling, the second mode at 1700 nm takes place. This is due to the quadrupole mode. This feature is clearly observable in the absorbance spectrum (Fig. 3.1f left, solid lines).

Right panels in the Fig. 3.1f shows the TH enhancement spectra as a function of the pump wavelength. Alongside this, each panel shows the corresponding absorbance curves. The top panel shows the uncoupled case, where the single peak in the nonlinear regime corresponds to the dipole resonance. In this case, an enhancement factor of around 100 is obtained at 1500 nm. When the coupling is increased (bottom panel), the second peak comes to existence with relatively small amplitude (around 30 at 1670 nm). This peak is relevant to the second mode of the linear spectrum. It is worth mentioning that the first peak in the TH enhancement at around 1520 nm remains almost unchanged. In this case, a sizable TH enhancement of about 365 at 1680 nm can be observed. This peak, which corresponds to the second mode, possesses a three times stronger signal, as compared to the higher energy mode at around 1500 nm. This study can clearly confirm the effect of coupling and phase relations in the linear regime on nonlinear response of complex plasmonic nanostructures.

In one more exciting research work, Zhang et al. have employed the coupling effect in other nonlinear interactions, e.g. four-wave mixing (FWM) at the nanoscale. They have reported a plasmonic nanocluster that supports two Fano resonances (FRs), a temporally oscillating superposition of two spatially coherent subradiant modes [76]. Such an interesting capability of plasmonic nanostructures holds significant potentials for realizing novel sensors, detectors, as well as new types of optoelectronic devices.

3.2.2 Vector Beam Excitation

The polarization state of the fundamental beam can also directly influence the nonlinear behaviour at the nanoscale. Recently, Bautista et al. [77] have demonstrated this effect by azimuthal oligomers consisting of several nanobars. Figure 3.2 demonstrates the second harmonic images of the radial oligomers, while excited with radial (see Fig. 3.2a) and azimuthal (see Fig. 3.2b) cylindrical vector beams. Corresponding simulations are shown in Figs. 3.2c, d, respectively. The SEM image of the corresponding sample is shown in Fig. 3.2e. The sample consists of oligomers with varying number of nanobars. As can be seen in Fig. 3.2a, c, for the case of radial excitation, most of the oligomers show the maximum intensity in the centre and a ring of varying intensities. The central intensity is caused by parallel excitation of all nanorods, when the illumination is symmetric, therefore, the transverse electric fields of the radial vector beams are oriented parallel to the long axes of all the constituent nanorods. However, when the oligomers are asymmetrically illuminated, the surrounding patterns exhibit lower intensity. Indeed, the ring-like nanorods cause n-fold rotational symmetries, which is more observable when the number of nanorods is fewer.

In contrast, for the case of azimuthal excitation, the authors observed a central minimum in the SHG scanning microscopy images for all cases (see Figs. 3.2b, d. In this case, the nanorods barely contribute to the overall SHG, the transverse electric fields of the azimuthal vector beams are oriented perpendicular to each individual nanorod. Similar to the radial excitation, the SHG intensities of the are impacted by

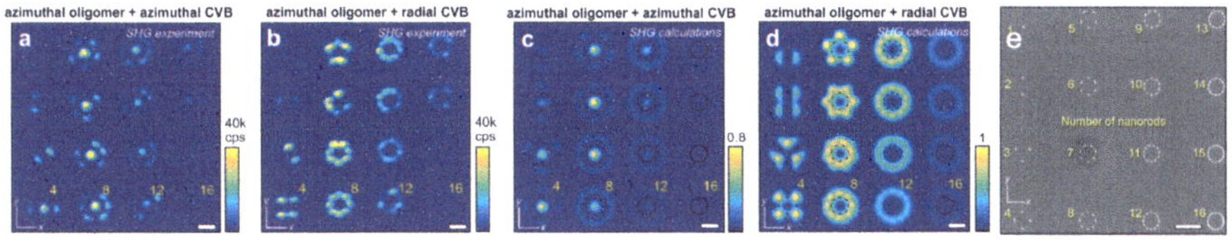

Fig. 3.2 **a** Experimentally measured second harmonic scanning images of the array of azimuthal oligomers with varying number of nanorods employing **a** azimuthal and **b** radial cylindrical vector beams. Scale bar is 1 μm. **c**, **d** Calculated corresponding images to a and b, respectfully, with excitation wavelength of 1060 nm. **e** Representative SEM image of an array azimuthal oligomers with increasing numbers of nanorods, as shown with marks. Scale bars are 2 μm. Adapted from [77].

the number of nanorods. This study demonstrates the potential of nonlinear engineering nanophotonics through the vectorial beam control.

3.2.3 Decoding Near-Field Distributions

A significant application of nonlinear spectroscopy in plasmonics is the ability to decode the near-field distribution. Rahmani et al. have recently shown this ability. They have shown that nonlinear spectroscopy can be used to detect the variation of near-field enhancement in isotropic nanostructures [63], e.g. oligomers. In isotropic nanostructures, the linear far-field properties are independent of the direction of the incident polarization. In other words, linear far-field spectroscopy does not provide any information about the asymmetric near-field distribution in such symmetric structures. However, Rahmani et al. have demonstrated a detectable variation of nonlinear signals from isotropic oligomers. They have monitored the SHG intensity versus the polarization rotation of the incident light. Subsequently, a meaningful correlation between the nonlinear response of symmetric oligomers and the polarisation angle of the incident pump has been revealed. Figure 3.3 shows the dependence of SHG from symmetric pentamers, on the pump polarization angle at two different wavelengths. The two wavelengths are chosen as the representative features of two distinct near-field distributions when rotating the polarization angle of the pump from $\theta = 0°$ to $\theta = 90°$ (see near-field intensity simulations in Fig. 3.3a). Figure 3.3b shows the measured SHG signal, in which significant fluctuations of the SHG intensity are observable. By rotating the pump polarization angle, the SHG signal varies from a minimum at $\theta = 45°$ to maximum at $\theta = 0°$ and $\theta = 90°$. This approach can be employed for detecting the symmetry classes of oligomers, as well as the symmetry classes of many molecules in nature. Moreover, this finding can enable the characterization of the near-field patterns in symmetric nanostructures, a challenging task due to the diffraction limits.

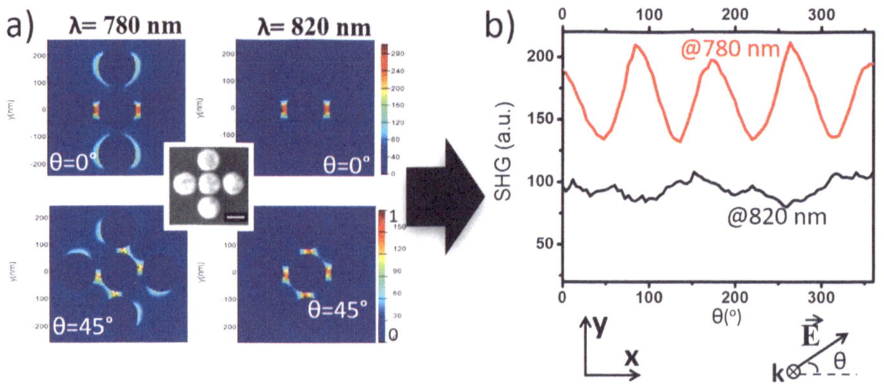

Fig. 3.3 **a** Numerically calculated near-field enhancement of a pentamer at 780 nm and 820 nm. **b** Experimentally measured second harmonic spectrum of the pentamer at 780 nm and 820 nm versus rotating azimuthal angle of the pump polarization. Adapted from [63]. ©2017 by the American Chemical Society

3.2.4 Hybrid Metallo-Dielectrics

Recently, it has been shown that one can combine metallic and dielectric/semiconductor nanoparticles to further enhance the nonlinear efficiency at the nanoscale. Several types of hybrid nanostructures, consisting of metallic particles embedded in dielectric and semiconductor media or vice-versa, have been realized in this respect. Indium tin oxide (ITO) particles within gold nanobars [53, 54, 78, 79] is an example of such efforts with promising outcomes. Aouani et al. have obtained a million-fold enhancement of third-harmonic generation (THG) from an ITO nanoparticle, after being decorated by two Au nanobars [54]. Other hybrid combinations include but are not limited to nanopatterned metallic films filled with GaAs [80], metal/dielectric core-shell nanoparticles [81], plasmonic ring filled by Germanium [82], AlGaAs [83] disks, etc.

3.3 Dielectric Nonlinear Metamaterials

3.3.1 Third-Order Nonlinearities from Si/Ge Nanostructures and Metasurfaces

Silicon (Si) and Germanium (Ge) are semiconductors of broad fundamental and technological interests. They have been continuously employed to realize unique functionalities in the modern nanophotonics field. Owing to their high refractive index, Si/Ge nanostructures can support strong Mie-type resonances of both electric and magnetic natures, offering novel performance in subwavelength manipulation and concentration of light. Due to their large third-order susceptibilities as well as large linear refractive index, Si- and Ge-based nanostructures form novel building blocks for third-order all-dielectric nonlinear nanophotonics. As compared to their metallic counterpart, Si and Ge-based nanostructures have several advantages: They have less absorption and lower laser damage threshold. So it allows for pumping at higher laser power to achieve higher nonlinear conversion efficiency; and also the electric field confined in Si- or Ge-based nanostructures is not limited to the surface. By exciting and engineering the Mie-type resonances supported by Si/Ge nanostructures, the electromagnetic field can penetrate inside the structures. This will significantly increase the effective light-matter interacting area, and facilitate the nonlinear processes [84]; Importantly, both Si and Ge materials are complementary-metal–oxide–semiconductor (CMOS) compatible, this will give great design and fabrication flexibility for linking to existing integrated photonic architectures [85].

3.3.1.1 Third-Order Nonlinear Process Based on Mie Resonances

In recent years, with the development of "all-dielectric nanophotonics", third-order nonlinear processes based on Si or Ge semiconductors have been intensively investigated. Specifically, Si and Ge Mie resonators have been the primary focus for investigating the nonlinear processes in recent years. For example, the fundamental magnetic dipole (MD) resonance supported by Si/Ge nanodisks features a strong circular displacement current excited inside the disk. It can significantly enhance the THG process [86]. Shcherbakov et al. have experimentally demonstrated that by exciting the resonance from a Si nanodisk, it is possible to get two orders of magnitude enhancement of THG as compared with that from unstructured bulk silicon slab [87] (Fig. 3.4a). Based on the excitation and interference between different Mie-type multipoles, researchers have been exploring various ways to enhance the near-fields of the pump to boost the nonlinear processes further. For example, by exciting electric dipole (ED) and toroidal dipole (TD) modes with the same scattering magnitude but out of phase, and overlapping them spectrally and spatially, it will create the nonradiating anapole state. The scattering contributions from ED and TD will cancel each other, and exhibit a strongly enhanced near-field distribution inside the nanostructure. Si/Ge nanodisks supporting anapole states have been used to investigate the nonlinear process, and can significantly improve the third-order nonlinear conversion efficiency of THG and FWM [82, 88–91]. Furthermore, the anapole excitation can be further improved by placing an Au nanoring outside the anapole resonator (Fig. 3.4b), or placing the anapole resonator on a metallic mirror (Fig. 3.4c). In such a case, the excitation of both ED and TD modes can be efficiently enhanced, resulting in a much stronger near-field enhancement. It has been shown that the third-harmonic radiation intensity from a dielectric anapole resonator located on a metallic mirror can be enhanced by two orders of magnitude as compared to a typical anapole resonator located on an insulating substrate [92].

By combining several individual resonators together to form complex nanoparticle arrangements such as dimers [93], trimers [94] or quadrumers [95, 96], it can

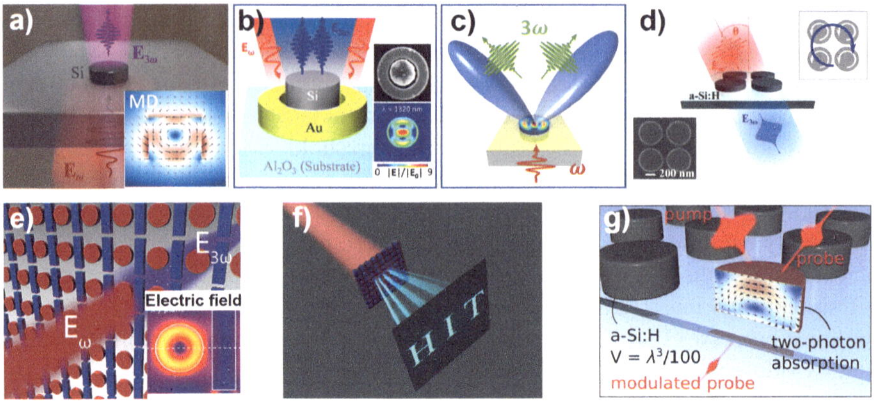

Fig. 3.4 **a** Illustration of THG process from Si nanodisk driven by MD resonance. Inset: Local electric field distribution at the MD resonance. **b** THG enhancement through a metal-dielectric hybrid nanostructure supporting an anapole state. Right upside: SEM image of the experimentally fabricated sample. Right downside: Calculated electric field distribution at the anapole state. **c** Schematic illustration of THG process from an anapole resonator on an Au mirror surface. **d** Scheme of the resonant THG in silicon quadrumers. Left downside: SEM image of the fabricated sample. **e** Schematic illustration of THG enhancement by a Fano resonant metasurfaces. Inset: Simulated electric near field distribution at the Fano resonance. **f** Nonlinear holographic image generation by a Si metasurface based on a THG process. **g** Ultrafast all-optical switching in subwavelength nonlinear silicon nanodisks. Panel **a** is adapted from [87] (©2014 by the American Chemical Society), panel **b** is adapted from [82] (©2017 by the American Chemical Society), panel **c** is adapted from [92] (Creative Commons Attribution License (CC BY) https://creativecommons.org/licenses/by/4.0/), panel **d** is adapted from [95] (©2016 by the American Chemical Society), panel **e** is adapted from [97] (©2015 by the American Chemical Society), panel **f** is adapted from [107] (©2018 by the American Chemical Society) and panel **g** is adapted from [108] (©2015 by the American Chemical Society)

support strong collective responses and Fano resonances with enhanced near-fields, which will benefit the nonlinear generation processes. For example, Shorokhov et al. observed a multifold enhancement of THG in dielectric quadrumers of silicon nanodisks supporting high-quality collective modes associated with the magnetic Fano resonance originating from the interplay between the collective optically induced magnetic responses of quadrumers and the individual magnetic responses of their constituent Si nanodisks [95] (Fig. 3.4d). Arrangement of Si/Ge nanoparticles to form nanoparticles array or metasurfaces can yield a more complicated optical responses [97, 98]. Driven by a high-quality Fano resonance in Si-based metasurfaces composed of disks and bars, Yang et al. have demonstrated highly efficient THG enhancement [97] (Fig. 3.4e).

Besides, based on the coupling between resonant nanoparticles in metasurfaces, topologically protected edge states [99, 100], bound state in the continuum [101–103], and other high-quality collective modes [104, 105] can be generated and supported to facilitate the third-order nonlinear processes in Si/Ge nanostructures and metasurfaces. Besides geometric engineering, different electric and magnetic modes can also be selectively excited by structured light, such as an azimuthally-polarized beam or a radially-polarized beam [106]. These modes can be employed to boost the third-order nonlinear processes significantly.

3.3.1.2 Nonlinear Wavefront Engineering

Another important perspective of Si/Ge-based nonlinear nanophotonics is nonlinear optical wavefront control. While linear wavefront control based on nanoparticles and metasurfaces has been widely studied for long times, nonlinear wavefront control using nanoparticles and metasurfaces have become more and more popular these days. It enables efficient generation of light at new frequencies in predefined shapes. Nonlinear metasurfaces have been a great tool to achieve this. Metasurfaces are composed of subwavelength particles, so-called meta-atoms. Arranging meta-atoms in different shapes and geometries enable the excitation of both linear and nonlinear multipoles of different polarization-dependence or polarization-independence. It enables engineering the spatial characteristics of the nonlinear interactions by locally modify the nonlinear responses from each meta-atoms, i.e. control of the nonlinear light amplitude, phase, polarization in the sub-wavelength regime, to achieve the design of nonlinear optical wavefronts of light. By engineering the polarization, amplitude, phase of the nonlinear emission waves locally, various nonlinear applications have been demonstrated from Si/Ge-based nanostructures and metasurfaces. Recently, Gao et al. have demonstrated efficient third-harmonic holography by use of a metasurfaces consisting of C-shaped Si nanoantennas [107]. The near-fields from the incident light response is enhanced by the excitation of the resonance at the pump frequency, whereas the THG signal is redistributed to the air gap region by higher-order resonances to reduce the absorption loss at the harmonic frequency (Fig. 3.4f). Besides, nonlinear vortex beams [109], nonlinear image tuning [102] and nonlinear multiplexed holography [107, 110] have been reported recently.

3.3.1.3 Ultrafast Optical Switching

Furthermore, due to the enhanced nonlinearities from Si or Ge nanoresonators, it is enabled all-optical modulation of light fields both spatially and spectrally at the nanoscale, realizing novel functionalities such as ultrafast all-optical switching [108, 111, 112]. This will further benefit applications such as optical interconnects, optical communication and computing. Pump-probe measurements have shown that the switching of the Si nanodisks can be governed by bandwidth-limited 65-fs long two-photon absorption being enhanced by a factor of 80 as compared to unstructured Si film (Fig. 3.4g) [108].

Owing to its vast range of possibilities and potential applications, the emerging field of nonlinear nanophotonics based on CMOS-compatible materials (such as Si, Ge, SiN, etc) has attracted increasing attention in recent years [113–116]. Si/Ge nanostructures and metasurfaces holds the promising ability of advanced field control functionalities, enhanced nonlinear responses, and meanwhile they can be easily linked to other integrated photonic architectures. It is expected that in the near future, together with the development of nanoscale fabrication techniques, tunable and active metamaterials, multiple optical functions could be integrated together onto nonlinear nanostructures to realize functional nonlinear on-chip elements. This will take a further step for developing and realizing efficient nonlinear photonic metadevices, enabling various applications in biosensing, nanolasing, all-optical signal processing and quantum nanophotonics area.

3.3.2 III–V Nonlinear Metasurfaces

Dielectric nanoresonators indeed offer a unique platform for the enhancement of light-matter interactions at the nanoscale. The ultimate goals in the field are to be able to tailor the nonlinear emission from such nanoresonators with high-level of control in directivity and efficiency, see Fig. 3.5a. For this purpose, the implementation of materials with non-centrosymmetric crystalline structure is highly beneficial, as such materials exhibit quadratic nonlinear response, which is ultimate, the strongest nonlinear term in the material polarisation. While metallic nanoresonators exhibit quadratic nonlinear response, this is predominantly based on surface effects and therefore weak. Therefore, implementing dielectric materials with the bulk nonlinear response, such as lithium niobate ($LiNbO_3$) and gallium arsenide (GaAs) for nonlinear nanoresonators and metasurfaces is of paramount importance.

3.3.2.1 Second Harmonic Generation in III–V Nanoresonators

Out of all natural materials the noncentrosymmetric crystalline structure of GaAs offers probably the strongest bulk quadratic nonlinearity with nonlinear susceptibility of $d = 100$ pm/V [121, 122]. The GaAs and its aluminium compounds also exhibit high transparency in a broad spectral window from 730 nm up to the far-infrared. This transparency is due to their direct electronic bandgap, tunable by varying the aluminium compositions and further preventing two-photon absorption at telecommunication wavelengths. Unfortunately, the use of GaAs comes with two main challenges: (i) the peculiar off-diagonal nature of the nonlinear susceptibility tensor [122] and the difficulties of fabrication on transparent substrates without defects in the crystalline structure [123].

Despite these peculiarities, the first theoretical studies on AlGaAs nanoresonators for SHG have predicted efficiencies reaching 10^{-3} with a pump intensity of 1 GW/cm^2 [117, 124–126], see Fig. 3.5b. These theoretical predictions have inspired to experiment with AlGaAs nanoresonators. Experiments with multiple resonators in a metasurface have quickly followed using structures fabricated through selective oxidation of the substrate [127–130] or through a transfer to a transparent substrate [118, 131], see Fig. 3.5c.

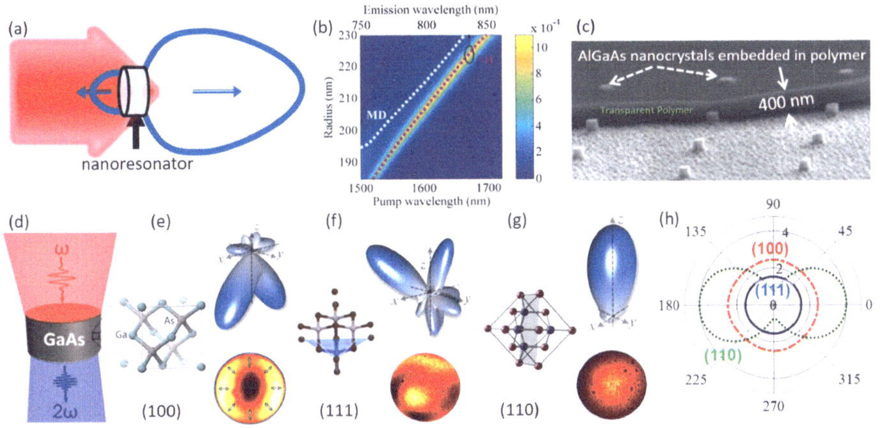

Fig. 3.5 Nonlinear quadratic nanoresonators: **a** Tailoring of emission characteristics and second harmonic directivity in forward and backward directions from quadratic nonlinear resonators. **b** SHG efficiency as a function of pump wavelength and nanodisk radius. The MD resonance and the resonant mode at the second harmonic wavelength are marked with a dotted white and magenta line, respectively. After [117]. **c** SEM image of the AlGaAs nanoresonators being transferred into a transparent polymer and subsequently onto a transparent substrate [118]. **d–g** Tailoring the second harmonic emission via crystalline orientation of the underlying material: **e** (100) crystalline orientation and the typical radiation patterns for a nanodisk of $h = 300$ nm, $r = 245$ nm; **f, g** (111) [119] and (110) crystalline orientation, respectively, with radiation patterns for a nanodisk of $h = 400$ nm, $r = 320$ nm. **h** Forward to backward ratio of the second harmonic emission for the three different crystalline orientations vs. the incident polarisation angle [120]. This figure is a collage of original images and other ones taken from the following references: [117, 118] (©2016 by the American Chemical Society), [119] (©2019 by the American Chemical Society) and [120] (©2020 by the American Chemical Society)

While the AlGaAs is a linearly isotropic material, the nonlinear properties of such nanoresonators depend strongly on the crystalline orientation. The nonlinear efficiency remains relatively similar for different crystalline orientations and is mainly defined by the proximity of the resonances to the excitation and the second harmonic wavelengths. The magnetic dipolar resonance has been found to strongly enhance the nonlinear conversion efficiency [117] (Fig. 3.5b), however, the presence of electric or magnetic anapole also provides strong near-fields enhancement and harmonic emission enhancement [132, 133].

The directionality and the radiation pattern of the second harmonic emission, however, are strongly affected by the crystalline orientation. Most of the AlGaAs metasurfaces and nanoresonators have been fabricated using a (100) crystalline orientation, which is common to VCSEL fabrication [134]. However, for this crystalline orientation the second harmonic emission comes in a doughnut shape [118], where no light is emitted at the direction parallel to the excitation direction, regardless of the nanoresonator size. This is a fundamental limitation and comes from the nature of the nonlinear susceptibility tensor. Typical radiation patterns and their projection (k-space image) are shown in Fig. 3.5d, e, f, g. For the case of (100) crystalline orientation the lack of normal direction of the second harmonic emission and the complex polarisation patterns (shown with arrows) inhibits the practical application of such resonators as nanoscale light sources.

A large body of work has been devoted to circumventing the lack of directional emission. These include the use of inclined beam excitation [124, 135], broken symmetry resonators [136], in-plane holographic gratings [137], and the interplay between Mie and Bragg resonances in AlGaAs metasurfaces [138]. However, the practical applicability of these approaches remains limited. Changing the crystalline orientation of the underlying material gives much bigger opportunities to retain the resonator geometry, while enabling emission normal to the resonator and control of the back-to-forward emission. For example, the use of (111) crystalline orientation [119] (Fig. 3.5f) allows for redirecting a significant portion of the second harmonic emission towards the normal direction of the disk axis. In addition, the efficiency of the SHG process becomes polarisation insensitive and does not depend on the position of the incident polarisation, which is of great practical advantage. However the radiation patterns is changing with the polarisation, as the nonlinearly excited multipoles change their relative phase.

The use of resonators with (110) crystalline orientation (Fig. 3.5g) probably holds the greatest potential for practical use. In this case, most of the energy of the SHG is directed towards the normal direction and this property does not significantly depend on the size of the nanoresonators. As such, these nanoantennas would be beneficial to combine in spatially-variant metasurfaces, having applications in nonlinear holography and ultra-thin light sources. Furthermore, the backward-forward directionality of the emission can be easily altered by varying the incident polarisation direction. As shown in Fig. 3.5h only in the case of (110) crystalline orientation one can vary the forward-backward directionality of emission, which is due to the nature of the excited multipoles. As such, the (110) III–V nanoresonators form an exciting platform for quadratic nonlinear metasurfaces.

3.3.2.2 Quadratic Nonlinear Metasurfaces

The combination of multiple nanoresonators of the same or different sizes (but at the same height) in an array forms a nonlinear dielectric metasurfaces. When excited by a pump beam, the emission of all nanoresonators interfere in the far-field to give rise to the far-field emission, Fig. 3.6a. Importantly, if the emission of each nanoresonator is zero in the normal direction, then the overall nonlinear emission of the metasurfaces at normal direction is also zero. This is the case for (100) crystalline orientation of AlGaAs nanoresonators, which is illustrated in Fig. 3.6a and was recently observed experimentally in [139]. Figure 3.6b shows the back-focal plane image of the SHG from a (100) AlGaAs metasurfaces, showing that the second harmonic emission collected by a high-numerical aperture objective is directed primarily to the four first-order diffraction orders, while the emission at the zeroth order is negligible.

Nevertheless, when using finite-size pump beams and high-numerical aperture collection objective one can detect the nonlinear emission, as done in the earlier

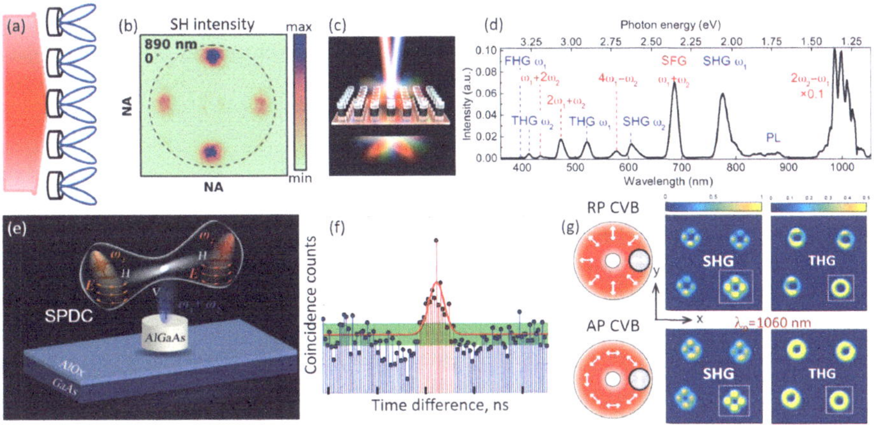

Fig. 3.6 Quadratic nonlinear metasurfaces. **a** The total nonlinear emission from a metasurface is a superposition of the emission from individual nanoresonators. **b** Back-focal plane image of the second harmonic pattern emitted from (100) AlGaAs metasurfaces, demonstrating the lack of zero emission order due to the symmetry of the nonlinear tensor. **c**, **d** Nonlinear mixing of two beams on a AlGaAs metasurfaces (c), demonstrating seven nonlinear processes (d), such as SHG, THG, 4th harmonic generation, sum frequency generation, FWM, six-wave mixing, and photo-luminescence induced by two-photon absorption. **e**, **f** Spontaneous parametric down-conversion in an AlGaAs nanoresonator (a) and the photon correlations measured due to the excitation with a pump beam at 785 nm. The central bin represents bi-photons due to the spontaneous parametric down-conversion, while the red Gaussian fit represents photons due to the thermal excitation of the semiconductor. The green band represents the statistical error of the measurement. **g** Second and TH scanning maps from AlGaAs nanoantennas (diameter of 585 nm) excited with cylindrical vector beams of radial (top tow) and azimuthal polarisation (bottom row). Image size: 9.75 μm × 9.75μm. Insets: calculated far-field SHG and THG scanning maps of the nanodisks using the corresponding cylindrical vector beam. Panel **b** is adapted from [139] (©2019 by the American Chemical Society), panels **c** and **d** are adapted from [140] (Creative Commons Attribution License (CC BY) https://creativecommons.org/licenses/by/4.0/), panels **e** and **f** are adapted from panel [141], and panel **g** is adapted from [133] with permission from The Royal Society of Chemistry

experimental works [117, 124, 125, 127, 129, 130, 142]. Following experiments [140] have explored this option to demonstrate multiple nonlinear mixing processes in a single metasurface, excited by two beams, see Fig. 3.6c. Seven nonlinear processes have been simultaneously observed, including SHG, THG, fourth harmonic generation, sum-frequency generation, FWM, six-wave mixing, and photo-luminescence induced by two-photon absorption (Fig. 3.6d). This experiment demonstrates the power of nonlinear metasurfaces as novel light sources.

It is worth noting that due to the peculiarity of the nonlinear tensor of III–V materials and the absorption in the visible spectral range, researchers have also explored the opportunities for integration of other materials platforms. Lithium niobate shows particular promises in this direction and linear metasurfaces for structural colours have already been fabricated [143] and single nanoresonators have been tested for second harmonic generation [144]. However, the difficulties in the fabrication techniques are still hampering this development. Recent theoretical suggestions on the use of monolithic substrates [145] might boost the future development of this materials platform.

3.3.2.3 Applications of Quadratic Nonlinear III–V Nanoresonators and Metasurfaces

The field of quadratic nonlinear metasurfaces and nanoantennas have made significant advances over the past years and the field is maturing to find novel applications in light sources. One of the most important applications includes the generation of entangled photons through the process of spontaneous parametric down-conversion in AlGaAs nanoresonators [141], as schematically depicted in Fig. 3.6e. The spontaneous parametric down-conversion process is strongly enhanced by the magnetic resonance of the antenna, allowing, for the first time, to measure correlated photons from a sub-wavelength nanocrystal. The measured coincidences (Fig. 3.6f) have very concise duration, however, go well above the statistical noise of the detectors.

Other applications include the generation of cylindrical vector beams [118] as well as their use for exciting the nanoantenna [133]. In a follow-up work [146] such excitation has been used in combination with a bound state in the continuum (BIC) mode to enhance further the nonlinear conversion per energy coupled into the resonator. Finally, the nonlinear up-conversion from infra-red to visible can be used to the conversion of infrared images to visible [147], therefore, opening the exciting new application in infrared imaging and vision.

3.4 Conclusions and Outlook

The field of nonlinear metamaterials continues to mature and brings a vast of practical applications. Nowadays, it is shifting towards planar and integrated metadevices. It includes tuning and switching functionalities which are achieved by structuring meta-atoms at the subwavelength scale. Many devices have been already demonstrated at microwave, terahertz, and visible frequency ranges by employing plasmonic and semiconductor materials, as well as graphene and liquid crystals. It is expected that many more practical metadevices will emerge in the near future, including acoustical, mechanical, and even quantum. The nonlinear response can be easier achieved there via electrostatic or optomechanical forces, leading to novel structures with an exotic response. One of the remaining challenges is the speed of the response. As we have covered in this chapter, there are several attempts to improve it by using semiconductor materials, but further investigations are required. Transmission modulation and the broadband nonlinear optical response are the key features in the next generation metadevice applications, such as terahertz-rate all-optical data processing and ultrafast optical limiters. Another challenge for quantum metamaterials is the ability of the active coherent control of quantum meta-atom states. It implies that different parts of the bulk media should exhibit similar collective properties. By achieving it, quantum metamaterials would allow for fast and high-precision image acquisition and processing. Moreover, at the fundamental level, they can be viewed as a natural test-bed for the study of the quantum-classical transition at the macroscopic scale.

Acknowledgements This work was supported by the Australian Research Council and UNSW Scientia Fellowship.

References

1. U. Leonhardt, Nat. Photonics **1**, 207 (2007)
2. V.G. Veselago, Sov. Phys. Uspekhi **10**, 509 (1968)
3. J.B. Pendry, Phys. Rev. Lett. **85**, 3966 (2000)
4. H. Lamb, Proc. Lond. Math. Soc. **1**, 473 (1904)
5. H.C. Pocklington, Nature **71**, 607 (1905)
6. L. Mandel'shtam, Zh. Eksp. Teor. Fiz. **15**, 475 (1945)
7. J.D. Joannopoulos, P. Villeneuve, S. Fan, Nature **386**, 143 (1997)
8. S.A. Maier, H.A. Atwater, J. Appl. Phys. **98**, 011101 (2005)
9. W. Rotman, IRE Trans. Antennas Propag. **10**, 82 (1962)
10. P.A. Belov, R. Marqués, S.I. Maslovski, I.S. Nefedov, M. Silveirinha, C.R. Simovski, S.A. Tretyakov, Phys. Rev. B **67**, 113103 (2003)
11. C.R. Simovski, P.A. Belov, Phys. Rev. E **70**, 046616 (2004)
12. S.A. Schelkunoff, H.T. Friis, *Antennas: Theory and Practice* (Wiley, New York, 1952)
13. J.B. Pendry, A.J. Holden, D.J. Robbins, W.J. Stewart, IEEE Trans. Microw. Theory Tech. **47**, 2075 (1999)
14. A.I. Kuznetsov, A.E. Miroshnichenko, Y.H. Fu, J. Zhang, B. Luk'yanchuk, Sci. Rep. **2**, 492 (2012)
15. A. Mackay, B. Sanz-Izquierdo, E.A. Parker, Forum Electromagn. Res. Methods Appl. Technol. **2**, 1 (2014)
16. S. Narayan, R.M. Jha, IEEE Antennas Propag. Mag. **57**, 135 (2015)
17. J.B. Pendry, D. Schurig, D.R. Smith, Science **312**, 1780 (2006)
18. D.R. Smith, D. Schurig, Phys. Rev. Lett. **90**, 5 (2003)
19. C. Holloway, M. Mohamed, E. Kuester, A. Dienstfrey, IEEE Trans. Electromagn. Compat. **47**, 853 (2005)
20. C.É. Kriegler, M.S. Rill, S. Linden, M. Wegener, IEEE J. Sel. Top. Quantum Electron. **16**, 367 (2010)
21. R. Ro, V. Varadan, V. Varadan, IEE Proc. H – Microw. Antennas Propag. **139**, 441 (1992)
22. M.C.K. Wiltshire, J.B. Pendry, J.V. Hajnal, J. Phys. Condens. Matter **21**, 292201 (2009)
23. M. Decker, M. Ruther, C.E. Kriegler, J. Zhou, C.M. Soukoulis, S. Linden, M. Wegener, Opt. Lett. **34**, 2501 (2009)
24. A. Serdyukov, I. Semchenko, S.A. Tretyakov, A. Sihvola, *Electromagnetics of Bi-anisotropic Materials: Theory and Applications* (Gordon and Breach Science Publishers, Amsterdam, 2001)
25. M.G. Silveirinha, Phys. Rev. B **75**, 115104 (2007)
26. M. Lapine, I.V. Shadrivov, Y.S. Kivshar, Rev. Mod. Phys. **86**, 1093 (2014)
27. P. Jung, A.V. Ustinov, S.M. Anlage, Supercond. Sci. Technol. **27**, 073001 (2014)
28. N. Lazarides, G. Neofotistos, G.P. Tsironis, Phys. Rev. B **91**, 054303 (2015)
29. S. Larouche, V. Radisic, Phys. Rev. A **97**, 043863 (2018)
30. X. Liu, S. Larouche, D.R. Smith, Opt. Commun. **410**, 53 (2018)

31. D. Filonov, Y. Kramer, V. Kozlov, B.A. Malomed, P. Ginzburg, Appl. Phys. Lett. **109**, 111904 (2016)
32. L. Liu, L. Wu, J. Zhang, Z. Li, B. Zhang, Y. Luo, Adv. Sci. **5**, 1800661 (2018)
33. A. Rose, D. Huang, D.R. Smith, Phys. Rev. Lett. **107**, 063902 (2011)
34. N. Segal, S. Keren-Zur, N. Hendler, T. Ellenbogen, Nat. Photonics **9**, 180 (2015)
35. W.R.C. Somerville, D.A. Powell, I.V. Shadrivov, Appl. Phys. Lett. **98**, 161111 (2011)
36. K.E. Hannam, D.A. Powell, I.V. Shadrivov, Y.S. Kivshar, Appl. Phys. Lett. **100**, 081111 (2012)
37. S.Y. Elnaggar, G.N. Milford, IEEE Trans. Antennas Propag. **66**, 481 (2018)
38. M. Lapine, I.V. Shadrivov, D.A. Powell, Y.S. Kivshar, Nat. Mater. **11**, 30 (2012)
39. M. Liu, D.A. Powell, I.V. Shadrivov, M. Lapine, Y.S. Kivshar, Nat. Commun. **5**, 4441 (2014)
40. A. Li, Z. Luo, H. Wakatsuchi, S. Kim, D.F. Sievenpiper, IEEE Access **5**, 27439 (2017)
41. I.V. Shadrivov, S.K. Morrison, Y.S. Kivshar, Opt. Express **14**, 9344 (2006)
42. Z. Luo, Q. Wang, X.G. Zhang, J.W. Wu, J.Y. Dai, L. Zhang, H.T. Wu, H.C. Zhang, H.F. Ma, Q. Cheng, T.J. Cui, Adv. Opt. Mater. **7**, 1970071 (2019)
43. I.V. Shadrivov, A.A. Sukhorukov, Y.S. Kivshar, A.A. Zharov, A.D. Boardman, P. Egan, Phys. Rev. E **69**, 016617 (2004)
44. L.J. Zhang, L. Chen, C.H. Liang, J. Electromagn. Waves Appl. **22**, 1031 (2008)
45. T. Ozawa, H.M. Price, A. Amo, N. Goldman, M. Hafezi, L. Lu, M.C. Rechtsman, D. Schuster, J. Simon, O. Zilberberg, I. Carusotto, Rev. Mod. Phys. **91**, 015006 (2019)
46. D.A. Dobrykh, A.V. Yulin, A.P. Slobozhanyuk, A.N. Poddubny, Y.S. Kivshar, Phys. Rev. Lett. **121**, 163901 (2018)
47. W. Withayachumnankul, D. Abbott, IEEE Photonics J. **1**, 99 (2009)
48. X. Zhao, J. Zhang, K. Fan, G. Duan, G.D. Metcalfe, M. Wraback, X. Zhang, R.D. Averitt, Photonics Res. **4**, A16 (2016)
49. H.R. Seren, J. Zhang, G.R. Keiser, S.J. Maddox, X. Zhao, K. Fan, S.R. Bank, X. Zhang, R.D. Averitt, Light Sci. Appl. **5**, e16078 (2016)
50. Y.C. Jun, I. Brener, J. Opt. **14**, 114013 (2012)
51. M. Kauranen, A.V. Zayats, Nat. Photonics **6**, 737 (2012)
52. V. Giannini, A.I. Fernández-Domínguez, S.C. Heck, S.A. Maier, Chem. Rev. **111**, 3888 (2011)
53. B. Metzger, M. Hentschel, T. Schumacher, M. Lippitz, X. Ye, C.B. Murray, B. Knabe, K. Buse, H. Giessen, Nano Lett. **14**, 2867 (2014)
54. H. Aouani, M. Rahmani, M. Navarro-Cía, S.A. Maier, Nat. Nanotechnol. **9**, 290 (2014)
55. B. Metzger, L. Gui, J. Fuchs, D. Floess, M. Hentschel, H. Giessen, Nano Lett. **15**, 3917 (2015)
56. J. Butet, P.F. Brevet, O.J. Martin, ACS Nano **9**, 10545 (2015)
57. O. Wolf, S. Campione, A. Benz, A.P. Ravikumar, S. Liu, T.S. Luk, E.A. Kadlec, E.A. Shaner, J.F. Klem, M.B. Sinclair, I. Brener, Nat. Commun. **6**, 7667 (2015)
58. M.W. Klein, C. Enkrich, M. Wegener, S. Linden, Science **313**, 502 (2006)
59. H. Aouani, M. Navarro-Cia, M. Rahmani, T.P. Sidiropoulos, M. Hong, R.F. Oulton, S.A. Maier, Nano Lett. **12**, 4997 (2012)
60. S.D. Gennaro, M. Rahmani, V. Giannini, H. Aouani, T.P. Sidiropoulos, M. Navarro-Cía, S.A. Maier, R.F. Oulton, Nano Lett. **16**, 5278 (2016)
61. F. Niesler, N. Feth, S. Linden, J. Niegemann, J. Gieseler, K. Busch, M. Wegener, Opt. Lett. **34**, 1997 (2009)
62. N. Panoiu, W. Sha, D. Lei, G. Li, J. Opt. **20**, 083001 (2018)
63. M. Rahmani, A.S. Shorokhov, B. Hopkins, A.E. Miroshnichenko, M.R. Shcherbakov, R. Camacho-Morales, A.A. Fedyanin, D.N. Neshev, Y.S. Kivshar, ACS Photonics **4**, 454 (2017)
64. S.A. Maier, *Plasmonics: Fundamentals and Applications* (Springer, Berlin, 2007)
65. M. Rahmani, T. Tahmasebi, Y. Lin, B. Lukiyanchuk, T. Liew, M. Hong, Nanotechnology **22**, 245204 (2011)
66. F. Lu, W. Zhang, L. Huang, S. Liang, D. Mao, F. Gao, T. Mei, J. Zhao, Opto-Electron. Adv. **1**, 180010 (2018)
67. L. Chen, Y. Zhou, M. Wu, M. Hong, Opto-Electron. Adv. **1**, 170001 (2018)
68. M. Rahmani, B. Luk'yanchuk, M. Hong, Laser Photonics Rev. **7**, 329 (2013)
69. W. Zhang, M. Rahmani, W. Niu, S. Ravaine, M. Hong, X. Lu, Sci. Rep. **5**, 8382 (2015)

70. F. Della Picca, R. Berte, M. Rahmani, P. Albella, J.M. Bujjamer, M. Poblet, E. Cortés, S.A. Maier, A.V. Bragas, Nano Lett. **16**, 1428 (2016)
71. P. Segovia, G. Marino, A.V. Krasavin, N. Olivier, G.A. Wurtz, P.A. Belov, P. Ginzburg, A.V. Zayats, Opt. Express **23**, 30730 (2015)
72. G. Marino, P. Segovia, A.V. Krasavin, P. Ginzburg, N. Olivier, G.A. Wurtz, A.V. Zayats, Laser Photonics Rev. **12**, 1700189 (2018)
73. P. Wang, A.V. Krasavin, M.E. Nasir, W. Dickson, A.V. Zayats, Nat. Nanotechnol. **13**, 159 (2018)
74. E.P. Franken, A.E. Hill, C.E. Peters, G. Weinreich, Phys. Rev. Lett. **7**, 118 (1961)
75. J. Krauth, T. Schumacher, J. Defrance, B. Metzger, M. Lippitz, T. Weiss, H. Giessen, M. Hentschel, ACS Photonics **6**, 2850 (2019)
76. Y. Zhang, F. Wen, Y.R. Zhen, P. Nordlander, N.J. Halas, Proc. Natl. Acad. Sci. USA **110**, 9215 (2013)
77. G. Bautista, C. Dreser, X. Zang, D.P. Kern, M. Kauranen, M. Fleischer, Nano Lett. **18**, 2571 (2018)
78. H. Aouani, M. Navarro-Cía, M. Rahmani, S.A. Maier, Adv. Opt. Mater. **3**, 1059 (2015)
79. H. Linnenbank, Y. Grynko, J. Förstner, S. Linden, Light Sci. Appl. **5**, e16013 (2016)
80. W. Fan, S. Zhang, N.C. Panoiu, A. Abdenour, S. Krishna, R. Osgood, K. Malloy, S. Brueck, Nano Lett. **6**, 1027 (2006)
81. Y. Pu, R. Grange, C.L. Hsieh, D. Psaltis, Phys. Rev. Lett. **104**, 207402 (2010)
82. T. Shibanuma, G. Grinblat, P. Albella, S.A. Maier, Nano Lett. **17**, 2647 (2017)
83. V.F. Gili, L. Ghirardini, D. Rocco, G. Marino, I. Favero, I. Roland, G. Pellegrini, L. Duò, M. Finazzi, L. Carletti, A. Locatelli, A. Lemaître, D. Neshev, C. De Angelis, G. Leo, M. Celebrano, Beilstein J. Nanotech. **9**, 2306 (2018)
84. D. Smirnova, Y.S. Kivshar, Optica **3**, 1241 (2016)
85. I. Staude, J. Schilling, Nat. Photonics **11**, 274 (2017)
86. D.A. Smirnova, A.B. Khanikaev, L.A. Smirnov, Y.S. Kivshar, ACS Photonics **3**, 1468 (2016)
87. M.R. Shcherbakov, D.N. Neshev, B. Hopkins, A.S. Shorokhov, I. Staude, E.V. Melik-Gaykazyan, M. Decker, A.A. Ezhov, A.E. Miroshnichenko, I. Brener, A.A. Fedyanin, Y.S. Kivshar, Nano Lett. **14**, 6488 (2014)
88. G. Grinblat, Y. Li, M.P. Nielsen, R.F. Oulton, S.A. Maier, Nano Lett. **16**, 4635 (2016)
89. G. Grinblat, Y. Li, M.P. Nielsen, R.F. Oulton, S.A. Maier, ACS Nano **11**, 953 (2016)
90. G. Grinblat, Y. Li, M.P. Nielsen, R.F. Oulton, S.A. Maier, ACS Photonics **4**, 2144 (2017)
91. R. Colom, L. Xu, L. Marini, F. Bedu, I. Ozerov, T. Begou, J. Lumeau, A.E. Miroshnishenko, D. Neshev, B.T. Kuhlmey, S. Palomba, N. Bonod, ACS Photonics **6**, 1295 (2019)
92. L. Xu, M. Rahmani, K.Z. Kamali, A. Lamprianidis, L. Ghirardini, J. Sautter, R. Camacho-Morales, H. Chen, M. Parry, I. Staude, G. Zhang, D. Neshev, A.E. Miroshnichenko, Light Sci. Appl. **7**, 44 (2018)
93. L. Wang, S. Kruk, L. Xu, M. Rahmani, D. Smirnova, A. Solntsev, I. Kravchenko, D. Neshev, Y.S. Kivshar, Nanoscale **9**, 2201 (2017)
94. M.R. Shcherbakov, A.S. Shorokhov, D.N. Neshev, B. Hopkins, I. Staude, E.V. Melik-Gaykazyan, A.A. Ezhov, A.E. Miroshnichenko, I. Brener, A.A. Fedyanin, Y.S. Kivshar, ACS Photonics **2**, 578 (2015)
95. A.S. Shorokhov, E.V. Melik-Gaykazyan, D.A. Smirnova, B. Hopkins, K.E. Chong, D.Y. Choi, M.R. Shcherbakov, A.E. Miroshnichenko, D.N. Neshev, A.A. Fedyanin, Y.S. Kivshar, Nano Lett. **16**, 4857 (2016)
96. M.K. Kroychuk, D.F. Yagudin, A.S. Shorokhov, D.A. Smirnova, I.I. Volkovskaya, M.R. Shcherbakov, G. Shvets, Y.S. Kivshar, A.A. Fedyanin, Adv. Opt. Mater. **7**, 1900447 (2019)
97. Y. Yang, W. Wang, A. Boulesbaa, I.I. Kravchenko, D.P. Briggs, A. Puretzky, D. Geohegan, J. Valentine, Nano Lett. **15**, 7388 (2015)
98. H. Liu, C. Guo, G. Vampa, J.L. Zhang, T. Sarmiento, M. Xiao, P.H. Bucksbaum, J. Vučković, S. Fan, D.A. Reis, Nat. Phys. **14**, 1006 (2018)
99. S. Kruk, A. Poddubny, D. Smirnova, L. Wang, A. Slobozhanyuk, A. Shorokhov, I. Kravchenko, B. Luther-Davies, Y.S. Kivshar, Nat. Nanotechnol. **14**, 126 (2019)

100. D. Smirnova, S. Kruk, D. Leykam, E. Melik-Gaykazyan, D.Y. Choi, Y.S. Kivshar, Phys. Rev. Lett. **123**, 103901 (2019)
101. K. Koshelev, Y. Tang, K. Li, D.Y. Choi, G. Li, Y.S. Kivshar, ACS Photonics **6**, 1639 (2019)
102. L. Xu, K. Zangeneh Kamali, L. Huang, M. Rahmani, A. Smirnov, R. Camacho-Morales, Y. Ma, G. Zhang, M. Woolley, D. Neshev, A.E. Miroshnichenko, Adv. Sci. **6**, 1802119 (2019)
103. L. Carletti, S.S. Kruk, A.A. Bogdanov, C. De Angelis, Y.S. Kivshar, Phys. Rev. Res. **1**, 023016 (2019)
104. K.I. Okhlopkov, P.A. Shafirin, A.A. Ezhov, N.A. Orlikovsky, M.R. Shcherbakov, A.A. Fedyanin, ACS Photonics **6**, 189 (2018)
105. S. Chen, M. Rahmani, K.F. Li, A. Miroshnichenko, T. Zentgraf, G. Li, D. Neshev, S. Zhang, ACS Photonics **5**, 1671 (2018)
106. E.V. Melik-Gaykazyan, S.S. Kruk, R. Camacho-Morales, L. Xu, M. Rahmani, K.Z. Kamali, A. Lamprianidis, A.E. Miroshnichenko, A.A. Fedyanin, D.N. Neshev, Y.S. Kivshar, ACS Photonics **5**, 728 (2017)
107. Y. Gao, Y. Fan, Y. Wang, W. Yang, Q. Song, S. Xiao, Nano Lett. **18**, 8054 (2018)
108. M.R. Shcherbakov, P.P. Vabishchevich, A.S. Shorokhov, K.E. Chong, D.Y. Choi, I. Staude, A.E. Miroshnichenko, D.N. Neshev, A.A. Fedyanin, Y.S. Kivshar, Nano Lett. **15**, 6985 (2015)
109. L. Wang, S. Kruk, K. Koshelev, I. Kravchenko, B. Luther-Davies, Y.S. Kivshar, Nano Lett. **18**, 3978 (2018)
110. B. Reineke, B. Sain, R. Zhao, L. Carletti, B. Liu, L. Huang, C. De Angelis, T. Zentgraf, Nano Lett. **19**, 6585 (2019)
111. K.M. Dani, Z. Ku, P.C. Upadhya, R.P. Prasankumar, S. Brueck, A.J. Taylor, Nano Lett. **9**, 3565 (2009)
112. S. Makarov, S. Kudryashov, I. Mukhin, A. Mozharov, V. Milichko, A. Krasnok, P. Belov, Nano Lett. **15**, 6187 (2015)
113. G. Li, S. Zhang, T. Zentgraf, Nat. Rev. Mater. **2**, 17010 (2017)
114. Y.S. Kivshar, Natl. Sci. Rev. **5**, 144 (2018)
115. A. Krasnok, M. Tymchenko, A. Alù, Mater. Today **21**, 8 (2018)
116. B. Sain, C. Meier, T. Zentgraf, Adv. Photonics **1**, 024002 (2019)
117. L. Carletti, A. Locatelli, O. Stepanenko, G. Leo, C. De Angelis, Opt. Express **23**, 26544 (2015)
118. R. Camacho-Morales, M. Rahmani, S. Kruk, L. Wang, L. Xu, D.A. Smirnova, A.S. Solntsev, A. Miroshnichenko, H.H. Tan, F. Karouta, S. Naureen, K. Vora, L. Carletti, C. De Angelis, C. Jagadish, Y.S. Kivshar, D.N. Neshev, Nano Lett. **16**, 7191 (2016)
119. J.D. Sautter, L. Xu, A.E. Miroshnichenko, M. Lysevych, I. Volkovskaya, D.A. Smirnova, R. Camacho-Morales, K. Zangeneh Kamali, F. Karouta, K. Vora, H.H. Tan, M. Kauranen, I. Staude, C. Jagadish, D.N. Neshev, M. Rahmani, Nano Lett. **19**, 3905 (2019)
120. L. Xu, G. Saerens, M. Timofeeva, D.A. Smirnova, I. Volkovskaya, M. Lysevych, R. Camacho-Morales, M. Cai, K. Zangeneh Kamali, L. Huang, F. Karouta, H.H. Tan, C. Jagadish, A.E. Miroshnichenko, R. Grange, D.N. Neshev, M. Rahmani, ACS Nano (2020). https://doi.org/10.1021/acsnano.9b07117
121. I. Shoji, T. Kondo, A. Kitamoto, M. Shirane, R. Ito, J. Opt. Soc. Am. B **14**, 2268 (1997)
122. M. Ohashi, T. Kondo, R. Ito, S. Fukatsu, Y. Shiraki, K. Kumata, S. Kano, J. Appl. Phys. **74**, 596 (1993)
123. M. Rahmani, G. Leo, I. Brener, A.V. Zayats, S.A. Maier, C. De Angelis, H. Tan, V.F. Gili, F. Karouta, R. Oulton, K. Vora, M. Lysevych, I. Staude, L. Xu, A.E. Miroshnichenko, C. Jagadish, D.N. Neshev, Opto-Electron. Adv. **1**, 180021 (2018)
124. L. Carletti, A. Locatelli, D. Neshev, C. De Angelis, ACS Photonics **3**, 1500 (2016)
125. L. Carletti, D. Rocco, A. Locatelli, C. De Angelis, V.F. Gili, M. Ravaro, I. Favero, G. Leo, M. Finazzi, L. Ghirardini, M. Celebrano, G. Marino, A.V. Zayats, Nanotechnology **28**, 114005 (2017)
126. M. Guasoni, L. Carletti, D. Neshev, C. De Angelis, IEEE J. Quantum Electron. **53**, 1 (2017)
127. V.F. Gili, L. Carletti, A. Locatelli, D. Rocco, M. Finazzi, L. Ghirardini, I. Favero, C. Gomez, A. Lemaître, M. Celebrano, C. De Angelis, G. Leo, Opt. Express **24**, 15965 (2016)

128. S. Liu, M.B. Sinclair, S. Saravi, G.A. Keeler, Y. Yang, J. Reno, G.M. Peake, F. Setzpfandt, I. Staude, T. Pertsch, I. Brener, Nano Lett. **16**, 5426 (2016)
129. S. Liu, G.A. Keeler, J.L. Reno, M.B. Sinclair, I. Brener, Adv. Opt. Mater. **4**, 1457 (2016)
130. L. Ghirardini, L. Carletti, V. Gili, G. Pellegrini, L. Duò, M. Finazzi, D. Rocco, A. Locatelli, C. De Angelis, I. Favero, M. Ravaro, G. Leo, A. Lemaître, M. Celebrano, Opt. Lett. **42**, 559 (2017)
131. S.S. Kruk, R. Camacho-Morales, L. Xu, M. Rahmani, D.A. Smirnova, L. Wang, H.H. Tan, C. Jagadish, D.N. Neshev, Y.S. Kivshar, Nano Lett. **17**, 3914 (2017)
132. J. Cambiasso, G. Grinblat, Y. Li, A. Rakovich, E. Cortés, S.A. Maier, Nano Lett. **17**, 1219 (2017)
133. R. Camacho-Morales, G. Bautista, X. Zang, L. Xu, L. Turquet, A. Miroshnichenko, H.H. Tan, A. Lamprianidis, M. Rahmani, C. Jagadish, D.N. Neshev, M. Kauranen, Nanoscale **11**, 1745 (2019)
134. F. Koyama, J. lightwave Technol. **24**, 4502 (2006)
135. L. Carletti, G. Marino, L. Ghirardini, V.F. Gili, D. Rocco, I. Favero, A. Locatelli, A.V. Zayats, M. Celebrano, M. Finazzi, G. Leo, C. De Angelis, D.N. Neshev, ACS Photonics **5**, 4386 (2018)
136. P.P. Vabishchevich, S. Liu, M.B. Sinclair, G.A. Keeler, G.M. Peake, I. Brener, ACS Photonics **5**, 1685 (2018)
137. L. Ghirardini, G. Marino, V.F. Gili, I. Favero, D. Rocco, L. Carletti, A. Locatelli, C. De Angelis, M. Finazzi, M. Celebrano, D.N. Neshev, G. Leo, Nano Lett. **18**, 6750 (2018)
138. G. Marino, C. Gigli, D. Rocco, A. Lemaître, I. Favero, C. De Angelis, G. Leo, ACS Photonics **6**, 1226 (2019)
139. F.J.F. Löchner, A.N. Fedotova, S. Liu, G.A. Keeler, G.M. Peake, S. Saravi, M.R. Shcherbakov, S. Burger, A.A. Fedyanin, I. Brener, T. Pertsch, F. Setzpfandt, I. Staude, ACS Photonics **5**, 1786 (2018)
140. S. Liu, P.P. Vabishchevich, A. Vaskin, J.L. Reno, G.A. Keeler, M.B. Sinclair, I. Staude, I. Brener, Nat. Commun. **9**, 2507 (2018)
141. G. Marino, A.S. Solntsev, L. Xu, V.F. Gili, L. Carletti, A.N. Poddubny, M. Rahmani, D.A. Smirnova, H. Chen, A. Lemaître, G. Zhang, A.V. Zayats, C.D. Angelis, G. Leo, A.A. Sukhorukov, D.N. Neshev, Optica **6**, 1416 (2019)
142. D. Rocco, V.F. Gili, L. Ghirardini, L. Carletti, I. Favero, A. Locatelli, G. Marino, D.N. Neshev, M. Celebrano, M. Finazzi, G. Leo, C. De Angelis, Photonics Res. **6**, B6 (2018)
143. B. Gao, M. Ren, W. Wu, H. Hu, W. Cai, J. Xu, Laser Photonics Rev. **13**, 1800312 (2019)
144. F. Timpu, J. Sendra, C. Renaut, L. Lang, M. Timofeeva, M.T. Buscaglia, V. Buscaglia, R. Grange, ACS Photonics **6**, 545 (2019)
145. L. Carletti, C. Li, J. Sautter, I. Staude, C.D. Angelis, T. Li, D.N. Neshev, Opt. Express **27**, 33391 (2019)
146. K. Koshelev, S. Kruk, E. Melik-Gaykazyan, J.H. Choi, A. Bogdanov, H.G. Park, Y.S. Kivshar, Science **367**, 288 (2020)
147. D.N. Neshev, in *Frontiers in Optics + Laser Science APS/DLS*. OSA Technical Digest (Optical Society of America, 2019), paper FM5B.1

Chapter 4
Nonlinearity and Discreteness: Solitons in Lattices

Boris A. Malomed

Abstract An overview is given of basic models combining discreteness in their linear parts (i.e., the models are built as dynamical lattices) and nonlinearity acting at sites of the lattices or between the sites. The considered systems include the Toda and Frenkel–Kontorova lattices (along with their dissipative versions), as well as equations of the discrete nonlinear Schrödinger and Ablowitz–Ladik types, and their combination in the form of the Salerno model. The interplay of discreteness and nonlinearity gives rise to a variety of states, most important ones being discrete solitons. Basic results for 1D and 2D discrete solitons are collected in the review, including 2D solitons with embedded vorticity, and some results concerning mobility of discrete solitons. Main experimental findings are overviewed too. Models of the semi-discrete type, and basic results for solitons supported by them, are also considered, in a brief form. Perspectives for the development of topics covered in the review are discussed throughout the text.

4.1 Introduction: Discretization of Continuum Models, and the Continuum Limit of Discrete Ones

Standard models of dynamical media are based on partial differential equations, typical examples being the nonlinear Schrödinger (NLS) equation for the mean-field complex wave function $\psi(x, y, z.t)$ in atomic Bose–Einstein condensates (BECs; in that case, the NLS equation is usually called the Gross–Pitaevskii equation (GPE) [1]), and the NLS equation for the envelope amplitude of the electromagnetic field in optical media [2]. In the scaled form, the NLS equation is

$$i\psi_t = -(1/2)\nabla^2\psi + g|\psi|^2\psi + U(x, y, z)\psi, \tag{4.1}$$

B. A. Malomed (✉)
Department of Physical Electronics, Faculty of Engineering, School of Electrical Engineering, and Center for Light-Matter Interaction, Tel Aviv University, 69978 Tel Aviv, Israel
e-mail: malomed@tauex.tau.ac.il

P. G. Kevrekidis et al. (eds.), *Emerging Frontiers in Nonlinear Science*, Nonlinear Systems and Complexity 32, https://doi.org/10.1007/978-3-030-44992-6_4

where $g = +1$ and -1 correspond, respectively, to the self-defocusing and focusing signs of the local cubic nonlinearity, and $U(x, y, z)$ is a real external potential. In the application to optics, the evolution variable t is replaced by coordinate z in the propagation direction, while original z is replaced by the temporal variable, $\tau = t - z/V_{\text{gr}}$, where t is time, and V_{gr} is the group velocity of the carrier wave [3]. In optics, the effective potential may be two-dimensional (2D), $-U(x, y)$ being a local variation of the refractive index in the transverse plane.

In many cases the potential represents a spatially periodic pattern, such as optical lattices (OLs) in BECs [4, 5], or photonic crystals which steer the propagation of light waves in optics [6]:

$$U_{\text{latt}}(x, y, z) = -\varepsilon\left[\cos(2\pi L/x) + \cos(2\pi L/y) + \cos(2\pi L/z)\right], \tag{4.2}$$

as well as its 2D and 1D reductions. A deep lattice potential, which corresponds to large ε, splits the continuous wave function into an array of "droplets" trapped in local potential wells, which are coupled by weak tunneling. Accordingly, in the framework of the *tight-binding approximation*, the NLS equation is replaced by a discrete NLS (DNLS) equation, which was derived, in the 1D form, for arrays of optical fibers [7–10] and arrays of plasmonic nanowires [11], as well as for BEC loaded in a deep OL trap [12]:

$$\begin{aligned} i\dot{\psi}_{l,m,n} = &-(1/2)\left[\left(u_{l+1,m,n} + \psi_{l-1,m,n} - 2\psi_{l,m,n}\right) + \left(\psi_{l,m+1,n} + \psi_{l,m-1,n} - 2\psi_{l,m,n}\right)\right. \\ &\left. + \left(\psi_{l,m,n+1} + \psi_{l,m,n-1} - 2\psi_{l,m,n}\right)\right] + g\left|\psi_{l,m,n}\right|^2 \psi_{l,m,n}, \end{aligned} \tag{4.3}$$

where the set of integer indices (l, m, n) replaces (x, y, z). DNLS equation (4.3) is often reduced to 2D and 1D forms. While it includes the linear coupling between the nearest neighbors, 1D lattices can be built in the form of zigzag chains, making it relevant to add couplings between the next-nearest neighbors [13, 14]. 2D lattices with similar additional coupling are known too [15].

As concerns the sign parameter, $g = \pm 1$, (4.3) admits flipping $+1 \leftrightarrow -1$ by means of the *staggering transformation* of the discrete wave function:

$$\psi_{l,m,n}(t) \equiv (-1)^{l+m+n} \exp(-6it)\, \tilde{\psi}^*_{l,m,n}(t) \tag{4.4}$$

where $*$ stands for the complex-conjugate expression, and in the 2D and 1D situations, $\exp(-6it)$ is replaced by $\exp(-4it)$ and $\exp(-2it)$, respectively.

It is well known that the 2D and 3D continuous NLS equation (4.1) with the self-focusing nonlinearity, i.e. $g < 0$, gives rise to the *critical* and *supercritical collapse*, respectively, i.e. appearance of singular solutions in the form of infinitely narrow and infinitely tall peaks, after a finite evolution time [2]. The discreteness arrests the collapse, replacing it by a *quasi-collapse* [16] when the width of the shrinking peak becomes comparable to the spacing of the DNLS lattice.

The DNLS equation and its extensions constitute a class of models with a large number of physical realizations, which have drawn much interest as subjects of mathematical studies as well [17]. The class also includes systems of coupled DNLS equations [18, 19].

The 1D continuous NLS equation without the external potential and with either sign of the nonlinearity, g, is integrable by means of the inverse-scattering transform [20–23], although it is nonintegrable in the 2D and 3D geometries. On the contrary to that, the 1D DNLS equation is not integrable, i.e. the direct discretization destroys the integrability [24, 25]. However, the continuous NLS equation admits another discretization in 1D, which leads to an integrable discrete model, viz. the Ablowitz–Ladik (AL) equation [26]:

$$i\dot{\psi}_n = -\left(\psi_{n+1} + \psi_{n-1}\right)\left(1 + \mu\left|\psi_n\right|^2\right), \tag{4.5}$$

where positive and negative values of the real nonlinearity coefficient, μ, correspond to the self-focusing and defocusing, respectively. Integrable discrete equations, such as the AL and Toda-lattice (see (4.9) below) ones, are exceptional models which play a fundamentally important role, providing exact solutions for discrete solitons and other dynamical states [27].

Considerable interest was also drawn to the nonintegrable combination of the AL and DNLS equations, in the form of the Salerno model (SM) [28], with an additional onsite cubic term, different from the intersite one in (4.5):

$$i\dot{\psi}_n = -\left(\psi_{n+1} + \psi_{n-1}\right)\left(1 + \mu\left|\psi_n\right|^2\right) - 2\left|\psi_n\right|^2\psi_n, \tag{4.6}$$

with the magnitude and sign of the onsite nonlinearity coefficient fixed by means of the rescaling and staggering transformation, respectively. The SM finds a physical realization in the context of the Bose–Hubbard model, i.e. BEC loaded in a deep OL, in the case when dependence of the intersite hopping rate on populations of the sites is taken into regard [29, 30].

While the above-mentioned DNLS, AL, and SM discrete systems are derived as the discretization of continuous NLS equations, one can look at this relation in the opposite direction: starting from discrete equations, one can derive their *continuum limit*. In particular, in the case of the SM equation (4.6), the continuum approximation is introduced by replacing the intersite combination of the discrete fields by a truncated Taylor's expansion,

$$\psi_n(t) \equiv e^{2it}\Psi(x,t), \qquad \Psi\left(x = n \pm 1, t\right) \approx \Psi\left(x = n\right) \pm \Psi_x\big|_{x=n} + (1/2)\Psi_{xx}\big|_{x=n}, \tag{4.7}$$

where $\Psi(x)$ is treated as a function of the continuous coordinate x, which coincides with n when it takes integer values. The substitution of this approximation in (4.6) leads to a generalized (nonintegrable) form of the 1D NLS equation [31]

$$i\Psi_t = -\left(1 + \mu\left|\Psi\right|^2\right)\Psi_{xx} - 2\left(1 + \mu\right)\left|\Psi\right|^2\Psi, \tag{4.8}$$

which amounts to the standard 1D NLS equation (4.1) with $g = +1$ and $U = 0$ in the case of $\mu = 0$.

The objective of this Chapter is to present an overview of basic discrete nonlinear models and dynamical states produced by them, chiefly in the form of bright solitons (self-trapped localized modes). Before proceeding to models based on equations of the DNLS, AL, and SM types, simpler ones, which were derived for chains of interacting particles, are considered in the next section. The paradigmatic model of the latter type is provided by the 1D Toda-lattice (TL) equation [32], written for coordinates $u_n(t)$ of particles with unit mass and exponential potential of interaction between adjacent ones:

$$\ddot{u}_n + e^{-(u_{n+1}-u_n)} - e^{-(u_n-u_{n-1})} = 0. \tag{4.9}$$

This equation can also be written for separations $r_n(t) \equiv u_{n+1}(t) - u_n(t)$ between the particles:

$$\ddot{r}_n + e^{-r_{n+1}} + e^{-r_{n-1}} - 2e^{-r_n} = 0. \tag{4.10}$$

Equation (4.10) is integrable [20], its continuum limit being the so-called "bad" Boussinesq equation [33], which is formally integrable too[1] [22]:

$$r_{tt} - r_{xx} - (1/12)r_{xxxx} + \left(r^2\right)_{xx} = 0 \tag{4.11}$$

Another famous, although not integrable, model of a chain of pairwise-interacting particles with coordinates $u_n(t)$, is the Fermi–Pasta–Ulam (FPU) system [34, 35]:

$$\ddot{u}_n = (u_{n+1} + u_{n-1} - 2u_n)\left[1 + \alpha\left(u_{n+1} - u_{n-1}\right)\right], \tag{4.12}$$

where α is a constant. This model was one of the first objects of numerical simulations performed in the context of fundamental research (back in 1953, published in 1955 [34], see also [36]). Later, it became known that a very essential contribution to the original FPU work was made by Mary Tsingou [37], therefore the model is also called Fermi–Pasta–Ulam–Tsingou system.

The initial objective of the original numerical FPU-Tsingou experiment was to observe the onset of ergodicity in the evolution governed by (4.12). A surprising result was that long simulations demonstrated a quasi-periodic evolution, without manifestations of ergodicity (i.e. without statistically uniform distribution of the energy between all degrees of freedom of the lattice system). Eventually, this perplexing result was explained (in the same paper [38] by N. Zabusky and M. Kruskal which had introduced word "soliton") by the fact that the continuum limit of (4.12) may be reduced (for unidirectional propagation of excitations in the continuum medium) to the Korteweg – de Vries equation, which, being integrable, does not feature ergodicity.

The next section briefly addresses, in addition to the TL, more complex models which combine the inter-particle interactions (taken in the linear approximation, unlike the exponential terms in (4.10)), and onsite potentials – most typically, in the

[1] "Bad" implies that (4.11) gives rise to an unstable dispersion relation.

form of $U = \varepsilon \sum_n (1 - \cos u_n)$, with $\varepsilon > 0$, which is the source of the nonlinearity in the corresponding Frenkel–Kontorova (FK) model. It was originally introduced as a model for dislocations in a crystalline lattice [39], and has found a large number of realizations in other physical settings [40].

This Chapter also addresses, in a brief form, other fundamental aspects of nonlinear discrete systems. These are discrete multidimensional models, semi-discrete ones, and experimental realizations of discrete media and bright solitons in them, chiefly in the realm of nonlinear optics. The experimental results are considered in a brief form too. Dissipative discrete nonlinear systems are partly addressed in this Chapter, as a systematic consideration of dissipative discrete systems is a subject for a separate review.

Because the length of the Chapter is limited, the presentation and bibliography are not aimed to be comprehensive; rather, particular results mentioned in sections following below are selected as examples which help to understand general principles supported by a large body of theoretical and experimental findings.

4.2 Excitations in Chains of Interacting Particles

4.2.1 The Toda Lattice

The TL equation (4.9) is characterized, first of all, by its linear spectrum. Looking for solutions to the linearized version of the equation in the form of "phonon modes", i.e. plane waves with an infinitesimal amplitude $u^{(0)}$, frequency χ and wavenumber p (which is constrained to the first Brillouin zone, $0 < p < 2\pi$),

$$(u_n)_{\text{linearized}} = u^{(0)} \exp(ipn - i\chi t), \tag{4.13}$$

it is easy to obtain the respective dispersion relation,

$$\chi = \pm 2 \sin(p/2). \tag{4.14}$$

Further, (4.13) produces phase velocities of the linear waves $V_{\text{ph}} = \chi/p$, which take values $|V_{\text{ph}}| < 1$.

Integrable equation (4.9) generates exact soliton solutions, which were first found in the original work of Toda [32]. The soliton represents a lattice deformation traveling at constant velocity c:

$$u_n = -\ln\left|\xi^{-2} - \frac{\xi^{-2} - 1}{1 + \xi^{2(n-ct)}}\right|, \quad c = \pm\frac{\xi^{-1} - \xi}{\ln(\xi^{-2})}, \tag{4.15}$$

where ξ is an arbitrary real parameter taking values $0 < \xi < 1$, the respective interval of the inverse velocities being

$$0 < |c|^{-1} < 1. \tag{4.16}$$

Note that this interval has no overlap with the above-mentioned range of the phase velocities of the linear modes, $|V_{\text{ph}}| < 1$, in accordance with the well-known principle that solitons may exist in *bandgaps* of linear spectra, i.e. in regions where linear waves do not exist.

Comparing values of the solution (4.15) at $n \to \pm\infty$, one concludes that the soliton carries compression of the TL by a finite amount, $\Delta u \equiv u_{n=+\infty} - u_{n=-\infty} = \ln(\xi^{-2})$, while a characteristic width of the soliton is $\Delta n \sim 1/\ln(\xi^{-2})$. Similar to other integrable systems [20–23], collisions between solitons do not affect their shapes and velocities, leading solely to finite shifts of the solitons' centers.

The limit of $\xi \to 0$ implies that the TL reduces to a chain of hard particles, which interact when they collide. Accordingly, the soliton's structure degenerates into a single fast moving particle, the propagation being maintained by periodically occurring collisions, as a result of which the moving particle comes to a halt, transferring its momentum to the originally quiescent one. In the opposite limit, $\xi \to 1$, soliton (4.15) becomes a very broad solution, traveling with the minimum velocity, $c \to 1$. As mentioned above, no TL solitons exists with velocities $|c| < 1$.

Equation (4.9) conserves the total momentum, $P = \sum_{-\infty}^{+\infty} \dot{u}_n$, and Hamiltonian (energy),

$$H_{\text{TL}} = \sum_{n=-\infty}^{+\infty} \left\{ \frac{1}{2}\dot{u}_n^2 + \left[e^{-(u_{n+1}-u_n)} - 1 \right] \right\}. \tag{4.17}$$

In fact, integrable equations, including (4.9), conserve an infinite number of dynamical invariants, the momentum and energy being the lowest-order ones in the infinite sequence [20]; however, higher-order invariants do not have a straightforward physical interpretation.

A realistic implementation of the TL includes friction forces with coefficient $\alpha > 0$, which should be compensated by an "ac" (time-periodic) driving force with amplitude ε and frequency ω [41–43]. The accordingly modified equation (4.9) is

$$\ddot{u}_n + e^{-(u_{n+1}-u_n)} - e^{(u_n-u_{n-1})} = -\alpha\dot{u}_n + \varepsilon q_n \cos(\omega t). \tag{4.18}$$

Here coefficients q_n may be realized as the charge of the particles, if the drive is applied by an ac electric field. Nontrivial coupling of the field to the TL dynamics is not possible if all the charges are identical, i.e. $q_n \equiv 1$. Indeed, in the latter case one can trivially eliminate the drive by defining $u_n(t) \equiv v_n(t) - \varepsilon\omega^{-2}\cos(\omega t)$, ending up with equation (4.18) for $v_n(t)$ with no drive. The simplest nontrivial coupling is provided by assuming $q_n = (-1)^n$, i.e. alternating positive and negative charges at neighboring sites of the TL [41]. A particular choice of the periodic pattern for q_n defines the respective size, a, of the cell of the ac-driven TL (in particular, $q_n = (-1)^n$ corresponds to $a = 2$).

The periodic passage of the soliton running through the lattice with velocity c, i.e. with temporal period $T = a/c$, may provide compensation of the friction losses if it resonates with the periodicity of the ac drive, which defines the spectrum of *resonant velocities* [41],

$$c_N = \pm\omega a / [2\pi (1 + 2N)], \tag{4.19}$$

where integer $N = 0, 1, 2, ...$ determines the order of the resonance.[2] Velocities c_N are relevant if they satisfy restriction $|c_N| > 1$ (see (4.16)), which implies $\omega > 2\pi/a$.

The progressive motion of solitons is actually supported by the drive whose strength, ε, exceeds a certain minimum (threshold) value, ε_{thr}, which is roughly proportional to the friction coefficient, α [42].

A specific class of dynamical chains with essentially nonlinear interaction between adjacent particles of a finite size (spheres) represents models of 1D granular media, in which spheres interact when they come in touch. It was demonstrated that such chains (in particular, those with the Hertz potential of the contact interaction [44]) support self-trapped states in the form of discrete breathers [45].

4.2.2 The Frenkel–Kontorova Model and Related Systems

A paradigmatic example of lattices which combine interactions between adjacent particles and the onsite potential acting on each particle is the FK model [40], which is the discretization of the commonly known sine-Gordon (sG) equation. In 1D, the sG equation for a real wave field u is [20–23]

$$u_{tt} - u_{xx} + \sin u = 0. \tag{4.20}$$

Elementary solutions to (4.20) are kinks

$$u_{\text{kink}} = \arctan\left[\exp\left(\sigma (x - ct) / \sqrt{1 - c^2}\right)\right], \tag{4.21}$$

with the velocity taking values $-1 < c < +1$ and the topological charge σ being

$$\sigma \equiv [u (x = +\infty) - u (x = -\infty)] / (2\pi) = \pm 1, \tag{4.22}$$

The discretization of (4.20) with stepsize h implies defining

$$x \equiv hn, u (x = hn) \equiv u_n, \tag{4.23}$$

and the replacement of the second derivative by its finite-difference counterpart:

$$u_{xx} \to h^{-2} (u_{n+1} + u_{n-1} - 2u_n). \tag{4.24}$$

---

[2] For velocities given by (4.19) with odd integer $1 + 2N$ replaced by an even one, $2N$, with $N = 1, 2, ...$, the transfer of energy from the drive to the moving soliton averages to zero.

The result is the FK model, which also includes the local friction with coefficient $\alpha \geq 0$, and an external force f_n, that may be time-dependent:

$$\ddot{u}_n - (1/h^2)(u_{n+1} + u_{n-1} - 2u_n) + \sin u_n = -\alpha \dot{u}_n + f_n(t). \tag{4.25}$$

The linearization of (4.25), with $\alpha = f_n = 0$, for phonon modes (4.13) gives rise to the following spectrum:

$$\chi^2 = 1 + (4/h^2)\sin^2(p/2), \tag{4.26}$$

cf. its counterpart (4.14) for the TL. The form of spectrum (4.26) implies that localized oscillatory states may exists in the inner and outer bandgaps, with frequencies $\chi^2 < 1$ and $\chi^2 > 1 + 4/h^2$, respectively.

In the connection to the linear spectrum, it is relevant to mention that considerable interest was recently drawn to specially designed discrete lattices whose spectrum includes a *flatband*, i.e. a degenerate branch of the $\chi(p)$ dependence in the form of $\chi = \text{const}$, as such systems admit the existence of localized discrete modes in the absence of nonlinearity [46, 47]. Effects of nonlinearity on localized states in flatband systems have been investigated too [47, 48].

Generally similar to the discrete sG lattice governed by (4.25) are models based on the discretization of Klein-Gordon equations. Typically, they feature the onsite cubic nonlinearity, the simplest model being [49]

$$\ddot{u}_n - (1/h^2)(u_{n+1} + u_{n-1} - 2u_n) - u_n + u_n^3 = 0. \tag{4.27}$$

The spectrum of the linearization of (4.27) is *unstable*, with $\chi^2 = -1 + (4/h^2)\sin^2(p/2)$ taking negative values. However, kink solutions, which connect constant values $u_n = \pm 1$ at $n \to \pm\infty$, are stable, as the constant nonzero background is stable against small perturbations. As concerns moving kinks, it is possible to construct a discrete model with a specially designed combination of nonlinear terms, which admits exact solutions for moving kinks with particular values of the velocity [50].

Even in the case of $\alpha = f_n = 0$, the discrete sG equation (4.25), unlike its continuum counterpart (4.20), is not integrable. Therefore, in the absence of the friction, the single dynamical invariant of (4.25) is the energy, provided that the driving force is time-independent:

$$E = \sum_{n=-\infty}^{+\infty} \left[(1/2)\dot{u}_n^2 + (1/2)h^{-2}(u_{n+1} - u_n)^2 + (1 - \cos u_n) - f_n u_n\right]. \tag{4.28}$$

Note that, treating u_n as per (4.23), and similarly defining $f_n \equiv f(x = hn)$, one can formally write the energy as in the continuum setting, in which the discreteness is introduced by means of a lattice of delta-functions with period h:

$$E = \int_{-\infty}^{+\infty} dx \sum_{n=-\infty}^{+\infty} \delta(x - hn) \left\{ \frac{\dot{u}_n^2}{2} + \frac{[u(x+h) - u(x)]^2}{2h^2} + (1 - \cos u) - f(x)u \right\}. \tag{4.29}$$

A fundamentally important concept in models of the FK type is the Peierls-Nabarro (PN) potential [51]. It is naturally defined in the quasi-continuum approximation, which implies that the lattice's spacing is much smaller than a characteristic size of the mode under the consideration, i.e. $h \ll 1$. In this limit, the mode may be considered in the continuum form – e.g. as $u_{\text{kink}}(x - \xi)$, with the central point, ξ, placed at an arbitrary position, and the PN potential is defined as the total energy, given by (4.29), considered as a function of ξ [52]. Then, using identity

$$\sum_{n=-\infty}^{+\infty} \delta(x - hn) \equiv \frac{1}{h} \sum_{m=-\infty}^{+\infty} \exp\left(i \frac{2\pi m}{h} x\right), \tag{4.30}$$

one obtains, in the lowest approximation, which is determined by the lowest harmonics in expression (4.30), with $m = \pm 1$, an exponentially small but, nevertheless, relevant result:

$$U_{\text{PN}}(\xi) = \frac{U_0}{2} \cos\left(\frac{2\pi\xi}{h}\right), \quad U_0 = \frac{(4\pi/h)^2}{\sinh\left(\pi^2/h\right)}. \tag{4.31}$$

Thus, the broad quasi-continuum mode tends to have its center pinned at any local minimum of the PN potential, $\xi = h\left((1/2) + N\right)$, with arbitrary integer N. The PN potential barrier, which separates neighboring minima, and thus creates an obstacle for free motion of kinks, is U_0. The PN barrier may be suppressed in FK lattices with a long-range intersite interaction added to the linear coupling between the nearest neighbors [53].

Unlike the TL solitons (4.15), which may only exist as moving states with velocities $|c| > 1$, the existence of quiescent FK kinks, pinned to local potential minima, is not predicated on the presence of the driving force. On the other hand, the motion of kinks is braked by friction, as well as by radiative losses, i.e. emission of lattice "phonons" by a kink moving through the lattice, the latter effect usually being much weaker than friction. As well as in the TL model, the motion of kinks can be supported by an ac drive, $f_n = (-1)^n \varepsilon \cos(\omega t)$, at the same resonant velocities as given by (4.19), with $a = 2$ [54].

A relevant physical realization of the FK model is provided by an array of coupled long Josephson junctions (JJs) [55, 56] (each junction is a narrow dielectric layer separating two bulk superconductors [57]). An accurate model of the array is provided by (4.25), where $f_n \equiv f$ represents the bias current applied to each junction, while α is the coefficient of Ohmic loss. Especially interesting is this version of the FK with periodic boundary conditions, which corresponds to the circular JJ array built of N junctions [58, 59], as it gives rise to resonant interaction between a kink (in terms of JJs, it is a *fluxon*, i.e. a quantum of the magnetic flux), moving at velocity c

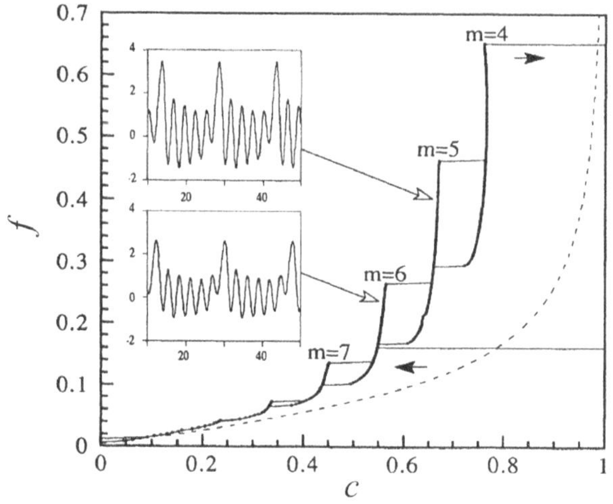

Fig. 4.1 The predicted relation between the velocity of the discrete kink, travelling in a ring-shaped FK lattice composed of $N = 10$ sites, subject to periodic boundary conditions, and a driving force. Numbers m, which label different vertical resonant steps, denote the "quantization" orders in (4.32). Arrows indicate directions of hysteretic jumps between the steps, and the dashed curve represents the $f(c)$ dependence in the continuum version of the system. Insets display dependences $\dot{u}_n$, which are proportional to the local voltage in the underlying JJ array, at the system's midpoint, corresponding to two points in the $f(c)$ (current-voltage) characteristic marked by arrows. Parameters in (4.25) are $h = 1$ and $\alpha = 0.1$. Reproduced from [58]

in the ring-shaped array, and phonon modes whose phase velocity χ/p, determined by the dispersion relation (4.26), may coincide with c. The periodicity of the array imposes the "quantization" condition on wavenumber p in (4.26),

$$p = (2\pi/hN)\,m, \quad m = 1, 2, 3, \ldots . \tag{4.32}$$

The analysis of the kink-phonon interaction leads to a dependence of the fluxon's velocity c on the driving force (current), f, in the form of resonant *Shapiro steps* [60] connected by hysteretic jumps, as shown in Fig. 4.1. This dependence predicts an experimentally observable current-voltage characteristic of the JJ system, as the voltage is proportional to c. The measured characteristic was found to be very close to the theoretical prediction [58].

Lastly, it is relevant to mention that the FK model also supports breathers, i.e. localized modes which are periodically oscillating functions of time [40, 61–63]. In the continuum limit, the breathers naturally carry over into the well-known exact breather solutions of the sG equation (4.20),

4.3 Nonlinear Schrödinger (NLS) Lattices

4.3.1 One-Dimensional (1D) Solitons

4.3.1.1 Fundamental States

DNLS equation (4.3) gives rise to discrete solitons, which cannot be represented by analytical solutions, but can be easily found in a numerical form. General properties of the soliton families can be understood by means of the variational approximation (VA). Results for solitons in models of the DNLS type are well known, being broadly represented in the literature [17]. Therefore, basic results for discrete NLS solitons are summarized here in a brief form. Most studies addressed the 1D version of (4.3), i.e.

$$i\dot{\psi}_n = -(1/2)\left(\psi_{n+1} + \psi_{n-1} - 2\psi_n\right) - |\psi_n|^2 \psi_n, \tag{4.33}$$

where the nonlinearity coefficient is fixed to be $g = -1$, which corresponds to the self-focusing sign of the onsite nonlinearity (recall that the sign of g may be flipped by means of the staggering transformation (4.4)). The DNLS equation conserves two dynamical invariants, viz. the total norm,

$$N = \sum_{n=-\infty}^{+\infty} |\psi_n|^2, \tag{4.34}$$

and Hamiltonian (energy),

$$H = \sum_{n=-\infty}^{+\infty} \left[(1/2)\,|\psi_n - \psi_{n-1}|^2 - (1/4)\,|\psi_n|^4\right]. \tag{4.35}$$

A fundamental property of the DNLS equation with self-attractive onsite nonlinearity is the modulational instability of its spatially homogeneous state [64].

Stationary solutions to (4.33) with real frequency ω are looked for in the usual form,

$$\psi_n(t) = e^{-i\omega t} u_n, \tag{4.36}$$

with real amplitudes u_n satisfying the discrete equation,

$$\omega u_n = -(1/2)\left(u_{n+1} + u_{n-1} - 2u_n\right) - u_n^3. \tag{4.37}$$

While (4.37) does not admit exact analytical solutions, the VA produces quite accurate approximations for discrete solitons. The VA is based on the Lagrangian, from which (4.37) can be derived by means of the variation with respect to the discrete field u_n:

$$L=\sum_{n=-\infty}^{+\infty}\left\{(1/4)\left[(u_n-u_{n-1})^2-u_n^4\right]-\omega u_n^2\right\}. \quad (4.38)$$

The use of the VA is based on a particular *ansatz*, i.e. a trial analytical expression which aims to approximate the solution [65]. The only discrete ansatz for which analytical calculations are feasible is represented by the exponential function [66–69]. In particular, an *onsite-centered* (OC) discrete soliton, i.e. one with a single maximum (which is placed, by definition, at site $n=0$) is approximated by

$$(u_n)_{\text{onsite}}=A\exp\left(-a|n|\right), \quad (4.39)$$

with $a>0$. The corresponding norm, calculated as per (4.34), is

$$N_{\text{ansatz}}=A^2\coth a.$$

Actually, ansatz (4.39) works well for strongly and moderately discrete solitons (see Fig. 4.2), but it is not appropriate for broad (quasi-continuum) modes, which are approximated by the commonly known soliton solution of the NLS equation (the 1D version of (4.1) with $U=0$),

$$\psi\left(x,t\right)=\eta\,\text{sech}\left(\eta\left(x-\xi\right)\right)\exp\left(i\eta^2 t\right), \quad (4.40)$$

with a large width, $\eta^{-1}\gg 1$, and central coordinate ξ.

For *intersite-centered* (IC) discrete solitons, with two symmetric maxima placed at two adjacent sites of the lattice, $n=0$ and $n=1$ (and a formal central point located between the sites, hence the name of these modes), an appropriate ansatz is

$$(u_n)_{\text{intersite}}=A\exp\left(-a|n-1/2|\right). \quad (4.41)$$

The substitution of ansatz (4.39) in Lagrangian (4.38) and straightforward calculations yield the following effective Lagrangian:

$$L_{\text{eff}}=(A^2/2)\coth(a/2)-(A^4/4)\coth\left(2a\right)-\omega A^2\coth a. \quad (4.42)$$

Then, for given $\omega<0$ (solitons with $\omega>0$ do not exist), the squared amplitude, A^2, and inverse width, a, of the discrete soliton are predicted by the Euler-Lagrange equations,

$$\frac{\partial L_{\text{eff}}}{\partial\left(A^2\right)}=\frac{\partial L_{\text{eff}}}{\partial a}=0. \quad (4.43)$$

The corresponding system of algebraic equations for A^2 and a can be easily solved numerically. A similar analysis was performed for the IC solitons, starting with ansatz (4.41). The VA produces quite accurate predictions for solitons of both types, see Fig. 4.2 and [70].

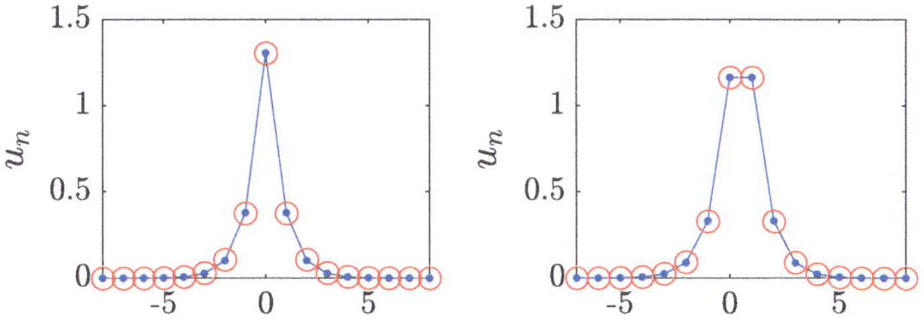

Fig. 4.2 Comparison of typical OC (left panel) and IC (right panel) 1D discrete solitons, obtained as numerical solutions of (4.37), shown by chains of blue dots, and their counterparts produced by the VA (shown by red open circles). In this figure, $\omega = -1$, see (4.36)

An extended version of VA for 1D discrete solitons was elaborated for nonstationary solutions, and compared to their numerically generated counterparts [66, 68]. Moreover, it was demonstrated that VA may be applied, in a more sophisticated form, even to a challenging problem of collisions of moving discrete solitons [67]. Further considerations addressed false instabilities, which are sometimes predicted by the nonstationary VA [71], and rigorous justification of the VA [72]. Finally, the VA and full numerical considerations demonstrate that the *entire family* of the OC discrete solitons is *stable*, while all the IC ones are *unstable* [17].

4.3.1.2 Mobility of 1D Discrete Solitons

The DNLS equation does not admit solutions for moving discrete solitons. Indeed, even in the quasi-continuum approximation, soliton (4.40) is running through the effective PN potential, which, for 1D DNLS modes, is

$$U_{\mathrm{PN}}(\xi) = -\frac{\pi^4}{3\sinh\left(\pi^2/\eta\right)}\cos\left(2\pi\xi\right), \tag{4.44}$$

cf. expression (4.31) for the PN barrier in the FK model. The periodic acceleration and deceleration of the quasi-continuous soliton moving across the PN potential gives rise to emission of small-amplitude "phonon" waves, i.e. losses which brake the motion. However, the emission effect is extremely weak in direct simulations of the DNLS equations, allowing the 1D discrete solitons to run indefinitely long [73]. On the other hand, discrete solitons in the 2D DNLS equation (see the following subsection) have no mobility. This is explained by the fact that the above-mentioned quasi-collapse effect [16] makes them very narrow modes strongly pinned to the underlying lattice.

The mobility of 1D discrete solitons in NLS lattices may be essentially enhanced by means of the *nonlinearity management* technique [74], i.e. replacing coefficient g in the 1D version of (4.3) by a combination of constant ("dc") and time-periodic "ac" terms [75]:

$$i\dot{u}_n + u_{n+1} + u_{n-1} - 2u_n + [g_{\rm dc} + g_{\rm ac}\sin(\omega t)]\,|u_n|^2 u_n = 0. \tag{4.45}$$

Similar to the situation for the damped driven TL, outlined above, discrete solitons may move across the lattice at special values of the velocity, determined by the resonance between the periodic passage of lattice sites by the soliton and periodically oscillating ac component of the nonlinearity coefficient in (4.45), cf. (4.19):

$$(c_{\rm res})_N^{(M)} = \frac{M\omega}{2\pi N}, \tag{4.46}$$

where integers N and M determine the order of the resonance. This prediction was corroborated by simulations of (4.45) [75].

4.3.1.3 Higher-Order Modes in the 1D DNLS Equation: Twisted Solitons and Bound States

In addition to the OC and IC solitons, which are fundamental states, (4.37) admits stable higher-order states in the form of *twisted modes*, which are subject to the antisymmetry condition, $u_n = -u_{1-n}$ [76]. Such states exist and are stable only in a strongly discrete form, vanishing in the continuum limit.

Stable discrete NLS solitons of the OC type may form bound states, which also represent higher-order modes of the DNLS equation. They are stable in the *out-of-phase* form, i.e. for opposite signs of the bound solitons [77, 78] (the same is true for 2D discrete solitons [79]). Stationary bound states do not exist either in the continuum limit, where bound states of NLS solitons are represented solely by periodically oscillating *breathers* [80].

4.3.2 Two-Dimensional (2D) Discrete Solitons and Solitary Vortices in Quiescent and Rotating Lattices

4.3.2.1 Static Lattices

The 2D cubic DNLS equation is a straightforward extension of the 1D equation (4.33). In particular, its stationary form is

$$\omega u_n = -(1/2)\left(u_{m+1,n} + u_{m-1,n} + u_{m,n+1} + u_{m,n-1} - 4u_{m,n}\right) - \left|u_{m,n}\right|^2 u_{m,n}, \tag{4.47}$$

cf. (4.37), where the stationary discrete wave function, $u_{m,n}$, may be complex. Fundamental-soliton solutions to (4.47) can also be predicted by means of the VA [81, 82] (see (4.53) below for the simplest 2D ansatz). More interesting in the 2D case are discrete solitons with *embedded vorticity*, which were introduced in [83] (see also [84]). Vorticity, alias topological charge, is defined as $\Delta\varphi/(2\pi)$, where

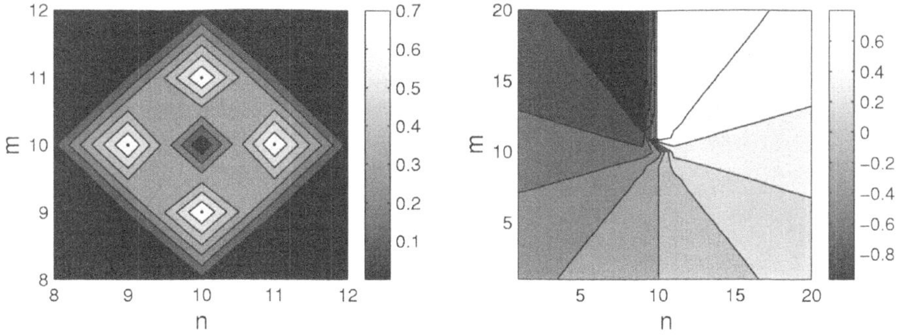

Fig. 4.3 A stable discrete vortex soliton with topological charge $S = 1$, produced by (4.47). The left and right panels show, respectively, distributions of the absolute value and phase of complex wave function $u_{m,n}$ in the plane of (m, n). Reproduced from [83]

$\Delta\varphi$ is a change of the phase of complex discrete wave function $u_{m,n}$, corresponding to any contour surrounding the vortex' pivot. Stability is an important issue for 2D discrete solitons, because it is commonly known that, in the continuum limit, the NLS equation in 2D gives rise solely to unstable solitons, including fundamental ones (usually called Townes' solitons [85]), which are unstable against the critical collapse, [2], and solitons with embedded vorticity [86], which are still more unstable [87].

A typical example of a stable discrete 2D soliton is displayed in Fig. 4.3. 2D fundamental and vortex solitons, with topological charges $S = 0$ and 1, remain stable at $-\omega > |\omega_{\mathrm{cr}}^{(S=0)}| \approx 0.50$ and $-\omega > |\omega_{\mathrm{cr}}^{(S=1)}| \approx 1.23$, respectively [83], while the higher-order localized discrete vortices with $S = 2$ and 4 are unstable, being replaced by stable modes, in the form of quadrupoles and octupoles [88]. Higher-order vortex solitons with $S = 3$ are stable only in a strongly discrete form, at $-\omega > |\omega_{\mathrm{cr}}^{(S=2)}| \approx 4.94$.

The theoretically predicted 2D discrete solitons with vorticity $S = 1$ were experimentally created in [89, 90], using a photorefractive crystal. Unlike uniform media of this type, where delocalized ("dark") optical vortices were originally produced [91, 92], these works made use of a very deep virtual photonic lattice as a quasi-discrete medium supporting nonlinear optical modes in light beams with extraordinary polarization (while the photonic lattice was induced by the interference of quasi-linear beams in the ordinary polarization). Intensity distributions observed in vortex solitons of the OC and IC types are displayed in Fig. 4.4.

Another interesting result demonstrated (and theoretically explained) in deep virtual photonic lattices is a possibility of periodic flipping of the topological charge of a vortex soliton initially created with $S = 2$ [93].

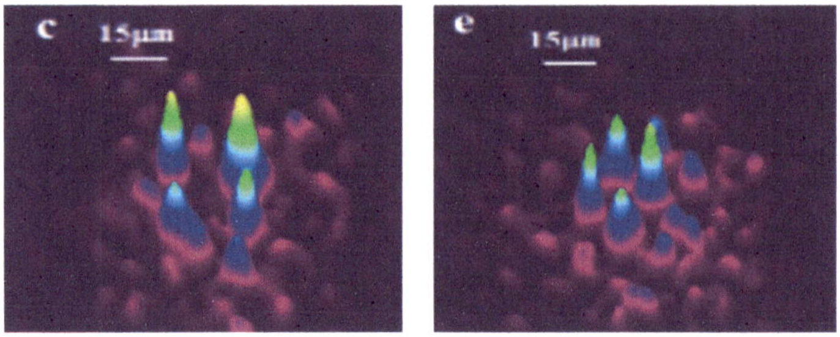

Fig. 4.4 Quasi-discrete optical solitons with vorticity $S = 1$, created in a bulk photorefractive crystal with an induced deep photonic lattice. The left and right panels display, respectively, OC and IC vortex solitons. Reproduced from [90]. ©2004 by the American Physical Society

4.3.2.2 Rotating Lattices

Dynamics of BEC loaded in OLs rotating at angular velocity Ω, as well as the propagation of light in a twisted nonlinear photonic crystal with pitch Ω, is modeled by the 2D version of (4.1), written in the rotating reference frame:

$$i\psi_t = -\left((1/2)\nabla^2 + \Omega\hat{L}_z\right)\psi - \varepsilon\left[\cos(kx) + \cos(ky)\right]\psi + g|\psi|^2\psi, \quad (4.48)$$

where $\hat{L}_z = i(x\partial_y - y\partial_x) \equiv i\partial_\theta$ is the operator of the z-component of the orbital momentum (θ is the angular coordinate in the (x, y) plane). In the tight-binding approximation, (4.48) is replaced by the following variant of the DNLS equation [94]:

$$\begin{aligned} i\dot{\psi}_{m,n} &= -(C/2)\left\{\left(\psi_{m+1,n} + \psi_{m-1,n} + \psi_{m,n+1} + \psi_{m,n-1} - 4\psi_{m,n}\right)\right. \\ &\left. -i\Omega\left[m\left(\psi_{m,n+1} - \psi_{m,n-1}\right) - n\left(\psi_{m+1,n} - \psi_{m-1,n}\right)\right]\right\} + g|\psi_{m,n}|^2\psi_{m,n}, \quad (4.49) \end{aligned}$$

where C is the intersite coupling constant. In [94], stationary solutions to (4.49) were looked for in the form of ansatz (4.36), fixing $\omega \equiv -1$ and varying C in (4.49) as a control parameter. Two species of localized states were thus constructed: off-axis fundamental discrete solitons, placed at distance R from the rotation pivot, and on-axis ($R = 0$) vortex solitons, with vorticities $S = 1$ and 2. At a fixed value of rotation frequency Ω, a stability interval for the fundamental soliton, $0 < C < C_{\max}^{(\mathrm{fund})}(R)$, monotonously shrinks with the increase of R, i.e. most stable are the discrete solitons with the center placed at the rotation pivot. Vortices with $S = 1$ are gradually destabilized with the increase of Ω (i.e. their stability interval, $0 < C < C_{\max}^{(\mathrm{vort})}(\Omega)$, shrinks). On the contrary, a remarkable finding is that vortex solitons with $S = 2$, which, as said above, are completely unstable in the usual DNLS equation with $\Omega = 0$, are *stabilized* by the rotation, in an interval $0 < C < C_{\mathrm{cr}}^{(S=2)}(\Omega)$, with $C_{\mathrm{cr}}^{(S=2)}(\Omega)$ growing as a function of Ω. In particular, $C_{\mathrm{cr}}^{(S=2)}(\Omega) \approx 2.5\Omega$ at small Ω [94].

4.3.3 Spontaneous Symmetry Breaking in Linearly-Coupled Lattices

A characteristic feature of many nonlinear *dual-core* systems, built of two identical linearly-coupled waveguides with intrinsic nonlinearity, is a *spontaneous-symmetry-breaking* (SSB) *bifurcation*, which destabilizes the symmetric ground state, with equal components in the coupled cores, and creates stable asymmetric ones, when the nonlinearity strength exceeds a critical value [95]. A system of linearly-coupled DNLS equations is a basic model for SSB in discrete settings. Its 2D form is [96]:

$$\begin{aligned} i\dot{\phi}_n &= -(1/4)\left(\phi_{m+1,n} + \phi_{m-1,n} + \phi_{m,n+1} + \phi_{m,n-1} - 4\phi_{m,n}\right) - \left|\phi_{m,n}\right|^2 \phi_{m,n} - K\psi_{m,n}\,, \\ i\dot{\psi}_n &= -(1/4)\left(\psi_{m+1,n} + \psi_{m-1,n} + \psi_{m,n+1} + \psi_{m,n-1} - 4\psi_{m,n}\right) - \left|\psi_{m,n}\right|^2 \psi_{m,n} - K\phi_{m,n}\,, \end{aligned} \tag{4.50}$$

where $\phi_{m,n}$ and $\psi_{m,n}$ are discrete fields, and $K > 0$ accounts for the linear coupling between them. Stationary states are looked for as $\left(\phi_{m,n}, \psi_{m,n}\right) = \exp\left(-i\omega t\right)\left(u_{m,n}, v_{m,n}\right)$, where the linear coupling makes it necessary to have identical frequency, ω, in both components. Real stationary fields are characterized by their norms,

$$E_{u,v} = \sum_{m,n=-\infty}^{+\infty} \left(u_{m,n}^2, v_{m,n}^2\right), \tag{4.51}$$

which define the asymmetry degree of the symmetry-broken states:

$$r = \left(E_u - E_v\right) / \left(E_u + E_v\right). \tag{4.52}$$

The present system can be analyzed by means of the VA, which is based on the simplest ansatz (cf. its 1D counterpart (4.39)):

$$\left(u_{m,n}, v_{m,n}\right) = (A, B)\exp\left[-a\left(|m| + |n|\right)\right], \tag{4.53}$$

with inverse width a and amplitudes, A and B, of the two components. The ansatz accounts for the SBB in the case of $A \neq B$. A typical example of a *stable* 2D discrete OC soliton is displayed in Fig. 4.5a, which corroborates accuracy of the VA. The full set of symmetric and asymmetric 2D discrete solitons is characterized, in Fig. 4.5b, by the dependence of asymmetry parameter r, defined in (4.52), on the total norm, $E \equiv E_u + E_v$ (see (4.51)). It is seen that the SSB bifurcation is one of a clearly *subcritical* type [97], with the two branches of broken-symmetry states originally going backward as unstable ones, and getting stable after passing the turning point. Accordingly, Fig. 4.5b demonstrates a considerable bistability area, where symmetric and asymmetric states coexist as ones stable against small perturbations.

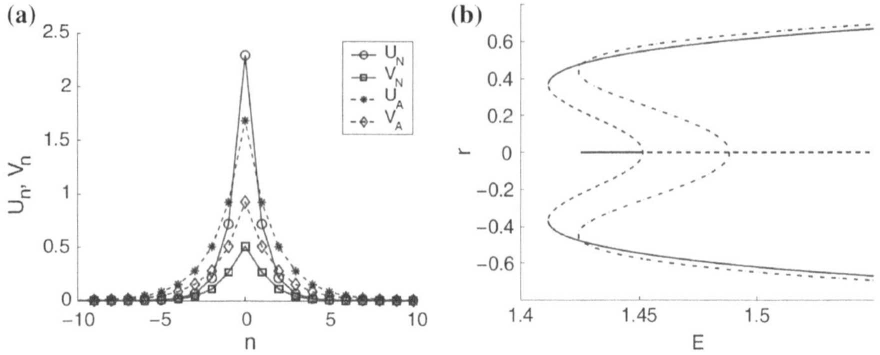

Fig. 4.5 **a** A stable 2D two-component discrete soliton with spontaneously broken symmetry between the components, generated by system (4.50). The 2D soliton, with the total norm $E \equiv E_u + E_v = 1.435$, is displayed by means of its 1D cross section. Symbols labelled (U_N, V_N) and (U_A, V_A) stand, respectively, for the components of the numerically constructed soliton and its analytical counterpart predicted by the VA based on ansatz (4.53). **b** Families of 2D discrete solitons of the OC type, generated by system (4.50), are shown by means of curves $r(E)$ (where r is the asymmetry parameter (4.52)). The dashed-dotted curve shows the VA prediction, while the solid and dashed ones depict stable and unstable solitons produced by the numerical solution. Reproduced from [96]

4.4 Ablowitz–Ladik and Salerno-Model Lattices

4.4.1 1D Models

1D models of AL and SM types, which are defined by (4.5) and (4.6), conserve the total norm, but its definition is different from the straightforward one, given by (4.34) for the DNLS equation; namely,

$$N_{\mathrm{AL,SM}} = (1/\mu) \sum_n \ln \left| 1 + \mu |\psi_n|^2 \right| \tag{4.54}$$

[26, 98]. The Hamiltonian of the AL and SM equations is also essentially different from the "naive" DNLS Hamiltonian given by (4.35). As found in the original work of Ablowitz and Ladik, the Hamiltonian of their model is

$$H_{\mathrm{AL}} = -\sum_n \left(\psi_n \psi_{n+1}^* + \psi_{n+1} \psi_n^* \right), \tag{4.55}$$

while for the SM, it is [98]

$$H_{\mathrm{SM}} = -\sum_n \left[\left(\psi_n \psi_{n+1}^* + \psi_{n+1} \psi_n^* \right) + (2/\mu) |\psi_n|^2 \right] + (2/\mu) N_{\mathrm{AL}}. \tag{4.56}$$

The price paid for ostensible "simplicity" of expression (4.55) is the complex form of the respective Poisson brackets, which determine the dynamical equations in terms of the Hamiltonian as $d\psi_n/dt = \{H, \psi_n\}$. For the AL and SM models, the Poisson brackets, written for a pair of arbitrary functions of the discrete field variables, $B\left(\psi_n, \psi_n^*\right), C\left(\psi_n, \psi_n^*\right)$, are

$$\{B, C\} = i\sum_n \left(\frac{\partial B}{\partial \psi_n}\frac{\partial C}{\partial \psi_n^*} - \frac{\partial B}{\partial \psi_n^*}\frac{\partial C}{\partial \psi_n}\right)\left(1 + \mu\,|\psi_n|^2\right). \tag{4.57}$$

As mentioned above, the continuum limit of the SM is represented by (4.8) [31]. This continuous equation conserves the total norm and Hamiltonian, written in terms of variables $\Psi(x)$ (see (4.7)), which are straightforward continuum counterparts of expressions (4.54)–(4.56):

$$(N_{\mathrm{AL}})_{\mathrm{cont}} = \frac{1}{\mu}\int_{-\infty}^{+\infty} dx\, \ln\left|1 + \mu|\Psi|^2\right|, \tag{4.58}$$

$$(H_{\mathrm{SM}})_{\mathrm{cont}} = \int_{-\infty}^{+\infty} dx\left[|\Psi_x|^2 - 2\left(\frac{1}{\mu} + 1\right)|\Psi|^2\right] + \frac{2}{\mu}(N_{\mathrm{AL}})_{\mathrm{cont}} \tag{4.59}$$

4.4.2 Discrete 1D Solitons

The AL equation (4.5) gives rise to an exact solution for solitons in the case of self-focusing nonlinearity, $\mu > 0$. Setting $\mu \equiv +1$ by means of rescaling, the exact solutions is

$$\psi_n(t) = (\sinh\beta)\,\mathrm{sech}\left[\beta(n - \xi(t))\right]\exp\left[i\alpha\left(n - \xi(t)\right) - i\varphi(t)\right], \tag{4.60}$$

where β and α are arbitrary real parameters that determine the soliton's amplitude, $A \equiv \sinh\beta$, its velocity, $V \equiv \dot{\xi} = 2\beta^{-1}(\sinh\beta)\sin\alpha$, and overall frequency $\Omega \equiv \dot{\varphi} = -2\left[(\cosh\beta)\cos\alpha + (\alpha/\beta)(\sinh\beta)\sin\alpha\right]$.

The existence of exact solutions for traveling solitons in this discrete system is a highly nontrivial property of the AL equation, which follows from its integrability. If the system is not integrable, motion of a discrete soliton through a lattice is hampered by emission of radiation, even if this effect may seem very weak in direct simulations [73]. On the other hand, there are some special discrete equations which are not integrable, but admit particular solutions for traveling solitons at exceptional values of the velocity, rather than at an arbitrary velocities, as in the case of the AL solitons [99, 100].

The stationary version of the SM, obtained by the substitution of the usual ansatz (4.36), with real u_n, in (4.6), is

$$\omega u_n = -\left(u_{n+1} + u_{n-1}\right)\left(1 + \mu u_n^2\right) - 2u_n^3, \tag{4.61}$$

cf. (4.37). Discrete solitons in the nonintegrable SM equation (4.6) with $\mu > 0$, i.e. with *noncompeting* intersite and onsite self-focusing nonlinearities, were investigated by means of numerical methods [98, 101, 102]. It has been demonstrated that the SM gives rise to static (and, sometimes, approximately mobile [101]) solitons at all positive values of μ.

Another possibility is to consider the SM with $\mu < 0$, which features *competing nonlinearities*, as the intersite cubic term, with coefficient $\mu < 0$ in (4.6), which accounts for nonlinear coupling between adjacent sites of the lattice, and the onsite term in (4.6) (the last term in that equation) represent, respectively, self-defocusing and focusing nonlinear interactions. It was found [31] that this version of the SM gives rise to families of quiescent discrete solitons, which are looked for in the usual form (4.36), with $\omega < 0$ and real amplitudes u_n, of two different types. One family represents ordinary discrete solitons, similar to those generated by the DNLS equation. Another family represents *cuspons,* featuring higher curvature of their profile at the center than exponential shapes. Examples of numerically found stable discrete solitons of these types are displayed in Fig. 4.6a. The border between the ordinary discrete solitons and cuspons is represented by a special discrete mode, in the form of a stable *peakon*, which is also shown in Fig. 4.6a.

The continuum limit of the SM with competing nonlinearities, given by (4.8) with $\mu < 0$, produces continuous solitons in the usual form, $\Psi = \exp(-i\omega t)\, U(x)$, in the frequency band $0 < -\omega < 1/|\mu| - 1$, provided that $|\mu| < 1$. At the edge of the soliton band, i.e. at $\omega = -(1/|\mu| - 1)$, (4.8) gives rise to an exact solution in the form of the continuous peakon $U_{\text{peakon}}(x) = \left(1/\sqrt{|\mu|}\right) \exp\left(-\sqrt{(1/|\mu|) - 1}|x|\right)$ [31]. For continuous solutions, the name of "peakon" implies a jump of the derivative at the central point, while cuspons do not exist in the continuum limit.

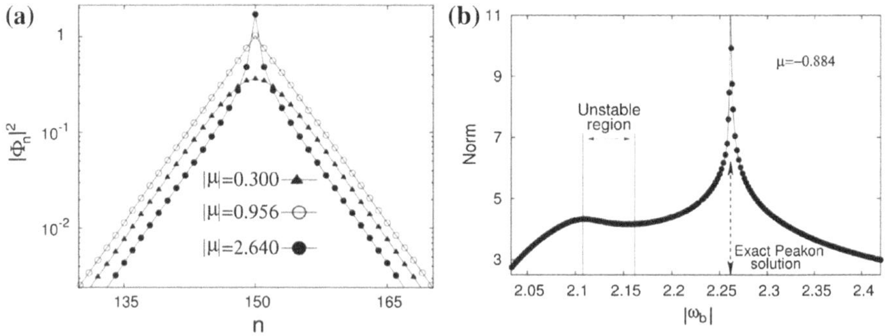

Fig. 4.6 **a** Examples of three different types of discrete solitons, shown on the logarithmic scale, which are produced by (4.61), i.e. the Salerno model, with competing nonlinearities ($\mu < 0$), at $\omega = -2.091$: an ordinary soliton for $\mu = -0.3$, a peakon for $\mu = -0.956$, and a cuspon for $\mu = -2.64$. In the figure, $|\Phi_n|$ has the same meaning as u_n in (4.61). **b** The norm of the discrete solitons in the Salerno model with competing nonlinearities, versus the frequency (denoted here ω_b, instead of ω), for $\mu = -0.884$. Reproduced from [31]

The stability analysis of discrete solitons produced by the SM with competing nonlinearities demonstrate that only a small subfamily of ordinary solitons is unstable, while all cuspons, including the peakon, are stable. For fixed $\mu = -0.884$, a typical situation for families of discrete solitons in the SM with competing nonlinearities is presented in Fig. 4.6b, which shows the norm (4.54) as a function of $|\omega|$. The plot clearly demonstrates that ordinary solitons and cuspons are separated by the peakon, as mentioned above. Except for the part of the ordinary-soliton family with the negative slope, $dN/d(|\omega|) < 0$, which is marked in Fig. 4.6b, the discrete solitons are stable. In particular, it is worthy to note that the cuspons and the peakon are completely stable modes. The instability of the segment of the family of ordinary discrete solitons with $dN/d(|\omega|) < 0$ exactly agrees with the prediction of the well-known Vakhitov–Kolokolov criterion [103]. On the other hand, it is seen from Fig. 4.6b that the Vakhitov–Kolokolov criterion, being valid for ordinary solitons, is actually *reversed* for cuspons [31].

As mentioned above, antisymmetric bound states of DNLS solitons are stable, while symmetric bound states are unstable [77, 78]. As shown in [31], the same is true for bound states of ordinary discrete solitons in the SM. However, a noteworthy finding is that, in the framework of the SM with competing nonlinearities, the situation is *exactly opposite* for the cuspons: their symmetric and antisymmetric bound states are stable and unstable, respectively [31].

4.4.3 The Two-Dimensional Salerno Model and Its Discrete Solitons

The 2D version of the SM was introduced in [104]. It is based on the following equation, cf. (4.6),

$$i\dot{\psi}_{n,m} = -\left[\left(\psi_{n+1,m} + \psi_{n-1,m}\right) + C\left(\psi_{n,m+1} + \psi_{n,m-1}\right)\right] \times \left(1 + \mu\left|\psi_{n,m}\right|^2\right) - 2\left|\psi_{n,m}\right|^2 \psi_{n,m}\,, \tag{4.62}$$

where real constant $C > 0$ accounts for a possible anisotropy of the 2D lattice. Similar to its 1D version, (4.62) conserves the norm and Hamiltonian, cf. (4.54) and (4.56),

$$(N_{\mathrm{AL}})_{2\mathrm{D}} = (1/\mu)\sum_{m,n} \ln\left|1 + \mu|\psi_{n,m}|^2\right|\,, \tag{4.63}$$

$$(H_{\mathrm{AL}})_{2\mathrm{D}} = -\sum_{n,m}\left[\left(\psi_{n,m}\psi^*_{n+1,m} + \psi_{n+1,m}\psi^*_{n,m}\right) + C\left(\psi_{n,m}\psi^*_{n,m+1} + \psi_{n,m+1}\psi^*_{n,m}\right)\right. \left. +(2/\mu)|\psi_{n,m}|^2\right] + (2/\mu)\left(N_{\mathrm{AL}}\right)_{2\mathrm{D}}\,. \tag{4.64}$$

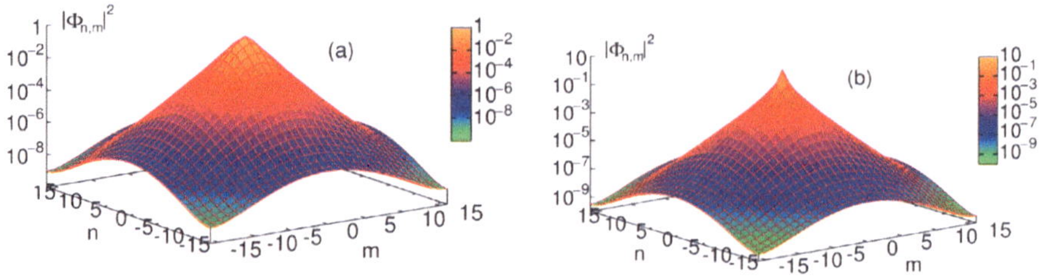

Fig. 4.7 Discrete solitons in the isotropic ($C = 1$) 2D Salerno model with competing nonlinearities [$\mu < 0$ in (4.62)], obtained for frequency $\omega = -4.22$: **a** a regular soliton at $\mu = -0.2$; **b** a cuspon at $\mu = -0.88$. Reproduced from [104]

The continuum limit of this model is a 2D continuous equation which is a straightforward extension of its 1D counterpart (4.8):

$$i\Psi_t + \left(1 + \mu |\Psi|^2\right)\left(\Psi_{xx} + \Psi_{yy}\right) + 2\left[(1 + C)\mu + 1\right]|\Psi|^2\Psi = 0. \qquad (4.65)$$

Note that term $\mu |\Psi|^2 \left(\Psi_{xx} + \Psi_{yy}\right)$ prevents the onset of the collapse in (4.65).

2D solitons are looked for in the same form as their 1D counterparts, $\psi_{mn}(t) = e^{-i\omega t}u_{mn}$, cf. (4.36). In the most interesting case of the competing nonlinearities, $\mu < 0$, the situation is similar to that outlined above for SM in 1D: there are ordinary discrete solitons, which have their stability and instability regions, and 2D cuspons, which are *entirely stable* in their existence region. Typical 2D solitons of both types are displayed in Fig. 4.7. Also similar to the 1D case, ordinary solitons and cuspons are separated by 2D peakons, which are stable. In addition to that, antisymmetric bound states of ordinary 2D discrete solitons, and symmetric complexes built of 2D cuspons, are stable, while the bound states with opposite parities are unstable, also like in the 1D model.

Along with the fundamental solitons, the 2D SM with the competing nonlinearities gives rise to vortex-soliton modes with narrow stability regions [104]. In the 2D SM with non-competing nonlinearities, unstable vortex solitons spontaneously transform into fundamental solitons, losing their vorticity (this is possible because the angular momentum is not conserved in the lattice system). The situation is essentially different in the 2D SM with competing nonlinearities, where unstable vortex modes transform into *vortical breathers*, i.e. persistently oscillating localized modes that keep the original vorticity.

4.5 A Brief Survey of Semi-discrete Systems

A topic which may be a subject for a separate major review, is semi-discrete systems, i.e. 2D setting which are discrete in one direction and continuous in the other. Accordingly, such systems can create semi-discrete solitons. A system of this type which was explored in detail is an array of optical fibers [105], modeled by a system of coupled NLS equations for amplitudes $u_n(z, \tau)$ of electromagnetic waves in individual fibers:

$$i\partial_z u_n + (1/2)D\partial_\tau^2 u_n + (\kappa/2)(u_{n+1} + u_{n-1} - 2u_n) + |u_n|^2 u_n = 0, \tag{4.66}$$

where D is the group-velocity-dispersion coefficient in each fiber, and $\kappa > 0$ is the coefficient of coupling between adjacent fibers in the array. It supports semi-discrete solitons in the case of anomalous dispersion, i.e. $D > 0$. A remarkable property of semi-discrete modes generated by (4.66) is their ability to stably move *across the array*, under the action of a *kick* applied to them at $z = 0$ [106]:

$$u_n(\tau) \rightarrow \exp(ian)\, u_n(\tau), \tag{4.67}$$

with real a. An example of such a moving mode is displayed in Fig. 4.8. This property may be compared to the above-mentioned mobility of 1D discrete solitons in the DNLS equation [73].

Similarly, quasi-discrete settings modeled by an extension of (4.66) with two transverse spatial coordinates, were used for the creation of spatiotemporal optical solitons ("light bullets") [107], as well as soliton-like transient modes with embedded vorticity [108]. Waveguides employed in those experiments feature a transverse hexagonal-lattice structure, written in bulk silica by means of optical technology.

A spatiotemporal vortex state (in the experiment, it is actually a transient one) in the bundle-like structure is presented by Fig. 4.9, which displays both numerically predicted and experimentally observed distributions of intensity of light in the transverse plane, together with a phase plate used in the experiment to embed the vorticity into the incident spatiotemporal pulse which was used to create the mode.

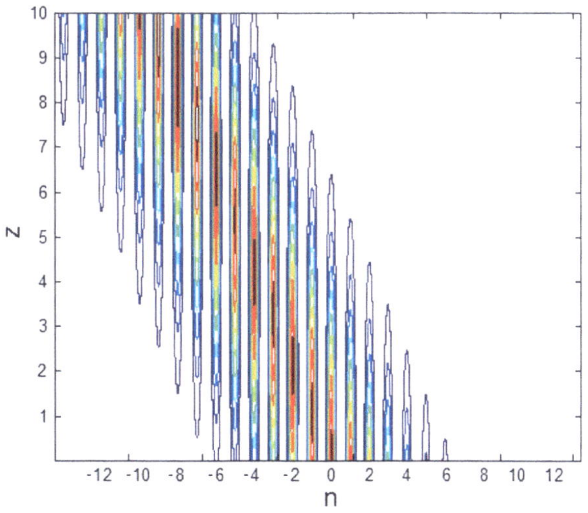

Fig. 4.8 An example of a semi-discrete spatiotemporal mode, generated by (4.66), which performs stable transverse motion under the action of the kick, defined according to (4.67), with $a = 1.5$. The cross section of the plot at any fixed z shows the distribution of power $|u_n(\tau)|^2$ for each n. Reproduced from [106]

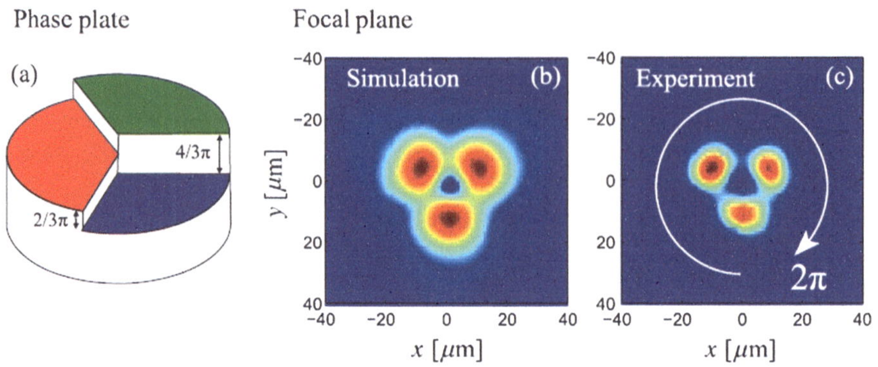

Fig. 4.9 A semi-discrete vortex soliton in a hexagonal quasi-discrete array of waveguides made in bulk silica. **a** A phase plate used for imprinting the vortex structure into the input beam. **b**, **c** Numerically simulated and experimentally observed (transient) intensity distributions in the transverse plane, with phase shifts $2\pi/3$ between adjacent peaks. Reproduced from [108].

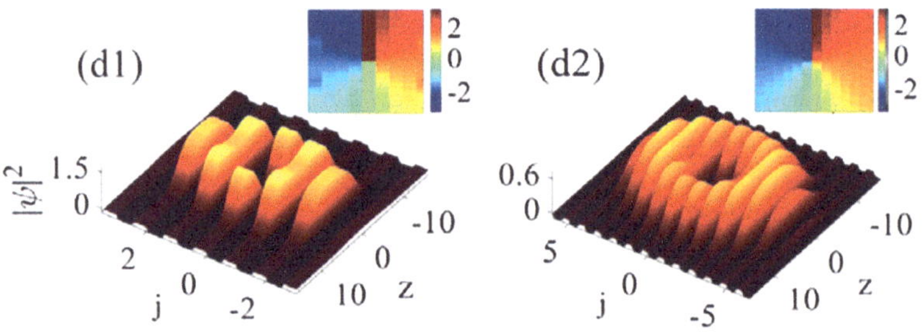

Fig. 4.10 Left and right panels display, respectively, examples of amplitude and phase profiles of stable OC and IC semi-discrete vortex solitons produced by (4.68). Reproduced from [109]

A new type of semi-discrete solitons was recently reported in [109], in the framework of an array of linearly coupled 1D GPEs, including the *Lee-Hung-Yang correction*, which represents an effect of quantum fluctuations around the mean-field states of a binary BEC [110, 111]. The system is

$$i\partial_t\psi_j = -(1/2)\partial_{zz}\psi_j - (1/2)\left(\psi_{j+1} - 2\psi_j + \psi_{j-1}\right) + g|\psi_j|^2\psi_j - |\psi_j|\psi_j, \tag{4.68}$$

where $\psi_j(z)$ is the mean-field wave function in the jth core, the self-attractive quadratic term represents the Lee-Hung-Yang effect, and $g > 0$ accounts for the mean-field self-repulsion. This system gives rise to many families of semi-discrete solitons, including a novel species of *semi-discrete vortex solitons*. Typical examples of such states are displayed in Fig. 4.10.

4.6 Conclusion

4.6.1 Summary of the Chapter

The interplay of discreteness and intrinsic nonlinearity in various physical media gives rise to a great variety of static and dynamical states. Among them, especially interesting are self-trapped ones in the form of discrete solitons. The present Chapter aims to briefly review basic theoretical models combining discreteness and nonlinearity, and basic results for discrete solitons produced by such models. Essential experimental findings are included too (in particular, those for 2D and 3D discrete solitons with embedded vorticity). In many cases, discreteness helps to produce states which either do not exist or are unstable in continuum counterparts of discrete settings. In particular, the 1D DNLS equation gives rise to stable bound states of fundamental solitons, and the 2D DNLS equation readily creates fundamental and vortex solitons, whose counterparts are completely unstable in the continuum. On the other hand, some properties which are obvious in the continuum limit, such as mobility of solitons, are problematic in discrete settings.

The work in this area is currently in progress, and new results may be expected. A promising direction is to generate discrete counterparts of complex continuous modes with intrinsic topological structures. Some results obtained in this direction have already been reported, such as discrete solitons in a system with spin-orbit coupling [112], sophisticated 3D discrete modes with embedded vorticity [113, 114], and discrete skyrmions [115]. A challenging task is the experimental realization of such results which, thus far, were only predicted in the theoretical form.

4.6.2 Topics Not Included in the Chapter

Due to length limitations, some essential models and methods are not considered here. One of them is the *anti-continuum limit*, which makes it possible to obtain "stems" for many families of discrete solitons by considering, at first, lattice models with no coupling between the sites. Using this approach, one can construct a great deal of modes, by formally putting together various solutions supported by non-interacting sites of the lattice. Then, the analysis allows one to identify solution branches that can be extended to small nonzero values of the intersite coupling. This method is efficient in constructing many families of discrete solitons in diverse models [61, 78, 84, 116].

Interaction of discrete solitons with local defects in the underlying lattice, as well as with interfaces and edges (surfaces, if the underlying lattice is two- or three-dimensional) is another vast area of theoretical and experimental studies. In particular, defects and surfaces may often help to create and stabilize localized modes which do not exist or are unstable in uniform lattices, such as Tamm [117] and topological-insulator [118, 119] states.

Large topics are solitons in discrete dissipative nonlinear systems, and in systems subject to the condition of the parity-time (PT) symmetry. In this Chapter, dissipative systems, which include friction and driving forces, are considered only in terms of the TL and FK models (in particular, for arrays of Josephson junctions, in Fig. 4.1). Other dissipative versions of TL models are known too, in the form of LC transmission lines for electric pulses. They support traveling discrete solitons, which have been produced in theoretical and experimental forms alike [120–122].

In other contexts, basic nonlinear dissipative models are represented by discrete complex Ginzburg–Landau equations, i.e. DNLS equations with complex coefficients in front of onsite linear and nonlinear terms, which account for dissipative losses and compensating gain [123]. These models give rise to discrete solitons which do not exist in continuous families, unlike the DNLS solitons, but rather as isolated *attractors* [124–128].

Systems with PT symmetry are dissipative models which share many properties with conservative ones. They include mutually symmetric spatially separated linear gain and loss elements [129–131]. This arrangement makes it natural to consider PT-symmetric systems with a discrete structure. Their experimental realization in optics [131] suggests to include the Kerr nonlinearity, thus opening the way to prediction of PT-symmetric discrete solitons [132, 133]. In particular, various species of stable 1D and 2D discrete solitons were predicted in chains of PT-symmetric elements [134–139]. An example of PT-symmetric solitons has been created experimentally in a similar discrete setting [140].

Acknowledgements I appreciate the invitation of the Editors of the volume *Emerging Frontiers in Nonlinear Science* to submit this Chapter. One of the Editors, Prof. J. Cuevas-Maraver, has provided a great deal of help in the course of preparing the manuscript. I would like to thank colleagues in collaboration with whom I have been working, in the course of a long time, on various topics related to the review: G.E. Astrakharchik, P. Beličev, A.R. Bishop, R. Blit, L.L. Bonilla, R. Carretero-González, Zhaopin Chen, Zhigang Chen, C. Chong, M. Cirillo, J. Cuevas-Maraver, J. D'Ambroise, F. K. Diakonos, S. V. Dmitriev, N. Dror, O. Dutta, A. Eckhard, L. M. Floría, D. J. Frantzeskakis, S. Fu, G. Gligorić, J. Gómez-Gardeñes, N. Grønbech-Jensen, L. Hadžievski, D. Herring, J. Hietarinta, P. Hauke, N.V. Hung, Y.V. Kartashov, T. Kapitula, D.J. Kaup, P.G. Kevrekidis, V.V. Konotop, T. Kuusela, M. Lewenstein, Y. Li, A. Maluckov, R.S. Osgood, N.C. Panoiu, I.E. Papachalarampus, J. Petrović, M.A. Porter, K.Ø. Rasmussen, H. Sakaguchi, L. Torner, M. Trippenbach, V. Ustinov, R.A. Van Gorder, M.I. Weinstein, and K. Zegadlo.

References

1. C. Pethick, H. Smith, *Bose-Einstein Condensation in Dilute Gases* (Cambridge University Press, Cambridge, 2002)
2. G. Fibich, *The Nonlinear Schrödinger Equation: Singular Solutions and Optical Collapse* (Springer, Heidelberg, 2015)
3. Y.S. Kivshar, G.P. Agrawal, *Optical Solitons: From Fibers to Photonic Crystals* (Academic, San Diego, 2003)
4. O. Morsch, M. Oberthaler, Rev. Mod. Phys. **78**, 179 (2006)

5. M.A. Porter, R. Carretero-González, P.G. Kevrekidis, B.A. Malomed, Chaos **15**, 015115 (2005)
6. M. Skorobogatiy, J. Yang, *Fundamentals of Photonic Crystal Guiding* (Cambridge University Press, Cambridge, 2008)
7. D.N. Christodoulides, R.I. Joseph, Opt. Lett. **13**, 794 (1988)
8. H.S. Eisenberg, Y. Silberberg, R. Morandotti, A.R. Boyd, J.S. Aitchison, Phys. Rev. Lett. **81**, 3383 (1998)
9. D.N. Christodoulides, F. Lederer, Y. Silberberg, Nature **424**, 817 (2003)
10. F. Lederer, G.I. Stegeman, D.N. Christodoulides, G. Assanto, M. Segev, Y. Silberberg, Phys. Rep. **463**, 1 (2008)
11. F. Ye, D. Mihalache, B. Hu, N.C. Panoiu, Phys. Rev. Lett. **104**, 106802 (2010)
12. A. Smerzi, A. Trombettoni, Phys. Rev. A **68**, 023613 (2003)
13. A. Szameit, R. Keil, F. Dreisow, M. Heinrich, T. Pertsch, S. Nolte, A. Tünnermann, Opt. Lett. **34**, 2838 (2009)
14. C. Chong, R. Carretero-González, B.A. Malomed, P.G. Kevrekidis, Phys. D **240**, 1205 (2011)
15. A. Szameit, T. Pertsch, S. Nolte, A. Tünnermann, F. Lederer, Phys. Rev. A **77**, 043804 (2008)
16. E.W. Laedke, K.H. Spatschek, S.K. Turitsyn, Phys. Rev. Lett. **73**, 1055 (1994)
17. P.G. Kevrekidis, *The Discrete Nonlinear Schrödinger Equation: Mathematical Analysis, Numerical Computations, and Physical Perspectives* (Springer, Berlin, 2009)
18. A. Locatelli, D. Modotto, D. Paloschi, C. De Angelis, Opt. Commun. **237**, 97 (2004)
19. G. Herring, P.G. Kevrekidis, B.A. Malomed, R. Carretero-González, D.J. Frantzeskakis, Phys. Rev. E **76**, 066606 (2007)
20. V.E. Zakharov, S.V. Manakov, S.P. Novikov, L.P. Pitaevskii, *Solitons: The Inverse Scattering Method* (Nauka Publishers, Moscow, 1980). [English translation: Consultants Bureau, New York, 1984]
21. M.J. Ablowitz, H. Segur, *Solitons and Inverse Scattering Method* (SIAM, Philadelphia, 1981)
22. F. Calogero, A. Degasperis, *Spectral Transform and Solitons: Tools to Solve and Investigate Nonlinear Evolution Equations* (North-Holland, New York, 1982)
23. A.C. Newell, *Solitons in Mathematics and Physics* (SIAM, Philadelphia, 1985)
24. M.J. Ablowitz, B.M. Herbst, SIAM, J. Appl. Math. **50**, 339 (1990)
25. D. Levi, M. Petrera, C. Scimiterna, Europhys. Lett. **84**, 10003 (2008)
26. M.J. Ablowitz, J.F. Ladik, J. Math. Phys. **17**, 1011 (1976)
27. Y.B. Suris, *The Problem of Integrable Discretization: Hamiltonian Approach* (Birkhauser, Basel, 2003)
28. M. Salerno, Phys. Lett. A **162**, 381–384 (1992)
29. O. Dutta, A. Eckardt, P. Hauke, B. Malomed, M. Lewenstein, New J. Phys. **13**, 023019 (2011)
30. O. Dutta, M. Gajda, P. Hauke, M. Lewenstein, D.-S. Luhmann, B.A. Malomed, T. Sowinski, J. Zakrzewski, Rep. Prog. Phys. **78**, 066001 (2015)
31. J. Gómez-Gardeñes, B.A. Malomed, L.M. Floría, A.R. Bishop, Phys. Rev. E **73**, 036608 (2006)
32. M. Toda, J. Phys. Soc. Jpn. **22**, 431 (1967)
33. R.S. Johnson, *A Modern Introduction to the Mathematical Theory of Water Waves* (Cambridge University Press, Cambridge, 1997)
34. E. Fermi, J. Pasta, S. Ulam, *Los Alamos Report LA-1940* (1955), p. 975
35. G. Gallavoti (ed.), *The Fermi-Pasta-Ulam Problem: A Status Report* (Springer, Berlin, 2008)
36. M.A. Porter, N.J. Zabusky, B. Hu, D.K. Campbell, Am. Sci. **97**, 214 (2009)
37. T. Dauxois, Phys. Today **6**, 55 (2008)
38. N.J. Zabusky, M.D. Kruskal, Phys. Rev. Lett. **15**, 240 (1965)
39. Y.I. Frenkel, T. Kontorova, J. Phys. **1**, 137 (1939)
40. Y.S. Kivshar, O.M. Braun, *The Frenkel-Kontorova Model: Concepts, Methods and Applications* (Springer, Berlin, 2004)
41. B.A. Malomed, Phys. Rev. A **45**, 4097 (1992)
42. T. Kuusela, J. Hietarinta, B.A. Malomed, J. Phys. A **26**, L21 (1993)

43. K.Ø. Rasmussen, B.A. Malomed, A.R. Bishop, N. Grønbech-Jensen, Phys. Rev. E **58**, 6695–6699 (1998)
44. K.L. Johnson, *Contact Mechanics* (Cambridge University Press, Cambridge, 1985)
45. N. Boechler, G. Theocharis, S. Job, P.G. Kevrekidis, M.A. Porter, C. Daraio, Phys. Rev. Lett. **104**, 244302 (2010)
46. O. Derzhko, J. Richter, M. Maksymenko, Int. J. Mod. Phys. B **29**, 1530007 (2015)
47. D. Leykam, A. Andreanov, S. Flach, Adv. Phys. X **3**, 1 (2018)
48. K. Zegadlo, N. Dror, N.V. Hung, M. Trippenbach, B.A. Malomed, Phys. Rev. E **96**, 012204 (2017)
49. I. Roy, S.V. Dmitriev, P.G. Kevrekidis, A. Saxena, Phys. Rev. E **76**, 026660 (2007)
50. S.V. Dmitriev, A. Khare, P.G. Kevrekidis, A. Saxena, L. Hadžievski, Phys. Rev. E **77**, 056603 (2008)
51. Y. Ishimori, T. Munakata, J. Phys. Soc. Jpn. **51**, 3367 (1982)
52. Y.S. Kivshar, D.K. Campbell, Phys. Rev. E **48**, 3077 (1993)
53. S.F. Mingaleev, Y.B. Gaididei, E. Majerníková, S. Shpyrko, Phys. Rev. E **61**, 4454 (2000)
54. L.L. Bonilla, B.A. Malomed, Phys. Rev. B **43**, 11539 (1991)
55. S. Hontsu, J. Ishii, J. Appl. Phys. **63**, 2021 (1988)
56. S. Sakai, P. Bodin, N.F. Pedersen, J. Appl. Phys. **73**, 2411 (1993)
57. G. Paternó, A. Barone, *Physics and Applications of the Josephson Effect* (Wiley, New York, 1982)
58. A.V. Ustinov, M. Cirillo, B.A. Malomed, Phys. Rev. B **47**, 8357(R) (1993)
59. S. Watanabe, H.S.J. van der Zant, S.H. Strogatz, T.P. Orlando, Phys. D **97**, 429 (1996)
60. S. Shapiro, Phys. Rev. Lett. **11**, 80 (1963)
61. S. Aubry, Phys. D **103**, 201 (1997)
62. J.L. Marín, F. Falo, P.J. Martínez, L.M. Floría, Phys. Rev. E **63**, 066603 (2001)
63. A.V. Savin, O.V. Gendelman, Phys. Rev. E **67**, 041205 (2003)
64. Y.S. Kivshar, M. Peyrard, Phys. Rev. A **46**, 3198 (1992)
65. B.A. Malomed, Progr. Optics **43**, 71 (2002)
66. B.A. Malomed, M.I. Weinstein, Phys. Lett. A **220**, 91 (1996)
67. I.E. Papacharalampous, P.G. Kevrekidis, B.A. Malomed, D.J. Frantzeskakis, Phys. Rev. E **68**, 046604 (2003)
68. D.J. Kaup, Math. Comput. Simulat. **69**, 322 (2005)
69. B.A. Malomed, D.J. Kaup, R.A. Van Gorder, Phys. Rev. E **85**, 026604 (2012)
70. J. Cuevas, G. James, P.G. Kevrekidis, B.A. Malomed, B. Sánchez-Rey, J. Nonlinear Math. Phy. **15**(Suppl. 3), 124 (2008)
71. R. Rusin, R. Kusdiantara, H. Susanto, J. Phys. A: Math. Theor. **51**, 475202 (2018)
72. C. Chong, D.E. Pelinovsky, G. Schneider, Phys. D **241**, 115 (2012)
73. D.B. Duncan, J.C. Eilbeck, H. Feddersen, J.A.D. Wattis, Phys. D **68**, 1 (1993)
74. B.A. Malomed, *Soliton Management in Periodic Systems* (Springer, New York, 2006)
75. J. Cuevas, B.A. Malomed, P.G. Kevrekidis, Phys. Rev. E **71**, 066614 (2005)
76. S. Darmanyan, A. Kobyakov, F. Lederer, J. Exp. Theor. Phys. **86**, 683 (1998)
77. T. Kapitula, P.G. Kevrekidis, B.A. Malomed, Phys. Rev. E **63**, 036604 (2001)
78. D.E. Pelinovsky, P.G. Kevrekidis, D.J. Frantzeskakis, Phys. D **212**, 1 (2005)
79. P.G. Kevrekidis, B.A. Malomed, A.R. Bishop, J. Phys. A: Math. Gen. **34**, 9615 (2001)
80. J. Satsuma, N. Yajima, Progr. Theor. Phys. Suppl. **55**, 284 (1974)
81. M.I. Weinstein, Nonlinearity **12**, 673 (1999)
82. C. Chong, R. Carretero-González, B.A. Malomed, P.G. Kevrekidis, Phys. D **238**, 126 (2009)
83. B.A. Malomed, P.G. Kevrekidis, Discrete vortex solitons. Phys. Rev. E **64**, 026601 (2001)
84. J. Cuevas, G. James, P.G. Kevrekidis, K.J.H. Law, Phys. D **238**, 1422 (2009)
85. R.Y. Chiao, E. Garmire, C.H. Townes, Phys. Rev. Lett. **13**, 479 (1964)
86. V.I. Kruglov, Y.A. Logvin, V.M. Volkov, J. Mod. Opt. **39**, 2277–2291 (1992)
87. B.A. Malomed, Phys. D **399**, 108 (2019)
88. P.G. Kevrekidis, B.A. Malomed, Z. Chen, D.J. Frantzeskakis, Phys. Rev. E **70**, 056612 (2004)

89. D.N. Neshev, T.J. Alexander, E.A. Ostrovskaya, Y.S. Kivshar, H. Martin, I. Makasyuk, Z.G. Chen, Phys. Rev. Lett. **92**, 123903 (2004)
90. J.W. Fleischer, G. Bartal, O. Cohen, O. Manela, M. Segev, J. Hudock, D.N. Christodoulides, Phys. Rev. Lett. **92**, 123904 (2004)
91. Z. Chen, M. Segev, D.W. Wilson, R.E. Muller, P.D. Maker, Phys. Rev. Lett. **78**, 2948 (1997)
92. Z. Chen, M.-F. Shih, M. Segev, D.W. Wilson, R.E. Muller, P.D. Maker, Opt. Lett. **22**, 1751 (1997)
93. A. Bezryadina, E. Eugenieva, Z. Chen, Opt. Lett. **31**, 2456 (2006)
94. J. Cuevas, B.A. Malomed, P.G. Kevrekidis, Phys. Rev. E **76**, 046608 (2007)
95. B.A. Malomed, in: *Nonlinear Dynamics: Materials, Theory and Experiments. 3rd Dynamics Days South America, Valparaiso, November 2014. Springer Proceedings in Physics*, ed. by M. Tlidi, M. Clerc, vol. 173 (Springer, Cham, 2016), p. 97
96. G. Herring, P.G. Kevrekidis, B.A. Malomed, R. Carretero-González, D.J. Frantzeskakis, Phys. Rev. E **76**, 066606 (2007)
97. G. Iooss, D.D. Joseph, *Elementary Stability Bifurcation Theory* (Springer, New York, 1980)
98. D. Cai, A.R. Bishop, N. Grønbech-Jensen, Phys. Rev. E **53**, 4131 (1996)
99. P.G. Kevrekidis, Phys. D **183**, 68 (2003)
100. O.F. Oxtoby, I.V. Barashenkov, Phys. Rev. E **76**, 036603 (2007)
101. D. Cai, A.R. Bishop, N. Grønbech-Jensen, Phys. Rev. E **56**, 7246 (1997)
102. S.V. Dmitriev, P.G. Kevrekidis, B.A. Malomed, D.J. Frantzeskakis, Phys. Rev. E **68**, 056603 (2003)
103. N.G. Vakhitov, A.A. Kolokolov, Radiophys. Quantum Electron. **16**, 783 (1973)
104. J. Gómez-Gardeñes, B.A. Malomed, L.M. Floría, A.R. Bishop, Phys. Rev. E **74**, 036607 (2006)
105. A.B. Aceves, C. De Angelis, A.M. Rubenchik, S.K. Turitsyn, Opt. Lett. **19**, 329 (1994)
106. R. Blit, B.A. Malomed, Phys. Rev. A **86**, 043841 (2012)
107. S. Minardi, F. Eilenberger, Y.V. Kartashov, A. Szameit, U. Röpke, J. Kobelke, K. Schuster, H. Bartelt, S. Nolte, L. Torner, F. Lederer, A. Tünnermann, T. Pertsch, Phys. Rev. Lett. **105**, 263901 (2010)
108. F. Eilenberger, K. Prater, S. Minardi, R. Geiss, U. Röpke, J. Kobelke, K. Schuster, H. Bartelt, S. Nolte, A. Tünnermann, T. Pertsch, Phys. Rev. X **3**, 041031 (2013)
109. X. Zhang, X. Xu, Y. Zheng, Z. Chen, B. Liu, C. Huang, B.A. Malomed, Y. Li, Phys. Rev. Lett. **123**, 133901 (2019)
110. D.S. Petrov, Phys. Rev. Lett. **115**, 155302 (2015)
111. D.S. Petrov, G.E. Astrakharchik, Phys. Rev. Lett. **117**, 100401 (2016)
112. P. Belicev, G. Gligorić, J. Petrović, A. Maluckov, L. Hadzievski, B.A. Malomed, J. Phys. B At. Mol. Opt. Phys. **48**, 065301 (2015)
113. P.G. Kevrekidis, B.A. Malomed, D.J. Frantzeskakis, R. Carretero-González, Phys. Rev. Lett. **93**, 080403 (2004)
114. R. Carretero-González, P.G. Kevrekidis, B.A. Malomed, D.J. Frantzeskakis, Phys. Rev. Lett. **94**, 203901 (2005)
115. P.G. Kevrekidis, R. Carretero-González, D.J. Frantzeskakis, B.A. Malomed, F.K. Diakonos, Phys. Rev. E **75**, 026603 (2007)
116. D. Chen, S. Aubry, G.P. Tsironis, Phys. Rev. Lett. **77**, 4776 (1996)
117. Y.S. Kivshar, Laser Phys. Lett. **5**, 703 (2008)
118. D.R. Gulevich, D. Yudin, D.V. Skryabin, I.V. Iorsh, I.A. Shelykh, Sci. Rep. **7**, 1780 (2017)
119. Y.V. Kartashov, D.V. Skryabin, Optica **3**, 1228 (2016)
120. D. Yemele, P. Marquie, J.M. Bilbault, Phys. Rev. E **68**, 016605 (2003)
121. K.T.V. Koon, J. Leon, P. Marquie, P. Tchofo-Dinda, Phys. Rev. E **75**, 066604 (2007)
122. E. Kengne, R. Vaillancourt, Int. J. Mod. Phys. B **23**, 1 (2009)
123. V. Hakim, W.J. Rappel, Phys. Rev. A **46**, 7347(R) (1992)
124. N.K. Efremidis, D.N. Christodoulides, Phys. Rev. E **67**, 026606 (2003)
125. K. Maruno, A. Ankiewicz, N. Akhmediev, Opt. Commun. **221**, 199 (2003)
126. N.K. Efremidis, D.N. Christodoulides, K. Hizanidis, Phys. Rev. A **76**, 043839 (2007)

127. C. Mejia-Cortes, J.M. Soto-Crespo, M.I. Molina, R.A. Vicencio, Dissipative vortex solitons in two-dimensional lattices, Phys. Rev. A **82**, 063818 (2010)
128. C. Mejia-Cortes, J.M. Soto-Crespo, R.A. Vicencio, M.I. Molina, Vortex solitons of the discrete Ginzburg-Landau equation, Phys. Rev. A **83**, 043837 (2011)
129. C.M. Bender, Rep. Prog. Phys. **70**, 947 (2007)
130. K.G. Makris, R. El-Ganainy, D.N. Christodoulides, Z.H, Musslimani, Phys. Rev. Lett. **100**, 103904 (2008)
131. C.E. Rüter, K.G. Makris, R. El-Ganainy, D.N. Christodoulides, M. Segev, D. Kip, Nat. Phys. **6**, 192 (2010)
132. V.V. Konotop, J. Yang, D. Zezyulin, Rev. Mod. Phys. **88**, 035002 (2016)
133. S.V. Suchkov, A.A. Sukhorukov, J. Huang, S.V. Dmitriev, C. Lee, Y.S. Kivshar, Laser Phot. Rev. **10**, 177 (2016)
134. V.V. Konotop, D.E. Pelinovsky, D.A. Zezyulin, Europhys. Lett. **100**, 56006 (2012)
135. C. Huang, C. Li, L. Dong, Opt. Express **21**, 3917–3925 (2013)
136. D. Leykam, V.V. Konotop, A.S. Desyatnikov, Opt. Lett. **38**, 372 (2013)
137. Z. Chen, J. Liu, S. Fu, Y. Li, B.A. Malomed, Opt. Express **22**, 29679 (2014)
138. D.E. Pelinovsky, D.A. Zezyulin, V.V. Konotop, J. Phys. A: Math. Gen. **47**, 085204 (2014)
139. J. D'Ambroise, P.G. Kevrekidis, B.A. Malomed, Phys. Rev. E **91**, 033207 (2015)
140. M. Wimmer, A. Regensburger, M.A. Miri, C. Bersch, D.N. Christodoulides, U. Peschel, Nat. Commun. **6**, 7782 (2015)

Chapter 5
Universal Relaxation in Quantum Systems

Kazuya Fujimoto and Masahito Ueda

Abstract We review universal relaxation dynamics in quantum systems by focusing on the Kibble–Zurek mechanism, coarsening dynamics, and non-thermal fixed points. The first two subjects have recently been applied to quantum systems to explore universal phenomena. The last one concerns universal scale-invariant relaxation dynamics in isolated quantum systems. We overview the theoretical foundations of the universal properties and recent experiments on these topics on the basis of a quantum spin model, the Gross–Pitaevskii theory, and kinetic equations.

5.1 Introduction

Universality plays a pivotal role in our understanding of various phenomena from a unified point of view. The renormalization-group analysis of critical phase transitions unveils universal aspects behind many different systems close to critical points [1]. This method is also applicable to out-of-equilibrium phenomena described by stochastic models [2]. The universal dynamics of critical phenomena has been explored in such diverse systems as active matter [3, 4] and condensed matter systems [5].

Recent experimental advances in atomic, molecular, and optical (AMO) systems enable one to study non-equilibrium phenomena in quantum systems. A number of theoretical and experimental studies have explored the fundamental aspects of many-body quantum systems [6, 7] such as many-body localization [8], thermal-

K. Fujimoto (✉)
Institute for Advanced Research, Nagoya University, Nagoya 464-8601, Japan
e-mail: kazuya_fujimoto@rover.nuap.nagoya-u.ac.jp

Department of Applied Physics, Nagoya University, Nagoya 464-8603, Japan

M. Ueda
Department of Physics, University of Tokyo, 7-3-1 Hongo,
Bunkyo-ku, Tokyo 113-0033, Japan
e-mail: ueda@cat.phys.s.u-tokyo.ac.jp

RIKEN Center for Emergent Matter Science (CEMS),
Wako, Saitama 351-0198, Japan

P. G. Kevrekidis et al. (eds.), *Emerging Frontiers in Nonlinear Science*, Nonlinear Systems and Complexity 32, https://doi.org/10.1007/978-3-030-44992-6_5

ization in integrable models [9], and the eigenstate thermalization hypothesis [10]. In particular, universal relaxation dynamics has attracted growing interest from the viewpoint of thermalization in isolated quantum systems. A prime example is the defect formation close to a quantum phase transition point, around which universal scaling relations arising from the Kibble–Zurek (KZ) mechanism [11, 12] hold for the number of defects and the correlation length. After the defect formation, the system often exhibits universal relaxation dynamics. This type of relaxation dynamics cannot be captured by linear approximations such as the Bogoliubov analysis, and nonlinear interactions must be taken into account to fully understand the emergence of universal dynamics.

In this chapter, we review the fundamentals and recent developments of the KZ mechanism, coarsening dynamics, and non-thermal fixed points (NTFPs). The first two topics have been studied in quantum spin models and quantum degenerate gases. The last one has emerged as a unified scenario to understand universal relaxation dynamics in isolated quantum systems. While the concept of the NTFP was originally proposed in the context of quantum field theory [13, 14], two experimental signatures of the NTFPs have recently been observed in ultracold atomic gases [15, 16].

5.2 Kibble–Zurek Mechanism

Universal dynamics emerges close to second-order phase transitions, around which the characteristic time and length scales diverge and adiabatic processes break down. In this diabatic regime, defect formation occurs due to the KZ mechanism [11, 12], which predicts a universal scaling relation for the number of defects such as kinks, vortices, and domains. This section reviews the KZ mechanism and discusses its manifestation in quantum systems.

5.2.1 *Universal Scaling Relation of the KZ Mechanism*

We consider quench dynamics across a second-order phase transition, and discuss a universal scaling relation between the quench rate τ_q and the number N_d of defects created in the process. Suppose that a d-dimensional system is characterized by a parameter λ, which can be controlled by an external field, and that the system undergoes a phase transition at λ_c. Then, the deviation from the transition point is measured by $\epsilon = (\lambda - \lambda_c)/\lambda_c$. Suppose that we quench ϵ as $\epsilon = t/\tau_q$ in a time interval $[-t_i, t_f]$, during which the system is rendered to cross the critical point from a disordered state to a symmetry-broken one. Here, $-t_i < 0$ and $t_f > 0$ are the initial and final times of the quench. Then, the local order is expected to grow as schematically illustrated in Fig. 5.1a. Near the critical point, the characteristic time t_c and the length scale ξ_c diverge as

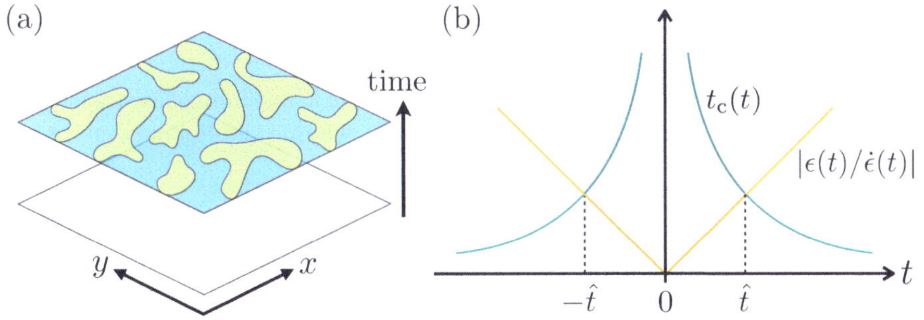

Fig. 5.1 **a** Schematic illustration of nucleation of domains after a second-order phase transition. Different color regions show different types of domains. **b** Schematic graph showing the freeze-out region. The green curves illustrate the divergence of the characteristic time $t_c(t)$ and the yellow lines show how the quench time scale $|\epsilon(t)/\dot{\epsilon}(t)|$ changes as a function of time. In the grey region $[-\hat{t}, \hat{t}]$, where $t_c(t) > |\epsilon(t)/\dot{\epsilon}(t)|$, the order parameter cannot adiabatically follow the change of the parameter

$$t_c = \frac{t_0}{|\epsilon|^{z\nu}}, \quad \xi_c = \frac{\xi_0}{|\epsilon|^{\nu}}, \tag{5.1}$$

where t_0 and ξ_0 are constants, and the critical exponents ν and z characterize the universality class of the phase transition. By comparing $t_c(t)$ with the quench time scale $|\epsilon(t)/\dot{\epsilon}(t)|$, we find that there is a freeze-out region $(-\hat{t} < t < \hat{t})$ in which the local order parameter cannot follow the quench process (see Fig. 5.1b). The freeze-out time $\hat{t}$ can be determined from $t_c(\hat{t}) = |\epsilon(\hat{t})/\dot{\epsilon}(\hat{t})|$, which gives $\hat{t} \propto \tau_q^{z\nu/(1+z\nu)}$. Outside of the region, the local order parameter can follow the change of the parameter $\epsilon(t)$, and thus the number of defects is determined by the physical quantities at $t = \hat{t}$ and obeys the following universal scaling law characterized by the critical exponents [12]:

$$N_d \propto \xi_c(\hat{t})^{D-d} \propto \tau_q^{\frac{(D-d)\nu}{1+z\nu}}, \tag{5.2}$$

where D is the dimension of the defect, e.g. $D = 0$ for a kink in a one-dimensional (1D) system. This is the universal KZ scaling relation.

The KZ mechanism was originally proposed by Kibble in the context of cosmology [11], and Zurek derived a universal scaling relation by applying the idea to the superfluid helium at finite temperature [12]. Recent AMO experiments enable one to study the KZ mechanism in quantum phase transitions, where the original argument has been extended to quantum fields [17–20].

5.2.2 KZ Mechanism in Quantum Systems

A prototypical model for describing the KZ mechanism in quantum phase transitions is the 1D transverse field Ising (TFI) model [17–20], whose Hamiltonian is given by

$$\hat{H}_{\mathrm{TFI}} = -J \sum_{i=1}^{L} \left(\hat{\sigma}_i^z \hat{\sigma}_{i+1}^z + g \hat{\sigma}_i^x \right), \tag{5.3}$$

where L is the number of lattice points, $\hat{\sigma}_i^\alpha$ $(\alpha = x, y, z)$ is the Pauli matrix at site i, $J > 0$ is the coupling constant, and g characterizes the strength of the transverse field. This model can be mapped to a free-fermion model via the Jordan–Wigner transformation [21]:

$$\hat{\sigma}_i^x = 1 - 2\hat{c}_i^\dagger \hat{c}_i, \tag{5.4}$$

$$\hat{\sigma}_i^z = -(\hat{c}_i + \hat{c}_i^\dagger) \prod_{j<i} (1 - 2\hat{c}_j^\dagger \hat{c}_j), \tag{5.5}$$

where $\hat{c}_i$ and $\hat{c}_i^\dagger$ are the fermion annihilation and creation operators. Substituting (5.4) and (5.5) into (5.3), we obtain the free-fermion model:

$$\hat{H}_{\mathrm{TFI}} = -J \sum_{i=1}^{L} \left(\hat{c}_i^\dagger \hat{c}_{i+1} + \hat{c}_{i+1} \hat{c}_i - g \hat{c}_i^\dagger \hat{c}_i + \frac{g}{2} + \mathrm{h.c.} \right), \tag{5.6}$$

where h.c. denotes the Hermitian conjugate of the preceding terms. Diagonalizing this Hamiltonian by means of the Bogoliubov transformation, we find that the TFI model undergoes a quantum phase transition at $g = 1$ between the ferromagnetic phase $(g < 1)$ and the paramagnetic one $(g > 1)$ [21]. Thus, we can test the universal KZ scaling relation for the quantum phase transition by quenching g and counting the number of magnetic kink defects $(D = 0)$.

In [18, 20], the control parameter g is linearly quenched as $g = -t/\tau_{\mathrm{q}}$ $(t \le 0)$, and the number density of excited kinks is given by

$$N_{\mathrm{k}} = \frac{1}{2L} \sum_{i=1}^{L} \langle 1 - \hat{\sigma}_{i+1}^z \hat{\sigma}_i^z \rangle, \tag{5.7}$$

where the brackets denote the quantum average over the state vector at $t = 0$. In [18], the number density is shown to obey

$$\lim_{L\to\infty} N_{\mathrm{k}} = \frac{1}{2\pi} \sqrt{\frac{\hbar}{2J\tau_{\mathrm{q}}}} \propto \tau_{\mathrm{q}}^{-1/2}. \tag{5.8}$$

This scaling relation is consistent with (5.2) because the 1D TFI model has $t_c \propto \Delta^{-1}$ and $\xi_c \propto \Delta^{-1}$ with the energy gap $\Delta \propto |\epsilon|$ and then the exponents ν and z are both equal to unity. Hence, the universal KZ scaling relation holds even in the quantum phase transition.

Quantum degenerate Bose gases offer yet another testbed for the KZ mechanism at zero temperature. Several theoretical works have predicted KZ scaling relations in multicomponent Bose-Einstein condensates (BECs) within the mean-field approximation [22–26]. Here we consider a binary BEC as an example. Let us consider a binary mixture composed of identical atoms with different internal states such as $|F=1, F_z=-1\rangle$ and $|F=2, F_z=0\rangle$ of ^{87}Rb with the hyperfine spin F and its z-component F_z [27, 28]. The binary BEC in a uniform system is well described by two macroscopic wave functions $\phi_m(\boldsymbol{r}, t)$ $(m=1,2)$ obeying the coupled Gross–Pitaevskii equations (GPEs):

$$i\hbar\frac{\partial}{\partial t}\phi_m(\boldsymbol{r}, t) = \frac{\delta E[\phi_m(\boldsymbol{r}, t), \phi_m(\boldsymbol{r}, t)^*]}{\delta\phi_m(\boldsymbol{r}, t)^*}, \tag{5.9}$$

where

$$E[\phi_m, \phi_m^*] = \int\left[\sum_{m=1}^{2}\left(\frac{\hbar^2}{2M}|\nabla\phi_m|^2 + \frac{g_0}{2}|\phi_m|^4\right) + g_1|\phi_1|^2|\phi_2|^2 - \hbar\Omega(\phi_1^*\phi_2 + \phi_2^*\phi_1)\right]\mathrm{d}\boldsymbol{r} \tag{5.10}$$

is the Gross–Pitaevskii energy functional. Here M is the atomic mass, g_0 and g_1 are the intra- and inter-species interaction strengths, and Ω is the Rabi coupling. The ground state is miscible or immiscible depending on the Rabi coupling Ω, and the energy gap of the excitation spectrum in the miscible region is given by $E_{\mathrm{gap}} \propto \sqrt{\Omega(\Omega-\Omega_c)}$ with the critical coupling Ω_c [23, 29]. In this case, the critical exponents are given by $\nu = 1/2$ and $z = 1$, and the KZ theory predicts $N_k \propto \tau_q^{-1/3}$ in the 1D quench dynamics from the miscible to immiscible phases. This scaling has been verified by numerical simulations [25].

5.2.3 Inhomogeneous KZ Mechanism

So far, we have considered the homogeneous systems, where the systems respect continuous or discrete space translation symmetry. Cold atom systems, however, are not usually homogeneous due to inhomogeneous trapping potentials. Thus, to observe the KZ scaling relation, modifications of the exponents due to inhomogeneities must be taken into account [30]. Here, the causality plays a crucial role as pointed out by Kibble and Volovik [31].

We illustrate the inhomogeneous KZ mechanism by using the TFI model with a spatially dependent coupling $J(i) = J(0) - ci^2$ where c is a constant [32]:

$$\hat{H}_{\rm ITFI} = -\sum_{i=1}^{L} J(i)\hat{\sigma}_i^z\hat{\sigma}_{i+1}^z - g(t)\sum_{i=1}^{L}\hat{\sigma}_i^x. \tag{5.11}$$

By assuming the local density approximation, the quantum phase transition occurs at the positions $\pm I_{\rm p}(t)$ satisfying $J(I_{\rm p}) = g(t)$, which means that kink defects can be generated in the region $[-I_{\rm p}(t), I_{\rm p}(t)]$. This implies that the region in the symmetry-broken phase spreads in time. According to the scaling analysis of [30], we can derive $I_{\rm p}(\hat{t}) \propto 1/\tau_{\rm q}$ at the freeze-out time $\hat{t}$ [32]. Combining this causality effect with the KZ mechanism, we estimate the defect number density as

$$N_{\rm k} \propto \frac{2I_{\rm p}(\hat{t})}{\xi_{\rm c}(\hat{t})} \propto \tau_{\rm q}^{-3/2}. \tag{5.12}$$

Thus, the exponent in the inhomogeneous system is different from that in the homogenous case (see (5.8)). This modification is numerically confirmed using the dynamical renormalization group [32]. In the context of trapped binary BECs, such a change in the exponent is numerically discussed in [25].

5.2.4 Experimental Test of the KZ Mechanism

Several AMO systems have provided playgrounds to test the KZ mechanism. A notable system is cold trapped ions, which can simulate quantum spin models by utilizing internal states of ions. Figure 5.2a shows an experimental setup to test the KZ mechanism of the TFI model by using the Landau-Zener crossing, and the observed kink number density after the quench is plotted in Fig. 5.2b. Cui et al. [33] found that the number density agrees well with the KZ scaling relation in (5.8). Cold trapped ions can be used to investigate symmetry breaking in ion crystals, in which defects of zig-zag configurations can be generated by ramping the transverse trapping

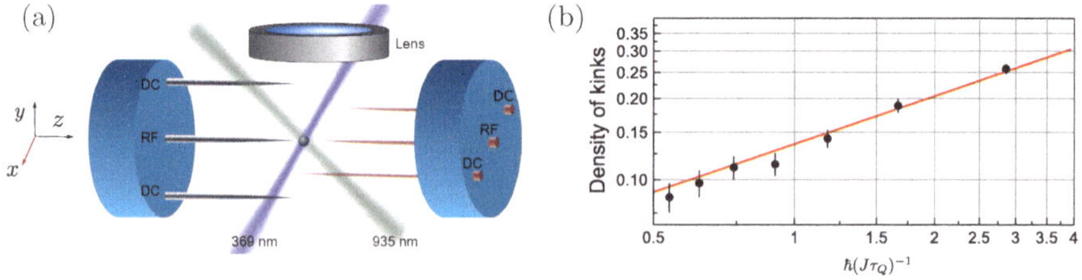

Fig. 5.2 **a** Experimental setup of the KZ mechanism in a system of cold trapped $^{171}Yb^+$ ions [33]. The system serves as a quantum simulator for the KZ scaling relation in (5.8). **b** Growth of the average kink density after the phase transition. The extracted exponent is 0.59 ± 0.03, which is consistent with (5.8). Reproduced from [33]. Creative Commons Attribution License (CC BY) https://creativecommons.org/licenses/by/4.0/

frequency [34–36]. The KZ scaling relation in this structural phase transition was observed [37, 38]. In quantum degenerate gases, on the other hand, multi-component BECs such as binary and spinor Bose gases exhibit second-order phase transitions, which are used to study the KZ mechanism. In [39], the universal scaling relation was observed in a ferromagnetic spinor Bose gas. The KZ mechanism at finite temperature was also reported in [40, 41].

5.3 Coarsening Dynamics

The previous section describes the universal aspects of defect-formation dynamics near critical points. In fact, universal dynamics could emerge far from or even without critical points. Such universal dynamics has been explored in the context of coarsening dynamics in classical systems [42, 43], and recent developments in AMO experiments have made it possible to observe the dynamics in quantum degenerate gases. This section describes the basics of coarsening dynamics on the basis of a few classical models, and then reviews recent related works in ultracold atomic gases.

5.3.1 Coarsening Dynamics in Classical Systems

Coarsening dynamics is a relaxation process accompanied by a scale-invariant growth of the order parameter [42, 43]. A prototypical example is merging of spin domains following the temperature quench in the Ising model, where the z-component of the spin vector characterizes the spin order. Here, by illustrating the concept of dynamical scaling, we discuss the universality classes of typical classical models.

5.3.1.1 Scale Invariance and Universality in Coarsening Dynamics

We denote the local order parameter by $\Phi(\boldsymbol{r}, t)$ and assume that the system respects translational symmetry and rotational symmetry. The growth of the order parameter is quantified by the correlation function $C(r, t)$ defined by

$$C(r, t) = \langle \Phi(\boldsymbol{x} + \boldsymbol{r}, t)\Phi(\boldsymbol{x}, t)\rangle, \tag{5.13}$$

where the brackets denote the ensemble average. If the system shows the scale-invariant relaxation dynamics, the correlation function obeys the dynamical scaling characterized by a certain function $F(x)$:

$$C(r, t) = F(r/L_{\mathrm{c}}(t)), \tag{5.14}$$

Table 5.1 $L_c(t)$ for the Ising and XY models

Model	Non-conserving	Conserving
1D Ising	$\log t$ [44]	$\log t$ [45]
1D XY	$t^{1/4}$ [46]	$t^{1/6}$ [42, 47, 48]
2D and 3D Ising	$t^{1/2}$ [49]	$t^{1/3}$ [49]
2D and 3D XY	$t^{1/2}$ [49–51]	$t^{1/4}$ [49, 52]

Table 5.2 $L_c(t)$ for the binary liquid model [42]

Regime	
Diffusive	$t^{1/3}$
Viscous hydrodynamic	t
Inertial hydrodynamic	$t^{2/3}$

where $L_c(t)$ is a time-dependent characteristic length. The striking feature of (5.14) is that the time dependence of $L_c(t)$ classifies the universality of the coarsening dynamics, depending on the fundamental properties of the system such as the spatial dimension and symmetry of the model. Tables 5.1 and 5.2 list several universality classes of classical coarsening dynamics. In the 1D Ising model, the coarsening dynamics is dominated by interacting domain walls. The strength of interactions decays exponentially as a function of the distance between two domain walls, leading to a logarithmic growth of $L_c(t)$. On the other hand, in the 2D and 3D Ising models, a domain wall has a curvature R and can move by itself with the velocity proportional to R^{-n} with some exponent n, leading to a power law growth of $L_c(t)$.

5.3.1.2 Classical Models in Coarsening Dynamics

To give a general picture of coarsening dynamics, we consider a system characterized by a multicomponent order parameter $\Phi_m(\boldsymbol{r}, t)$ $(m = 1, 2, \ldots)$. The coarsening dynamics is often described by two equations of motion: one is an order-parameter non-conserving model and the other is an order-parameter conserving one. Here, by conserving, we mean that a spatial integral of the order parameter is independent of time:

$$\frac{\mathrm{d}}{\mathrm{d}t} \int \Phi_m(\boldsymbol{r}, t) \mathrm{d}\boldsymbol{r} = 0. \tag{5.15}$$

In the non-conserving model, the order parameter $\Phi_m(\boldsymbol{r}, t)$ obeys the Ginzburg–Landau (GL)-type equation [42, 43]:

$$\frac{\partial}{\partial t} \Phi_m(\boldsymbol{r}, t) = -\frac{\delta F[\boldsymbol{\Phi}(\boldsymbol{r}, t)]}{\delta \Phi_m(\boldsymbol{r}, t)}, \tag{5.16}$$

where $F[\boldsymbol{\Phi}(\boldsymbol{r}, t)]$ is a model-dependent free-energy functional. In the Ising and XY models, the functionals are given by

$$F_{\text{Ising}}[\Phi] = \int \left(|\nabla \Phi|^2 + a(\Phi^2 - 1)^2 \right) \mathrm{d}\boldsymbol{r}, \tag{5.17}$$

$$F_{\text{XY}}[\boldsymbol{\Phi}] = \int \left(|\nabla \boldsymbol{\Phi}|^2 + a(|\boldsymbol{\Phi}|^2 - 1)^2 \right) \mathrm{d}\boldsymbol{r}, \tag{5.18}$$

where a is a coupling constant. Here, $\boldsymbol{\Phi} = (\Phi_1, \Phi_2)$ is the vector order parameter consisting of two scalar functions Φ_1 and Φ_2. We note that in the GL model the energy decreases in time and the integral of $\Phi(\boldsymbol{r}, t)$ is not conserved:

$$\frac{\mathrm{d}}{\mathrm{d}t} F[\boldsymbol{\Phi}(\boldsymbol{r}, t)] = - \int \sum_m \left(\frac{\delta F[\boldsymbol{\Phi}(\boldsymbol{r}, t)]}{\delta \Phi_m(\boldsymbol{r}, t)} \right)^2 \mathrm{d}\boldsymbol{r} \leq 0, \tag{5.19}$$

$$\frac{\mathrm{d}}{\mathrm{d}t} \int \Phi_m(\boldsymbol{r}, t) d\boldsymbol{r} = - \int \frac{\delta F[\boldsymbol{\Phi}(\boldsymbol{r}, t)]}{\delta \Phi_m(\boldsymbol{r}, t)} \mathrm{d}\boldsymbol{r}. \tag{5.20}$$

Because of (5.20), the GL equation is a non-conserving model.

An order-parameter conserving system is modeled by

$$\frac{\partial}{\partial t} \Phi_m(\boldsymbol{r}, t) = \nabla^2 \frac{\delta F[\boldsymbol{\Phi}(\boldsymbol{r}, t)]}{\delta \Phi_m(\boldsymbol{r}, t)}, \tag{5.21}$$

which is known as the Cahn–Hilliard (CH)-type equation [42, 43]. In this case, we obtain

$$\frac{\mathrm{d}}{\mathrm{d}t} F[\boldsymbol{\Phi}(\boldsymbol{r}, t)] = - \int \sum_m \left(\nabla \frac{\delta F[\boldsymbol{\Phi}(\boldsymbol{r}, t)]}{\delta \Phi_m(\boldsymbol{r}, t)} \right)^2 \mathrm{d}\boldsymbol{r} \leq 0, \tag{5.22}$$

$$\frac{\mathrm{d}}{\mathrm{d}t} \int \Phi_m(\boldsymbol{r}, t) \mathrm{d}\boldsymbol{r} = 0. \tag{5.23}$$

Thus, the integral of $\Phi_m(\boldsymbol{r}, t)$ is constant, while the energy decreases in time. The existence of the conserved quantity drastically changes the relaxation dynamics and lead to different universality classes. We summarize the universality classes for the Ising and XY models in Table 5.1, which shows that the exponents in the conserving models are smaller than those in the non-conserving one except for the 1D Ising model.

A binary liquid model has also been investigated and shows quite different dynamics. Here, we summarize the time dependence of $L_c(t)$ in Table 5.2, which has three exponents depending on which terms are dominant.

5.3.2 Coarsening Dynamics in Quantum Degenerate Systems

We now review coarsening dynamics in quantum degenerate systems. Many theoretical works focus on segregating multi-component BECs and discuss the universality classes in comparison with the classical models [53–62]. Here, we discuss the coarsening dynamics of a ferromagnetic spin-1 BEC.

5.3.2.1 Coarsening Dynamics in a Spin-1 BEC

The spin-1 BEC considered here is well described by three macroscopic wave functions ψ_m with the magnetic quantum number $m = 1, 0, -1$ [63, 64]. Within the mean-field approximation, we can derive the GPE given by

$$i\hbar\frac{\partial}{\partial t}\psi_m(\boldsymbol{r},t) = \frac{\delta E[\psi_m(\boldsymbol{r},t), \psi_m(\boldsymbol{r},t)^*]}{\delta\psi_m(\boldsymbol{r},t)^*}, \tag{5.24}$$

where

$$E[\psi_m, \psi_m^*] = \int\left[\sum_{m=-1}^{1}\left(\frac{\hbar^2}{2M}|\nabla\psi_m|^2 + qm^2|\psi_m|^2\right) + \frac{g_{\mathrm{d}}}{2}\rho^2 + \frac{g_{\mathrm{s}}}{2}\boldsymbol{F}^2\right]\mathrm{d}\boldsymbol{r} \tag{5.25}$$

is the spin-1 Gross–Pitaevskii energy functional. Here, q is the coefficient of the quadratic Zeeman energy, g_{d} is the spin-independent interaction strength, g_{s} is the spin-dependent one, $\rho = \sum_m |\psi_m|^2$ is the total particle-number density, and $\boldsymbol{F} = \sum_{m,n}\psi_m^*(\hat{\boldsymbol{F}})_{mn}\psi_n$ is the spin density vector with the spin-1 matrix vector $\hat{\boldsymbol{F}}$. The spin-dependent interaction is assumed to be ferromagnetic ($g_{\mathrm{s}} < 0$). Then, the ground state depends on the value of q as shown in Fig. 5.3. The polar state $\psi_0 = \sqrt{\rho_0}$ is nonmagnetic, while the ferromagnetic state is fully magnetized ($\psi_1 = \sqrt{\rho_0}$ or $\psi_{-1} =$

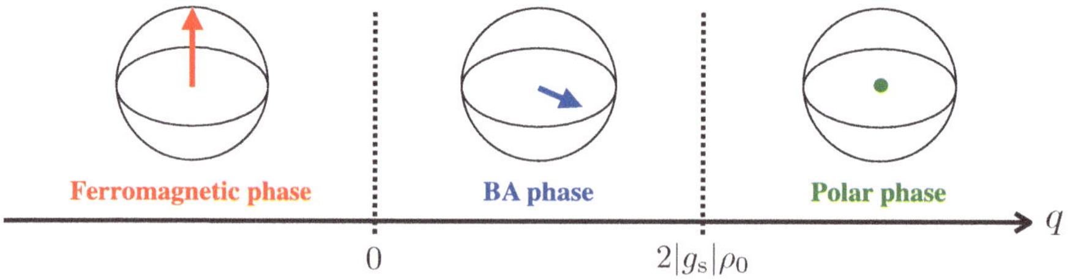

Fig. 5.3 Phase diagram of the ground states of the spin-1 ferromagnetic BEC as a function of the quadratic Zeeman energy q. The ground states are obtained with the mean-field approximation [63, 64], where ρ_0 denotes the uniform particle-number density. The ferromagnetic state is fully magnetized, the broken-axisymmetry state is transversely magnetized, and the polar state is nonmagnetic

$\sqrt{\rho_0}$). Here, ρ_0 is the uniform particle-number density. The broken-axisymmetry state has only transverse magnetization. Thus, through the quench from the polar phase to the ferromagnetic phase, we can investigate the coarsening dynamics similar to the Ising-type dynamics in ultracold atomic gases.

We might expect that the universality class for the Ising-type spin-1 BEC should belong to the Ising class in Table 5.1. However, this needs to be carefully examined due to the conservation law respected by (5.24). In contrast to the classical models discussed above, the GPE conserves the total energy:

$$\frac{\mathrm{d}}{\mathrm{d}t}E[\psi_m, \psi_m^*] = \int \sum_{m=-1}^{1} \left(\frac{\delta E}{\delta \psi_m} \frac{\partial \psi_m}{\partial t} + \frac{\delta E}{\delta \psi_m^*} \frac{\partial \psi_m^*}{\partial t} \right) \mathrm{d}\boldsymbol{r} = 0. \tag{5.26}$$

Thus, the domain dynamics should be different from that in the dissipative classical Ising models.

To consider the universality class in more detail, we follow Williamson and Blakie [55, 56] to rewrite (5.24) in terms of the hydrodynamic description [65–67]. First, we derive the continuity equation for the total particle-number density ρ by using (5.24):

$$\frac{\partial}{\partial t}\rho + \nabla \cdot (\rho \boldsymbol{v}) = 0, \tag{5.27}$$

where the velocity $\boldsymbol{v}$ is given by

$$\boldsymbol{v} = \frac{\hbar}{2Mi\rho} \sum_{m=-1}^{1} \left(\psi_m^* \nabla \psi_m - \psi_m \nabla \psi_m^* \right). \tag{5.28}$$

Assuming the fully magnetized state and the uniform density, we can derive

$$\frac{\partial}{\partial t}\boldsymbol{v} + (\boldsymbol{v} \cdot \nabla)\boldsymbol{v} = -\frac{1}{M\rho_0}\nabla P, \tag{5.29}$$

where the pressure is $P = g_\mathrm{s} F_z^2/2$. Here, we neglect the third derivatives of the spin density vectors because in the late stage of the dynamics the large-scale structures become dominant and the terms with high-order derivatives are expected to be negligible. If the pressure satisfies the Young-Laplace equation, the pressure difference across a domain wall is inversely proportional to the domain-wall curvature R. Using this fact and applying the scale transformations ($\boldsymbol{r} \to s^p \boldsymbol{r}, t \to st$) with a constant s and the exponent p to (5.29), we find that the equation remains invariant at $p = 2/3$. This implies that the correlation length obeys $L_\mathrm{c}(t) \propto t^{2/3}$, which agrees with the time evolution in the inertial hydrodynamic region of the binary liquid as shown in Table 5.2. By comparing the derivation given here with that in the binary liquid [42] we find that the coarsening mechanism in the Ising-type BEC is almost identical to the classical binary liquid in the hydrodynamic region. Williamson and Blakie numeri-

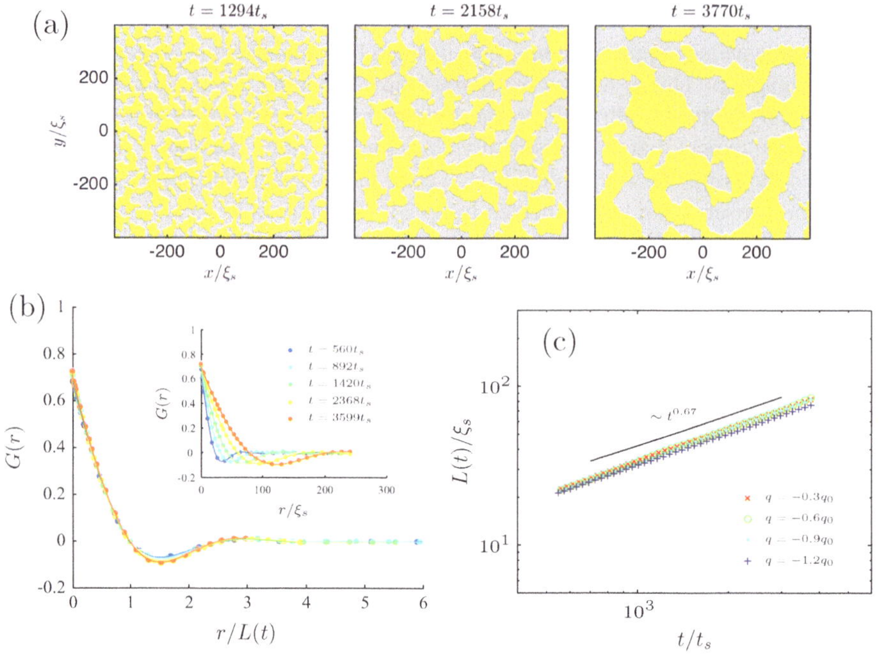

Fig. 5.4 Numerical simulations of the ferromagnetic spin-1 Gross–Pitaevskii equation in two dimensions [56]. **a** Spatial spin distribution $F_z(\boldsymbol{r}, t)$ at three different times. **b** Time evolution of the spin correlation function $G(r, t)$. The inset shows the raw numerical data. In the main panel the abscissa is normalized by the correlation length $L(t)$. All data points collapse into a single curve, indicating the dynamical scaling. **c** Time evolution of $L(t)$ for several different values of quadratic Zeeman coupling q. All data points again collapse into a single line which indicates that the correlation length obeys the 2/3 power law. Reproduced from [56]. ©2016 by the American Physical Society

cally studied the coarsening dynamics as shown in Fig. 5.4a, and found the dynamical scaling of the spin correlation function $G(r, t) = \langle F_z(\boldsymbol{r} + \boldsymbol{x}, t) F_z(\boldsymbol{x}, t)\rangle / \rho_0^2$ and the 2/3 power law (see Figs. 5.4b and c). Recently, several theoretical works including scalar and binary BECs have found growth laws other than the 2/3 power law [53, 57–62].

5.3.2.2 Experiments in BECs

Ultracold atomic gases have a relatively long lifetime, and thus enable one to investigate the coarsening dynamics [68, 69]. Guzman et al. [68] studied the relaxation dynamics in a ferromagnetic spinor Bose gas, and observed the growth of a large-scale spin texture. In a binary BEC, De et al. [69] observed the generation of spin domains by quenching the Zeeman energy shift, and the number of domains is found

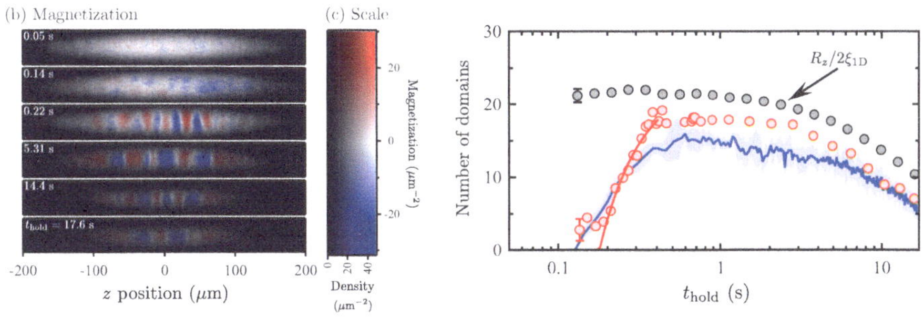

Fig. 5.5 Experimental observation of the spin-domain relaxation dynamics of binary BECs [69]. (Left) Time evolution of the magnetization. Spin domains are generated by the quench of the Zeeman energy shift. (Right) Time evolution of the number of spin domains. The number decreases over time, which implies the coarsening and growth of the spin order. The red circles show the experimental data, and the blue solid curve shows the result of the numerical simulation based on the stochastic GPE. The gray circles show $R_z/2\xi_{\rm 1D}$ with the system size $2R_z$ and the averaged domain size $4\xi_{\rm 1D}$. Reproduced from [69]. Creative Commons Attribution License (CC BY) https://creativecommons.org/licenses/by/3.0/

to decrease in time as shown in Fig. 5.5. The experimental result is well described by the stochastic Gross–Pitaevskii model; however, the universal aspects of the coarsening dynamics, e.g. dynamical scaling, remains to be investigated.

5.4 Non-thermal Fixed Points

Emergence of universal relaxation dynamics does not necessarily require second-order phase transitions as described in Sect. 5.3. Then, it is natural to ask how we can understand the universal aspects from a unified point of view. A notable scenario is a non-thermal fixed point (NTFP), which was originally proposed in the context of quantum field theory [13, 14]. Recently, it has been applied to ultracold atomic gases, and several theoretical and experimental studies have uncovered scale-invariant relaxations related to the NTFP in turbulent decay [70–72] and coarsening dynamics [73–75].

5.4.1 Concept of NTFPs

The NTFP concerns the scale-invariant relaxation dynamics that emerges in isolated quantum systems [13, 14]. Let us consider quench dynamics characterized by a two-point correlation function $G(\boldsymbol{k}, t)$ such as the momentum distribution. When the scale invariance emerges, the correlation function exhibits the dynamical scaling:

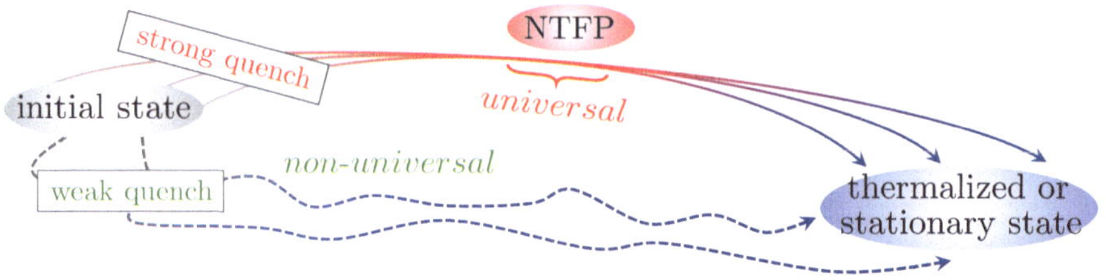

Fig. 5.6 Schematic illustration of universal and non-universal relaxations. When the system is strongly quenched, the spatio-temporal scale invariance emerges in the relaxation process before a thermalized or stationary state is reached. In the course of the dynamics, the system passes through a particular scale-invariant point, which is referred to as a non-thermal fixed point (NTFP). A weak quench may not lead to such universal relaxation dynamics. Reproduced from [74]

$$G(\boldsymbol{k}, t) = s^{\alpha/\beta} G(s\boldsymbol{k}, s^{-1/\beta} t), \tag{5.30}$$

where s is a scaling parameter and the exponents α and β characterize the universality of the dynamics. Substituting $s = t^{\beta}$ into (5.30), we find that the correlation function can be expressed by a single scaling function $f(x)$ as

$$G(\boldsymbol{k}, t) = t^{\alpha} f(t^{\beta}\boldsymbol{k}), \tag{5.31}$$

where $f(t^{\beta}\boldsymbol{k}) = G(t^{\beta}\boldsymbol{k}, 1)$. When the control parameter is strongly quenched, the system often shows the dynamical scaling before reaching a stationary state. This situation is schematically illustrated in Fig. 5.6, where the system passes through a NTFP en route to a thermalized or stationary state. On the other hand, when the initial quench is not strong enough, the system may not show such universal dynamics.

The universal relaxation dynamics was originally discussed using the kinetic equation for the momentum distribution $n(\boldsymbol{k}, t)$ in quantum field theory, and subsequent works argued that NTFPs manifest themselves in the relaxation dynamics in ultracold atomic gases [70–75]. In the following, we discuss some specified examples.

5.4.2 NTFPs in the Scalar Gross–Pitaevskii Model

We consider a one-component BEC, which is well described by a single macroscopic wavefunction $\psi(\boldsymbol{r}, t)$ obeying the scalar GPE [27, 28]:

$$i\hbar\frac{\partial}{\partial t}\psi(\boldsymbol{r}, t) = -\frac{\hbar^2}{2M}\nabla^2\psi(\boldsymbol{r}, t) + U|\psi(\boldsymbol{r}, t)|^2\psi(\boldsymbol{r}, t), \tag{5.32}$$

where U characterizes the strength of interaction. Numerical simulations of (5.32) with random initial noises [76, 77] can show dynamical scaling of the momentum distribution defined by

$$n(\boldsymbol{k},t) = |\bar{\psi}(\boldsymbol{k},t)|^2, \tag{5.33}$$

where $\bar{\psi}(\boldsymbol{k},t)$ is the Fourier component of the macroscopic wavefunction. Here, we discuss the universal relaxation dynamics in two dimensions [73].

We note that the conservation law respected by (5.32) leads to a scaling relation for the exponents α and β unique to the NTFP. The GPE conserves the total particle number $\int n(\boldsymbol{k},t)\mathrm{d}\boldsymbol{k}$. Substituting the scaling form $n(\boldsymbol{k},t) = t^{\alpha} f(t^{\beta}\boldsymbol{k})$ into the conservation law, we find

$$\int n(\boldsymbol{k},t)\mathrm{d}\boldsymbol{k} = \int t^{\alpha} f(t^{\beta}\boldsymbol{k})\mathrm{d}\boldsymbol{k} \propto t^{\alpha-d\beta}. \tag{5.34}$$

Requiring that the total particle number be conserved, we find the scaling relation

$$\alpha = d\beta \tag{5.35}$$

with the spatial dimension d.

By numerically solving (5.32) for highly excited initial states with many quantized vortices, Karl and Gasenzer [73] have demonstrated that the exponents α and β have two different sets of values depending on the initial vortex number:

$$(\alpha,\beta) = \begin{cases} (1.10, 0.56) & \text{(Gaussian NTFP);} \\ (0.402, 0.193) & \text{(anomalous NTFP),} \end{cases} \tag{5.36}$$

both of which are consistent with the scaling relation (5.35). The former is called the Gaussian NTFP because the free-particle dispersion relation $\omega \propto k^2$ gives $\beta = 1/2$ as in dynamic critical phenomena [2]. Figure 5.7 shows the numerical results for the anomalous NTFP, in which the initial state has many quantized vortices with the winding number of six. Then, the time evolution of the density distribution in the left panel of Fig. 5.7 shows the relaxation of density fluctuations. The momentum distribution exhibits the dynamical scaling as shown in the right panel of Fig. 5.7, where the exponents (5.36) of the anomalous NTFP can make all data points collapse to a single curve. This is a hallmark of the NTFP. For initial states with more quantized vortices, the power-law exponents are confirmed to approach the Gaussian NTFP.

5.4.3 NTFPs and Kinetic Equations

The idea of NTFPs can be understood on the basis of the kinetic theory [13, 14, 80–82]. We consider a quantum model such as the O(N) model, and derive an effective kinetic equation for the momentum distribution $N_{\mathrm{kin}}(\boldsymbol{k},t)$ of quasiparticles by employing quantum field theoretic methods:

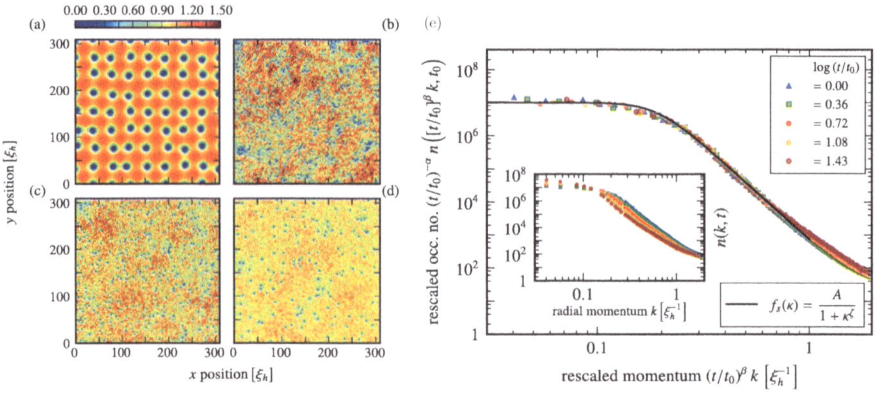

Fig. 5.7 Numerical study on the NTFP in the two-dimensional Bose gas [73]. Figures **a–d** show the time evolution of the density distributions calculated from the GPE. The initial state (a) has many quantized vortices with the winding number of six. Such highly quantized vortices suffer dynamical instabilities and develop complicated density distributions in Figs. **b–d**. Figure **e** shows the dynamical scaling of the momentum distribution at different times. The inset shows the raw numerical data, and in the main panel the abscissa and the ordinate are normalized by the corresponding power-law functions of time, from which the exponents $\alpha = 0.402 \pm 0.05$ and $\beta = 0.193 \pm 0.05$ are obtained. Reproduced from [73].

$$\frac{\partial}{\partial t} N_{\text{kin}}(\boldsymbol{k}, t) = I[N_{\text{kin}}](\boldsymbol{k}, t), \tag{5.37}$$

where $I[N_{\text{kin}}](\boldsymbol{k}, t)$ is the collision integral that depends on the original Hamiltonian. The investigation of (5.37) uncovers universal properties of the non-equilibrium dynamics. Note that a similar analysis has been employed in the context of weak wave turbulence [78, 79] to show that several steady turbulent states are characterized by the power-law behavior of $N_{\text{kin}}(\boldsymbol{k})$.

Solving the kinetic equation (5.37) under the assumption of the dynamical scaling (5.31), we can seek for the power-law exponents α and β. The obtained time- and scale-invariant state is originally identified with the NTFP of (5.37). We follow [80] to assume that the momentum distribution is isotropic and expressed by

$$N_{\text{kin}}(k, t) = t^\alpha h(t^\beta k) \tag{5.38}$$

with $h(t^\beta k) = N_{\text{kin}}(t^\beta k, 1)$. We assume that the collision integral has the following scaling property:

$$I[N_{\text{kin}}](k, t) = s^{-\mu} I[N_{\text{kin}}](sk, s^{-1/\beta} t), \tag{5.39}$$

which is a natural extension of (5.30) and characterized by the exponent μ depending on theoretical models and approximations. In a weakly interacting bosonic system, a perturbative calculation gives $\mu = 2d - z - 3\alpha/\beta$ with the dynamical exponent z, which is defined by $\epsilon(sk) = s^z\epsilon(k)$ with the energy spectrum $\epsilon(k)$ for quasiparticles. Substituting $s = t^\beta$ into (5.39), we have

$$I[N_{\mathrm{kin}}](k,t) = t^{-\beta\mu} I[h](t^\beta k). \tag{5.40}$$

Substituting (5.38) and (5.40) into (5.37), we obtain

$$t^{\alpha-1}\left(\alpha + \beta q\frac{\partial}{\partial q}\right)h(q)\Big|_{q=t^\beta k} = t^{-\beta\mu} I[h](t^\beta k). \tag{5.41}$$

Requiring the scale invariance of (5.41), we obtain the scaling relation:

$$\alpha = 1 - \beta\mu. \tag{5.42}$$

Furthermore, the energy conservation and the particle-number conservation give other scaling relations. When the system shows a particle turbulent cascade, the particle-number conservation gives $\alpha = \beta d$. On the other hand, when an energy turbulent cascade emerges, the energy conservation gives $\alpha = \beta(d + z)$. Using these relations and (5.42), we can determine the exponents α and β. While we have considered the unidirectional self-similar transport, in which either particle turbulent cascade or energy turbulent cascade occurs, a bi-directional self-similar transport is also possible. Details of non-relativistic bosonic models are described in [80–82].

5.4.4 Experimental Investigation of NTFPs

Universal relaxation dynamics passing through a NTFP has recently been observed in a quenched spin-1 ferromagnetic Bose gas in the 1D system [15]. Employing the quench protocol from the polar phase to the broken-axisymmetry phase (see Fig. 5.3), Prüfer et al. observed scale-invariant relaxation dynamics of the transverse component of the spin density vector after an initial growth of the transverse spins, which is triggered by quench-induced instability. To investigate the relaxation dynamics from the viewpoint of the NTFP, they measured the spin correlation function given by

$$f_\theta(k,t) = \langle|\theta(k,t)|^2\rangle, \tag{5.43}$$

where $\theta(k,t)$ is the Fourier transform of the azimuth angle of the spin density vector. They find that the correlation function showed the dynamical scaling $f_\theta(k,t) = s^{\alpha/\beta} f_\theta(sk, s^{-1/\beta}t)$ over 5 sec, which is the hallmark of the NTFP.

Erne et al. [16] reported a similar dynamical scaling in the 1D Bose gas, in which a sudden temperature quench by evaporative cooling promotes relaxation. They observed the dynamical scaling of the momentum distribution $n(k,t)$. Using a random soliton model [83], they analyzed the measured distribution and the scaling exponents.

5.5 Concluding Remarks

We have reviewed the universal relaxation in several quantum systems. First, we have discussed the KZ mechanism characterized by second-order phase transitions by using the TFI model and the GPE. Second, we have addressed the universal aspects of the coarsening dynamics that emerges even without or far from the critical point and the universality classes in dissipative classical models and the spin-1 Gross–Pitaevskii model. Finally, we have reviewed recent studies of the NTFP on the basis of the scalar GPE, the kinetic equation, and the experimental observations. In these non-equilibrium phenomena, interactions between quasiparticles in a kinetic equation and topological excitations in coarsening dynamics play key roles, and thus nonlinear effects are essential in the emergence of the universal relaxation dynamics.

In contrast to the vast amount of research performed on classical relaxation processes, the study on the relaxation and coarsening dynamics in the quantum regime has only recently attracted much attention due to the experimental progress in ultracold atomic gases. While we have reviewed some of the remarkable achievements in this direction, many fundamental problems remain to be investigated. For example, what happens if the system relaxes to a turbulent state? Will the system be attracted to a special NTFP? What happens if excitations are non-Abelian? Such a case can be studied for biaxial nematic and cyclic phases of spin-2 BECs. It is also of great interest to explore the cases in which the final relaxed state is topologically nontrivial (e.g. topological order) or non-thermal (e.g. many-body localization). What is the role of the Lieb-Robinson bound and the entanglement entropy for these exotic cases? With these and other interesting questions, we are led to the conclusion that the field of quantum relaxation and coarsening dynamics is still in its infancy and the exploration of this new research arena will contribute to the deepening and widening of our fundamental understanding of quantum, many-body and statistical physics.

References

1. J. Cardy, *Scaling and Renormalization in Statistical Physics* (Cambridge University Press, Cambridge, 1996)
2. P.C. Hohenberg, B.I. Halperin, Rev. Mod. Phys. **49**, 435 (1977)
3. J. Toner, Y. Tu, Phys. Rev. Lett. **75**, 4326 (1995)
4. J. Toner, Y. Tu, Phys. Rev. E **58**, 4828 (1998)
5. U. Täuber, *Critical Dynamics* (Cambridge University Press, Cambridge, 2014)
6. A. Polkovnikov, K. Sengupta, A. Silva, M. Vengalattore, Rev. Mod. Phys. **83**, 863 (2011)
7. J. Eisert, M. Friesdorf, C. Gogolin, Nat. Phys. **11**, 124 (2015)
8. R. Nandkishore, D.A. Huse, Annu. Rev. Condens. Matter Phys. **6**, 15 (2015)
9. T. Kinoshita, T. Wenger, D.S. Weiss, Nature **440**, 900 (2006)

10. L. D'Alessio, Y. Kafri, A. Polkovnikov, M. Rigol, Adv. Phys. **65**, 239 (2016)
11. T.W.B. Kibble, J. Phys. A: Math. Gen. **9**, 1387 (1976)
12. W.H. Zurek, Nature **317**, 505 (1985)
13. J. Berges, A. Rothkopf, J. Schmidt, Phys. Rev. Lett. **101**, 041603 (2008)
14. J. Berges, G. Hoffmeister, Nucl. Phys. B **813**, 383 (2009)
15. M. Prüfer, P. Kunkel, H. Strobel, S. Lannig, D. Linnemann, C. Schmied, J. Berges, T. Gasenzer, M.K. Oberthaler, Nature **563**, 217 (2018)
16. S. Erne, R. Bücker, T. Gasenzer, J. Berges, J. Schmiedmayer, Nature **563**, 225 (2018)
17. W.H. Zurek, U. Dorner, P. Zoller, Phys. Rev. Lett. **95**, 105701 (2005)
18. J. Dziarmaga, Phys. Rev. Lett. **95**, 245701 (2005)
19. A. Polkovnikov, Phys. Rev. B **72**, 161201(R) (2005)
20. A. del Campo, Phys. Rev. Lett. **121**, 200601 (2018)
21. S. Sachdev, *Quantum Phase Transitions* (Cambridge University Press, Cambridge, 2011)
22. H. Saito, Y. Kawaguchi, M. Ueda, Phys. Rev. A **76**, 043613 (2007)
23. C. Lee, Phys. Rev. Lett. **102**, 070401 (2009)
24. B. Damski, W.H. Zurek, New J. Phys. **11**, 063014 (2009)
25. J. Sabbatini, W.H. Zurek, M.J. Davis, Phys. Rev. Lett. **107**, 230402 (2011)
26. T. Świsłocki, E. Witkowska, J. Dziarmaga, M. Matuszewski, Phys. Rev. Lett. **110**, 045303 (2013)
27. C.J. Pethick, H. Smith, *Bose-Einstein Condensation in Dilute Gases* (Cambridge University Press, Cambridge, 2008)
28. L.P. Pitaevskii, S. Stringari, *Bose-Einstein Condensation and Superfluidity* (Oxford University Press, Oxford, 2016)
29. P. Tommasini, E.J.V. de Passos, A.F.R. de Toledo Piza, M.S. Hussein, E. Timmermans, Phys. Rev. A **67**, 023606 (2003)
30. A. del Campo, T.W.B. Kibble, W.H. Zurek, J. Phys.: Condens. Matter **25**, 404210 (2013)
31. T.W.B. Kibble, G.E. Volovik, JETP Lett. **65**, 102 (1997)
32. F.J. Gómez-Ruiz, A. del Campo, Phys. Rev. Lett. **122**, 080604 (2019)
33. J. Cui, Y. Huang, Z. Wang, D. Cao, J. Wang, W. Lv, L. Luo, A. del Campo, Y. Han, C. Li, G. Guo, Sci. Rep. **6**, 33381 (2016)
34. A. del Campo, G. De Chiara, G. Morigi, M.B. Plenio, A. Retzker, Phys. Rev. Lett. **105**, 075701 (2010)
35. G.D. Chiara, A. del Campo, G. Morigi, M.B. Plenio, A. Retzker, New J. Phys. **12**, 115003 (2010)
36. M. Mielenz, J. Brox, S. Kahra, G. Leschhorn, M. Albert, T. Schaetz, H. Landa, B. Reznik, Phys. Rev. Lett. **110**, 133004 (2013)
37. J. Roßnagel, G. Jacob, C. Degünther, S.T. Dawkins, U.G. Poschinger, R. Nigmatullin, A. Retzker, M.B. Plenio, F. Schmidt-Kaler, K. Singer, Nat. Commun. **4**, 2290 (2013)
38. K. Pyka, J. Keller, H.L. Partner, R. Nigmatullin, T. Burgermeister, D.M. Meier, K. Kuhlmann, A. Retzker, M.B. Plenio, W.H. Zurek, A. del Campo, T.E. Mehlstäubler, Nat. Commun. **4**, 2291 (2013)
39. M. Anquez, B.A. Robbins, H.M. Bharath, M. Boguslawski, T.M. Hoang, M.S. Chapman, Phys. Rev. Lett. **116**, 155301 (2016)
40. G. Lamporesi, S. Donadello, S. Serafini, F. Dalfovo, G. Ferrari, Nat. Phys. **9**, 656 (2013)
41. N. Navon, A.L. Gaunt, R.P. Smith, Z. Hadzibabic, Science **347**, 167 (2015)
42. A.J. Bray, Adv. Phys. **43**, 357 (1994)
43. A. Onuki, *Phase Transition Dynamics* (Cambridge University Press, Cambridge, 2002)
44. K. Kawasaki, T. Nagai, Physica A **121**, 175 (1983)
45. T. Kawakatsu, T. Munakata, Prog. Theor. Phys. **74**, 11 (1985)
46. T.J. Newman, A.J. Bray, M.A. Moore, Phys. Rev. B **42**, 4514 (1990)
47. M. Mondello, N. Goldenfeld, Phys. Rev. E **47**, 2384 (1993)
48. M. Rao, A. Chakrabarti, Phys. Rev. E **49**, 3727 (1994)
49. A.D. Rutenberg, A.J. Bray, Phys. Rev. E **51**, 5499 (1995)

50. A.N. Pargellis, P. Finn, J.W. Goodby, P. Panizza, B. Yurke, P.E. Cladis, Phys. Rev. A **46**, 7765 (1992)
51. B. Yurke, A.N. Pargellis, T. Kovacs, D.A. Huse, Phys. Rev. E **47**, 1525 (1993)
52. S. Puri, A.J. Bray, F. Rojas, Phys. Rev. E **52**, 4699 (1995)
53. K. Damle, S.N. Majumdar, S. Sachdev, Phys. Rev. A **54**, 5037 (1996)
54. J. Hofmann, S.S. Natu, S.D. Sarma, Phys. Rev. Lett. **113**, 095702 (2014)
55. L.A. Williamson, P.B. Blakie, Phys. Rev. Lett. **116**, 025301 (2016)
56. L.A. Williamson, P.B. Blakie, Phys. Rev. A **94**, 023608 (2016)
57. K. Kudo, Y. Kawaguchi, Phys. Rev. A **88**, 013630 (2013)
58. K. Kudo, Y. Kawaguchi, Phys. Rev. A **91**, 053609 (2015)
59. A. Bourges, P.B. Blakie, Phys. Rev. A **95**, 023616 (2017)
60. L.M. Symes, P.B. Blakie, Phys. Rev. A **96**, 013602 (2017)
61. L.A. Williamson, P.B. Blakie, Phys. Rev. Lett. **119**, 255301 (2017)
62. K. Fujimoto, R. Hamazaki, M. Ueda, Phys. Rev. Lett. **120**, 073002 (2018)
63. K. Kawaguchi, M. Ueda, Phys. Rep. **520**, 253 (2012)
64. D.M. Stamper-Kurn, M. Ueda, Rev. Mod. Phys. **85**, 1191 (2013)
65. A. Lamacraft, Phys. Rev. A **77**, 063622 (2008)
66. K. Kudo, Y. Kawaguchi, Phys. Rev. A **82**, 053614 (2010)
67. E. Yukawa, M. Ueda, Phys. Rev. A **86**, 063614 (2012)
68. J. Guzman, G.-B. Jo, A.N. Wenz, K.W. Murch, C.K. Thomas, D.M. Stamper-Kurn, Phys. Rev. A **84**, 063625 (2011)
69. S. De, D.L. Campbell, R.M. Price, A. Putra, B.M. Anderson, I.B. Spielman, Phys. Rev. A **89**, 033631 (2014)
70. B. Nowak, D. Sexty, T. Gasenzer, Phys. Rev. B **84**, 020506(R) (2011)
71. J. Schole, B. Nowak, T. Gasenzer, Phys. Rev. A **86**, 013624 (2012)
72. M. Karl, B. Nowak, T. Gasenzer, Phys. Rev. A **88**, 063615 (2013)
73. M. Karl, T. Gasenzer, New J. Phys. **19**, 093014 (2017)
74. K. Fujimoto, R. Hamazaki, M. Ueda, Phys. Rev. Lett. **122**, 173001 (2019)
75. C.-M. Schmied, M. Prüfer, M.K. Oberthaler, T. Gasenzer, Phys. Rev. A **99**, 033611 (2019)
76. P.B. Blakie, A.S. Bradley, M.J. Davis, R.J. Ballagh, C.W. Gardiner, Adv. Phys. **57**, 363 (2008)
77. A. Polkovnikov, Ann. Phys. **325**, 1790 (2010)
78. V.E. Zakharov, V.S. L'vov, G. Falkovich, *Kolmogorov Spectra of Turbulence I: Wave Turbulence* (Springer, Berlin, 1992)
79. S. Nazarenko, *Wave Turbulence*. Lecture Notes in Physics, vol. 825 (Springer, Heidelberg, 2011)
80. I. Chantesana, A.P. Orioli, T. Gasenzer, Phys. Rev. A **99**, 043620 (2019)
81. C. Scheppach, J. Berges, T. Gasenzer, Phys. Rev. A **81**, 033611 (2010)
82. A.N. Mikheev, C. Schmied, T. Gasenzer, Phys. Rev. A **99**, 063622 (2019)
83. M. Schmidt, S. Erne, B. Nowak, D. Sexty, T. Gasenzer, New J. Phys. **14**, 075005 (2012)

Chapter 6
Nonlinearity + Networks: A 2020 Vision

Mason A. Porter

Abstract I briefly survey several fascinating topics in networks and nonlinearity. I highlight a few methods and ideas, including several of personal interest, that I anticipate to be especially important during the next several years. These topics include temporal networks (in which a network's entities and/or their interactions change in time), stochastic and deterministic dynamical processes on networks, adaptive networks (in which a dynamical process on a network is coupled to dynamics of network structure), and network structure and dynamics that include "higher-order" interactions (which involve three or more entities in a network). I draw examples from a variety of scenarios, including contagion dynamics, opinion models, waves, and coupled oscillators.

6.1 Introduction

Network analysis is one of the most exciting areas of applied and industrial mathematics [1–4]. It is at the forefront of numerous and diverse applications throughout the sciences, engineering, technology, and the humanities. The study of networks combines tools from many areas of mathematics, including graph theory, linear algebra, probability, statistics, optimization, statistical mechanics, scientific computation, and nonlinear dynamics.

In this chapter, I give a short overview of popular and state-of-the-art topics in network science. My discussions of these topics, which I draw preferentially from ones that relate to nonlinear and complex systems, will be terse, but I will cite many review articles and highlight specific research papers for those who seek more details. This chapter is not a review or even a survey; instead, I give my perspective on the short-term and medium-term future of network analysis in applied mathematics for 2020 and beyond.

M. A. Porter (✉)
Department of Mathematics, University of California, Los Angeles, CA 90095, USA
e-mail: mason@math.ucla.edu

P. G. Kevrekidis et al. (eds.), *Emerging Frontiers in Nonlinear Science*, Nonlinear Systems and Complexity 32, https://doi.org/10.1007/978-3-030-44992-6_6

My presentation proceeds as follows. In Sect. 6.2, I review a few basic concepts from network analysis. In Sect. 6.3, I discuss the dynamics of networks in the form of time-dependent ("temporal") networks. In Sect. 6.4, I discuss dynamical processes—both stochastic and deterministic—on networks. In Sect. 6.5, I discuss adaptive networks, in which there is coevolution of network structure and a dynamical process on that structure. In Sect. 6.6, I discuss higher-order structures (specifically, hypergraphs and simplicial complexes) that aim to go beyond the standard network paradigm of pairwise connections. I conclude with an outlook in Sect. 6.7.

6.2 Background on Networks

In its broadest form, a network encodes the connectivity patterns and connection strengths in a complex system of interacting entities [1]. The most traditional type of network is a graph $G = (V, E)$ (see Fig. 6.1**a**), where V is a set of "nodes" (i.e. "vertices") that encode entities and $E \subseteq V \times V$ is a set of "edges" (i.e. "links" or "ties") that encode the interactions between those entities. However, recent uses of the term "network" have focused increasingly on connectivity patterns that are more general than graphs [5]: a network's nodes and/or edges (or their associated weights) can change in time [6, 7] (see Sect. 6.3), its nodes and edges can include annotations [8], a network can include multiple types of edges and/or multiple types of nodes [9, 10], a network can have associated dynamical processes [11] (see Sects. 6.3, 6.4, and 6.5), a network can include memory [12], a network's connections can occur between an arbitrary number of entities [13, 14] (see Sect. 6.6), and so on.

Associated with a graph is an adjacency matrix $\mathbf{A}$ with entries a_{ij}. In the simplest scenario, edges either exist or they don't. If edges have directions, $a_{ij} = 1$ when there is an edge from entity j to entity i and $a_{ij} = 0$ when there is no such edge. When $a_{ij} = 1$, node i is "adjacent" to node j (because we can reach i directly from j), and the associated edge is "incident" from node j and to node i. The edge from j to i is an "out-edge" of j and an "in-edge" of i. The number of out-edges of a node is its "out-degree", and the number of in-edges of a node is its "in-degree". For an undirected network, $a_{ij} = a_{ji}$, and the number of edges that are attached to a node is the node's "degree". One can assign weights to edges to represent connections with different strengths (e.g. stronger friendships or larger transportation capacity) by defining a function $w : E \longrightarrow \mathbb{R}$. In many applications, the weights are nonnegative, although several applications [15] (such as in international relations) incorporate positive, negative, and zero weights. In some applications, nodes can also have self-edges (i.e. nodes can be adjacent to themselves) and multi-edges (i.e. adjacent nodes can have multiple edges between them). The spectral properties of adjacency (and other) matrices give important information about their associated graphs [1, 16]. For undirected networks, it is common to exploit the beneficent property that all eigenvalues of symmetric matrices are real.

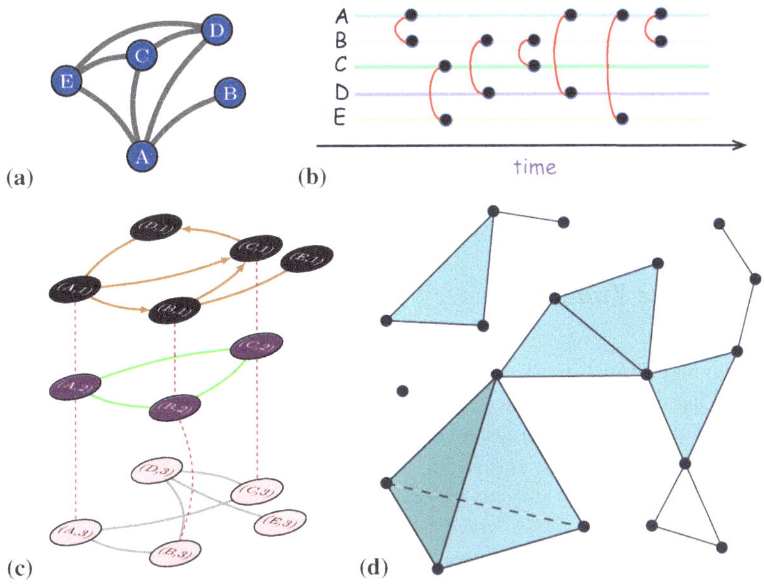

Fig. 6.1 Several types of network structures: (**a**) a graph, (**b**) a temporal network, (**c**) a multilayer network, and (**d**) a simplicial complex. (I drew panels (**a**) and (**c**) using TIKZ- NETWORK, which is by Jürgen Hackl and is available at https://github.com/hackl/tikz-network. Panel (**b**) is inspired by Fig. 1 of [6]. Panel (**d**), which is in the public domain, was drawn by Wikipedia user Cflm001 and is available at https://en.wikipedia.org/wiki/Simplicial_complex

6.3 Time-Dependent Networks

Traditional studies of networks consider time-independent structures, but most networks evolve in time. For example, social networks of people and animals change based on their interactions, roads are occasionally closed for repairs and new roads are built, and airline routes change with the seasons and over the years. To study such time-dependent structures, one can analyze "temporal networks". See [6, 7] for reviews and [17, 18] for edited collections.

The key idea of a temporal network is that networks change in time, but there are many ways to model such changes, and the time scales of interactions and other changes play a crucial role in the modeling process. There are also other important modeling considerations. To illustrate potential complications, suppose that an edge in a temporal network represents close physical proximity between two people in a short time window (e.g. with a duration of two minutes). It is relevant to consider whether there is an underlying social network (e.g. the friendship network of mathematics Ph.D. students at UCLA) or if the people in the network do not in general have any other relationships with each other (e.g. two people who happen to be visiting a particular museum on the same day). In both scenarios, edges that represent close physical proximity still appear and disappear over time, but indirect connections (i.e.

between people who are on the same connected component, but without an edge between them) in a time window may play different roles in the spread of information. Moreover, network structure itself is often influenced by a spreading process or other dynamics, as perhaps one arranges a meeting to discuss a topic (e.g. to give me comments on a draft of this chapter). See my discussion of adaptive networks in Sect. 6.5.

6.3.1 Discrete Time

For convenience, most work on temporal networks employs discrete time (see Fig. 6.1**b**). Discrete time can arise from the natural discreteness of a setting, discretization of continuous activity over different time windows, data measurement that occurs at discrete times, and so on.

6.3.1.1 Multilayer Representation of Temporal Networks

One way to represent a discrete-time (or discretized-time) temporal network is to use the formalism of "multilayer networks" [9, 10]. One can also use multilayer networks to study networks with multiple types of relations, networks with multiple subsystems, and other complicated networked structures.

A multilayer network $M = (V_M, E_M, V, \mathtt{L})$ (see Fig. 6.1**c**) has a set V of nodes—these are sometimes called "physical nodes", and each of them corresponds to an entity, such as a person—that have instantiations as "state nodes" (i.e. node-layer tuples, which are elements of the set V_M) on layers in $\mathtt{L}$. One layer in the set $\mathtt{L}$ is a combination, through the Cartesian product $L_1 \times \cdots \times L_d$, of elementary layers. The number d indicates the number of types of layering; these are called "aspects". A temporal network with one type of relationship has one type of layering, a time-independent network with multiple types of social relationships also has one type of layering, a multirelational network that changes in time has two types of layering, and so on. The set of state nodes in M is $V_M \subseteq V \times L_1 \times \cdots \times L_d$, and the set of edges is $E_M \subseteq V_M \times V_M$. The edge $((i, \alpha), (j, \beta)) \in E_M$ indicates that there is an edge from node j on layer β to node i on layer α (and vice versa, if M is undirected). For example, in Fig. 6.1**c**, there is a directed intralayer edge from (A,1) to (B,1) and an undirected interlayer edge between (A,1) and (A,2). The multilayer network in Fig. 6.1**c** has three layers, $|V| = 5$ physical nodes, $d = 1$ aspect, $|V_M| = 13$ state nodes, and $|E_M| = 20$ edges. To consider weighted edges, one proceeds as in ordinary graphs by defining a function $w : E_M \longrightarrow \mathbb{R}$. As in ordinary graphs, one can also incorporate self-edges and multi-edges.

Multilayer networks can include both intralayer edges (which have the same meaning as in graphs) and interlayer edges. The multilayer network in Fig. 6.1**c** has 4 directed intralayer edges, 10 undirected intralayer edges, and 6 undirected interlayer edges. In most studies that have employed multilayer representations of temporal networks, researchers have included interlayer edges only between state nodes in contiguous layers and only between state nodes that are associated with the same entity (see Fig. 6.1**c**). However, this restriction is not always desirable (see [19] for an example), and one can envision interlayer couplings that incorporate ideas like time horizons and interlayer edge weights that decay over time. For convenience, many researchers have used undirected interlayer edges in multilayer analyses of temporal networks, but it is often desirable for such edges to be directed to reflect the arrow of time [20]. The sequence of network layers, which constitute time layers, can represent a discrete-time temporal network at different time instances or a continuous-time network in which one bins (i.e. aggregates) the network's edges to form a sequence of time windows with interactions in each window.

Each d-aspect multilayer network with the same number of nodes in each layer has an associated adjacency tensor $\mathcal{A}$ of order $2(d + 1)$. For unweighted multilayer networks, each edge in E_M is associated with a 1 entry of $\mathcal{A}$, and the other entries (the "missing" edges) are 0. If a multilayer network does not have the same number of nodes in each layer, one can add empty nodes so that it does, but the edges that are attached to such nodes are "forbidden". There has been some research on tensorial properties of $\mathcal{A}$ [21] (and it is worthwhile to undertake further studies of them), but the most common approach for computations is to flatten $\mathcal{A}$ into a "supra-adjacency matrix" $\mathbf{A}_M$ [9, 10], which is the adjacency matrix of the graph G_M that is associated with M. The entries of the diagonal blocks of $\mathbf{A}_M$ correspond to intralayer edges, and the entries of its off-diagonal blocks correspond to interlayer edges.

6.3.1.2 Centrality, Clustering, and Large-Scale Network Structures

Following a long line of research in sociology [22], two important ingredients in the study of networks are examining (1) the importances ("centralities") of nodes, edges, and other small network structures and the relationship of measures of importance to dynamical processes on networks and (2) the large-scale organization of networks [1, 23].

Studying central nodes in networks is useful for numerous applications, such as ranking Web pages, football teams, or physicists [24]. It can also help reveal the roles of nodes in networks, such as nodes that experience high traffic or help bridge different parts of a network [1, 23]. Mesoscale features can impact network function and dynamics in important ways. Small subgraphs called "motifs" may appear frequently in some networks [25], perhaps indicating fundamental structures such as feedback loops and other building blocks of global behavior [26]. Various types of larger-scale network structures, such as dense "communities" of nodes [27, 28] and core–periphery structures [29, 30], are also sometimes related to dynamical modules (e.g. a set of synchronized neurons) or functional modules (e.g. a set of proteins that are important for a certain regulatory process) [31]. A common way to study large-scale structures[1] is inference using statistical models of random networks, such as stochastic block models [33]. Much recent research has generalized the study of large-scale network structure to temporal and multilayer networks [9, 18, 34].

Various types of centrality—including betweenness centrality [35, 36], Bonacich and Katz centrality [37, 38], communicability [39], PageRank [40, 41], and eigenvector centrality [42, 43]—have been generalized to temporal networks using a variety of approaches. Such generalizations make it possible to examine how node importances change over time as network structure evolves.

In recent work, my collaborators and I used multilayer representations of temporal networks to generalize eigenvector-based centralities to temporal networks [20, 44].[2] One computes the eigenvector-based centralities of nodes for a time-independent network as the entries of the "dominant" eigenvector, which is associated with the largest positive eigenvalue (by the Perron–Frobenius theorem, the eigenvalue with the largest magnitude is guaranteed to be positive in these situations) of a centrality matrix $C(\mathbf{A})$. Examples include eigenvector centrality (by using $C(\mathbf{A}) = \mathbf{A}$) [46], hub and authority scores[3] (by using $C(\mathbf{A}) = \mathbf{A}\mathbf{A}^T$ for hubs and $\mathbf{A}^T\mathbf{A}$ for authorities) [47], and PageRank [24].

Given a discrete-time temporal network in the form of a sequence of adjacency matrices $\mathbf{A}^{(t)} \in \mathbb{R}^{N \times N}$ for $t \in \{1, \ldots, T\}$, where $a_{ij}^{(t)}$ denotes a directed edge from entity j to entity i in time layer t, we construct a "supracentrality matrix" $\mathbb{C}(\omega)$, which couples the centrality matrices $C(\mathbf{A}^{(t)})$ of the individual time layers. We then compute the dominant eigenvector of $\mathbb{C}(\omega)$, where ω is an interlayer coupling strength.[4] In [20, 44], a key example was the ranking of doctoral programs in the mathematical sciences (using data from the Mathematics Genealogy Project [48]), where an edge from one institution to another arises when someone with a Ph.D. from the first institution supervises a Ph.D. student at the second institution. By calculating time-dependent centralities, one can study how the rankings of mathematical-sciences doctoral programs change over time and the dependence of such rankings on the value of ω. Larger values of ω impose more ranking consistency across time, so centrality trajectories over time are less volatile for larger ω [20, 44].

Multilayer representations of temporal networks have been very insightful in the detection of communities and how they split, merge, and otherwise evolve over time. Numerous methods for community detection—including inference via stochastic block models [49], maximization of objective functions (especially "modularity") [50], and methods based on random walks and bottlenecks to their traversal of a network [51, 52]—have been generalized from graphs to multilayer networks. They

---

[1]There are recent theoretical advances on examining network structure amidst rich but noisy data [32], and it is important for research on both network structure and dynamics to explicitly consider such scenarios.

[2]There is also much research on generalizing centralities (including eigenvector-based centralities [45]) to other types of multilayer networks, such as multiplex networks [9, 34].

[3]Nodes that are good authorities tend to have good hubs that point to them, and nodes that are good hubs tend to point to good authorities.

[4]A major open problem in multilayer network analysis is the measurement and/or inference of values of ω (and generalizations of it in the form of coupling tensors) [10].

have yielded insights in a diverse variety of applications, including brain networks [53], granular materials [54], political voting networks [50, 55], the spread of infectious diseases [56], and ecology and animal behavior [57, 58]. To assist with such applications, there are efforts to develop and analyze multilayer random-network models that incorporate rich and flexible structures [59], such as diverse types of interlayer correlations.

6.3.1.3 Activity-Driven Models

Activity-driven (AD) models of temporal networks [60] are a popular family of generative models that encode instantaneous time-dependent descriptions of network dynamics through a function called an "activity potential", which gives a mechanism to generate connections and characterizes the interactions between entities in a network. An activity potential encapsulates all of the information about the temporal network dynamics of an AD model, making it tractable to study dynamical processes (such as ones from Sect. 6.4) on networks that are generated by such a model. It is also common to compare the properties of networks that are generated by AD models to those of empirical temporal networks [18].

In the original AD model of Perra et al. [60], one considers a network with N entities, which we encode by the nodes. We suppose that node i has an activity rate $a_i = \eta x_i$, which gives the probability per unit time to create new interactions with other nodes. The scaling factor η ensures that the mean number of active nodes per unit time is $\eta \langle x \rangle N$, where $\langle x \rangle = \frac{1}{N} \sum_{i=1}^{N} x_i$. We define the activity rates such that $x_i \in [\epsilon, 1]$, where $\epsilon > 0$, and we assign each x_i from a probability distribution $F(x)$ that can either take a desired functional form or be constructed from empirical data. The model uses the following generative process:

- At each discrete time step (of length Δt), start with a network G_t that consists of N isolated nodes.
- With a probability $a_i \Delta t$ that is independent of other nodes, node i is active and generates m edges, each of which attaches to other nodes uniformly (i.e. with the same probability for each node) and independently at random (without replacement). Nodes that are not active can still receive edges from active nodes.
- At the next time step $t + \Delta t$, we delete all edges from G_t, so all interactions have a constant duration of Δt. We then generate new interactions from scratch. This is convenient, as it allows one to apply techniques from Markov chains.

Because entities in time step t do not have any memory of previous time steps, $F(x)$ encodes the network structure and dynamics.

The AD model of Perra et al. [60] is overly simplistic, but it is amenable to analysis and has provided a foundation for many more general AD models, including ones that incorporate memory [61]. In Sect. 6.6.4, I discuss a generalization of AD models to simplicial complexes [62] that allows one to study instantaneous interactions that involve three or more entities in a network.

6.3.2 Continuous Time

Many networked systems evolve continuously in time, but most investigations of time-dependent networks rely on discrete or discretized time. It is important to undertake more analysis of continuous-time temporal networks.

Researchers have examined continuous-time networks in a variety of scenarios. Examples include a compartmental model of biological contagions [63], a generalization of Katz centrality to continuous time [38], generalizations of AD models (see Sect. 6.3.1.3) to continuous time [64, 65], and rankings in competitive sports [66].

In a recent paper [67], my collaborators and I formulated a notion of "tie-decay networks" for studying networks that evolve in continuous time. We distinguished between *interactions*, which we modeled as discrete contacts, and *ties*, which encode relationships and their strengths as a function of time. For example, perhaps the strength of a tie decays exponentially after the most recent interaction. More realistically, perhaps the decay rate depends on the weight of a tie, with strong ties decaying more slowly than weak ones. One can also use point-process models like Hawkes processes [68] to examine similar ideas from a node-centric perspective.

Suppose that there are N interacting entities, and let $\mathbf{B}(t)$ be the $N \times N$ time-dependent, real, non-negative matrix with entries $b_{ij}(t)$ that encode the tie strength from entity j to entity i at time t for each j and i. In [67], we made the following simplifying assumptions:

1. As in [69], ties decay exponentially when there are no interactions: $\frac{\mathrm{d}b_{ij}}{\mathrm{d}t} = -\alpha b_{ij}$, where $\alpha \geq 0$ is the decay rate.
2. If two entities interact at time $t = \tau$, the strength of the tie between them grows instantaneously by 1.

See [70] for a comparison of various choices, including those in [67, 69], for tie evolution over time.

In practice (e.g. in data-driven applications), one obtains $\mathbf{B}(t)$ by discretizing time, so let's suppose that there is at most one interaction in each time step of length Δt. This occurs, for example, in a Poisson process. Such time discretization is common in the simulation of stochastic dynamical systems, such as in Gillespie algorithms [11, 71, 72]. Consider an $N \times N$ matrix $\mathbf{A}(t)$ in which $a_{ij}(t) = 1$ if node j interacts with node i at time t and $a_{ij}(t) = 0$ otherwise. For a directed network, $\mathbf{A}(t)$ has exactly one nonzero entry in each time step when there is an interaction and no nonzero entries when there isn't one. For an undirected network, because of the symmetric nature of interactions, there are exactly two nonzero entries in time steps that include an interaction. We write

$$\mathbf{B}(t + \Delta t) = e^{-\alpha \Delta t}\mathbf{B}(t) + \mathbf{A}(t + \Delta t) . \tag{6.1}$$

Equivalently, if interactions between entities occur at times $\tau^{(\ell)}$ such that $0 \leq \tau^{(0)} < \tau^{(1)} < \cdots < \tau^{(T)}$, then at time $t \geq \tau^{(T)}$, we have

$$\mathbf{B}(t) = \sum_{k=0}^{T} e^{-\alpha(t-\tau^{(k)})}\mathbf{A}(\tau^{(k)})\,. \tag{6.2}$$

In [67], my coauthors and I generalized PageRank [24, 73] to tie-decay networks. One nice feature of our tie-decay PageRank is that it is applicable not just to data sets, but also to data streams, as one updates the PageRank values as new data arrives. By contrast, one problematic feature of many methods that rely on multilayer representations of temporal networks is that one needs to recompute everything for an entire data set upon acquiring new data, rather than updating prior results in a computationally efficient way.

6.4 Dynamical Processes on Networks

A dynamical process can be discrete, continuous, or some mixture of the two; it can also be either deterministic or stochastic. It can take the form of one or several coupled ordinary differential equations (ODEs), partial differential equations (PDEs), maps, stochastic differential equations, and so on.

A dynamical process requires a rule for updating the states of its dependent variables with respect one or more independent variables (e.g. time), and one also has initial conditions and/or boundary conditions. To formalize a dynamical process on a network, one needs a rule for how to update the states of the network's nodes and/or edges.

The nodes (of one or more types) of a network are connected to each other in nontrivial ways by one or more types of edges. This leads to a natural question: How does nontrivial connectivity between nodes affect dynamical processes on a network [11]? When studying a dynamical process on a network, the network structure encodes which entities (i.e. nodes) of a system interact with each other and which do not. If desired, one can ignore the network structure entirely and just write out a dynamical system. However, keeping track of network structure is often a very useful and insightful form of bookkeeping, which one can exploit to systematically explore how particular structures affect the dynamics of particular dynamical processes.

Prominent examples of dynamical processes on networks include coupled oscillators [74, 75], games [76], and the spread of diseases [77, 78] and opinions [79, 80]. There is also a large body of research on the control of dynamical processes on networks [81, 82].

Most studies of dynamics on networks have focused on extending familiar models—such as compartmental models of biological contagions [77] or Kuramoto phase oscillators [75]—by coupling entities using various types of network structures, but it is also important to formulate new dynamical processes from scratch, rather than only studying more complicated generalizations of our favorite models. When trying to illuminate the effects of network structure on a dynamical process, it is often insightful to provide a baseline comparison by examining the process on a convenient ensemble of random networks [11].

6.4.1 An Illustrative Example: A Threshold Model of a Social Contagion

A simple, but illustrative, dynamical process on a network is the Watts threshold model (WTM) of a social contagion [11, 79]. It provides a framework for elucidating how network structure can affect state changes, such as the adoption of a product or a behavior, and for exploring which scenarios lead to "virality" (in the form of state changes of a large number of nodes in a network).

The original WTM [83], a binary-state threshold model that resembles bootstrap percolation [84], has a deterministic update rule, so stochasticity can come only from other sources (see Sect. 6.4.2). In a binary state model, each node is in one of two states; see [85] for a tabulation of well-known binary-state dynamics on networks. The WTM is a modification of Mark Granovetter's threshold model for social influence in a fully-mixed population [86]. See [87, 88] for early work on threshold models on networks that developed independently from investigations of the WTM. Threshold contagion models have been developed for many scenarios, including contagions with multiple stages [89], models with adoption latency [90], models with synergistic interactions [91], and situations with hipsters (who may prefer to adopt a minority state) [92].

In a binary-state threshold model such as the WTM, each node i has a threshold R_i that one draws from some distribution. Suppose that R_i is constant in time, although one can generalize it to be time-dependent. At any time, each node can be in one of two states: 0 (which represents being inactive, not adopted, not infected, and so on) or 1 (active, adopted, infected, and so on). A binary-state model is a drastic oversimplification of reality, but the WTM is able to capture two crucial features of social systems [93]: interdependence (an entity's behavior depends on the behavior of other entities) and heterogeneity (because nodes with different threshold values behave differently). One can assign a seed number or seed fraction of nodes to the active state, and one can choose the initially active nodes either deterministically or randomly.

The states of the nodes change in time according to an update rule, which can either be synchronous (such that it is a map) or asynchronous (e.g. as a discretization of continuous time) [11]. In the WTM, the update rule is deterministic, so the choice of synchronous versus asynchronous updating affects only how long it takes to reach a steady state; it does not affect the steady state itself. With a stochastic update rule, the synchronous and asynchronous versions of ostensibly the "same" model can behave in drastically different ways [94]. In the WTM on an undirected network, to update the state of a node, one compares its fraction s_i/k_i of active neighbors (where s_i is the number of active neighbors and k_i is the degree of node i) to the node's threshold R_i. An inactive node i becomes active (i.e. it switches from state 0 to state 1) if $s_i/k_i \geq R_i$; otherwise, it stays inactive. The states of the nodes in the WTM are monotonic, in the sense that a node that becomes active remains active forever. This feature is convenient for deriving accurate approximations for the global behavior of the WTM using branching-process approximations [11, 85] and when analyzing the behavior of the WTM using tools such as persistent homology [95].

6.4.2 *Stochastic Processes*

A dynamical process on a network can take the form of a stochastic process [1, 11]. There are several possible sources of stochasticity: (1) choice of initial condition, (2) choice of which nodes or edges to update (when considering asynchronous updating), (3) the rule for updating nodes or edges, (4) the values of parameters in an update rule, and (5) the selection of particular networks from a random-graph ensemble (i.e. a probability distribution on graphs). Some or all of these sources of randomness can be present when studying dynamical processes on networks. It is desirable to compare the sample mean of a stochastic process on a network to an ensemble average (i.e. to an expectation over a suitable probability distribution).

Prominent examples of stochastic processes on networks include percolation [96], random walks [97], compartment models of biological contagions [77, 78], bounded-confidence models of the dynamics of continuous-valued opinions [98], and other opinion and voter models [11, 79, 80, 99].

6.4.2.1 Example: A Compartmental Model of a Biological Contagion

Compartmental models of biological contagions are a topic of intense interest in network science [1, 11, 77, 78]. A compartment represents a possible state of a node; examples include susceptible, infected, zombified, vaccinated, and recovered. An update rule determines how a node changes its state from one compartment to another. One can formulate models with as many compartments as desired [100], but investigations of how network structure affects dynamics typically have employed examples with only two or three compartments [77, 78]. The formulation and analysis of compartmental models on networks have played a crucial role—and have influenced both human behavior and public policy—during the current coronavirus disease 2019 (COVID-19) pandemic. For one nice example of such important work, see [101].

Researchers have studied various extensions of compartmental models, including contagions on multilayer and temporal networks [9, 34, 102], metapopulation models on networks for simultaneously studying network connectivity and subpopulations with different characteristics [103], non-Markovian contagions on networks for exploring memory effects [104], and explicit incorporation of individuals with essential societal roles (e.g. health-care workers) [105]. As I discuss in Sect. 6.4.4, one can also examine coupling between biological contagions and the spread of information (e.g. "awareness") [106, 107]. One can also use compartmental models to study phenomena, such as the dissemination of ideas on social media [108] and forecasting of political elections [109], that are much different from the spread of diseases.

One of the most prominent examples of a compartmental model is a susceptible–infected–recovered (SIR) model, which has three compartments. Susceptible nodes are healthy and can become infected, and infected nodes can eventually recover. The steady state of the basic SIR model on a network is related to a type of bond percolation [110–113]. There are many variants of SIR models and other compartmental models on networks [77]. See [114] for an illustration using susceptible–infected–susceptible (SIS) models.

Suppose that an infection is transmitted from an infected node to a susceptible neighbor at a rate of λ. The probability of a transmission event on one edge between an infected node and a susceptible node in an infinitesimal time interval $\mathrm{d}t$ is $\lambda\,\mathrm{d}t$. Assuming that all infection events are independent, the probability that a susceptible node with s infected neighbors becomes infected (i.e. that it transitions from the S compartment to the I compartment, which represents both being infected and being infective) in a time step of duration $\mathrm{d}t$ is

$$1 - (1 - \lambda\,\mathrm{d}t)^s \to \lambda\, s\,\mathrm{d}t \quad \text{as} \quad \mathrm{d}t \to 0\,. \tag{6.3}$$

If an infected node recovers at a constant rate of μ, the probability that it switches from state I to state R (the recovered compartment) in an infinitesimal time interval $\mathrm{d}t$ is $\mu\,\mathrm{d}t$.

6.4.3 Deterministic Dynamical Systems

When there is no source of stochasticity, a dynamical process on a network is "deterministic". A deterministic dynamical system can take the form of a system of coupled maps, ODEs, PDEs, or something else. As with stochastic systems, the network structure encodes which entities of a system interact with each other and which do not.

There are numerous interesting deterministic dynamical systems on networks—just incorporate nontrivial connectivity between entities into your favorite deterministic model—although it is worth noting that some stochastic aspects (e.g. choosing parameter values from a probability distribution or sampling choices of initial conditions) can arise in these models.

6.4.3.1 Example: Coupled Oscillators

For concreteness, let's consider the popular setting of coupled oscillators. Each node in a network is associated with an oscillator, and we want to examine how network structure affects the collective behavior of the coupled oscillators.

It is common to investigate various forms of synchronization (a type of coherent behavior), such that the rhythms of the oscillators adjust to match each other (or to match a subset of the oscillators) because of their interactions [115]. A variety of methods, such as "master stability functions" [116], have been developed to study the local stability of synchronized states and their generalizations [11, 74], such as cluster synchrony [117]. Cluster synchrony, which is related to work on "coupled-cell networks" [26], uses ideas from computational group theory to find synchronized sets of oscillators that are not synchronized with other sets of synchronized oscillators. Many studies have also examined other types of states, such as "chimera states" [118], in which some oscillators behave coherently but others behave incoherently. (Analogous phenomena sometimes occur in mathematics departments.)

A ubiquitous example is coupled Kuramoto oscillators on a network [74, 75, 119], which is perhaps the most common setting for exploring and developing new methods for studying coupled oscillators. (In principle, one can then build on these insights in studies of other oscillatory systems, such as in applications in neuroscience [120].) Coupled Kuramoto oscillators have been used for modeling numerous phenomena, including jetlag [121] and singing in frogs [122]. Indeed, a "Snowbird" (SIAM) conference on applied dynamical systems would not be complete without at least several dozen talks on the Kuramoto model. In the Kuramoto model, each node i has an associated phase $\theta_i(t) \in [0, 2\pi)$. In the case of "diffusive" coupling between the nodes,[5] the dynamics of the ith node is governed by the equation

$$\dot{\theta}_i := \frac{\mathrm{d}\theta_i}{\mathrm{d}t} = \omega_i + \sum_{j=1}^{N} b_{ij} a_{ij} \sin(\theta_j - \theta_i), \qquad i \in \{1, \dots, N\}, \tag{6.4}$$

where one typically draws the natural frequency ω_i of node i from some distribution $g(\omega)$, the scalar a_{ij} is an adjacency-matrix entry of an unweighted network, b_{ij} is the coupling strength that is experienced by oscillator i from oscillator j (so $b_{ij}a_{ij}$ is an element of an adjacency matrix $\mathbf{W}$ of a weighted network), and $f_{ij}(y) = \sin(y)$ is the coupling function, which depends only on the phase difference between oscillators i and j because of the diffusive nature of the coupling.

Once one knows the natural frequencies ω_i, the model (6.4) is a deterministic dynamical system, although there have been studies of coupled Kuramoto oscillators with additional stochastic terms [123]. Traditional studies of (6.4) and its generalizations draw the natural frequencies from some distribution (e.g. a Gaussian or a compactly supported distribution), but some studies of so-called "explosive synchro-

[5] In this case, linearization yields Laplacian dynamics, which is closely related to a random walk on a network [97].

nization" (in which there is an abrupt phase transition from incoherent oscillators to synchronized oscillators) have employed deterministic natural frequencies [119, 124]. The properties of the frequency distribution $g(\omega)$ have a significant effect on the dynamics of (6.4). Important features of $g(\omega)$ include whether it has compact support or not, whether it is symmetric or asymmetric, and whether it is unimodal or not [75, 125].

The model (6.4) has been generalized in numerous ways. For example, researchers have considered a large variety of coupling functions f_{ij} (including ones that are not diffusive) and have incorporated an inertia term $\ddot{\theta}_i$ to yield a second-order Kuramoto oscillator at each node [75]. The latter generalization is important for studies of coupled oscillators and synchronized dynamics in electric power grids [126]. Another noteworthy direction is the analysis of Kuramoto model on "graphons" (see, for example, [127]), an important type of structure that arises in a suitable limit of large networks.

6.4.4 Dynamical Processes on Multilayer Networks

An increasingly prominent topic in network analysis is the examination of how multilayer network structures—multiple system components, multiple types of edges, co-occurrence and coupling of multiple dynamical processes, and so on—affect qualitative and quantitative dynamics [9, 34, 102]. For example, perhaps certain types of multilayer structures can induce unexpected instabilities or phase transitions in certain types of dynamical processes?

There are two categories of dynamical processes on multilayer networks: (1) a single process can occur on a multilayer network; or (2) processes on different layers of a multilayer network can interact with each other [102]. An important example of the first category is a random walk, where the relative speeds and probabilities of steps within layers versus steps between layers affect the qualitative nature of the dynamics. This, in turn, affects methods (such as community detection [51, 52]) that are based on random walks, as well as anything else in which diffusion is relevant [128, 129]. Two other examples of the first category are the spread of information on social media (for which there are multiple communication channels, such as Facebook and Twitter) and multimodal transportation systems [130]. For instance, a multilayer network structure can induce congestion even when a system without coupling between layers is decongested in each layer independently [131]. Examples of the second category of dynamical process are interactions between multiple strains of a disease and interactions between the spread of an infectious disease and the spread of information [106, 107, 132]. Many other examples have been studied [34], including coupling between oscillator dynamics on one layer and a biased random walk on another layer (as a model for neuronal oscillations coupled to blood flow) [133].

Numerous interesting phenomena can occur when dynamical systems, such as spreading processes, are coupled to each other [106]. For example, the spreading of one disease can facilitate infection by another [134], and the spread of awareness about a disease can inhibit the spread of the disease itself (e.g. if people stay home when they are sick) [135]. Interacting spreading processes can also exhibit other fascinating dynamics, such as oscillations that are induced by multilayer network structures in a biological contagion with multiple modes of transmission [136] and novel types of phase transitions [102].

A major simplification in most work thus far on dynamical processes on multilayer networks is a tendency to focus on toy models. For example, a typical study of coupled spreading processes may consider a standard (e.g. SIR) model on each layer, and it may draw the connectivity pattern of each layer from the same standard random-graph model (e.g. an Erdős–Rényi model or a configuration model). However, when studying dynamics on multilayer networks, it is particular important in future work to incorporate heterogeneity in network structure and/or dynamical processes. For instance, diseases spread offline but information spreads both offline and online, so investigations of coupled information and disease spread ought to consider fundamentally different types of network structures for the two processes.

6.4.5 *Metric Graphs and Waves on Networks*

Network structures also affect the dynamics of PDEs on networks [137–141]. Interesting examples include a study of a Burgers equation on graphs to investigate how network structure affects the propagation of shocks [138] and investigations of reaction–diffusion equations and Turing patterns on networks [140, 142]. The latter studies exploited the rich theory of Laplacian dynamics on graphs (and concomitant ideas from spectral graph theory) [16, 97] and examined the addition of nonlinear terms to Laplacians on various types of networks (including multilayer ones).

A mathematically oriented thread of research on PDEs on networks has built on ideas from so-called "quantum graphs" [137, 143] to study wave propagation on networks through the analysis of "metric graphs". Metric graphs differ from the usual "combinatorial graphs", which in other contexts are usually called simply "graphs".[6] As in a combinatorial graph, a metric graph has nodes and edges, but now each edge e also has an associated positive length $l_e \in (0, \infty]$. For many experimentally relevant scenarios (e.g. in models of circuits of quantum wires [144]), there is a natural embedding into space, but metric graphs that are not embedded in space are also appropriate for some applications.

As the nomenclature suggests, one can equip a metric graph with a natural metric. If a sequence $\{e_j\}_{j=1}^{m}$ of edges forms a path, the length of the path is $\sum_j l_{e_j}$. The distance $\rho(v_1, v_2)$ between two nodes, v_1 and v_2, is the minimum path length between them. We place coordinates along each edge, so we can compute a distance between points x_1 and x_2 on a metric graph even when those points are not located at nodes.

[6]Combinatorial graphs, and more general combinatorial objects, are my main focus in this chapter. This subsection is an exception.

Traditionally, one assumes that the infinite ends (which one can construe as "leads" at infinity, as in scattering theory) of infinite edges have degree 1. It is also traditional to assume that there is always a positive distance between distinct nodes and that there are no finite-length paths with infinitely many edges. See [143] for further discussion.

To study waves on metric graphs, one needs to define operators, such as the negative second derivative or more general Schrödinger operators. This exploits the fact that there are coordinates for all points on the edges—not only at the nodes themselves, as in combinatorial graphs. When studying waves on metric graphs, it is also necessary to impose boundary conditions at the nodes [143].

Many studies of wave propagation on metric graphs have considered generalizations of nonlinear wave equations, such as the cubic nonlinear Schrödinger (NLS) equation [145] and a nonlinear Dirac equation [146]. The overwhelming majority of studies of metric graphs (with both linear and nonlinear waves) have focused on networks with a very small number of nodes, as even small networks yield very interesting dynamics. For example, Marzuola and Pelinovsky [147] analyzed symmetry-breaking and symmetry-preserving bifurcations of standing waves of the cubic NLS equation on a dumbbell graph (with two rings attached to a central line segment and Kirchhoff boundary conditions at the nodes). Kairzhan et al. [148] studied the spectral stability of half-soliton standing waves of the cubic NLS equation on balanced star graphs. Sobirov et al. [149] studied scattering and transmission at nodes of sine–Gordon solitons on networks (e.g. on a star graph and a small tree).

A particularly interesting direction for future work is to study wave dynamics on large metric graphs. This will help extend investigations, as in ODEs and maps, of how network structures affect dynamics on networks to the realm of linear and nonlinear waves. One can readily formulate wave equations on large metric graphs by specifying relevant boundary conditions and rules at each junction. For example, Joly et al. [150] recently examined wave propagation of the standard linear wave equation on fractal trees. Because many natural real-life settings are spatially embedded (e.g. wave propagation in granular materials [54, 151] and traffic-flow patterns in cities), it will be particularly valuable to examine wave dynamics on (both synthetic and empirical) spatially-embedded networks [152]. Therefore, I anticipate that it will be very insightful to undertake studies of wave dynamics on networks such as random geometric graphs, random neighborhood graphs, and other spatial structures. A key question in network analysis is how different types of network structure affect different types of dynamical processes [11], and the ability to take a limit as model synthetic networks become infinitely large (specifically, a thermodynamic limit) is crucial for obtaining many key theoretical insights.

6.5 Adaptive Networks

Dynamics of networks and dynamics on networks do not occur in isolation; instead, they are coupled to each other. Researchers have studied the coevolution of network structure and the states of nodes and/or edges in the context of "adaptive networks" (which are also known as "coevolving networks") [153, 154]. Whether it is sensible to study a dynamical process on a time-independent network, a temporal network with frozen (or no) node or edge states, or an adaptive network depends on the relative time scales of the dynamics of a network's structure and the dynamics of the states of the nodes and/or edges of the network. See [11] for a brief discussion.

Models in the form of adaptive networks provide a promising mechanistic approach to simultaneously explain both structural features (e.g. degree distributions and temporal features (e.g. burstiness) of empirical data [155]. Incorporating adaptation into conventional models can produce extremely interesting and rich dynamics, such as the spontaneous development of extreme states in opinion models [156].

Most studies of adaptive networks that include some analysis (i.e. that go beyond numerical computations) have employed rather artificial adaption rules for adding, removing, and rewiring edges. This is relevant for mathematical tractability, but it is important to go beyond these limitations by considering more realistic types of adaptation and coupling between network structure (including multilayer structures, as in [157]) and the states of nodes and edges.

6.5.1 Contagion Models

When people are sick, they stay home from work or school. People also form and remove social connections (both online and offline) based on observed opinions and behaviors. To study these ideas using adaptive networks, researchers have coupled models of biological and social contagions with time-dependent networks [11, 79].

An early example of an adaptive network of disease spread is the SIS model in Gross et al. [158]. In this model, susceptible nodes sometimes rewire their incident edges to "protect themselves". Suppose that we have an N-node network with a constant number of undirected edges. Each node is either susceptible (i.e. in state S) or infected (i.e. in state I). At each time step, and for each edge—so-called "discordant edges"—between nodes in different states, the susceptible node becomes infected with probability λ. For each discordant edge, with some probability κ, the incident susceptible node breaks the edge and rewires to some other susceptible node. (Multi-edges and self-edges are not allowed.) This is called a "rewire-to-same" mechanism in the language of some adaptive opinion models [159, 160]. In each time step, infected nodes can also recover to become susceptible again.

Gross et al. [158] studied how the rewiring probability affects the "basic reproduction number", a scalar that measures how many infections occur, on average, from one infected node in a population in which all other nodes are susceptible to infection [77, 78, 100]. The basic reproduction number determines the size of a critical infection probability λ_* that is necessary to maintain a stable epidemic (as determined traditionally using linear stability analysis). A high rewiring rate can significantly increase λ_* and thereby significantly reduce the prevalence of a contagion. Although results like these are perhaps intuitively clear, other studies of contagions on adaptive networks have yielded potentially actionable (and arguably nonintuitive) insights. For example, Scarpino et al. [105] demonstrated using an adaptive compartmental model (along with some empirical evidence) that the spread of a disease can accelerate when individuals with essential societal roles (e.g. health-care workers) become ill and are replaced with healthy individuals.

6.5.2 Opinion Models

Another type of model with many interesting adaptive variants are opinion models [11, 80], especially in the form of generalizations of classical voter models [99].

Voter dynamics were first considered in the 1970s by Clifford and Sudbury [161] as a model for species competition, and the dynamical process that they introduced was dubbed "the voter model"[7] shortly thereafter by Holley and Liggett [162]. Voter dynamics are fun and are popular to study [99], although it is questionable whether it is ever possible to genuinely construe a voter model as a model of voters [163].

Holme and Newman [164] undertook an early study of a rewire-to-same adaptive voter model. Inspired by their research, Durrett et al. [159] compared the dynamics from two different types of rewiring in an adaptive voter model. In each variant of their model, one considers an N-node network and supposes that each node is in one of two states. The network structure and the node states coevolve. Pick an edge uniformly at random. If this edge is discordant, then with probability $1 - \kappa$, one of its incident nodes adopts the opinion state of the other. Otherwise, with complementary probability κ, a rewiring action occurs: one removes the discordant edge, and one of the associated nodes attaches to a new node either through a rewire-to-same mechanism (choosing uniformly at random among the nodes with the same opinion state) or through a "rewire-to-random" mechanism (choosing uniformly at random among all nodes). As with the adaptive SIS model of [158], self-edges and multi-edges are not allowed.

The models in [159] evolve until there are no discordant edges. There are several key questions. Does the system reach a consensus (in which all nodes are in the same state)? If so, how long does it take to converge to consensus? If not, how many opinion clusters (each of which is a connected component, perhaps interpretable as an "echo chamber", of the final network) are there at steady state? How long does it take to reach this state? The answers and analysis are subtle; they depend on the initial network topology, the initial conditions, and the specific choice of rewiring rule. As with other adaptive network models, researchers have developed some nonrigorous theory (e.g. using mean-field approximations and their generalizations) on adaptive voter models with simplistic rewiring schemes, but they have struggled to extend these ideas to models with more realistic rewiring schemes. There are very few

[7] There are several variants of "the" voter model, depending on choices such as whether one selects nodes or edges at random, that have substantively different qualitative dynamics [11, 97].

mathematically rigorous results on adaptive voter models, although there do exist some, under various assumptions on initial network structure and edge density [165].

Researchers have generalized adaptive voter models to consider more than two opinion states [166] and more general types of rewiring schemes [167]. As with other adaptive networks, analyzing adaptive opinion models with increasingly diverse types of rewiring schemes (ideally with a move towards increasing realism) is particularly important. In [160], Yacoub Kureh and I studied a variant of a voter model with nonlinear rewiring (in our model, the probability that determines whether a node rewires or adopts a neighbor's opinion is a function of how well it "fits in" with the nodes in its neighborhood), including a "rewire-to-none" scheme to model unfriending and unfollowing in online social networks. It is also important to study adaptive opinion models with more realistic types of opinion dynamics. A promising example is adaptive generalizations of bounded-confidence models (see the introduction of [98] for a brief review of bounded-confidence models), which have continuous opinion states, with nodes interacting either with nodes or with other entities (such as media [168]) whose opinion is sufficiently close to theirs. A recent numerical study examined an adaptive bounded-confidence model [169]; this is an important direction for future investigations.

6.5.3 Synchronization of Adaptive Oscillators

It is also interesting to examine how the adaptation of oscillators—including their intrinsic frequencies and/or the network structure that couples them to each other—affects the collective behavior (e.g. synchronization) of a network of oscillators [75]. Such ideas are useful for exploring mechanistic models of learning in the brain (e.g. through adaptation of coupling between oscillators to produce a desired limit cycle [170]).

One nice study is a paper by Skardal et al. [171], who examined an adaptive model of coupled Kuramoto oscillators as a toy model of learning. First, we write the Kuramoto system as

$$\frac{\mathrm{d}\theta_i}{\mathrm{d}t} = \omega_i + \sum_{j=1}^{N} f_{ij}(\theta_j - \theta_i)\,, \qquad i \in \{1, \ldots, N\}\,, \tag{6.5}$$

where f_{ij} is a 2π-periodic function of the phase difference between oscillators i and j. The function f_{ij} incorporates the matrix elements b_{ij} and a_{ij} from (6.4). One way to examine adaptation is to define an "order parameter" r_i (which, in its traditional form, quantifies the amount of coherence of the coupled Kuramoto oscillators [75]) for the ith oscillator by

$$r_i = \sum_{j=1}^{N} b_{ij} a_{ij} e^{\mathrm{i}\theta_j}\,, \quad i \in \{1, \ldots, N\}$$

and to consider the following dynamical system:

$$\frac{\mathrm{d}\theta_i}{\mathrm{d}t} = \omega_i + \lambda_D^{-1}\,\mathrm{Im}(z_i e^{-\mathrm{i}\theta_i})\,, \quad \tau\frac{\mathrm{d}z_i}{\mathrm{d}t} = r_i - z_i\,, \quad T\frac{\mathrm{d}b_{ij}}{\mathrm{d}t} = \alpha + \beta\,\mathrm{Re}(r_i z_i^*) - b_{ij}\,, \tag{6.6}$$

where $\mathrm{Re}(\zeta)$ denotes the real part of a quantity ζ and $\mathrm{Im}(\zeta)$ denotes its imaginary part. In the model (6.6), λ_D denotes the largest positive eigenvalue of the adjacency matrix $\mathbf{A}$, the variable $z_i(t)$ is a time-delayed version of r_i with time parameter τ (with $\tau \to 0$ implying that $z_i \to r_i$), and z_i^* denotes the complex conjugate of z_i. One draws the frequencies ω_i from some distribution (e.g. a Lorentz distribution, as in [171]), and we recall that b_{ij} is the coupling strength that is experienced by oscillator i from oscillator j. The parameter T gives an adaptation time scale, and $\alpha \in \mathbb{R}$ and $\beta \in \mathbb{R}$ are parameters (which one can adjust to study bifurcations). Skardal et al. [171] interpreted scenarios with $\beta > 0$ as "Hebbian" adaptation (see [172]) and scenarios with $\beta < 0$ as anti-Hebbian adaptation, as they observed that oscillator synchrony is promoted when $\beta > 0$ and inhibited when $\beta < 0$.

6.6 Higher-Order Structures and Dynamics

Most studies of networks have focused on networks with pairwise connections, in which each edge (unless it is a self-edge, which connects a node to itself) connects exactly two nodes to each other. However, many interactions—such as playing games, coauthoring papers and other forms of collaboration, and horse races—often occur between three or more entities of a network. To examine such situations, researchers have increasingly studied "higher-order" structures in networks, as they can exert a major influence on dynamical processes.

6.6.1 *Hypergraphs*

Perhaps the simplest way to account for higher-order structures in networks is to generalize from graphs to "hypergraphs" [1]. Hypergraphs possess "hyperedges" that encode a connection between an arbitrary number of nodes, such as between all coauthors of a paper. This allows one to make important distinctions, such as between a k-clique (in which there are pairwise connections between each pair of nodes in a set of k nodes) and a hyperedge that connects all k of those nodes to each other, without the need for any pairwise connections.

One way to study a hypergraph is as a "bipartite network", in which nodes of a given type can be adjacent only to nodes of another type. For example, a scientist can be adjacent to a paper that they have written [173] and a legislator can be adjacent to a committee on which they sit [174]. It is important to generalize ideas from graph theory to hypergraphs, such as by developing models of random hypergraphs [8, 175, 176].

6.6.2 Simplicial Complexes

Another way to study higher-order structures in networks is to use "simplicial complexes" [13, 177, 178]. A simplicial complex is a space that is built from a union of points, edges, triangles, tetrahedra, and higher-dimensional polytopes (see Fig. 6.1**d**). Simplicial complexes are topological spaces that one can use to approximate other topological spaces and thereby capture some of their properties.

A *p-dimensional simplex* (i.e. a *p-simplex*) is a p-dimensional polytope that is the convex hull of its $p + 1$ vertices (i.e. nodes). Any $\tilde{p}$-dimensional subset (with $\tilde{p} < p$) of a p-simplex that is itself a simplex is a *face* of the p-simplex. A *simplicial complex* K is a set of simplices such that (1) every face of a simplex from S is also in S and (2) the intersection of any two simplices $\sigma_1, \sigma_2 \in S$ is a face of both σ_1 and σ_2. An increasing sequence $\emptyset = K_0 \subseteq K_1 \subseteq K_2 \subseteq \cdots \subseteq K_l = K$ of simplicial complexes forms a *filtered simplicial complex*; each K_i is a *subcomplex*. As discussed in [13] and references therein, one can examine the homology of each subcomplex. In studying the homology of a topological space, one computes topological invariants that quantify features of different dimensionalities [178]. One studies the "persistent homology" of a filtered simplicial complex to quantify the topological structure of a data set (e.g. a point cloud) across multiple scales of such data. The goal of such "topological data analysis" is to measure the "shape" of data in the form of connected components, "holes" of various dimensionality, and so on [13]. From the perspective of network analysis, this yields insight into types of large-scale structure that complement traditional ones (such as community structure). See [179] for a friendly, nontechnical introduction to topological data analysis.

A natural goal is to generalize ideas from network analysis to simplicial complexes. Important efforts include generalizing configuration models of random graphs [180] to random simplicial complexes [181, 182]; generalizing well-known network growth mechanisms, such as preferential attachment [183]; and developing geometric notions, like curvature, for networks [184]. An important modeling issue when studying higher-order network data is the question of when it is more appropriate (or convenient) to use the formalisms of hypergraphs or simplicial complexes.

The computation of persistent homology has yielded insights into a diverse set of models and applications in network science and complex systems. Examples include granular materials [54, 185], functional brain networks [177, 186], quantification of "political islands" in voting data [187], percolation theory [188], contagion dynamics [95], swarming and collective behavior [189], chaotic flows in ODEs and PDEs [190], diurnal cycles in tropical cyclones [191], and mathematics education [192]. See the introduction to [13] for pointers to numerous other applications.

Most uses of simplicial complexes in network science and complex systems have been in the context of topological data analysis (especially the computation of persistent homology) and its applications [13, 14, 193]. In my discussion, however, I focus instead on a somewhat different (and increasingly popular) topic: the generalization of dynamical processes on and of networks to simplicial complexes to study the effects of higher-order interactions on network dynamics. Simplicial structures influence the collective behavior of the dynamics of coupled entities on networks (e.g. they can lead to novel bifurcations and phase transitions), and they provide a natural approach to analyze p-entity interaction terms, including for $p \geq 3$, in dynamical systems. Existing work includes research on linear diffusion dynamics (in the form of Hodge Laplacians, such as in [194]) and generalizations of a variety of other popular types of dynamical processes on networks.

6.6.3 Coupled Phase Oscillators with p-Body Interactions with $p \geq 3$

Given the ubiquitous study of coupled Kuramoto oscillators [75], a sensible starting point for exploring the impact of simultaneous coupling of three or more oscillators on a system's qualitative dynamics is to study a generalized Kuramoto model. For example, to include both two-entity ("two-body") and three-entity interactions in a model of coupled oscillators on networks, we write [195]

$$\dot{\mathbf{x}}_i = \boldsymbol{\omega}_i(\mathbf{x}_i) + \sum_{j,k} \mathbf{f}_{ijk}(\mathbf{x}_i, \mathbf{x}_j, \mathbf{x}_k)\,, \tag{6.7}$$

where $\boldsymbol{\omega}_i$ describes the intrinsic dynamics of oscillator i and the three-oscillator interaction term $\mathbf{f}_{ijk}$ can also encompass two-oscillator interaction terms $\mathbf{f}_{ij}(\mathbf{x}_i, \mathbf{x}_j)$.

An example of N coupled Kuramoto oscillators with three-body interactions is [195]

$$\begin{aligned}\dot{\theta}_i = \omega_i &+ \frac{1}{N}\sum_j \left[a_{ij}\sin(\theta_{ji} + \alpha_{1ij}) + b_{ij}\sin(2\theta_{ji} + \alpha_{2ij})\right] \\ &+ \frac{1}{N^2}\sum_{j,k}\left[c_{ijk}\sin(\theta_{ji} + \alpha_{3ijk})\cos(\theta_{ki} + \alpha_{4ijk})\right]\,,\end{aligned} \tag{6.8}$$

where we draw the coefficients a_{ij}, b_{ij}, c_{ijk}, α_{1ij}, α_{2ij}, α_{3ijk}, and α_{4ijk} from various probability distributions. Including three-body interactions leads to a large variety of intricate dynamics, and I anticipate that incorporating the formalism of simplicial complexes will be very helpful for categorizing the possible dynamics.

In the last few years, several other researchers have also studied Kuramoto models with three-body interactions [196–198]. A recent study [198], for example, discovered a continuum of abrupt desynchronization transitions with no counterpart in abrupt synchronization transitions. There have been mathematical studies of coupled oscillators with interactions of three or more entities using methods such as normal-form theory [199] and coupled-cell networks [26].

An important point, as one can see in the above discussion (which does not employ the mathematical formalism of simplicial complexes), is that one does not necessarily need to explicitly use the language of simplicial complexes to study interactions between three or more entities in dynamical systems. Nevertheless, I anticipate that explicitly incorporating the formalism of simplicial complexes will be useful both for studying coupled oscillators on networks and for other dynamical systems. In upcoming studies, it will be important to determine when this formalism helps illuminate the dynamics of multi-entity interactions in dynamical systems and when simpler approaches suffice.

6.6.4 Social Dynamics and Simplicial Complexes

Several recent papers have generalized models of social dynamics by incorporating higher-order interactions [62, 200–202]. For example, perhaps somebody's opinion is influenced by a group discussion of three or more people, so it is relevant to consider opinion updates that are based on higher-order interactions. Some of these papers use some of the terminology of simplicial complexes, but it is mostly unclear (except perhaps for [201]) how the models in those papers take advantage of the associated mathematical formalism, so arguably it often may be unnecessary to use such language. Nevertheless, these models are very interesting and provide promising avenues for further research.

Petri and Barrat [62] generalized activity-driven models to simplicial complexes. Such a simplicial activity-driven (SAD) model generates time-dependent simplicial complexes, on which it is desirable to study dynamical processes (see Sect. 6.4), such as opinion dynamics, social contagions, and biological contagions.

The simplest version of the SAD model is defined as follows.

- Each node i has an activity rate a_i that we draw independently from a distribution $F(x)$.
- At each discrete time step (of length Δt), we start with N isolated nodes. Each node i is active with a probability of $a_i \Delta t$, independently of all other nodes. If it is active, it creates a $(p-1)$-simplex (forming, in network terms, a clique of p nodes) with $p-1$ other nodes that we choose uniformly and independently at random (without replacement). One can either use a fixed value of p or draw p from some probability distribution.
- At the next time step, we delete all edges, so all interactions have a constant duration. We then generate new interactions from scratch.

This version of the SAD model is Markovian, and it is desirable to generalize it in various ways (e.g. by incorporating memory or community structure).

Iacopini et al. [200] recently developed a simplicial contagion model that generalizes an SIS process on a graph. Consider a simplicial complex K with N nodes, and associate each node i with a state $x_i(t) \in \{0, 1\}$ at time t. If $x_i(t) = 0$, node i is part of the susceptible compartment S; if $x_i(t) = 1$, it is part of the infected compartment I. The density of infected nodes at time t is $\rho(t) = \frac{1}{N}\sum_{i=1}^{N} x_i(t)$. Suppose that there are D parameters $\varpi_1, \ldots, \varpi_D$ (with $D \in \{1, \ldots, N-1\}$), where ϖ_d represents the probability per unit time that a susceptible node i that participates in a d-dimensional simplex σ is infected from each of the faces of σ, under the condition that all of the other nodes of the face are infected. That is, ϖ_1 is the probability per unit time that node i is infected by an adjacent node j via the edge (i, j). Similarly, ϖ_2 is the probability per unit time that node i is infected via the 2-simplex (i, j, k) in which both j and k are infected, and so on. The recovery dynamics, in which an infected node i becomes susceptible again, proceeds as in the SIR model that I discussed in Sect. 6.4.2, except that the node is again susceptible to the contagion. One can envision numerous interesting generalizations of this model (e.g. ones that are inspired by ideas that have been investigated in contagion models on graphs).

6.7 Outlook

The study of networks is one of the most exciting and rapidly expanding areas of mathematics, and it touches on myriad other disciplines in both its methodology and its applications. Network analysis is increasingly prominent in numerous fields of scholarship (both theoretical and applied), it interacts very closely with data science, and it is important for a wealth of applications.

My focus in this chapter has been a forward-looking presentation of ideas in network analysis. My choices of which ideas to discuss reflect their connections to dynamics and nonlinearity, although I have also mentioned a few other burgeoning areas of network analysis in passing. Through its exciting combination of graph theory, dynamical systems, statistical mechanics, probability, linear algebra, scientific computation, data analysis, and many other subjects—and through a comparable diversity of applications across the sciences, engineering, and the humanities—the mathematics and science of networks has plenty to offer researchers for many years.

Acknowledgements I thank Jesús Cuevas-Maraver, Panos Kevrekidis, and Avadh Saxena for the invitation to write this book chapter. I acknowledge financial support from the National Science Foundation (grant number 1922952) through the Algorithms for Threat Detection (ATD) program. I thank Mariano Beguerisse Díaz, Manlio De Domenico, James Gleeson, Petter Holme, Renaud Lambiotte, and Hiroki Sayama for their helpful comments. I am particularly grateful to Heather Brooks, Michelle Feng, Panos Kevrekidis, and Alice Schwarze for their thorough and insightful comments on drafts of this chapter.

References

1. M.E.J. Newman, *Networks*, 2nd edn. (Oxford University Press, Oxford, UK, 2018)
2. M.A. Porter, S.D. Howison, The role of network analysis in industrial and applied mathematics (2017), arXiv:1703.06843
3. M.A. Porter, G. Bianconi, Eur. J. App. Math. **27**, 807 (2016)
4. F. Bullo, *Lectures on Network Systems*, 1.3 edn. (Kindle Direct Publishing, 2019), http://motion.me.ucsb.edu/book-lns
5. R. Lambiotte, M. Rosvall, I. Scholtes, Nat. Phys. **15**, 313 (2019)
6. P. Holme, J. Saramäki, Phys. Rep. **519**, 97 (2012)
7. P. Holme, Eur. Phys. J. B **88**, 234 (2015)
8. P.S. Chodrow, A. Mellor, App. Netw. Sci. **5**, 9 (2020)
9. M. Kivelä, A. Arenas, M. Barthelemy, J.P. Gleeson, Y. Moreno, M.A. Porter, J. Complex Netw. **2**, 203 (2014)
10. M.A. Porter, Notices Am. Math. Soc. **65**, 1419 (2018)
11. M.A. Porter, J.P. Gleeson, *Dynamical Systems on Networks: A Tutorial*, vol. 4 in Frontiers in Applied Dynamical Systems: Reviews and Tutorials (Springer, Cham, Switzerland, 2016)
12. M. Rosvall, A. Esquivel, A. Lancichinetti, J. West, R. Lambiotte, Nat. Commun. **5**, 4630 (2014)
13. N. Otter, M.A. Porter, U. Tillmann, P. Grindrod, H.A. Harrington, Eur. Phys. J. Data Sci. **6**, 17 (2017)
14. A. Patania, F. Vaccarino, G. Petri, Eur. Phys. J. Data Sci. **6**, 7 (2017)
15. V.A. Traag, P. Doreian, A. Mrvar, in *Advances in Network Clustering and Blockmodeling*, ed. by P. Doreian, V. Batagelj, A. Ferligoj (Wiley, Hoboken, NJ, USA, 2020), p. 225
16. P. Van Mieghem, *Graph Spectra for Complex Networks* (Cambridge University Press, Cambridge, UK, 2013)
17. P. Holme, J. Saramäki (eds.), *Temporal Networks* (Springer, Heidelberg, Germany, 2013)
18. P. Holme, J. Saramäki (eds.), *Temporal Network Theory* (Springer, Cham, Switzerland, 2019)
19. E. Valdano, L. Ferreri, C. Poletto, V. Colizza, Phys. Rev. X **5**, 021005 (2015)
20. D. Taylor, M.A. Porter, P.J. Mucha, in *Temporal Network Theory*, ed. by P. Holme, J. Saramäki (Springer, Cham, Switzerland, 2019), p. 325
21. M. De Domenico, A. Solé-Ribalta, E. Cozzo, M. Kivelä, Y. Moreno, M.A. Porter, S. Gómez, A. Arenas, Phys. Rev. X **3**, 041022 (2013)
22. I. de Sola Pool, M. Kochen, Soc. Netw. **1**, 5 (1978)
23. S. Wasserman, K. Faust, *Social Network Analysis: Methods and Applications* (Cambridge University Press, Cambridge, UK, 1994)
24. D.F. Gleich, SIAM Rev. **57**, 321 (2015)
25. R. Milo, S. Shen-Orr, S. Itzkovitz, N. Kashtan, D. Chklovskii, U. Alon, Science **298**, 824 (2002)
26. M. Golubitsky, I. Stewart, A. Török, SIAM. J. Appl. Dyn. Syst. **4**, 78 (2005)
27. M.A. Porter, J.-P. Onnela, P.J. Mucha, Notices Am. Math. Soc. **56**, 1082 (2009)
28. S. Fortunato, D. Hric, Phys. Rep. **659**, 1 (2016)
29. P. Csermely, A. London, L.Y. Wu, B. Uzzi, J. Complex Netw. **1**, 93 (2013)
30. P. Rombach, M.A. Porter, J.H. Fowler, P.J. Mucha, SIAM Rev. **59**, 619 (2017)
31. H. Simon, Proc. Am. Philos. Soc. **106**, 467 (1962)
32. M.E.J. Newman, Nat. Phys. **14**, 542 (2018)
33. T.P. Peixoto, in *Advances in Network Clustering and Blockmodeling*, ed. by P. Doreian, V. Batagelj, A. Ferligoj (Wiley, Hoboken, NJ, USA, 2020), p. 289
34. A. Aleta, Y. Moreno, Annu. Rev. Condens. Matter Phys. **10**, 45 (2019)
35. J. Tang, M. Musolesi, C. Mascolo, V. Latora, V. Nicosia, in *SNS '10—Proceedings of the 3rd Workshop on Social Network Systems*. Paris (Association for Computing Machinery, New York City, NY, USA, 2010), Article No. 3
36. H. Kim, J. Tang, R. Anderson, C. Mascolo, Comput. Netw. **56**, 983 (2012)

37. K. Lerman, R. Ghosh, J.H. Kang, in *MLG'10—Proceedings of the 8th Workshop on Mining and Learning with Graphs*. Washington D.C. (Association for Computing Machinery, New York City, NY, USA, 2010), p. 70
38. P. Grindrod, D.J. Higham, Proc. R. Soc. A **470**, 20130835 (2014)
39. P. Grindrod, D.J. Higham, SIAM Rev. **55**, 118 (2013)
40. D. Walker, H. Xie, K.K. Yan, S. Maslov, J. Stat. Mech.: Theory Exp. **2007**, P06010 (2007)
41. R.A. Rossi, D.F. Gleich, in *WAW 2018—Proceedings of the 15th International Workshop on Algorithms and Models for the Web Graph*. Moscow (Springer, Heidelberg, Germany, 2012), p. 126
42. S. Praprotnik, V. Batagelj, Ars Math. Contemp. **11**, 11 (2015)
43. J. Flores, M. Romance, J. Comput. Appl. Math. **330**, 1041 (2018)
44. D. Taylor, S.A. Myers, A. Clauset, M.A. Porter, P.J. Mucha, Multiscale Model Simul. **15**, 537 (2017)
45. D. Taylor, M.A. Porter, P.J. Mucha, Tunable eigenvector-based centralities for multiplex and temporal networks (2019), arXiv:1904.02059
46. P. Bonacich, J. Math. Sociol. **2**, 113 (1972)
47. J. Kleinberg, J. ACM **46**(5), 604 (1999)
48. The Mathematics Genealogy Project (2020), http://www.genealogy.ams.org. Accessed 27 Jan 2020
49. T.P. Peixoto, M. Rosvall, Nat. Commun. **8**, 582 (2017)
50. P.J. Mucha, T. Richardson, K. Macon, M.A. Porter, J.-P. Onnela, Science **328**(5980), 876 (2010)
51. M. De Domenico, A. Lancichinetti, A. Arenas, M. Rosvall, Phys. Rev. X **5**, 011027 (2015)
52. L.G.S. Jeub, M.W. Mahoney, P.J. Mucha, M.A. Porter, Netw. Sci. **5**, 144 (2017)
53. M. Vaiana, S. Muldoon, Resolution limits for detecting community changes in multilayer networks (2018), arXiv:1803.03597
54. L. Papadopoulos, M.A. Porter, K.E. Daniels, D.S. Bassett, J. Complex Netw. **6**, 485 (2018)
55. J. Moody, P.J. Mucha, Netw. Sci. **1**, 119 (2013)
56. M. Sarzynska, E.A. Leicht, G. Chowell, M.A. Porter, J. Complex Netw. **4**, 363 (2016)
57. S. Pilosof, M.A. Porter, M. Pascual, S. Kéfi, Nat. Ecol. Evol. **1**, 0101 (2017)
58. K.R. Finn, M.J. Silk, M.A. Porter, N. Pinter-Wollman, Anim. Behav. **149**, 7 (2019)
59. M. Bazzi, L.G.S. Jeub, A. Arenas, S.D. Howison, M.A. Porter, Phys. Rev. Res. **2**, 023100 (2020)
60. N. Perra, B. Gonçalves, R. Pastor-Satorras, A. Vespignani, Sci. Rep. **2**, 469 (2012)
61. L. Zino, A. Rizzo, M. Porfiri, SIAM, J. Appl. Dyn. Syst. **17**, 2830 (2018)
62. G. Petri, A. Barrat, Phys. Rev. Lett. **121**, 228301 (2018)
63. E. Valdano, M.R. Fiorentin, C. Poletto, V. Colizza, Phys. Rev. Lett. **120**, 068302 (2018)
64. L. Zino, A. Rizzo, M. Porfiri, Phys. Rev. Lett. **117**, 228302 (2016)
65. L. Zino, A. Rizzo, M. Porfiri, J. Complex Netw. **5**, 924 (2017)
66. S. Motegi, N. Masuda, Sci. Rep. **2**, 904 (2012)
67. W. Ahmad, M.A. Porter, M. Beguerisse-Díaz, Tie-decay temporal networks in continuous time and eigenvector-based centralities (2019), arXiv:1805.00193
68. P.J. Laub, T. Taimre, P.K. Pollett, Hawkes processes (2015), arXiv:1507.02822
69. E.M. Jin, M. Girvan, M.E.J. Newman, Phys. Rev. E **64**, 046132 (2001)
70. X. Zuo, M.A. Porter, Models of continuous-time networks with tie decay, diffusion, and convection (2019), arXiv:1906.09394
71. R. Erban, S.J. Chapman, P.K. Maini, A practical guide to stochastic simulations of reaction-diffusion processes (2017), arXiv:0704.1908
72. C.L. Vestergaard, M. Génois, PLoS Comput. Biol. **11**, e1004579 (2015)
73. S. Brin, L. Page, Comput. Netw. **30**, 107 (1998)
74. A. Arenas, A. Díaz-Guilera, J. Kurths, Y. Moreno, C. Zhou, Phys. Rep. **469**, 93 (2008)
75. F.A. Rodrigues, T.K.DM. Peron, P. Ji, J. Kurths, Phys. Rep. **469**, 1 (2016)
76. M.O. Jackson, Y. Zenou, in *Handbook of Game Theory*, vol. 4, ed. by P. Young, S. Zamir (Elsevier, Amsterdam, The Netherlands, 2014), p. 95

77. I.Z. Kiss, J.C. Miller, P.L. Simon, *Mathematics of Epidemics on Networks: From Exact to Approximate Models* (Springer, Cham, Switzerland, 2017)
78. R. Pastor-Satorras, C. Castellano, P. Van Mieghem, A. Vespignani, Rev. Mod. Phys. **87**, 925 (2015)
79. S. Lehmann, Y.-Y. Ahn, *Complex Spreading Phenomena in Social Systems: Influence and Contagion in Real-World Social Networks* (Springer, Cham, Switzerland, 2018)
80. C. Castellano, S. Fortunato, V. Loreto, Rev. Mod. Phys. **81**, 591 (2009)
81. Y.Y. Liu, A.L. Barabási, Rev. Mod. Phys. **88**, 035006 (2016)
82. A. Motter, Chaos **25**, 097501 (2015)
83. D.J. Watts, Proc. Natl. Acad. Sci. USA **99**, 5766 (2002)
84. J. Chalupa, P.L. Leath, G.R. Reich, J. Phys. C **12**, L31 (1979)
85. J.P. Gleeson, Phys. Rev. X **3**, 021004 (2013)
86. M. Granovetter, Am. J. Sociol. **83**, 1420 (1978)
87. T.W. Valente, *Network Models of the Diffusion of Innovations* (Hampton Press, New York City, NY, USA, 1995)
88. D. Kempe, J. Kleinberg, É. Tardos, in *KDD '03—Proceedings of the Ninth ACM SIGKDD International Conference on Knowledge Discovery and Data Mining*. Washington D.C., August 2003 (Association for Computing Machinery, New York City, NY, USA, 2003), p. 137
89. S. Melnik, J.A. Ward, J.P. Gleeson, M.A. Porter, Chaos **23**, 013124 (2013)
90. S.W. Oh, M.A. Porter, Chaos **28**, 033101 (2018)
91. J.S. Juul, M.A. Porter, Chaos **28**, 013115 (2018)
92. J.S. Juul, M.A. Porter, Phys. Rev. E **99**, 022313 (2019)
93. P. Oliver, G. Marwell, R. Teixeira, Am. J. Sociol. **91**, 522 (1985)
94. P.G. Fennell, S. Melnik, J.P. Gleeson, Phys. Rev. E **94**, 052125 (2016)
95. D. Taylor, F. Klimm, H.A. Harrington, M. Kramár, K. Mischaikow, M.A. Porter, P.J. Mucha, Nat. Commun. **6**, 7723 (2015)
96. A.A. Saberi, Phys. Rep. **578**, 1 (2015)
97. N. Masuda, M.A. Porter, R. Lambiotte, Phys. Rep. **716–717**, 1 (2017)
98. X.F. Meng, R.A. Van Gorder, M.A. Porter, Phys. Rev. E **97**, 022312 (2018)
99. S. Redner, C. R. Phys. **20**, 275 (2019)
100. F. Brauer, C. Castillo Chavez, *Mathematical Models in Population Biology and Epidemiology*, 2nd edn. (Springer, Heidelberg, Germany, 2012)
101. A. Arenas, W. Cota, J. Gomez-Gardenes, S. Gómez, C. Granell, J.T. Matamalas, D. Soriano-Panos, B. Steinegger, A mathematical model for the spatiotemporal epidemic spreading of COVID19, medRxiv (2020), https://doi.org/10.1101/2020.03.21.20040022
102. M. De Domenico, C. Granell, M.A. Porter, A. Arenas, Nat. Phys. **12**, 901 (2016)
103. V. Colizza, R. Pastor-Satorras, A. Vespignani, Nat. Phys. **3**, 276 (2007)
104. P. Van Mieghem, R. Van de Bovenkamp, Phys. Rev. Lett. **110**, 108701 (2013)
105. S.V. Scarpino, A. Allard, L. Hébert-Dufresne, Nat. Phys. **12**, 1042 (2016)
106. Z. Wang, Z.X.W.M.A. Andrews, L. Wang, C.T. Bauch, Phys. Life Rev. **15**, 1 (2015)
107. S. Funk, M. Salathé, V.A.A. Jansen, J. R. Soc. Interface **7**, 1247 (2010)
108. S. Goel, A. Anderson, J. Hofman, D.J. Watts, Manag. Sci. **62**, 180 (2016)
109. A. Volkening, D.F. Linder, M.A. Porter, G.A. Rempala, Forecasting elections using compartmental models of infection (2019), arXiv:1811.01831
110. P. Grassberger, Math. Biosci. **63**, 157 (1983)
111. E. Kenah, J.M. Robins, Phys. Rev. E **76**, 036113 (2007)
112. P. Trapman, Theor. Popul. Biol. **71**, 160 (2007)
113. L. Hébert-Dufresne, O. Patterson-Lomba, G.M. Goerg, B.M. Althouse, Phys. Rev. Lett. **110**, 108103 (2013)
114. S. Morita, Sci. Rep. **6**, 22506 (2016)
115. A. Pikovsky, M. Rosenblum, Scholarpedia **2**, 1459 (2007)
116. L.M. Pecora, T.L. Carroll, Phys. Rev. Lett. **80**, 2109 (1998)

117. L.M. Pecora, F. Sorrentino, A.M. Hagerstrom, T.E. Murphy, R. Roy, Nat. Commun. **5**, 4079 (2014)
118. M.J. Panaggio, D.M. Abrams, Nonlinearity **28**, R67 (2015)
119. R.M. D'Souza, J. Gómez-Gardeñes, J. Nagler, A. Arenas, Adv. Phys. **68**, 123 (2019)
120. P. Ashwin, S. Coombes, R. Nicks, J. Math. Neurosci. **6**, 2 (2016)
121. Z. Lu, K. Klein-Cardeña, S. Lee, T.M. Antonsen, M. Girvan, E. Ott, Chaos **26**, 094811 (2016)
122. K. Ota, I. Aihara, T. Aoyagi, Royal Soc. Open Sci. **7**, 191693 (2020)
123. G.A. Gottwald, Chaos **27**(10), 101103 (2017)
124. S. Boccaletti, J.A. Almendral, S. Guan, I. Leyva, Z. Liu, I. Sendiña-Nadal, Z. Wang, Y. Zou, Phys. Rep. **660**, 1 (2016)
125. S.H. Strogatz, Phys. D **143**, 1 (2000)
126. D. Witthaut, M. Timme, New J. Phys. **14**, 083036 (2012)
127. G.S. Medvedev, SIAM. J. Math. Anal. **46**, 2743 (2014)
128. M. De Domenico, A. Solé-Ribalta, S. Gómez, A. Arenas, Proc. Natl. Acad. Sci. USA **111**, 8351 (2014)
129. J.M. Buldú, M.A. Porter, Netw. Neurosci. **2**(4), 418 (2018)
130. R. Gallotti, M. Barthelemy, Sci. Rep. **4**, 6911 (2014)
131. A. Solé-Ribalta, S. Gómez, A. Arenas, Phys. Rev. Lett. **116**, 108701 (2016)
132. S. Funk, S. Bansal, C.T. Bauch, K.T.D. Eames, W.J. Edmunds, A.P. Galvani, P. Klepac, Epidemics **10**, 21 (2015)
133. V. Nicosia, P.S. Skardal, A. Arenas, V. Latora, Phys. Rev. Lett. **118**, 138302 (2017)
134. J. Sanz, C.Y. Xia, S. Meloni, Y. Moreno, Phys. Rev. X **4**, 041005 (2014)
135. C. Granell, S. Gómez, A. Arenas, Phys. Rev. Lett. **111**, 128701 (2013)
136. K.A. Jacobsen, M.G. Burch, J.H. Tien, G.A. Rempala, J. Biol. Dyn. **12**, 746 (2018)
137. S. Gnutzmann, U. Smilansky, Adv. Phys. **55**, 527 (2006)
138. E. Mones, N.A.M. Araújo, T. Vicsek, H.J. Herrmann, Sci. Rep. **4**, 4949 (2014)
139. Y. Ide, H. Izuhara, T. Machida, Phys. A **457**, 331 (2016)
140. M. Asllani, D.M. Busiello, T. Carletti, D. Fanelli, G. Planchon, Phys. Rev. E **90**, 042814 (2014)
141. S. Contemori, F.D. Patti, D. Fanelli, F. Miele, Phys. Rev. E **93**, 032317 (2016)
142. N.E. Kouvaris, S. Hata, A. Díaz-Guilera, Sci. Rep. **5**, 10840 (2015)
143. P. Kuchment, Waves Random Complex **14**, S107 (2004)
144. Quantum wire (Wikipedia, The Free Encyclopedia) (2020), https://en.wikipedia.org/w/index.php?title=Quantum_wire&oldid=925324634. Accessed 27 Jan 2020
145. D. Noja, Phil. Trans. R. Soc. A **372**, 20130002 (2014)
146. K.K. Sabirov, D.B. Babajanov, D.U. Matrasuolv, P.G. Kevrekidis, J. Phys. A: Math. Theor. **51**, 435203 (2018)
147. J.L. Marzuola, D.E. Pelinovsky, Appl. Math. Res. Express **2016**, 98 (2016)
148. A. Kairzhan, D.E. Pelinovsky, R.H. Goodman, SIAM. J. Appl. Dyn. Syst. **18**, 1723 (2019)
149. Z. Sobirov, D. Babajanov, D. Matrasulov, K. Nakamura, H. Uecker, Europhys. Lett. **115**, 50002 (2016)
150. P. Joly, M. Kachanovska, A. Semin, Netw. Heterog. Media **14**, 205 (2019)
151. A. Leonard, L. Ponson, C. Daraio, J. Mech. Phys. Solids **73**, 103 (2014)
152. M. Barthelemy, *Morphogenesis of Spatial Networks* (Springer, Cham, Switzerland 2018)
153. T. Gross, B. Blasius, J. R. Soc. Interface **5**, 259 (2007)
154. H. Sayama, I. Pestov, J. Schmidt, B.J. Bush, C. Wong, J. Yamanoi, T. Gross, Comput. Math. Appl. **65**, 1645 (2013)
155. T. Aoki, L.E.C. Rocha, T. Gross, Phys. Rev. E **93**, 040301 (2016)
156. H. Sayama, R. Sinatra, Phys. Rev. E **91**, 032809 (2015)
157. R. Berner, J. Sawicki, E. Schöll, Phys. Rev. Lett. **124**, 088301 (2020)
158. T. Gross, C.J.D. D'Lima, B. Blasius, Phys. Rev. Lett. **96**, 208701 (2006)
159. R.T. Durrett, J.P. Gleeson, A.L. Lloyd, P.J. Mucha, F. Shi, D. Sivakoff, J.E.S. Socolar, C. Varghese, Proc. Natl. Acad. Sci. USA **109**, 3682 (2012)

160. Y.H. Kureh, M.A. Porter, Fitting in and breaking up: A nonlinear version of coevolving voter models. Phys. Rev. E, in press (2020), arXiv:1907.11608
161. P. Clifford, A. Sudbury, Biometrika **60**, 581 (1973)
162. R.A. Holley, T.M. Liggett, Ann. Probab. **3**, 643 (1975)
163. J. Fernández-Gracia, K. Suchecki, J.J. Ramasco, M. San Miguel, V.M. Eguíluz, Phys. Rev. Lett. **112**, 158701 (2014)
164. P. Holme, M.E.J. Newman, Phys. Rev. E **74**, 056108 (2006)
165. R. Basu, A. Sly, Ann. Appl. Probab. **27**, 1235 (2017)
166. F. Shi, P.J. Mucha, R.T. Durrett, Phys. Rev. E **88**, 062818 (2013)
167. N. Malik, F. Shi, H.W. Lee, P.J. Mucha, Chaos **26**, 123112 (2016)
168. H.Z. Brooks, M.A. Porter, Phys. Rev. Res. **2**, 023041 (2020)
169. M. Brede, Complexity **2019**, 1486909 (2019)
170. J.P. Stroud, M.A. Porter, G. Hennequin, T.P. Vogels, Nat. Neurosci. **21**, 1774 (2018)
171. P.S. Skardal, D. Taylor, J.G. Restrepo, Phys. D **267**, 27 (2014)
172. Y. Choe, in *Encyclopedia of Computational Neuroscience*, ed. by D. Jaeger, R. Jung (Springer, Heidelberg, Germany, 2014)
173. M.E.J. Newman, Phys. Rev. E **64**, 016131 (2001)
174. M.A. Porter, P.J. Mucha, M.E.J. Newman, C.M. Warmbrand, Proc. Natl. Acad. Sci. USA **102**, 7057 (2002)
175. G. Ghoshal, V. Zlatić, G. Caldarelli, M.E.J. Newman, Phys. Rev. E **79**, 066118 (2009)
176. P.S. Chodrow, Configuration models of random hypergraphs (2019), arXiv:1902.09302
177. C. Giusti, R. Ghrist, D.S. Bassett, J. Comput. Neurosci. **41**, 1 (2016)
178. R. Ghrist, *Elementary Applied Topology*. CreateSpace Independent Publishing Platform (Edition 1.0, 2014), https://www.math.upenn.edu/~ghrist/notes.html. Accessed 27 January 2020
179. C.M. Topaz, Topological Data Analysis. DSWeb: The Dynamical Systems Web (2016), https://dsweb.siam.org/The-Magazine/Article/topological-data-analysis-1. Accessed 27 Jan 2020
180. B.K. Fosdick, D.B. Larremore, J. Nishimura, J. Ugander, SIAM Rev. **60**, 315 (2018)
181. O. Bobrowski, M. Kahle, J. Appl. Comput. Topol. **1**, 331 (2018)
182. O.T. Courtney, G. Bianconi, Phys. Rev. E **93**, 062311 (2016)
183. G. Bianconi, C. Rahmede, Phys. Rev. E **93**, 032315 (2016)
184. A. Samal, R.P. Sreejith, J. Gu, S. Liu, E. Saucan, J. Jost, Sci. Rep. **8**, 8650 (2018)
185. M. Kramár, A. Goullet, L. Kondic, K. Mischaikow, Phys. D **283**, 37 (2014)
186. A.E. Sizemore, J.E. Phillips-Cremins, R. Ghrist, D.S. Bassett, Netw. Neurosci. **3**, 656 (2019)
187. M. Feng, M.A. Porter, Persistent homology of geospatial data: A case study with voting. SIAM Rev., in press (2020), arXiv:1902.05911
188. L. Speidel, H.A. Harrington, S.J. Chapman, M.A. Porter, Phys. Rev. E **98**, 012318 (2018)
189. C.M. Topaz, L. Ziegelmeier, T. Halverson, PLoS ONE **10**, e0126383 (2015)
190. G. Yalnız, N.B. Budanur, Chaos **30**, 033109 (2020)
191. S. Tymochko, E. Munch, J. Dunion, K. Corbosiero, R. Torn, Using persistent homology to quantify a diurnal cycle in hurricanes (2019), arXiv:1902.06202
192. N.H. Christianson, A. Sizemore Blevins, D.S. Bassett, Architecture and evolution of semantic networks in mathematics texts (2019), arXiv:1908.04911
193. V. Salnikov, D. Cassese, R. Lambiotte, Eur. J. Phys. **40**, 014001 (2018)
194. M.T. Schaub, A.R. Benson, P. Horn, G. Lippner, A. Jadbabaie, SIAM Rev. **62**, 353 (2020)
195. T. Tanaka, T. Aoyagi, Phys. Rev. Lett. **106**, 224101 (2011)
196. M. Komarov, A. Pikovsky, Phys. Rev. Lett. **110**, 134101 (2013)
197. M. Komarov, A. Pikovsky, Phys. Rev. E **92**, 020901(R) (2015)
198. P.S. Skardal, A. Arenas, Phys. Rev. Lett. **122**, 248301 (2019)
199. C. Bick, P. Ashwin, A. Rodrigues, Chaos **26**, 094814 (2016)
200. I. Iacopini, G. Petri, A. Barrat, V. Latora, Nat. Commun. **10**, 2485 (2019)
201. L. Horstmeyer, C. Kuehn, Phys. Rev. E **101**, 022305 (2020)
202. L. Neuhäuser, A. Mellor, R. Lambiotte, Phys. Rev. E **101**, 032310 (2020)

Chapter 7
Integrability and Nonlinear Waves

Mark J. Ablowitz

Abstract Major developments in the study of nonlinear wave equations were the discovery of the soliton, its connection to the eigenvalue of a linear operator and solutions of the underlying equations by the inverse scattering transform. Inverse scattering transform provides the solution to the initial value problem of a class of nonlinear equations. In this article the background, key ideas and methodology of inverse scattering transform are discussed in connection with some well-known physically important nonlinear equations including the Korteweg–de Vries and nonlinear Schrödinger equations. More recently a new class of nonlocal nonlinear Schrödinger type equations have been found to be integrable; they arise from new symmetries in the associated scattering problem that had not been previously known. The solution of these novel systems is also discussed in this review. Other methods of solution of soliton/integrable equations and comments regarding the future are also included.

7.1 Introduction

The topic of this chapter, integrability and nonlinear waves, has an interesting history. In fact, on both topics separately, integrability and nonlinear waves, one can write lengthy histories. What is surprising is that these topics were found to be intimately related in the 1960s. This is where we shall begin. In the mid 1960s M. Kruskal learned about the so-called Fermi–Pasta–Ulam (FPU) problem which is a nonlinear spring mass system (see Fig. 7.1) with force law:

$$F(\Delta) = -k(\Delta + \alpha\ \Delta^2),\ k, \alpha \text{ const.}$$

We refer the reader to [1] for additional background discussion about the FPU problem. This spring mass system had shown an unusual property, periodic initial con-

M. J. Ablowitz (✉)
Department of Applied Mathematics, University of Colorado, Campus Box 526,
Boulder, CO 80309-0526, USA
e-mail: mark.ablowitz@colorado.edu

P. G. Kevrekidis et al. (eds.), *Emerging Frontiers in Nonlinear Science*, Nonlinear Systems and Complexity 32, https://doi.org/10.1007/978-3-030-44992-6_7

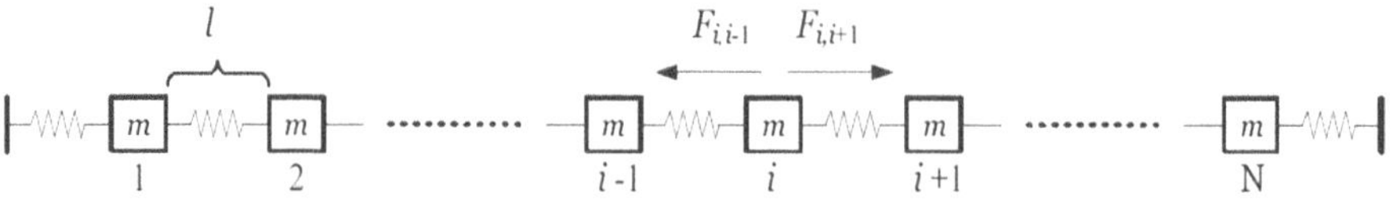

Fig. 7.1 Fermi–Pasta–Ulam mass-spring system

ditions nearly recurred after some time. To understand this recurrence phenomena Zabusky and Kruskal [2] took the continuum limit and they obtained the following equation

$$u_t + uu_x + \delta^2 u_{xxx} = 0, u(x, 0) = \cos(\pi x) \tag{7.1}$$

governing unidirectional waves; here δ^2 is related to α. Numerical computations were carried out in [2]; it was found that (7.1) like the FPU problem exhibited near recurrence of initial states. The explanation of this recurrence was attributed to the way special solutions of this equation interacted. In order to simplify the discussion, (7.1) can be transformed to the following normalized form

$$u_t + 6uu_x + u_{xxx} = 0 \tag{7.2}$$

Equation (7.2) has the following special localized solution called a solitary wave

$$u_s(x, t) = 2\kappa^2 \text{sech}^2 \kappa(x - 4\kappa^2 t - x_0) \tag{7.3}$$

where κ, x_0 are constant. Remarkably Zabusky and Kruskal found that two such waves, with different amplitudes, big one (amplitude: κ_1) to the left of the small one (amplitude: κ_2), maintained the same amplitude (and speed) after they interacted (see Fig. 7.2) they termed these localized waves *solitons*. Later in this Chapter we discuss the mathematical underpinnings of this concept.

As Kruskal and Zabusky found out, (7.2) was one that was studied many years earlier. Namely in 1895 in the context of small amplitude, shallow water waves D. Korteweg and his student de Vries [3] derived this equation. In physical units they found the following unidirectional equation

$$\frac{1}{\sqrt{gh}}\eta_t + \eta_x + \frac{3}{2h}\eta\eta_x + \frac{h^2}{2}\left(\frac{1}{3} - \hat{T}\right)\eta_{xxx} = 0 \tag{7.4}$$

where η is the wave elevation, g gravity, h mean depth, $\hat{T}$ normalized surface tension ($\hat{T} = \frac{T}{\rho g h^2}$, ρ density). This equation can be transformed to the normalized (7.2) which we term the Korteweg–de Vries (KdV) equation. Korteweg and de Vries found a special periodic solution in terms of Jacobian elliptic functions which they termed the cnoidal wave. A special limit of the cnoidal wave is the solitary wave. When $\frac{1}{3} - \hat{T} > 0$ (most water waves have 'small'surface tension) the solitary wave is one of elevation. When $\frac{1}{3} - \hat{T} < 0$ ('large' $\hat{T}$ can occur with high surface tension fluids like

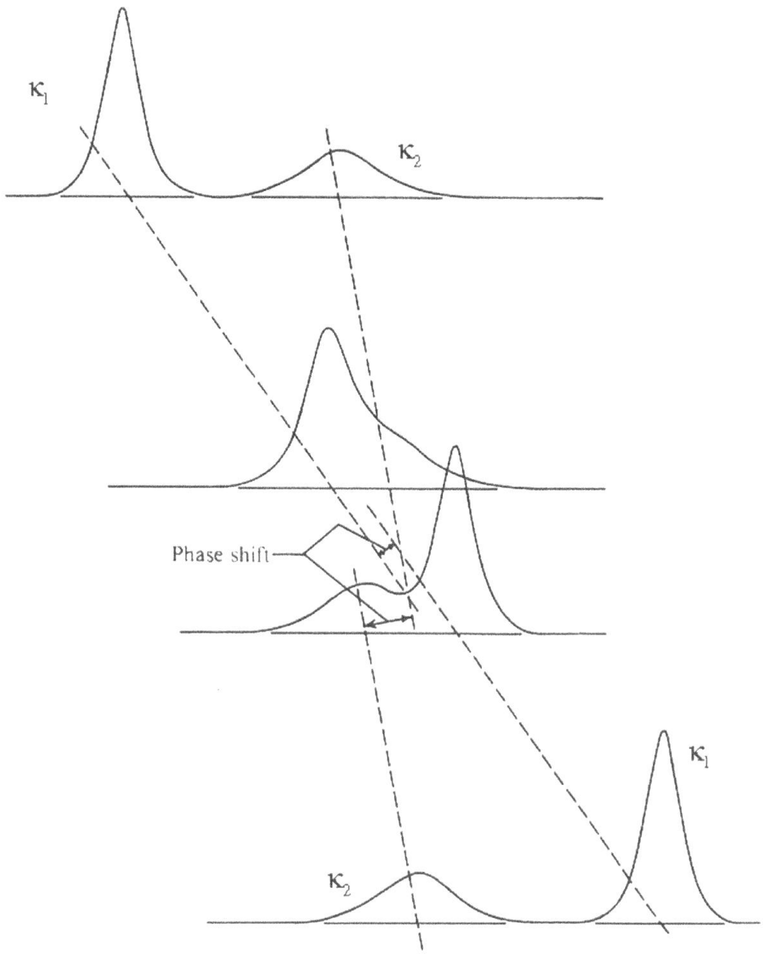

Fig. 7.2 Typical soliton interaction in the KdV equation. Here, $\kappa_1 > \kappa_2$

Mercury) the solitary wave is one of depression. Both have been observed, though solitary waves of elevation is the standard case. The solitary wave of depression was only observed in 2002 [4]. We remark that shallow water waves was heavily studied by Boussinesq in the 1870s cf. [5]. Boussinesq was primarily interested in two directional waves and found solitary waves for such equations; in his works he did mention that one could find unidirectional equations such as the KdV equation.

We remark that the name soliton has resonated within the broader physical science and engineering community. To many researchers a soliton only means a localized wave, namely this is what a solitary wave was originally called. So, many researchers do not use the term soliton in the context of how it was originally derived; i.e. solitons having special interaction properties and being associated with special equations such as (7.2). In this paper when we refer to a soliton we mean one that has special properties.

Solitary waves of elevation were first observed by John Scott Russell in 1834 while riding on horseback next to the union canal. He discusses this in his report to the British Association for the Advancement of Science in 1844 [6] where he described a "rounded smooth [...] well-defined heap of water". He made the following often quoted remark: "Such, in the month of August 1834, was my first chance interview with that singular and beautiful phenomenon ...". Though he termed this the "Great Wave of Translation", it later became known as the solitary wave. Russell also described experiments that he carried out on these solitary waves. Nevertheless Russell's work was controversial and not widely believed. It wasn't until Boussinesq and Korteweg–de Vries research with the underlying mathematics that explained Russell's observation was the controversy put to rest.

Earlier in the 1960s applied mathematicians realized that the KdV equation was universal. It always arises in weakly dispersive media with quadratically small amplitude cf. [7]. Soon after solitons were computationally discovered M. Kruskal and his postdoctoral associate R. Miura began studying conservation laws. The idea was that somehow conservation laws might explain the soliton interaction property. This proved to be the tip of the iceberg. They began finding numerous conservation laws associated with the KdV equation (7.2) and another closely aligned equation the so-called modified KdV (mKdV) equation

$$v_t - 6v^2 v_x + v_{xxx} = 0. \tag{7.5}$$

They found an infinite number of conservation laws to these equations [8] and the following transformation [9] between solutions of the mKdV and KdV equations

$$u(x,t) = v^2(x,t) - v_x(x,t). \tag{7.6}$$

At first they thought that the above transformation might be similar to the explicit Cole-Hopf transformation associated with the Burgers equation cf. [7]; since one can view this transformation as a Riccati equation for v in terms of u they linearized the transformation via $u(x,t) = v_x(x,t)/v(x,t)$ to find

$$v_{xx}(x,t) + u(x,t)v = 0 \tag{7.7}$$

This did not lead to any useful result regarding linearization. Then they argued that since the KdV equation is Galilean invariant one should be able to add a constant to $u(x,t)$ without loss of generality. Hence they were led to the time independent Schrödinger equation

$$\mathcal{L}_s v = v_{xx}(x,t) + u(x,t)v = \lambda v(x,t). \tag{7.8}$$

In [10] a second operator was found that was compatible with the KdV equation which we write in the equivalent form below

$$v_t(x,t) = \mathcal{M}v = (\gamma + u_x(x,t))v(x,t) - (4\lambda + 2u(x,t))v(x,t). \tag{7.9}$$

We note that from (7.8) λv converts to $(\partial_x^2 + u)v$; hence equivalently there are three derivatives in (7.9). Taking the time derivative of $\mathcal{L}_s v$ and two x−derivatives of $v_t(x,t)$ shows that $\lambda_t = 0$ if and only if the KdV equation is satisfied. KdV is said to be an isospectral flow associated with the time independent Schrödinger equation (7.8). This allowed Gardner, Greene, Kruskal and Miura to develop a method of solution (i.e. a linearization) to the KdV equation for rapidly decaying data [10]. Moreover solitons were associated with eigenvalues of the time independent Schrödinger equation. No question, this was a stunning and remarkable result.

Shortly afterwards Lax [11] generalized these ideas. He introduced two operators

$$\mathcal{L}v(x,t) = \lambda v(x,t), \quad v_t(x,t) = \mathcal{M}v(x,t) \tag{7.10}$$

Taking the time derivative of the first equation and using the second equation Lax found

$$\mathcal{L}_t + [\mathcal{L}, \mathcal{M}] = 0, \quad \lambda_t = 0 \tag{7.11}$$

where the bracket represents the commutator $[\mathcal{L}, \mathcal{M}] = \mathcal{L}\mathcal{M} - \mathcal{M}\mathcal{L}$. This result implies that whenever the formula (7.11) can be associated with a nonlinear evolution equation, it is an isospectral flow associated with $\mathcal{L}$.

The next important development was by Zakharov and Shabat [12] who critically used Lax ideas to find two compatible operators associated with the nonlinear Schrödinger (NLS) equation

$$iq_t + q_{xx} + 2q^2q^* = 0 \tag{7.12}$$

where $*$ represents the complex conjugate. This allowed Zakharov and Shabat to linearize the NLS equation for rapidly decaying data. Like the KdV equation, the NLS equation is universal. It occurs whenever one has weak nonlinearity and a slowly varying wave envelope.

Learning about the Zakharov and Shabat work Ablowitz, Kaup, Newell and Segur (AKNS) [13] generalized the approach in [12]. They introduced the operators

$$v_x(x,t) = \mathcal{L}v(x,t) = \begin{pmatrix} -ik & q(x,t) \\ r(x,t) & ik \end{pmatrix} v(x,t), \quad k = \text{const.} \tag{7.13}$$

$$v_t(x,t) = \mathcal{M}v(x,t), \qquad \mathcal{M} = \mathcal{M}[q, r; \lambda], \quad 2 \times 2 \text{ matrix} \tag{7.14}$$

Note from (7.13–7.14): Taking the t-derivative of the first equation and setting it equal to the x-derivative of the second equation leads to equations for the components of $\mathcal{M}$. In [13] methods to find compatible operators were developed. This led to *classes* of nonlinear equations which are compatible with these operators. In general these equations are evolution equations for both $r(x,t)$ and $q(x,t)$. The general evolution operator that was found to be associated with these equations is sometimes referred to as a recursion operator.

When there are symmetries between $r(x,t)$ and $q(x,t)$ we find one scalar equation. Below we list the best known of these equations and the symmetries between $r(x,t)$ and $q(x,t)$ that these equations satisfy.

- *NLS equation*:

$$r(x,t) = \mp q^*(x,t) \in \mathbb{C}: \quad iq_t + q_{xx} \pm 2q^2q^* = 0 \tag{7.15}$$

- *KdV equation*:

$$r(x,t) = -1: \quad q_t + q_{xxx} + 6qq_x = 0 \tag{7.16}$$

- *mKdV equation*:

$$r(x,t) = \mp q(x,t) \in \mathbb{R}: \quad q_t + q_{xxx} \pm 6q^2q_x = 0 \tag{7.17}$$

- *sine-Gordon equation* :

$$r(x,t) = -q(x,t) = u_x(x,t)/2 \in \mathbb{R}: \quad u_{xt} = \sin u \tag{7.18}$$

- *sinh-Gordon equation*:

$$r(x,t) = q(x,t) = u_x(x,t)/2 \in \mathbb{R}: \quad u_{xt} = \sinh u \tag{7.19}$$

Each of these equations have a different $\mathcal{M}$ operator. AKNS used this formulation to construct a method of solution/linearization to all of these equations and they found that solitons correspond to eigenvalues of the operator $\mathcal{L}$. They termed this method the Inverse Scattering Transform (IST) cf. [13, 14].

Soon after the above results were found, many new classes of nonlinear evolution equations were obtained; some of the physically interesting ones are listed below cf. [14, 15].

From matrix AKNS type systems the *vector NLS equation* was found

$$i\mathbf{q}_t + \mathbf{q}_{xx} + ||\mathbf{q}||^2\mathbf{q} = 0 \tag{7.20}$$

where $\mathbf{q} = (q_1, q_2, ...q_n)$, $||\mathbf{q}||^2 = \sum_j |q_j|^2$; similarly the *three wave interaction equations*

$$q_{1,t} + c_1 q_{1,x} = \gamma_1 q_2^* q_3^* : \quad 1, 2, 3 \text{ cyclic}, \quad c_1, \gamma_1 \text{ const.} \tag{7.21}$$

were solved by IST.[1]

Discrete nonlinear equations were solved by IST; two of the most interesting are the *Toda lattice equation*

---

[1]By solved we mean linearized in terms of integral equations.

$$q_{n,tt} = e^{-(q_n - q_{n-1})} - e^{-(q_{n+1} - q_n)} \tag{7.22}$$

and the *integrable discrete NLS equation*

$$iq_{n,t} = q_{n+1} + q_{n-1} \pm |q_n|^2 (q_{n+1} + q_{n-1}) \tag{7.23}$$

There were also nonlinear integro-differential equations; a well-known one is the *Benjamin–Ono equation*

$$q_t + qq_x + \mathcal{H}(q_{xx}) = 0 \tag{7.24}$$

where $\mathcal{H}q(x) = \frac{1}{\pi}\mathcal{P}\int_{-\infty}^{\infty} \frac{q(x')}{x'-x} dx'$ is the Hilbert transform.

There were also two space one time dimensional systems solved via IST; two important ones are the *Kadomtsev–Petviashvili (KP) equations*

$$\partial_x (q_t + 6qq_x + q_{xxx}) \pm 3q_{yy} = 0 : \quad \text{KPI (-), KPII (+)} \tag{7.25}$$

and the *Davey–Stewartson (DS) equation*

$$\begin{aligned} &iq_t + \frac{1}{2}\left(\gamma q_{xx} + q_{yy}\right) = \left(\phi - \sigma |q|^2\right) q \\ &\phi_{xx} - \gamma \phi_{yy} = 2\left(\sigma |q|^2\right)_{xx} \end{aligned} \tag{7.26}$$

where $\gamma = \pm 1, \sigma = \pm 1$.

Indeed there is also a four dimensional system that is integrable: *the self-dual Yang–Mills system* cf. [15].

The above list is only representative. There are many, many more. Most of these equations were derived/solved by IST in the 1970–1980s.

In a seemingly unrelated development Bender and Boettcher [16] found that the linear Schrödinger equation

$$iq_t + q_{xx} + V(x)q = 0, \quad \text{with } V(x) = V^*(-x) \tag{7.27}$$

could have real spectra. They proposed this symmetry as a different way to understand/construct quantum mechanical waves. A few years later optics researchers [17, 18] found that these ideas could be used to construct new optical wave modes which were subsequently observed in laboratory experiments [19]. This has become a very active field of research in optics.

This motivated Ablowitz and Musslimani to look for new nonlinear equations with this symmetry. They found [20] the following integrable NLS equation with PT symmetry

$$iq_t(x,t) + q_{xx}(x,t) + V(x,t;q)q(x,t) = 0 \quad \text{with V}(x,t;q) = \pm 2q(x,t)q^*(-x,t) \tag{7.28}$$

where we see that $V(x,t;q) = V^*(-x,t;q)$; alternatively the above equation, which we have termed the *PT-NLS equation*, can be written in the form

$$iq_t + q_{xx} \pm 2q^2(x,t)q^*(-x,t) = 0 \tag{7.29}$$

Remarkably this equation is associated with the AKNS eigenvalue problem (7.13) with the symmetry $r(x,t) = \mp q^*(-x,t)$. This was the first time anyone had found nonlinear equations with this symmetry. In 2016 Ablowitz and Musslimani [21] developed the inverse scattering transform for this PT-NLS equation and found how the associated symmetries in scattering space were found in IST.

In fact there are a huge number of nonlocal equations associated with these and similar symmetries. A few integrable equations with PT symmetry are listed below.

- *Integrable PT-symmetric vector NLS equation*

$$i\mathbf{q}_t + \mathbf{q}_{xx} + \mathbf{q}(x,t)\cdot\mathbf{q}^*(-x,t)\mathbf{q} = 0 \tag{7.30}$$

  where $\mathbf{q} = (q_1, q_2, ...q_N)$
- *Integrable PT-symmetric discrete NLS equation*

$$iq_{n,t} = q_{n+1} + q_{n-1} \pm q_n(t)q^*_{-n}(t)(q_{n+1} + q_{n-1}) \tag{7.31}$$

  The IST associated with this PT-DNLS equation was outlined in [22].
- *Integrable nonlocal PT-symmetric DS equation*

$$\begin{aligned} &iq_t + \frac{1}{2}\left(\gamma q_{xx} + q_{yy}\right) = \left(\phi - \sigma q(\mathbf{x},t)q^*(-\mathbf{x},t)\right)q \\ &\phi_{xx} - \gamma\phi_{yy} = 2\left(q(\mathbf{x},t)q^*(-\mathbf{x},t)\right)_{xx} \end{aligned} \tag{7.32}$$

  where $\gamma = \pm 1, \sigma = \pm 1;\ \mathbf{x} = (x,y)$; see [23].

In 2017 Ablowitz and Musslimani [24] went further and found many more new equations; in this case the equations are nonlocal in both space and time. Two other nonlocal equations of NLS type with their associated q, r symmetries are the following:

- *Reverse space-time (RST) NLS equation*

$$iq_t + q_{xx} \pm 2q^2(x,t)q(-x,-t) = 0 \tag{7.33}$$

- *Reverse time (RT) NLS equation*

$$iq_t + q_{xx} \pm 2q^2(x,t)q(x,-t) = 0, \quad r(x,t) = \mp q(x,-t) \tag{7.34}$$

Indeed, all of the NLS equations that are associated with the AKNS scattering system where the time operator $\mathcal{M}$ in (7.14) is a quadratic polynomial in powers of k; e.g. the (1, 1) component of $\mathcal{M}$ is given by $\mathcal{M}_{1,1} = 2ik^2 + iq(x,t)r(x,t)$, the two functions (potentials) $q(x,t), r(x,t)$ satisfy the coupled system

$$iq_t(x,t) = q_{xx}(x,t) - 2r(x,t)q^2(x,t) \tag{7.35}$$

$$-ir_t(x,t) = r_{xx}(x,t) - 2q(x,t)r^2(x,t) \tag{7.36}$$

Said differently, the classical NLS equation (7.12) with the symmetry $r(x,t) = \mp q^*(x,t)$, the PT-NLS equation (7.29) with the symmetry $r(x,t) = \mp q^*(-x,t)$, the RST-NLS equation (7.33) with the symmetry $r(x,t) = \mp q(-x,-t)$ and the RT-NLS equation (7.34) with the symmetry $r(x,t) = \mp q^*(x,-t)$, *all satisfy the coupled system (7.35)–(7.36).*

Interestingly, the coupled system (7.35)–(7.36) and hence all these nonlocal NLS reductions can be derived by asymptotic reductions of physical systems. In [25] these systems were derived from small amplitude quasi-monochromatic reductions of the KdV, nonlinear Klein-Gordon and water wave equations; indeed this reduction procedure leads to the conclusion that the coupled system (7.35)–(7.36) is universal.

In the literature nonlocal systems and generalizations have been considered from many points of view including how the solution of these nonlocal equations depend on long distance interactions are related to physical phenomena. Recently Yang [26] showed that the 2×2 vector Manakov system [27] has a nonlocal reduction of similar type to those discussed here; this Manakov system has numerous physical applications. Given the large number of new nonlocal systems found it is expected that many further physically based nonlocal systems will be obtained.

When we consider the time operator $\mathcal{M}$ in (7.14) as different polynomial functions of k different equations result from the AKNS procedure. For example when the time operator $\mathcal{M}$ is cubic we find the *nonlocal RST-mKdV equation*

$$q_t + q_{xxx} \pm 6q(x,t)q(-x,-t)q_x(x,t) = 0, \quad r(x,t) = \mp q(-x,-t) \tag{7.37}$$

and when the time operator $\mathcal{M}$ is inverse power of k we find the *nonlocal RST sine/sinh-Gordon equations*

$$q_{xt} + 2s(x,t)q(x,t) = 0, s_x(x,t) = \pm(q(x,t)q(-x,-t))_t, r(x,t) = \mp q(-x,-t), \tag{7.38}$$

We remark that the standard sine/sinh-Gordon equations arise when $r(x,t) = \mp q(x,t)$; in this case there is a further reduction to sine/sinh Gordon equations by taking $q(x,t) = \mp u_x(x,t)/2$ cf. (7.18)–(7.19). There are no apparent further reductions in the nonlocal RST case.

At this point we have seen that there are many integrable nonlinear wave equations both local and nonlocal.

The basic methodology of the inverse scattering transform will be discussed in the next section. Details and background information can be found in the following references [14, 15, 28–31]. The reason we discuss IST for rapidly decaying data in detail is the following.

1. Solitons are related to eigenvalues of the $\mathcal{L}$.
2. From given initial data one can determine the number of eigenvalues and hence the number of solitons.

3. Phase shifts of solitons can be determined from initial data.
4. Long time asymptotic structure of the solution can be obtained from initial data; e.g. for the KdV equation (7.2) the solitons move to the right away from where localized initial data is given. The central portion of the solution decays like $O(1/t^{2/3})$ in the region the where the localized initial conditions were prescribed. This portion is dominated by the self-similar solution which is related to the second Painlevé equation. There is also a collisionless shock layer where the data decays like $O((\log(t)/t)^{2/3})$ and an oscillatory region to the left which decays like $O(1/t^{1/2})$; its amplitude was determined in [14, 32]; the phase of the solution in the oscillatory region was determined a little later [33].
5. For the NLS and mKdV equations the long time amplitude and phase was found in [14, 27, 32–34] and later in [35] where a valuable steepest descent Riemann–Hilbert approach was developed.
6. Underlying Hamiltonian structures and action-angle variables associated with these equations were found.
7. An infinite number of conserved quantities/conservation laws can be determined.

There are other valuable approaches to understand solutions to integrable problems. But only IST for rapidly decaying data (and also data whose amplitude decays rapidly to constants at infinity) can do all of the above.

We also note that the periodic boundary value problem has attracted considerable attention cf. [14, 29, 36, 37]. These elegant algebro-geometric methods enable one to find multi-phase (also called finite gap) solutions to a range of nonlinear wave equations. We remind the reader that the single phase solution of the KdV equation was first obtained by Korteweg and de Vries [3]; they termed this solution as the cnoidal wave a limit of which is the solitary wave solution.

Later we will briefly discuss what we refer to as direct methods. These techniques allow one to obtain special solutions such as multi-soliton solutions via algebraic methods. They do not connect to the initial value problem.

7.2 Inverse Scattering Transform

Next we will outline the method of solution associated with second order systems (7.13)–(7.14); as mentioned above, the method is termed the Inverse Scattering Transform (IST). We note that while here we will concentrate on 2×2 systems, higher oder, discrete and multi-dimensional systems can also be analyzed.

The scheme is given by the following diagram

$$\begin{array}{ccc} q(x,0),\, r(x,0) & \xrightarrow{\text{Direct Scattering}} & \mathcal{L}: S(k,0) \\ & & \Big\downarrow {\scriptstyle t:\ \text{time evolution: } \mathcal{M}} \\ q(x,t),\, r(x,t) & \xleftarrow{\text{Inverse Scattering (RH)}} & S(k,t) \end{array}$$

Importantly we supplement the IST scheme with additional symmetry conditions in physical space such as $r(x,t) = \mp q^*(x,t), r(x,t) = \mp q^*(-x,t), r(x,t) = \mp q(-x,-t), \ldots$. We will see that these physical space symmetry conditions require symmetry conditions on scattering data.

7.2.1 Direct Scattering

For potentials q, r decaying sufficiently fast at infinity (i.e. functions $q, r \in L^1$), linearly independent eigenfunctions are defined by the following boundary conditions, $k \in \mathbb{R}$ (note: in this section, for simplicity, the time variable t is suppressed):

$$\phi(x,k) \sim \begin{pmatrix} 1 \\ 0 \end{pmatrix} e^{-ikx}, \quad \bar{\phi}(x,k) \sim \begin{pmatrix} 0 \\ 1 \end{pmatrix} e^{ikx} \quad \text{as } x \to -\infty \tag{7.39a}$$

$$\psi(x,k) \sim \begin{pmatrix} 0 \\ 1 \end{pmatrix} e^{ikx}, \quad \bar{\psi}(x,k) \sim \begin{pmatrix} 1 \\ 0 \end{pmatrix} e^{-ikx} \quad \text{as } x \to +\infty \tag{7.39b}$$

Note: $\bar{\phi}(x,k)$ is NOT the complex conjugate of $\phi(x,k)$, etc.; we use * for complex conjugation instead.

Since $\mathcal{L}$ is a second order operator we have

$$\phi(x,k) = a(k)\bar{\psi}(x,k) + b(k)\psi(x,k) \tag{7.40a}$$

$$\bar{\phi}(x,k) = \bar{a}(k)\psi(x,k) + \bar{b}(k)\bar{\psi}(x,k) \tag{7.40b}$$

Hence the 'scattering data': $a(k), \bar{a}(k), b(k), \bar{b}(k)$ satisfy:

$$a(k) = W(\phi, \psi), \quad \bar{a}(k) = W(\bar{\psi}, \bar{\phi}) \tag{7.41a}$$

$$b(k) = W(\bar{\psi}, \phi), \quad \bar{b}(k) = W\left(\bar{\phi}, \psi\right) \tag{7.41b}$$

where the Wronskian of u and v is given by

$$W(u,v) = u^{(1)}v^{(2)} - u^{(2)}v^{(1)}$$

By formulating appropriate integral equations, one can prove that the functions

$$M(x,k) = \phi(x,k)e^{ikx}, \quad N(x,k) = \psi(x,k)e^{-ikx}$$

are analytic and bounded in the upper half plane (UHP) and

$$\bar{M}(x,k) = \bar{\phi}(x,k)e^{-ikx}, \quad \bar{N}(x,k) = \bar{\psi}(x,k)e^{ikx}$$

are analytic and bounded in the lower half plane (LHP) cf. [14, 28].

From the first of equations (7.41) we can show that $a(k)$, $\bar{a}(k)$ are analytic in the UHP/LHP respectively.

Further analysis shows that the required scattering data we will need in order to reconstruct the potentials q, r are the 'reflection' coefficients

$$\rho(k) = b(k)/a(k), \bar{\rho}(k) = \bar{b}(k)/\bar{a}(k) : \text{ reflection coef.}$$

and certain information about so-called discrete spectrum discussed below.

(i) Location of the zero's of $a(k_j) = 0 : j = 1, 2...J$ and $\bar{a}(\bar{k}_j)$, $j = 1, 2...\bar{J}$

k_j and $\bar{k}_j$ are the eigenvalues. Proper (generic) eigenvalues correspond to square integrable (L^2) bound states which are simple and do not lie on the real axis.

At $a(k_j) = 0, k_j = \xi_j + i\eta_j, \eta_j > 0, j = 1, 2, ...J$ we have the following linear dependence

$$\phi_j(x) = b_j\psi_j(x) \text{ where } \phi_j(x) = \phi(x, k_j), \ \psi_j(x) = \psi(x, k_j); \tag{7.42}$$

Similarly at $\bar{a}(\bar{k}_j) = 0, \bar{k}_j = \bar{\xi}_j - i\bar{\eta}_j, \ \bar{\eta}_j > 0, j = 1, 2, ...\bar{J}$ we also have linear dependence

$$\bar{\phi}_j(x) = \bar{b}_j\bar{\psi}_j(x) \text{ where } \bar{\phi}_j(x) = \bar{\phi}(x, \bar{k}_j), \ \bar{\psi}_j(x) = \bar{\psi}(x, \bar{k}_j). \tag{7.43}$$

To reconstruct q, r we will also need the so-called 'norming constants'

(ii) $C_j = b_j/a'(k_j), j = 1, 2, ...J; \ \ \bar{C}_j = \bar{b}_j/\bar{a}'(\bar{k}_j), \ \ j = 1, 2, ...\bar{J}$, here $a'(k_j) = da(k)/dk)(k = k_j)$.

Summary: for the inverse problem for q, r, and linearization required scattering data:

$$\mathcal{S}(k) = \left\{\rho(k), \{k_j, C_j\}, j = 1, 2, ..., J; \ \bar{\rho}(k), \ \{\bar{k}_j, \bar{C}_j\}, j = 1, 2, ...\bar{J}\right\}$$

where $\rho(k) = b(k)/a(k)$, $\bar{\rho}(k) = \bar{b}(k)/\bar{a}(k)$ reflection coefficients and discrete spectra:

$$k_j : a(k_j) = 0, j = 1, 2...J; \ \bar{k}_j : \bar{a}(\bar{k}_j) = 0, j = 1, 2, ...\bar{J} \ : \ \text{ eigenvalues}$$

$$C_j = b_j/a'(k_j), j = 1, 2...J; \bar{C}_j = \bar{b}_j/\bar{a}'(\bar{k}_j), j = 1, 2, ...\bar{J} \ : \ \text{ norming constants}$$

7.2.2 Symmetries in Scattering Space

In this section we will discuss symmetries associated with the classical NLS and PT-NLS equations. Additional information and symmetries can be found in our papers [21, 24].

When $r(x) = \sigma q^*(x), \sigma = \mp 1$ e.g. classical NLS equation (7.12) we find

$$\overline{\psi}(x,k) = P_1 \psi^*(x,k^*), \;\; \overline{\phi}(x,t,k) = P_1^T \phi^*(x,k^*)$$

where

$$P_1 = \begin{pmatrix} 0 & 1 \\ \sigma & 0 \end{pmatrix}$$

Thus we have that the analytic functions in the UHP $\phi(x,k)$, $\psi(x,k)$ are related to analytic functions in the LHP $\bar{\psi}, \bar{\phi}$. This relationship is the analog of the Schwarz symmetry principle for these eigenfunctions. From this key relationship we can derive symmetries of the scattering data

$$\bar{a}(k) = a^*(k^*), \bar{k}_j = k_j^*, \bar{C}_j(t) = -C_j^*(t), \;\; j = 1, 2, ...J, \;\; \bar{b}(k,t) = \sigma b^*(k,t) \tag{7.44}$$

Thus the data we need is reduced by a factor of two. We also note that we can show that only when $\sigma = -1$ there are discrete eigenvalues.

When $r(x) = \sigma q^*(-x), \sigma = \mp 1$ e.g. PT-NLS equation (7.29) we find

$$\psi(x,k) = P_2 \phi^*(-x,-k^*), \;\; \overline{\psi}(x,k) = P_2^T \overline{\phi}^*(-x,-k^*)$$

$$P_2 = \begin{pmatrix} 0 & -\sigma \\ 1 & 0 \end{pmatrix}$$

In this case we have that $\psi(x,t,k)$ is related to $\phi(-x,t,-k^*)$ (both analytic in the UHP) and $\bar{\psi}(x,t,k)$ is related to $\bar{\phi}(-x,t,-k^*)$ (both analytic in the LHP). From this we can find the following symmetries of the scattering data.

$$a(k,t) = a^*(-k^*,t), \;\; \bar{a}(k,t) = \bar{a}^*(-k^*,t); \tag{7.45}$$

hence the pair: $k_j; -k_j^*, j = 1, 2..., J$, are eigenvalues and similarly the pair $\bar{k}_j; -\bar{k}_j^*, \; j = 1, 2..., J$ are eigenvalues.

Note: k_j is unrelated to $\bar{k}_j$; i.e. they 'live in separate worlds'. We also find

$$\bar{b}(k,t) = \sigma b^*(-k,t), \;\; k \in \mathbb{R}$$

When $k_j = i\eta_j, \bar{k}_j = -i\bar{\eta}_j$ there is a simple relation among parts of the norming constants; namely:

$$C_j(0) = b_j / a'(kj), \;\; b_j = e^{i\theta_j}, \theta_j \in \mathbb{R}, \;\; j = 1, 2..., J$$

$$\bar{C}_j(0) = \bar{b}_j / \bar{a}'(\bar{k}_j), \;\; \bar{b}_j = e^{i\bar{\theta}_j}, \bar{\theta}_j \in \mathbb{R}, \;\; j = 1, 2..., J$$

In this case we need to compute $a'(k_j), \bar{a}'(\bar{k}_j)$ via the trace formulae. The trace formulae determine $a(k), \bar{a}(k)$ in terms of the eigenvalues $k_j, \bar{k}_j$ and $b(k), \bar{b}(k)$; these formulae are given by

For $\mathrm{Im}(k) \geq 0$:

$$a(k) = \prod_{m=1}^{J} \left(\frac{k - k_m}{k - \bar{k}_m} \right) \exp \left\{ \frac{1}{2\pi i} \int_{-\infty}^{+\infty} \frac{\log\left(1 + b(\xi)\bar{b}(\xi)\right)}{\xi - k} d\xi \right\} \tag{7.46}$$

For $\mathrm{Im}(k) \leq 0$:

$$\bar{a}(k) = \prod_{m=1}^{J} \left(\frac{k - \bar{k}_m}{k - k_m} \right) \exp \left\{ \frac{1}{2\pi i} \int_{-\infty}^{+\infty} \frac{\log\left(1 + b(k)\bar{b}(k)\right)}{\xi - k} d\xi \right\} \tag{7.47}$$

We also note that we can show that only when $\sigma = -1$ for the case discussed above there are discrete eigenvalues. We also see that the symmetry $r(x) = \sigma q^*(x)$ leads to different scattering symmetries than does $r(x) = \sigma q^*(-x)$.

7.2.3 *Inverse Scattering*

From the scattering equations (7.40) dividing the first/second by $a(k)/\bar{a}(k)$ and multiplying by e^{ikx}/e^{-ikx} respectively and using the definitions of $M(x, t, k)$, $N(x, t, k)$ and $\bar{M}(x, k)$, $\bar{N}(x, t, k)$ (here we reinsert the time dependence which was omitted in prior sections) we have

$$\frac{M(x, t, k)}{a(k, t)} = \bar{N}(x, t, k) + \rho(k)e^{2ikx} N(x, t, k) \tag{7.48a}$$

$$\frac{\bar{M}(x, t, k)}{\bar{a}(k, t)} = N(x, t, k) + \bar{\rho}(k)e^{-2ikx} \bar{N}(x, t, k) \tag{7.48b}$$

Since we have that $M(x, k)$, $a(k)$, $N(x, k)$ are analytic in the UHP and $\bar{M}(x, k)$, $\bar{a}(k)$, $\bar{N}(x, k)$ are analytic in the LHP the above system (7.48) define a Riemann–Hilbert (RH) problem. Since the system is coupled we cannot solve it in closed form, as we can do with scalar RH problems, we can nevertheless find integral equations that govern the solution [38]. Furthermore these linear integral equations can be related to Gel'fand-Levitan–Marchenko type integral equations.

We define the projectors: if $f(\zeta)$, $\zeta \in \mathbb{R}$, is an integrable function the projection operators are

$$P^{\pm}[f](k) = \frac{1}{2\pi i} \int_{-\infty}^{+\infty} \frac{f(\zeta)}{\zeta - (k \pm i0)} d\zeta.$$

Thus if f_+ (respectively f_-) is analytic in the upper (respectively lower) k-plane and $f_\pm(k) \to 0$ as $|k| \to \infty$ for $\mathrm{Im}(k) > 0$ (respectively $\mathrm{Im}(k) < 0$), then

$$P^\pm\left[f_\pm\right] = \pm f_\pm, \qquad P^\pm\left[f_\mp\right] = 0$$

$P^\pm$ are usually referred to as projection operators into the upper/lower half k-planes.

Then subtracting the poles at $k_j/\bar{k}_j$ from each equation and taking the $P^\pm$ projectors we find the following system of coupled integral equations for $N(x,k)$, $\bar{N}(x,k)$ [28]

$$\overline{N}(x,t,k) = \begin{pmatrix}1\\0\end{pmatrix} + \sum_{j=1}^{J} \frac{C_j(t)e^{2ik_jx}N(x,t,k_j)}{k-k_j} + \frac{1}{2\pi i}\int_{-\infty}^{+\infty} \frac{\rho(\zeta,t)e^{2i\zeta x}N(x,t,\zeta)}{\zeta-(k-i0)}d\zeta\,, \tag{7.49}$$

$$N(x,t,k) = \begin{pmatrix}0\\1\end{pmatrix} + \sum_{j=1}^{J} \frac{\overline{C}_j(t)e^{-2i\bar{k}_jx}\overline{N}(x,t,\bar{k}_j)}{k-\bar{k}_j} - \frac{1}{2\pi i}\int_{-\infty}^{+\infty} \frac{\overline{\rho}(\zeta,t)e^{-2i\zeta x}\overline{N}(x,t,\zeta)}{\zeta-(k+i0)}d\zeta\,. \tag{7.50}$$

In order to obtain a closed system we substitute $k = \bar{k}_\ell$ and $k = k_\ell$ in (7.49) and (7.50) respectively. This yields an additional linear algebraic integral system of equations. Together these equations govern the inverse scattering problem for the eigenfunctions $N(x,t,k)$ and $\overline{N}(x,t,k)$.

To reconstruct the potentials for all time, $q(x,t)$, $r(x,t)$, we compare the asymptotic expansions of (7.49) and (7.50) to the asymptotic expansions of the eigenfunctions and find cf. [28]

$$q(x,t) = 2i\sum_{\ell=1}^{J} \bar{C}_\ell(t)e^{-2i\bar{k}_\ell x}\bar{N}_1(x,\bar{k}_\ell) + \frac{1}{\pi}\int_{-\infty}^{\infty} \bar{\rho}(\xi,t)e^{-2i\xi x}\bar{N}_1(x,t,\xi)d\xi,\,. \tag{7.51}$$

$$r(x,t) = -2i\sum_{\ell=1}^{J} C_\ell(t)e^{2ik_\ell x}N_2(x,k_\ell) + \frac{1}{\pi}\int_{-\infty}^{\infty} \rho(\xi,t)e^{-2i\xi x}N_2(x,t,\xi)d\xi,\,. \tag{7.52}$$

where $N(x,t,k)$ is a 2×1 vector: $N(x,t,k) = [N_1(x,t,k), N_2(x,t,k)]^t$ (with ^{t} representing transpose) and similarly for $\overline{N}(x,t,k)$.

In the case with zero reflection coefficient, i.e. $\rho(t) = \overline{\rho}(t) = 0$ the resulting *algebraic system* is given by

$$\overline{N}(x,t,\bar{k}_\ell) = \begin{pmatrix}1\\0\end{pmatrix} + \sum_{j=1}^{J} \frac{C_j(t)e^{2ik_jx}N(x,t,k_j)}{\bar{k}_\ell-k_j}\,, \tag{7.53}$$

$$N(x,t,k_\ell) = \begin{pmatrix} 1 \\ 0 \end{pmatrix} + \sum_{j=1}^{\overline{J}} \frac{\overline{C}_j(t)e^{-2i\overline{k}_j x}\overline{N}(x,t,\overline{k}_j)}{k_\ell - \overline{k}_j} . \tag{7.54}$$

These equations yield reflectionless eigenfunctions and from (7.51)–(7.52) reflectionless potentials (with $\rho(k,t) = \overline{\rho}(k,t) = 0$) when time is omitted or when time is inserted the soliton solutions. Next we discuss the time dependence.

7.2.4 Time Dependence of Scattering Data

When time is included in (7.49)–(7.50) we need to solve the coupled integral/algebraic equations in order to find the eigenfunctions $N(x,t,k)$, $\overline{N}(x,t,k)$; cf. [39] for a rigorous discussion of these and related issues. The time dependence of $N(x,t,k)$, $\bar{N}(x,t,k)$ and the potentials $q(x,t)$ and $r(x,t)$ in (7.51) and (7.52) are encoded in the reflection coefficients $\rho(k,t)$, $\bar{\rho}(k,t)$ the eigenvalues and norming constants $C_j(t)$ and $\overline{C}_j(t)$. Their time evolution is derived from the compatible equations (7.13) and (7.14). The space, time and space-time nonlocal NLS, mKdV and sG equations belong to the same 'hierarchy', i.e. they all originate from the same scattering problem (7.13) with different time operators $\mathcal{M}$ which in turn determines the time evolution of the scattering data and norming constants. For the problems we are considering, we find the following time dependence cf. [28]:

$$a(k,t) = a(k,0), \ \bar{a}(k,t) = \bar{a}(k,0),$$

i.e. $a(k)$ and $\bar{a}(k)$ and its zeros, which we have denoted by $k_j, \bar{k}_j, \ j = 1, 2...J$ respectively, are constant in time. For classical NLS and nonlocal NLS problems

$$b^{\mathrm{NLS}}(k,t) = b(k,0)e^{-4ik^2 t},$$

$$\bar{b}^{\mathrm{NLS}}(k,t) = \bar{b}(k,0)e^{4ik^2 t}, k \in \mathbb{R}$$

$$C_j^{\mathrm{NLS}}(t) = C_j(0)e^{-4ik_j^2 t} , \tag{7.55}$$

$$\overline{C}_j^{\mathrm{NLS}}(t) = \overline{C}_j(0)e^{4i\overline{k}_j^2 t}. \tag{7.56}$$

Here, k_j and $\overline{k}_j$ are also referred to as soliton eigenvalues and $C_j(0)$, $\bar{C}_j(0)$ are the norming constants. For mKdV and nonlocal mKdV problems

$$b^{\mathrm{mKdV}}(k,t) = b(k,0)e^{8ik^3 t},$$

$$\bar{b}^{\mathrm{mKdV}}(k,t) = \bar{b}(k,0)e^{-8ik^3 t}, k \in \mathbb{R}$$

$$C_j^{\mathrm{mKdV}}(t) = C_j(0)e^{8ik_j^3 t}\,, \tag{7.57}$$

$$\overline{C}_j^{\mathrm{mKdV}}(t) = \overline{C}_j(0)e^{-8i\overline{k}_j^3 t}, \tag{7.58}$$

7.2.5 Soliton Solutions

As indicated above pure soliton solutions arise when $\rho(k,t) = \overline{\rho}(k,t) = 0$. This means pure solitons are obtained from the linear algebraic system (7.53)–(7.54). In the case of $J = 1$, i.e. a one soliton solution, we find

$$N_2(x,t) = \bar{N}_1(x,t) = \frac{1}{1 + \frac{C_1(t)\bar{C}_1(t)}{(k_1-\bar{k}_1)^2 e^{2i(k_1-\bar{k}_1)x}}}\,. \tag{7.59}$$

Hence the reflectionless potentials/one soliton solutions from (7.51)–(7.52) are given by

$$q(x,t) = \frac{2ie^{-2i\bar{k}_1 x}\bar{C}_1(t)}{1 + \frac{C_1(t)\bar{C}_1(t)}{(k_1-\bar{k}_1)^2 e^{2i(k_1-\bar{k}_1)x}}}\,, \tag{7.60}$$

$$r(x,t) = -\frac{2ie^{2ik_1 x}C_1(t)}{1 + \frac{C_1(t)\bar{C}_1(t)}{(k_1-\bar{k}_1)^2 e^{2i(k_1-\bar{k}_1)x}}}\,. \tag{7.61}$$

For classical NLS substituting its time dependence and symmetries with $J = 1, k_1 = \xi + i\eta$ we find

$$q_{\mathrm{NLS}}(x,t) = 2\eta \operatorname{sech}(2\eta(x - 4\xi t - x_0))e^{-2i\xi x + 4i(\xi^2-\eta^2)t - i\psi_0}\,, \tag{7.62}$$

where $e^{2\eta x_0} = |C_1(0)|/(2\eta)$, $\psi_0 = \arg(C_1(0)) - \pi/2$. We also note that the above symmetries in scattering space imply that $r(x,t)$ given by (7.61) automatically satisfy the physical symmetry $r(x,t) = \mp q^*(x,t)$.

Using the same ideas the one soliton solution associated with the PT-NLS equation is given by

$$q_{\mathrm{PT}}(x,t) = \frac{2(\eta+\bar{\eta})e^{i\bar{\theta}}e^{-2\bar{\eta}x - 4i\bar{\eta}^2 t}}{1 - e^{i(\theta+\bar{\theta})}e^{-2(\eta+\bar{\eta})x + 4i(\eta^2-\bar{\eta}^2 t)}}\,. \tag{7.63}$$

Alternatively this 1-soliton solution (7.63) can be written in the form

$$q_{\mathrm{PT}}(x,t) = \frac{(\eta+\bar{\eta})e^{i(\bar{\theta}-\theta-\pi)/2}e^{-(\bar{\eta}-\eta)x}e^{-2i(\eta^2+\bar{\eta}^2)t}}{\cosh\left[(\eta+\bar{\eta})x - 2i(\eta^2-\bar{\eta}^2)t - i(\theta+\bar{\theta}+\pi)/2\right]}\,. \tag{7.64}$$

Next we make some remarks.

- The solution $q(x,t)$ given in (7.63) is doubly periodic in time with periods given by $T_1 = \frac{\pi}{2\bar{\eta}^2}$ and $T_2 = \frac{\pi}{2(\eta^2-\bar{\eta}^2)}$.
- The intensity $|q(x,t)|^2$ breathes in time with period given by $T = \frac{\pi}{2(\eta^2-\bar{\eta}^2)}$
- The solution (7.63) can develop a singularity in finite time. Indeed, at the origin ($x = 0$) the solution (7.64) becomes singular when

$$t_n = \frac{2n\pi - (\theta + \bar{\theta})}{4(\eta^2 - \bar{\eta}^2)}, n \in \mathbb{Z}\,. \tag{7.65}$$

- The solution (7.63) is characterized by two important time scales: the singularity time scale and the periodicity of 'breathing'.
- An interesting feature of this solution of (7.63) is that it can be defined after singularity has developed; i.e. it has a pole in time and it can be avoided in the complex time plane; i.e. the solution is of Painlevé type.
- We recall that not all members of the one-soliton family develop a singularity at finite time. Namely, if one let $\eta = \bar{\eta} \equiv \eta$ in (7.63) then we arrive at the following well behaved soliton solution of the nonlocal $\mathcal{PT}$-symmetric NLS equation (7.29)

$$q(x,t) = 2\eta \text{sech}[2\eta x - i\theta]e^{-4i\eta^2 t}\,, \tag{7.66}$$

  where η and θ are arbitrary real constants.
- This singularity requires having an eigenvalue which in turn requires having a sufficiently large L^1 norm cf. [14].
- It should be stressed that another important nonlinear equation exhibits a singularity in finite time; i.e. the $2+1$, $3+1$ dimensional NLS equations. In order for a singularity to occur the L^2 norm has to be sufficiently large. cf. [7, 40]. But $2+1$, $3+1$ dimensional NLS equations are not known/expected to be integrable.

In [24] the other nonlocal one soliton solutions to the RST-NLS and RT-NLS equations are obtained.

Next we make a remark about soliton solutions to the real valued nonlocal RST-mKdV equation (7.37). In this case we have $r(x,t) = \sigma q(-x,-t),\ q \in \mathbb{R}$. For this case

$$C_1(0) = i(\eta + \bar{\eta})b_1, \bar{C}_1(0) = -i(\eta + \bar{\eta})\bar{b}_1\,, \tag{7.67}$$

but now with

$$b_1 = \pm 1, \bar{b}_1 = \pm 1.$$

Thus the only difference from PT-NLS is that now we require $\theta, \bar{\theta} = 0, \pi$. Therefore, in this case there are only two free real parameters η, $\bar{\eta}$ and for the real nonlocal mKdV equation (7.37) the one soliton solution is given by

$$q(x,t) = \frac{2\gamma_1(\eta + \bar{\eta})e^{-2\bar{\eta}x + 8\bar{\eta}^3 t}}{1 + \gamma_2 e^{-2\eta x + 8\eta^3 t - 2\bar{\eta}x + 8\bar{\eta}^3 t}}\,, \tag{7.68}$$

where $\gamma_j = \pm 1$, $j = 1, 2$. If we set $\gamma_1 = \gamma_2 = 1$ then the solution can be written in the following form

$$q(x,t) = \frac{(\eta + \bar{\eta})e^{\eta(x-4\eta^2 t)}e^{-\bar{\eta}(x-4\bar{\eta}^2 t)}}{\cosh\left[(\eta(x - 4\eta^2 t) + \bar{\eta}(x - 4\bar{\eta}^2)t\right]} \,. \tag{7.69}$$

This solution is not singular. When $\eta = \bar{\eta}$ the solution reduces to the well-known solution of the real mKdV equation

$$q(x,t) = \frac{2\eta}{\cosh\left[(2\eta(x - 4\eta^2 t)\right]} \,. \tag{7.70}$$

A singular solution to the nonlocal mKdV equation occurs when $\gamma_2 = -1$. This seems to be the case for these classes of nonlocal problems: there are singular and nonsingular solution sectors.

7.2.6 Extensions

The methods associated with rapidly decaying data can be generalized to find solutions to nonlinear wave equations whose data does not decay at infinity. Among the most interesting situations is the NLS, integrable discrete NLS and vector NLS equations with data that rapidly decays to constant magnitude cf. [31, 41–43]; recently these ideas were used to analyze the nonlocal NLS equation [44]. The KdV with step boundary values was also analyzed [45–49]. An interesting use of these ideas was the consideration of a soliton moving into a step-like initial condition [50]. This leads to the novel concept of a pseudo-embedded eigenvalues/solitons. In this situation one has a soliton that becomes trapped by the step and although it seems there is an eigenvalue, it is not the case. For problems like those above one can obtain detailed information about the initial value solution to the wave equation.

7.3 Direct Methods

Often one does not require the solution of the initial value problem or even a connection to initial value problems. In this case direct methods of solution might be sufficient. To do this it turns out that there is a range of useful alternatives.

- Bilinear and related methods were pioneered by R. Hirota cf. [51]; see also [14]. In this approach one transforms the original nonlinear equation to a convenient form e.g. a bilinear equation, where one can find multi-soliton solutions as a sum of exponential functions. This method was enhanced by the KP hierarchy reduction methods cf. [52]. Noteworthy is the study of line soliton solutions of the KP equation cf. [53–55] and the observation of these structures in water waves on shallow flat beaches [56].

- Another technique is to use a Darboux transformation [57]. These ideas go back to geometric studies of Darboux and others in the 1800s. The method is to construct a seed solution, typically from the vacuum, of the linear compatible pair. Using this first solution, we can develop further solutions in an iterative fashion.
- Bäcklund transformations are also a valuable approach [58]. In a sense they are similar to Darboux transformations except the iterative solutions are developed from compatible nonlinear systems (see also [14]).
- Direct integral equation or Riemann–Hilbert/Dbar methods (cf. [15]). Here one postulates an integral equation cf. [14, 59] or the dressing method using a Riemann–Hilbert/Dbar formulation [60, 61] motivated typically by inverse scattering methods. By applying suitable operators one can find nonlinear equations and use the integral equation or Riemann–Hilbert/Dbar method to obtain solutions. It appears that of the above direct methods currently the ones most frequently used by researchers to obtain solutions are bilinear/related techniques and Darboux transformations.

7.4 Outlook

This Chapter involves integrability and nonlinear waves. The field of integrable equations is considerably broader than this, comprising numerous other areas of research including: symmetries, methods to determine linear compatible systems, classification of integrable systems of certain types, studies of wide classes of discrete integrable systems; Painlevé type equations both continuous and discrete. There are journals that are closely related to studies in these areas. To cover such a large scope/territory would require a considerably longer Chapter than this.

It was truly remarkable that so many interesting/important nonlinear wave equations have been identified and solved via IST. Some of these have been mentioned earlier: e.g. KdV, modified KdV, NLS, vector NLS, three wave interaction equations, Toda lattice, integrable discrete NLS, Benjamin–Ono, Kadomtsev–Petviashvili equation, Davey–Stewartson system etc. There are many others that have not been discussed here; e.g. the intermediate long wave equation, sine-Hilbert equation, loop-soliton system, Landau-Lifshitz system, Heisenberg ferromagnetic system, self-induced transparency system, two dimensional Einstein equations, Benney system [14, 15], Camassa–Holm equation [62] and many others. The methods used to understand nonlinear wave equations via IST were new and very different from those used earlier which were usually direct/explicit transformations such as the Hopf-Cole transformation cf. [7] or hodograph type transformations cf. [14]. IST and its related mathematics including the compatibility of two linear systems (i.e. Lax pairs), solitons being related to eigenvalues of a linear operator, the connection to direct/inverse scattering which in turn involves linear integral equations and Riemann–Hilbert problems, led to many new and unexpected results.

Furthermore the deep relationship of integrable nonlinear waves/IST to PT-symmetric problems (e.g. the PT-NLS equation) discovered in 2013–2018 [20–22, 24, 25] was also unanticipated. Its connection to new physical space/hence new scattering space symmetries of the AKNS system (7.13 with $r(x,t) = \mp q(-x,t)$) led to large classes of new integrable nonlinear wave systems being identified and solved/linearized via IST. After so many years with so many people studying IST/integrability and related matters it was surprising to find relatively simple equations connected to the well-known AKNS linear compatible system. Now one must ask: what other interesting/important nonlinear waves systems will be identified? It is natural to expect there will be more of them; the only issue is when will they be found and what is an appropriate starting point. For the nonlocal equations discussed here the surprise was i) they could be found by a modification of the AKNS and higher order AKNS and $(2+1)$ AKNS formalism; (ii) they are 'simple' equations; (iii) there are so many new nonlocal systems; (iv) nevertheless the scattering theory and solution process has numerous nontrivial differences from the known 'classical' AKNS systems

In terms of IST there are open questions of how far can the method be developed including: rigorous functional analytic/geometric analysis, incorporating slower decay than currently allowed in the theory (thereby incorporating algebraically slow decaying solitons), finding new types of solutions bounded on the real line cf. [63, 64], new methods to analyze for $2+1$ dimensional systems which tend to different types of nonzero data at infinity, analyzing different types of boundary conditions e.g. the semi-infinite problem [65], random initial/boundary data. Although methods to solve periodic initial boundary value problems have been heavily studied there is still a need to put these theories in a form where applied researchers can more easily use them and to develop hopefully more explicit/effective methods to solve periodic initial value problems for a large range of nonlinear wave problems. Finding solutions to equations which are 'close' to, i.e. perturbations of integrable systems is important; continuing development of effective perturbation methods for one and two-dimensional systems is an important goal. There are interesting IST type formulations and formal solutions associated with vector fields and dispersionless equations like the dispersionless KP, Heavenly equations and $2+1$ dimensional Toda equation [66–68] and a number of others. These problems should be further analyzed and better understood; here are a number of open questions that remain.

It is also natural to expect other novel classes of eigenvalue problems to be found that can be used to create new compatible linear systems and new types of solutions.

There is considerable interest in discrete systems. We briefly discussed semi-discrete equations above cf. [28]. But there are well-known integrable ordinary and partial difference evolution equations related to compatible Lax pairs were discovered early on cf. [14]. Over the years many more have been identified and studied cf. [69]. In some cases the IST has been developed but in many cases their relationship/solution via IST remains unexplored.

A goal that still seems elusive at this point in time is to find 'simple' three space, one time $(3+1)$ dimensional systems that can be effectively solved by IST. To date potential integrable $3+1$ dimensional systems have been complicated and highly nonlocal; cf. [15] for a discussion of this issue. What is clear from this analysis is that the type of Lax pair needed to integrate/linearize local $3+1$ dimensional systems will need to be quite different from its $1+1$, $2+1$ counterparts. There is a four dimensional system that is integrable: the self-dual Yang–Mills system. But here the natural way to study this system is as $4+0$ or $2+2$ dimensional systems. Interestingly the general self-dual Yang–Mills system (general underlying group structure) has the virtue of being, in a sense, universal. From it one can derive many (all?) known integrable systems cf. [15]. There is also a multidimensional sine-Gordon equation of geometric significance that can be connected to a hierarchy of one-dimensional linear problems. But there are constraints on this system that make it similar to one dimensional problems than a 'N +1' dimensional system cf. [15].

7.5 Conclusion

In this Chapter the study of integrability and nonlinear waves was surveyed. Given the necessary space constraints only certain aspects of this field were discussed: namely the inverse scattering transform (IST) as a tool to investigate important nonlinear wave systems. A brief history/background underlying this method (IST) was given including the motivation from the Fermi–Pasta–Ulam problem, numerical discovery of solitons, connection of solitons to eigenvalues of linear operators associated with rapidly decaying data and direct/inverse scattering/linearization of the kdV equation by Gardner, Greene, Kruskal, Miura [10], Lax pairs, [11] solutions of the NLS equation by Zakharov and Shabat [12], solutions of NLS, mKdV, sine/sinh-Gordon, KdV equations and many more by Ablowitz, Kaup, Newell and Segur [13]. Surprisingly after many years new nonlocal classes of integrable nonlinear systems by finding new symmetries of the AKNS linear scattering problem [20]. Many physically interesting nonlinear wave equations solvable by IST were mentioned cf. [15].

The methodology of IST applied to equations associated with second order linear operators is an area where IST can be effectively used was surveyed. Some of the equations that are amenable to this approach are the KdV, mKdV, NLS, sine/sinh Gordon, PT-NLS, RST-NLS, RST-mKdV equations and many more. Extensions of the IST technique were briefly mentioned as were different methods to find solutions to integrable nonlinear wave equations. The outlook for the future was considered and discussed.

Acknowledgements MJA was partially supported by NSF under Grant No. DMS-1712793.

References

1. M.D. Kruskal, in *Nonlinear Wave Motion*. AMS Lectures in Applied Mathematics, vol. 15, ed. by A.C. Newell (American Mathematical Society, Providence, 1974), p. 61
2. N.J. Zabusky, M.D. Kruskal, Phys. Rev. Lett. **15**, 240 (1965)
3. D.J. Korteweg, G. de Vries, Philos. Mag. **39**, 442 (1895)
4. E. Falcon, C. Laroche, S. Fauve, Phys. Rev. Lett. **89**, 204501 (2002)
5. J. Boussinesq, Mem. Pres. Acad. Sci. Paris Ser. **2**(23), 1 (1877)
6. J.S. Russell, in *Proceedings of the 14th meeting of the British Association for the Advancement of Science*, ed. by J. Murray (London, 1844), p. 311
7. M.J. Ablowitz, *Nonlinear Dispersive Waves, Asymptotic Analysis and Solitons* (Cambridge University Press, Cambridge, 2011)
8. R.M. Miura, C.S. Gardner, M.D. Kruskal, J. Math. Phys. **9**, 1204 (1968)
9. R.M. Miura, J. Math. Phys. **9**, 1202 (1968)
10. C.S. Gardner, J.M. Greene, M.D. Kruskal, R.M. Miura, Phys. Rev. Lett. **19**, 1095 (1967)
11. P.D. Lax, Commun. Pure Appl. Math **21**, 467 (1968)
12. V.E. Zakharov, A.B. Shabat, Sov. Phys., J. Exp. Theor. Phys. **34**, 62 (1972)
13. M.J. Ablowitz, D.J. Kaup, A.C. Newell, H. Segur, Stud. Appl. Math. **53**, 249 (1974)
14. M.J. Ablowitz, H. Segur, *Solitons and Inverse Scattering Transform*. Studies in Applied Mathematical, vol. 4 (SIAM, Philadelphia, 1981)
15. M.J. Ablowitz, P.A. Clarkson, *Solitons, Nonlinear Evolution Equations and Inverse Scattering* (Cambridge University Press, Cambridge, 1991)
16. C.M. Bender, S. Boettcher, Phys. Rev. Lett. **80**, 5243 (1998)
17. K.G. Makris, R. El Ganainy, D.N. Christodoulides, Z.H. Musslimani, Phys. Rev. Lett. **100**, 103904 (2008)
18. Z.H. Musslimani, K.G. Makris, R. El Ganainy, D.N. Christodoulides, Phys. Rev. Lett. **100**, 030402 (2008)
19. C.E. Rüter, K.G. Makris, R. El-Ganainy, D.N. Christodoulides, M. Segev, D. Kip, Nat. Phys. **6**, 192 (2010)
20. M.J. Ablowitz, Z.H. Musslimani, Phys. Rev. Lett. **110**, 064105 (2013)
21. M.J. Ablowitz, Z.H. Musslimani, **29**, 915 (2016)
22. M.J. Ablowitz, Z.H. Musslimani, Phys. Rev. E **90**, 032912 (2014)
23. A.S. Fokas, Nonlinearity **29**, 319 (2016)
24. M.J. Ablowitz, Z.H. Musslimani, Stud. Appl. Math. **139**, 7 (2017)
25. M.J. Ablowitz, Z.H. Musslimani, J. Phys. A: Math. Theor. **52**, 15LT02 (2019)
26. J. Yang, Phys. Rev. E **98**, 042202 (2018)
27. S.V. Manakov, Sov. Phys. JETP **38**, 693 (1974)
28. M.J. Ablowitz, B. Prinari, A.D. Trubatch, *Discrete and Continuous Nonlinear Schrödinger Systems* (Cambridge University Press, Cambridge, 2004)
29. S.P. Novikov, S.V. Manakov, L.P. Pitaevskii, V.E. Zakharov, *Theory of Solitons. The inverse Scattering Method* (Plenum, New York, 1984)
30. F. Calogero, A. Degasperis, *Spectral Transform and Solitons I* (North Holland, Amsterdam, 1982)
31. L.D. Faddeev, L.A. Takhtajan, *Hamiltonian Methods in the Theory of Solitons* (Springer, Berlin, 1987)
32. M.J. Ablowitz, H. Segur, Stud. Appl. Math. **57**, 13 (1977)
33. H. Segur, M.J. Ablowitz, Phys. D **3**, 165 (1981)
34. V.E. Zakharov, S.V. Manakov, J. Exp. Theor. Phys. **44**, 106 (1976)
35. P. Deift, X. Zhou, Bull. Am. Math. Soc. **26**, 119 (1992)
36. I.M. Krichever, Funct. Anal. Appl. **11**, 12 (1977)
37. E.D. Belekolos, A.I. Bobenko, V.Z. Enol'skii, A.R. Its, V.B. Matveev, *Algebro-geometrical Approach to Nonlinear Integrable Equations* (Springer, Berlin, 1994)
38. M.J. Ablowitz, A.S. Fokas, *Complex Variables: Introduction and Applications*, 2nd edn. (Cambridge University Press, Cambridge, 2003)

39. R. Beals, R.R. Coifman, Commun. Pure Appl. Math. **38**, 29 (1985)
40. G. Fibich, *The Nonlinear Schrödinger Equation, Singular Solutions and Collapse* (Springer, Heidelberg, 2015)
41. V.E. Zakharov, A.B. Shabat, Sov. Phys. JETP **37**, 823 (1973)
42. M.J. Ablowitz, G. Biondini, B. Prinari, Inverse Prob. **23**, 1711 (2007)
43. B. Prinari, M.J. Ablowitz, G. Biondini, J. Math. Phys. **47**, 063508 (2006)
44. M.J. Ablowitz, X.-D. Luo, Z.H. Musslimani, J. Math. Phys. **59**, 0011501 (2018)
45. V.S. Buslaev, V.N. Fomin, Vestnik Leningrad Univ. **17**, 56 (1962)
46. A. Cohen, Commun. Part. Differ. Eq. **9**, 751 (1984)
47. A. Cohen, T. Kappeler, Indiana U. Math. J. **34**, 127 (1985)
48. T. Aktosun, J. Math. Phys. **40**, 5289 (1999)
49. I. Egorova, Z. Gladka, T. Luc Lange, G. Teschl, Zh. Mat. Fiz. Anal. Geom. **11**, 123 (2015)
50. M.J. Ablowitz, X.-D. Luo, J.T. Cole, J. Math. Phys. **59**, 091406 (2018)
51. R. Hirota, *The Direct Method in Soliton Theory* (Cambridge University Press, Cambridge, 2004)
52. T. Miwa, M. Jimbo, E. Date, *Solitons, Differential Equations, Symmetries and Infinite Dimensional Lie Algebras* (Cambridge University Press, Cambridge, 2000)
53. G. Biondini, Y. Kodama, J. Phys. A: Math. Gen. **36**, 10529 (2003)
54. S. Chakravarty, Y. Kodama, Stud. Appl. Math. **123**, 83 (2009)
55. Y. Kodama, *Solitons in Two-Dimensional Shallow Water* (SIAM, Philadelphia, 2018)
56. M.J. Ablowitz, D.E. Baldwin, Phys. Rev. E **86**, 036305 (2012)
57. V.B. Matveev, M.A. Salle, *Darboux Transformation and Solitons* (Springer, Berlin, 1991)
58. C. Rogers, W.K. Schief, *Bäcklund and Darboux Transformations, Geometry and Modern Applications in Soliton Theory* (Cambridge University Press, Cambridge, 2002)
59. F.W. Nijhoff, G.R.W. Quispel, H.W. Capel, Phys. Lett. A **97**, 125 (1983)
60. V.E. Zakharov, A.B. Shabat, Funct. Anal. Appl. **13**, 166 (1979)
61. V.E. Zakharov, S.V. Manakov, Funct. Anal. Appl. **19**, 89 (1985)
62. R. Camassa, D.D. Holm, Phys. Rev. Lett. **71**, 1661 (1993)
63. S. Dyachenko, D. Zakharov, V. Zakharov, Phys. D **333**, 148 (2016)
64. D. Zakharov, V. Zakharov, Generalized primitive potentials, arXiv:1907.05038
65. A.S. Fokas, *A Unified Approach to Boundary Value Problems* (SIAM, Philadelphia, 2007)
66. S.V. Manakov, P.M. Santini, JETP Lett. **83**, 462 (2006)
67. S.V. Manakov, P.M. Santini, Phys. Lett. A. **359**, 613 (2006)
68. S.V. Manakov, P.M. Santini, J. Phys. A; Math. Theor. **42**, 095203 (2009)
69. J. Hietarinta, N. Joshi, F.W. Nijhoff, *Discrete Systems and Integrability* (Cambridge University Press, Cambridge, 2016)

Chapter 8
Nonequilibrium Phenomena in Nonlinear Lattices: From Slow Relaxation to Anomalous Transport

Stefano Iubini, Stefano Lepri, Roberto Livi, Antonio Politi, and Paolo Politi

Abstract This chapter contains an overview of the effects of nonlinear interactions in selected problems of non-equilibrium statistical mechanics. Most of the emphasis is put on open setups, where energy is exchanged with the environment. With reference to a few models of classical coupled anharmonic oscillators, we review anomalous but general properties such as extremely slow relaxation processes, or non-Fourier heat transport.

S. Iubini
Dipartimento di Fisica e Astronomia, Università di Padova, via F. Marzolo 8 I-35131, Padova, Italy
e-mail: stefano.iubini@unipd.it

S. Iubini · S. Lepri (✉) · R. Livi · P. Politi
Consiglio Nazionale delle Ricerche, Istituto dei Sistemi Complessi via Madonna del Piano 10, I-50019 Sesto Fiorentino, Italy
e-mail: stefano.lepri@isc.cnr.it

R. Livi
e-mail: roberto.livi@unifi.it

P. Politi
e-mail: paolo.politi@isc.cnr.it

S. Lepri · R. Livi · P. Politi
Istituto Nazionale di Fisica Nucleare, Sezione di Firenze, via G. Sansone 1 I-50019, Sesto Fiorentino, Italy

R. Livi
Dipartimento di Fisica e Astronomia, Università di Firenze, via G. Sansone 1 I-50019, Sesto Fiorentino , Italy

A. Politi
Institute for Pure and Applied Mathematics & SUPA University of Aberdeen, Aberdeen AB24 3UE, Scotland, United Kingdom
e-mail: a.politi@abdn.ac.uk

P. G. Kevrekidis et al. (eds.), *Emerging Frontiers in Nonlinear Science*, Nonlinear Systems and Complexity 32, https://doi.org/10.1007/978-3-030-44992-6_8

8.1 Introduction

The title of this chapter contains two negations, *nonequilibrium* and *nonlinearity*, which signal a double source of difficulties. First, at variance with equilibrium statistical mechanics, there is no general approach to describe the evolution of a generic system far from equilibrium. Second, nonlinear forces notoriously have to be handled with care.

From the fundamental point of view, nonlinear interactions are essential for the theoretical foundations of irreversible processes: a derivation of phenomenological relations (like for instance Fourier's law) from microscopic dynamics is indeed one of the challenges of mathematical physics. On the other hand, understanding the role of nonlinearity, low-dimensionality, long-range interactions, disorder etc. may help developing innovative ideas for nanoscale thermal management with possible future applications like controlling the heat fluxes in small devices built on molecular junctions, carbon nanotubes, polymers and nano-structured materials [1].

This chapter aims at illustrating the combined effect of nonlinear interactions on relaxation and transport: since this is an exceedingly vast topic, we focus mainly on selected specific issues. In particular, we wish to review through some examples (mostly relying on numerical simulations) of how relaxation and transport are affected by nonlinear interactions in systems of classical nonlinear oscillators. This class of model represents a large variety of different physical problems like atomic vibrations in crystals and molecules or field modes in optics or acoustics.

The chapter is organized as follows. For concreteness, we discuss mostly one-dimensional arrays of classical oscillators, that are reviewed in Sect. 8.2. For later purposes, their equilibrium thermodynamics is recalled in Sect. 8.3. Section 8.4 deals with the typical time-scales of relaxation to a steady state in the presence of a dissipation applied to the boundaries and discusses how nonlinear localization can significantly affect the process. This is the most detailed section, since the topic is still open and we have preferred to add some recent details for the sake of clarity. In Sect. 8.5 we address the issue of nonequilibrium steady states for chains in contact with different thermal reservoirs. We recall there how nonlinear interactions affect fluctuations of conserved quantities and conspire to yield energy superdiffusion. Finally, in Sect. 8.6 we outline possible future developments.

8.2 Classical Coupled Nonlinear Oscillators: Basic Models

A vast number of micro- and mesoscopic models have been introduced and studied to understand nonequilibrium dynamics. Many of them involve some form of stochasticity [2]. Here we concentrate on open Hamiltonian models described by unidimensional arrays of N classical nonlinear oscillators. Two families of models are reviewed: (i) separable systems characterized by kinetic and potential energy; (ii) non-separable ones such as the Discrete Nonlinear Schrödinger equation.

The first class is generally characterized by the Hamiltonian

$$H = \sum_{n=1}^{N} \left[\frac{p_n^2}{2m} + U(q_n) + V(q_{n+1} - q_n) \right], \tag{8.1}$$

where q_n and p_n denote position and momentum of the point–like particles; m is their mass while, the potential $V(x)$ accounts for the nearest-neighbour interactions between consecutive particles, and finally, the on-site potential $U(q_n)$ accounts for the possible interaction with an external environment (e.g. a substrate).

The corresponding evolution equations are

$$m\ddot{q}_n = -U'(q_n) - F(r_n) + F(r_{n-1}), \quad n = 1, \ldots, N, \tag{8.2}$$

where $r_n = q_{n+1} - q_n$, $F(x) = -V'(x)$, and the prime denotes a derivative with respect to the argument. If q_n represents a longitudinal position, then $L = \sum_{n=1}^{N-1} r_n$ represents the total length of the chain, which, in the case of fixed boundary conditions (b.c.) and for $U = 0$, is a constant of motion. Conversely, if the particles are confined in a simulation "box" of length L with periodic b.c., we have $q_{n+N} = q_n + L$. Alternatively, one can adopt a lattice interpretation whereby the (discrete) position is $x_n = an$ (where a is the lattice spacing), while q_n is a transversal displacement. Thus, the chain length is obviously equal to Na.

For isolated systems, the Hamiltonian (8.1) is a constant of motion. If the pinning potential U is constant, the total momentum $P = \sum_{n=1}^{N} p_n$ is conserved, as well. Since we are interested in heat transport, one can set $P = 0$ (i.e. we assume to work in the center–of–mass reference frame) without loss of generality. As a result, the relevant state variables of microcanonical equilibrium are the specific energy (i.e. the energy per particle) $h = H/N$ and the elongation $\ell = L/N$ (i.e. the inverse of the particle density).

An important subclass is the one in which V is quadratic, which can be regarded as a discretization of the Klein–Gordon field: relevant examples are the Frenkel–Kontorova [3, 4] and "ϕ^4" models [5, 6] which, in suitable units, correspond to $U(y) = 1 - \cos(y)$ and $U(y) = y^2/2 + y^4/4$, respectively. Another toy model that has been studied in some detail is the ding-a-ling system [7], where U is quadratic and the nearest-neighbor interactions are replaced by elastic collisions.

8.2.1 The Fermi–Pasta–Ulam–Tsingou Chain

In this context, the most paradigmatic example is the Fermi–Pasta–Ulam-Tsingou (FPUT) model, with $U(q_n) = 0$ and

$$V(r_n) = \frac{k_2}{2}(r_n - a)^2 + \frac{k_3}{3}(r_n - a)^3 + \frac{k_4}{4}(r_n - a)^4, \tag{8.3}$$

introduced in a widely acknowledged seminal work [8] in nonlinear dynamics. It is well known that the initial goal of the study was to demonstrate that a generic nonlinear interaction should eventually drive an isolated mechanical system with many degrees of freedom, towards an equilibrium state characterised by energy equipartition among normal modes. Actually, in the following decades the related problem of steady state transport was also considered [9–11].

Following the notation of the original work [8], the coupling terms k_3 and k_4 are denoted by α and β, respectively; historically this model is sometimes referred to as the "FPU-$\alpha\beta$" model. In the absence of the cubic nonlinearity ($k_3 = 0$), the system is referred to as "FPU-β" model. Notice that upon introducing the displacement $u_n = q_n - na$ from the equilibrium position, r_n can be rewritten as $u_{n+1} - u_n + a$, so that the lattice spacing a disappears from the equations.

8.2.2 The Discrete Nonlinear Schrödinger Equation

The Discrete Nonlinear Schrödinger (DNLS) equation has been widely investigated in various domains of physics as a prototype model for the propagation of nonlinear excitations [12–14]. Originally, it was proposed to describe electronic transport in biomolecules [15] and later for nonlinear wave propagation in photonic or phononic crystals [16, 17] as well as in ultra-cold atom gases in optical lattices [18].

The system (in its dimensionless form) is described by the Hamiltonian

$$H = \sum_{n=1}^{N} \left(|z_n|^4 + z_n^* z_{n+1} + z_n z_{n+1}^*\right) \tag{8.4}$$

where the complex variables z_n and $-iz_n^*$ ($n = 1, \ldots, N$) are canonical variables. The Hamilton equations $\dot{z}_n = -\partial H/\partial(iz_n^*)$ are written as

$$i\dot{z}_n = -2|z_n|^2 z_n - z_{n-1} - z_{n+1} \, . \tag{8.5}$$

Sometimes it is convenient to decompose z_n into real and imaginary components: $z_n = (p_n + iq_n)/\sqrt{2}$; this way q_n and p_n are standard conjugate canonical variables. Besides the Hamiltonian, the system admits a second constant of motion, namely the total norm $A = \sum_{n=1}^{N} |z_n|^2$ which, depending on the physical context, can be interpreted as the gas particle number, optical power, etc. At variance with its continuum counterpart, the DNLS is non-integrable: it typically displays a chaotic dynamics.

As other oscillator models, including the FPUT one, the DNLS equation admits a set of localized solutions, the so-called *discrete breathers* (DB) [14, 19], characterized by a large amplitude on a single site, $|z_n|^2 \gg s^2$, where s^2 is the amplitude of the surrounding background. In the limit of $s \ll 1$, when perturbative calculations can be carried out, the long term stability has been discussed in full detail [20]. We later on show that breather stability is an important issue also in physical setups, where the background is fully chaotic.

8.2.3 The Coupled Rotors Model

Another interesting system is the coupled rotors chain described by the equations of motion

$$\dot{q}_n = p_n \,, \quad \dot{p}_n = \sin(q_{n+1} - q_n) - \sin(q_n - q_{n-1}) \,. \tag{8.6}$$

This model is sometimes referred to as the Hamiltonian version of the XY spin chain. It is a sort of intermediate model between standard oscillator chains (notice that here the "position" q_n is an angle) and the DNLS equation, once we think of the variable z_n as composed of amplitude and phase. In fact, it can be shown that in some limit the DNLS reduces to a rotor chain [21].

8.3 Equilibrium

Equilibrium thermodynamics of the above models can be calculated by standard means. For the Hamiltonian (8.1) without pinning ($U = 0$) this can be accomplished straightforwardly by computing the partition function in the isobaric ensemble [22]. For models like Klein–Gordon and DNLS lattices, the computation requires the use of transfer integral methods [23].

The case of the DNLS is of particular interest: a thermodynamic equilibrium state is in fact specified by two intensive parameters, the mass density $a = A/N \geq 0$ and the energy density $h = H/N$ or equivalently by the conjugate variables μ (chemical potential) and β (inverse temperature). The equilibrium phase-diagram in the (a, h) plane [23] is bounded from below by the $(T = 0)$ ground–state line $h = a^2 - 2a$ corresponding to a uniform state with constant amplitude and constant phase–differences $z_n = \sqrt{a}e^{i(\mu t + \pi n)}$, with $\mu = 2(a - 1)$. States below this curve are not physically accessible.

The positive–temperature region lies between the ground–state line and the infinite-temperature ($\beta = 0$) line, given by $h = 2a^2$. In this limit, the grand–canonical equilibrium distribution becomes proportional to $\exp(\beta\mu A)$, where the finite (negative) product $\beta\mu$ implies a diverging chemical potential. Equilibrium states at infinite temperature are therefore characterized by an exponential distribution of the amplitudes, $P(|z_n|^2) = a^{-1}e^{-|z_n|^2/a}$ and random phases. Finally, states above the $\beta = 0$ line belong to the so–called negative–temperature region [23, 24]. From a thermodynamic point of view, the presence of states at absolute negative temperature is a consequence of the entropy being a decreasing function of the internal energy. Very recently it has been found that this region is characterized by inequivalence of statistical ensembles [25]. While just above the $\beta = 0$ line the microcanonical partition function can be computed explicitly by large-deviation techniques, the grandcanonical partition function is undefined, due to the presence of a branch-cut singularity in the complex β-plane. Moreover, the microcanonical ensemble predicts the presence of a first-order phase transition from a thermalized phase, below the $\beta = 0$ line, to a condensed phase, above the $\beta = 0$ line. This situation represents a typical scenario of broken ergodicity in the negative-temperature phase, induced by a condensation phenomenon, due to the spontaneous formation and coalescence of DB.

Since the DNLS Hamiltonian (8.4) is not separable, one cannot determine temperature and chemical potential via the standard molecular-dynamics tools. It is necessary to make use of the microcanonical definition provided in [26]. The general expressions are nonlocal and rather involved; we refer to [27] for details and the related bibliography. Alternatively, one can determine the relations $a(T, \mu)$, $h(T, \mu)$ numerically, by putting the system in interaction with an external reservoir that imposes T and μ and by measuring the corresponding equilibrium densities.[1]

8.4 Relaxation

In this section we review and discuss relaxation dynamics, namely how the steady state is reached starting from nonequilibrium initial conditions. Generally speaking one may distinguish between two cases:

(i) relaxation to thermal equilibrium (energy equipartition) from a particular initial state in the *isolated* (microcanonical) setup;
(ii) evolution towards a steady state in an *open* setup, whereby the system is allowed to exchange energy, momentum etc. with the environment, composed of one (or more) reservoirs.

A typical example of case (i) is the numerical experiment discussed in the original FPUT paper, a problem deeply related to the validity of the ergodic hypothesis. In the following, we will not dwell further on the FPUT problem: the interested reader can look at some recent literature [30–32]. Here, we focus more on boundary-induced relaxation phenomena with a particular emphasis given to the DNLS, for the existence of a negative-temperature region.

[1]The actual implementation of a reservoir for the DNLS is less straightforward than for usual oscillator models [28]. Two main strategies have been proposed: the first is a Monte-Carlo dynamics [27] whereby the reservoir performs random perturbations δz_1 of, say, the state variable z_1 that are accepted or rejected according to a grand-canonical Metropolis cost-function $\exp[-\beta(\Delta H - \mu \Delta A)]$, where ΔH and ΔA are respectively the variations of energy and mass produced by δz_1. Between successive interactions with the environment the dynamics is Hamiltonian and can be integrated by symplectic algorithms [29]. Another approach is based on a Langevin dynamics with a dissipation designed in such a way that equilibrium corresponds to the grand-canonical measure [24].

8.4.1 Localization by Boundary Cooling

A first numerical evidence of slow-relaxation induced by the spontaneous emergence of localized inhomogeneities has been provided by Tsironis and Aubry [33], who discussed the chain dynamics in the presence of a nonlinear pinning potential. The system, initially prepared in a thermalized state at some given temperature T, was put in contact with a cold (zero temperature) heat bath, by adding a damping term on a few boundary particles. The chain eventually converges to a quasi-stationary state, where a residual amount of energy is kept under the form of a few isolated DBs. It has been later argued that the energy relaxation obeys a stretched exponential law in time [34, 35].

The rotor model is yet another interesting example where boundary dissipation leads to a slow relaxation. In [36] it was found that for long enough times the energy decreases very slowly, according to a typical stretched-exponential law

$$E(t) = E(0)\exp(-(t/\tau)^{\sigma})$$

with $\sigma < 1$ (typically $\sigma \approx 0.5$) and τ being some characteristic timescale. The occurrence of slow dynamics has been associated with a progressive destruction of localized excitations (the so-called rotobreathers) and energy release at the boundaries. At very long times a residual quasi-stationary state is again observed: a finite fraction of the initial energy is stored into a single rotobreather and remains constant over the rest of the simulation.

Recently, some mathematical insight on the origin of such slow process has been proposed: it has been argued that the dissipation rate may become arbitrarily small in certain physical regimes due to the decoupling of non-resonant terms, as it happens in KAM problems [37–39].

A renewed interest in the problem of boundary cooling was provided by the proposal of implementing it as a technique to localize Bose-Einstein condensates in optical lattices [40]. Referring to the DNLS model, stationary and traveling localized states were generated by removing atoms at the optical-lattice ends. Regimes of stretched-exponential decay for the number of atoms trapped in the lattice were clearly identified by numerical simulations. Further studies showed that the dynamics of dissipated energy exhibits a characteristic avalanche behavior [41].

8.4.2 Dynamical Freezing of Relaxation to Equilibrium

In this section we discuss the relaxation process of a localized excitation (DB) in the DNLS model (see (8.4)–(8.5)). For positive temperatures, DBs have a negligible probability to arise at equilibrium and yet, as commented above, there are ways to grow them (e.g. by boundary dissipation). Therefore, it is important to understand their relaxation process. Simulations have been performed by superposing a large-

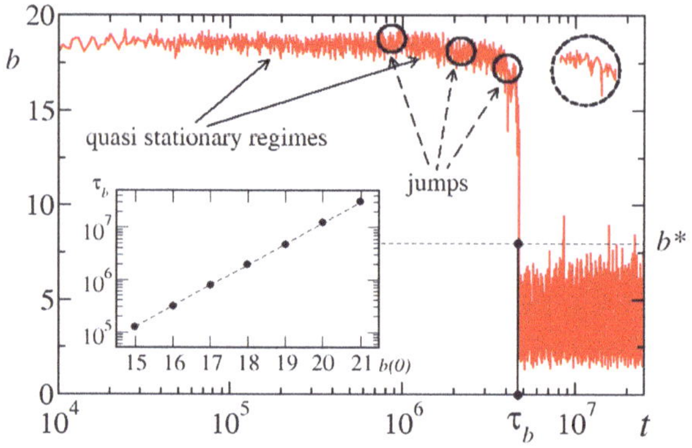

Fig. 8.1 The relaxation process of a breather through the time dependence of its mass, $b(t)$. The abrupt character of the process allows to define a threshold value b^* and to determine the relaxation time τ_b as the shortest time satisfying the condition $b(\tau_b) = b^*$. In the top right circle we zoom in a jump of the mass. The inset shows the exponential increase of τ_b with the initial mass of the breather: the dashed line is a fit, giving $\tau_b \approx e^{\alpha b(0)}$, with $\alpha = 0.91 \pm 0.01$. The breather is initially at the centre of a chain of $N = 31$ sites

mass DB at the origin, $n = 0$, to an otherwise equilibrium configuration for the background (i.e. $-N \leq n \leq N$). The temperature $T = 10$ and the chemical potential $\mu = -6.4$ are imposed by connecting the chain boundaries to suitable Langevin baths.

The main results are reported in Fig. 8.1, where we can see the time dependence of the breather mass $b = |z_0|^2$ in a typical simulation (notice the logarithmic time scale). The decay process is fairly abrupt and we can identify quasi-stationary regimes, separated by jumps. Because of the abrupt decay of $b(t)$ we can define the relaxation time τ_b by setting a threshold b^* (dashed horizontal time). In the inset, we show how τ_b depends on the initial mass of the breather, namely exponentially. In the rest of this section we will discuss the meaning of this numerical finding and some related results.

The rapid increase of τ_b with $b(0)$ suggests that the larger the mass of the breather, the weaker the relaxation mechanism. This is not surprising, because the natural frequency of a breather of mass b is $\omega = 2b$, so the coupling term between the breather and the nearest neighbors becomes negligible on the typical time scale of the background (which is of order one). However, a rotational frequency proportional to the breather mass cannot justify by itself an exponentially slow decay process. Before trying any theoretical explanation, it is necessary to gain more insight on the relaxation process.

From Fig. 8.1 we can see that the *laminar* regime preceding the final breather disruption is approximately quasi-stationary, with no specific drift (except for the final part, where the mass clearly tends to decrease). It is, therefore, tempting to characterize this regime in terms of a suitable diffusion coefficient.[2] The next question is the identification of an optimal variable to characterize the hypothetical diffusion process. The mass $b(t)$ is too noisy to extract reliable estimates. A principal component analysis performed on the triplet of variables z_{-1}, z_0, z_1 suggests to use $\tilde{Q} \simeq E_b^{1/4}$ [42], where $E_b = |z_0|^4 + \frac{1}{2}[z_0^*(z_1 + z_{-1}) + \text{c.c.}]$ is the breather energy (see (8.4)). Accordingly we define $D_{\tilde{Q}} = \langle [\tilde{Q}^2(t+\tau) - \tilde{Q}^2(t)]^2 \rangle / \tau$. The evaluation of $D_{\tilde{Q}}$ for breathers of increasing height (see Fig. 4 of [42]) shows that the diffusion coefficient decreases exponentially with $b(0)$, in agreement with the direct evidence of an almost frozen dynamics. On the basis of fluctuation-dissipation considerations, we should also expect a drift $v \approx D_{\tilde{Q}}/T$ and, accordingly, an exponential increase of the decay time. However, while the exponential increase is confirmed by the simulations, we do not see a clear evidence of a (downward) drift.

Figure 8.1 rather suggests that the relaxation is accompanied and perhaps caused by sporadic jumps. This statement is supported by Fig. 8.2, where we plot the average transfer of energy between the breather and the background in a given interval of time, as a function of the initial mass of the neighbouring site, $a_1(0)$ (suitably rescaled). There is an evident, narrow peak accompanied by some pinnacles. The peak appears at a value very close to the analytical threshold for the existence of symmetric bound states (dimers) between two neighbouring sites, $\sqrt{a_1^*(0)} = \sqrt{b(0)} - \sqrt{2}$.

The very emergence of such bound states is attested by the inset of the same figure, where we plot the time dependence of the mass of the breather $b(t)$ along with the mass $a_1(t)$ of a neighboring site. When it happens that $a_1 > a_1^*$ (see the dashed line) the two sites are strongly coupled together and rotate with the same frequency. This bound state eventually dissolves, with the net result that the "post-dimer" breather has lost some energy with respect to the "pre-dimer" one.

Analytical, non-rigorous considerations allow evaluating the typical time scale of such phenomena as the expectation time for a background fluctuation in one of the two neighbouring sites to become larger than the threshold a_1^*. In fact, from the high-T equilibrium distribution $P(\theta)$ to have a mass θ [23], we can approximately estimate the breather lifetime $\tau_b \approx 1/P(a_1^*)$, which has the asymptotic expression $\tau_b \approx \exp(\beta b^2)$ for a diverging breather mass b. This expression implies a superexponential growth of the relaxation time with the mass, suggesting that the asymptotic mechanism for breather decay might not be dimer formation (τ_b seems to increase exponentially), but current simulation data does not allow to exclude it either.

As a matter of fact, the additional peaks appearing in Fig. 8.2 suggest the existence of other mechanisms and the special setup we are going to discuss allows one to conclude that relaxation occurs even if dimer formation is suppressed. We direct reader's attention to the inset of Fig. 8.3, where we consider the relaxation process of the very same breather in different conditions. The full line is the standard DNLS model, i.e. a curve fully similar to that plotted in Fig. 8.1. The dotted line has been determined with a unidirectional coupling between breather and background: the former feels the latter but not the other way around. In practice the breather is coupled

[2] Bounded fluctuations would be possible only in the presence of an attractor, but this is a Hamiltonian system.

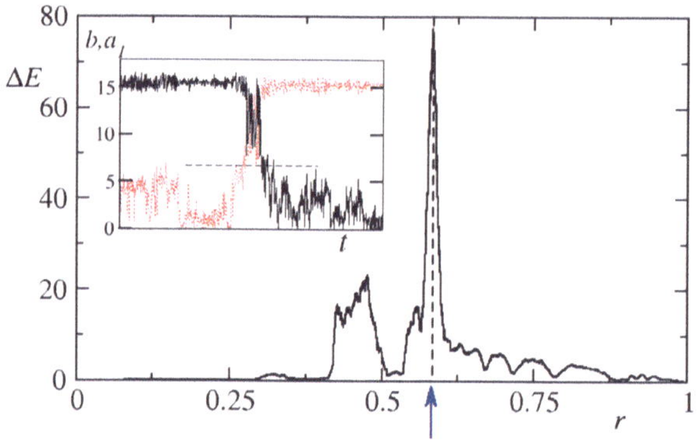

Fig. 8.2 The energy lost or gained on average by the breather after a fixed amount of time, as a function of the initial mass of the neighbouring site, $a_1(0)$ (more precisely, of the ratio $r = a_1(0)/b(0)$). The sharp peak corresponds to the condition $\sqrt{a_1(0)} = \sqrt{b(0)} - \sqrt{2}$, which allows the formation of a symmetric bound state (dimer) between the breather and its neighbour. In the inset we plot the time dependence of the mass in these two sites along with such threshold value (horizontal dashed line)

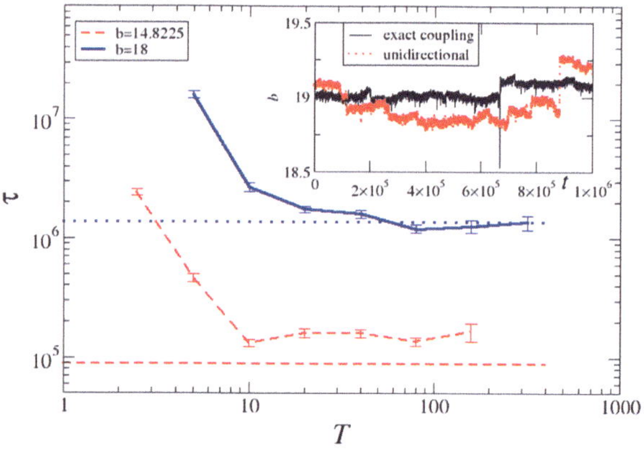

Fig. 8.3 The relaxation time of a breather as a function of the temperature T, for two different initial masses. The horizontal lines represent the analytical, asymptotic values as deduced by an approximate approach (see the main text) where the breather feels the background but not the other way around. In the inset we compare the time evolution of the mass of the same initial breather, with exact and unidirectional coupling

to a background whose evolution is independent of the breather itself. The result is a qualitatively similar evolution of the breather mass.

Since unidirectional coupling makes impossible the rising of a symmetric bound state, in this case dimers cannot play any role in DB relaxation. On the other hand, resonance mechanisms would not be affected by the unidirectional character of the coupling and therefore they would persist, but the approximate expression for the relaxation time would still give a superexponential growth.

Unidirectional coupling is useful also to discuss the temperature dependence of the relaxation time, $\tau_b(T)$. The curves plotted in Fig. 8.3 refer to different initial masses of the breather and the central result we focus on is the asymptotic behavior of $\tau_b(T)$, which seems to reach a constant. In other words, the relaxation time of a finite breather is expected to be finite even at infinite temperature. The horizontal lines have been determined analytically with the same criterion discussed here above, taking the limit $T \to \infty$ but also allowing $|\mu| = T/a$ to diverge so that the density of mass, a, is kept constant (simulations have been performed in such a setup).

So far, we have mentioned two potentially interesting processes: (i) the diffusion occurring during the laminar (quasi-stationary) regime; (ii) the jumps, both visible in Fig. 8.1. The diffusion coefficient turns out to decrease exponentially with b and the probability of dimer or resonances formation (related to the jumps) also decreases exponentially or even superexponentially. It is evident that in the presence of several relaxation channels, a frozen dynamics may appear if and only if all mechanisms are exponentially slow.

However, it is not clear at all *why* dynamics is almost frozen. A former paper by some of the present authors [42]—a first attempt in this direction—suggests that $\tilde{Q}$ (the quantity derived from the principal component analysis and used to derive the diffusion coefficient) is the approximate expression of an adiabatic invariant, which might be broken by jumps. We have tried to determine the adiabatic invariant perturbatively, by computing higher orders, but the attempt has not been successful.

We wish to stress that understanding the relaxation process of a breather at positive temperature is a useful 'if not necessary' step to later understand the negative T phase from a dynamical point of view.

8.4.3 Role of Negative Temperatures

In the previous section we have seen that relaxation to equilibrium may be very slow in the DNLS, in the positive temperature region of the (a, h) plane. As mentioned in Sect. 8.3, there exists a second region ($h > 2a^2$) where the absolute temperature is expected to be negative (on the basis of microcanonical arguments). Relaxation phenomena in this region turn out to be a rather controversial issue, not yet fully settled.

Entropic arguments [43, 44] suggest that all the excess energy, which cannot be stored in a homogeneous background for $h > 2a^2$, should eventually concentrate into a single breather. In fact, this is precisely what happens in a simplified, purely stochastic version of the DNLS, where it has been shown that multiple breathers progressively merge through a non conventional coarsening process [45, 46].

On the other hand, coalescence has not been observed in molecular dynamics simulations slightly above the $\beta = 0$ line (e.g. for $a = 1$ and $h = 2.4$). On the contrary, it looks like a sort of stationary regime sets in, characterized by a small breather density, where DBs spontaneously form and then die, after some typical lifetime [24], see Fig. 8.4. This is due to the presence of a finite interaction (hopping) energy: the

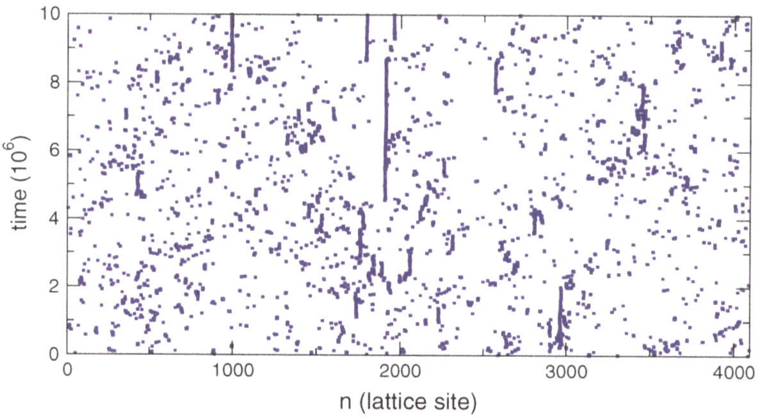

Fig. 8.4 Evolution of the local amplitude for DNLS in a negative temperature state, the dots correspond to points where $|z_n|^2 > 10$. Microcanonical simulation with $a = 1$ and $h = 2.4$, $N = 4192$

background can store excess energy in the phase differences of neighbouring sites and this implies that breathers can spontaneously nucleate. These findings have been recently confirmed by more extensive simulations [47] which suggest the existence of a finite region in the (a, h)-plane, where the dynamics covers a subregion of the available phase space, doing so in an "ergodic manner" (i.e. no coarsening).

Further, independent evidence of a relatively stable negative-temperature regime comes from the nonequilibrium simulations performed in [48], where a DNLS chain was put in contact on one side with a positive temperature heat bath, while on the other, with a pure dissipator. Depending on the temperature value of the first heat bath, an extended portion of the chain settles in a regime characterized by a position-dependent negative temperature, a flux of mass and energy, without being accompanied by the onset of breathers.

Finally, recent statistical-mechanics calculations [25] suggest that slightly above the critical $\beta = 0$ line, strong finite-size effects are to be expected, which might affect the interpretation of the numerical simulations.

8.5 Transport

Let us now turn our attention to nonequilibrium steady states that emerge, at long enough times, when the system is in contact with two (or more) heat reservoirs operating at different temperatures. Generally speaking, several methods, based on both deterministic and stochastic (Langevin or Monte-Carlo) algorithms, have been proposed [28]. A complementary approach is based on linear-response theory, which amounts to computing the equilibrium correlation function of currents. In principle, this task can be accomplished in any equilibrium ensemble, the microcanonical one being the most natural choice.

8.5.1 Anomalous Energy Transport

The main results, emerging from a long series of works, can be summarized as follows. Models of the form (8.2) with $U(q) = 0$ typically display *anomalous* transport and relaxation features. Said differently, Fourier's law *does not hold*: the kinetics of energy carriers is so correlated that they are able to propagate *faster* than in the standard (diffusive) case. We refer the reader to existing review papers [1, 28, 49, 50] for a more comprehensive description. Here we just mention how this anomalous behavior manifests itself in the simulations.

- The finite-size heat conductivity $\kappa(L)$ diverges in the limit of a large system size $L \to \infty$ as, $\kappa(L) \propto L^\gamma$ [51], i.e. the heat transport coefficient is ill-defined in the thermodynamic limit.
- The equilibrium correlation function of the total energy current J displays a non-integrable long-time tail $\langle J(t)J(0)\rangle \propto t^{-(1-\delta)}$, with $0 \le \delta < 1$ [52, 53]. Accordingly, the Green–Kubo formula yields an infinite value of the conductivity.
- Energy perturbations propagate superdiffusively [54, 55]: a local perturbation of the energy spreads, while its variance broadens in time as $\sigma^2 \propto t^\beta$, with $\beta > 1$.
- Temperature profiles in the nonequilibrium steady states are nonlinear, even for vanishing applied temperature gradients. Typically they are the solution of a *fractional heat equation* [56, 57].

There is a large body of numerical evidence that the above features occur generically in 1D, whenever the conservation of energy, momentum and length holds. This is related to the existence of long-wavelength (Goldstone) modes (an acoustic phonon branch in the linear spectrum of (8.2) with $U = 0$) that are very weakly damped. Indeed, it is sufficient to add external (e.g. substrate) forces, to make all anomalies disappear and restore Fourier's law.

8.5.2 Universality and the Kardar–Parisi–Zhang Equation

The nonlinear fluctuating hydrodynamics approach is able to justify and predict several universal features of anomalous transport in anharmonic chains [22, 58]. The main entities are the random fields describing deviations of the conserved quantities with respect to their stationary values. The role of fluctuations is taken into account by renormalization group or some kind of self-consistent theory.

The main theoretical insight is the intimate relation between the anharmonic chain and one of the most important equations in nonequilibrium statistical physics, the celebrated Kardar–Parisi–Zhang (KPZ) equation, originally introduced in the (seemingly unrelated) context of surface growth [59]. The KPZ equation for the stochastic field $h(x, t)$ in one spatial dimension reads

$$\frac{\partial h}{\partial t} = \nu \frac{\partial^2 h}{\partial x^2} + \frac{\kappa}{2} \left(\frac{\partial h}{\partial x} \right)^2 + \eta. \tag{8.7}$$

where $\eta(x, t)$ represents a Gaussian white noise with $\langle \eta(x, t)\eta(x', t') \rangle = 2D\delta(x - x')\delta(t - t')$ and ν, κ, D are the relevant parameters. It has been shown [22, 58] that large-scale dynamical properties of anharmonic chains are in the same dynamical universality class as (8.7). Loosely speaking, we can represent the displacement field as the superposition of counter-propagating plane waves, modulated by an envelope that is ruled, at large scales, by (8.7). As a consequence, correlations of observables display in the hydrodynamic limit *anomalous dynamical scaling*. For instance, the dynamical structure factor $S(k, \omega)$ of the particle displacement shows for $k \to 0$ two sharp peaks at $\omega = \pm\omega_{\max}(k)$ that correspond to the propagation of sound modes and for $\omega \approx \pm\omega_{\max}$ behaves as

$$S(k, \omega) \sim f_{\mathrm{KPZ}} \left(\frac{\omega \pm \omega_{\max}}{\lambda_s k^{3/2}} \right). \tag{8.8}$$

Remarkably, the scaling function f_{KPZ} is universal and known exactly, while λ_s is a model-dependent parameter. The main point is that the *dynamical exponent* $z = 3/2$ is different from $z = 2$ expected for a standard diffusive process.

Most of the predictions have been successfully tested for several models. For a chain of coupled anharmonic oscillators with three conserved quantities like the FPUT chains, such theoretical predictions have been successfully compared with the numerics [60, 61]. Other positive tests have been reported in [62]. One further prediction is that the FPUT-β model should belong to a different (non-KPZ) universality class, as previously suggested by the numerics [63, 64].

8.5.3 Coupled Transport

As known from irreversible thermodynamics, when there are more conserved quantities, the corresponding currents can be coupled: in the linear response regime, transport is described by Onsager coefficients. The best-known example is that of thermoelectricity, whereby useful electric work can be extracted in the presence of temperature gradients [65–67].

In the present context, the simplest example is the one-dimensional rotor model, (8.6), that admits two conserved quantities (energy and angular momentum), two associated currents, and only one relevant thermodynamic parameter, the temperature T. In this case [45] one can easily introduce the interaction with two reservoirs by fixing the average angular momenta ω_0, ω_1 and kinetic temperatures T_0, T_1 at the chain ends. This can be obtained by adding the Langevin term $\gamma(\omega_0 - \dot{q}_1) +$

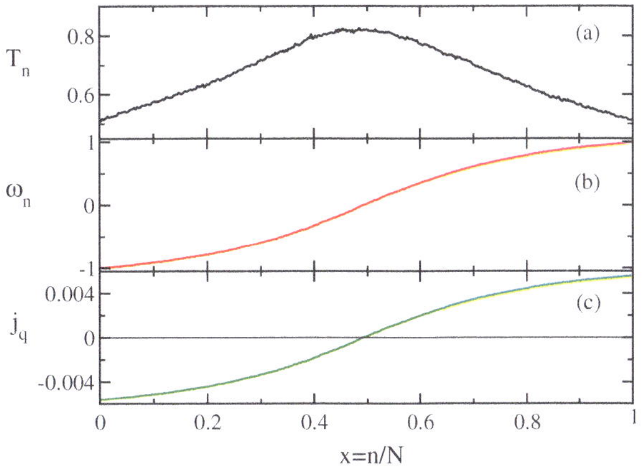

Fig. 8.5 Simulation of the rotor chain with $N = 400$ particles, in contact at its boundaries with two heat baths at temperature $T_0 = T_1 = 0.5$ and in the presence of torques $\omega_0 = -1$ and $\omega_1 = 1$: **a** temperature profile; **b** frequency (chemical potential) profile; **c** local heat flux. Adapted from [68]

$\sqrt{2\gamma T_0}\,\xi$ to the equation of motion (8.6) of the leftmost rotor: γ defines the coupling strength with the bath and ξ is a Gaussian white random noise with zero mean and unit variance. An analogous term, with ω_1 and T_1 replacing ω_0 and T_0, is added to the equation of motion of the rightmost rotor.

To illustrate the peculiarities of this setup, Fig. 8.5 reports frequency and temperature profiles [68] in a case where only an angular momentum gradient is applied, i.e. $T_1 = T_0$ and $\omega_0 = -1$ and $\omega_1 = 1$. The temperature profile T_n is non-monotonic [69] as a consequence of the coupling with the momentum flux j_p imposed by the torque at the boundaries, although, in the end, the energy flux j_h vanishes for symmetry reasons. By recalling that $j_h = j_q + \omega j_p$ we see that the heat flux $j_q = -\omega j_p$ varies along the chain being everywhere proportional to the frequency, so that it is negative in the left part and positive in the right side (this is again consistent with symmetry considerations). Thus heat is generated in the central hotter region, where the temperature is higher and transported towards the two edges. The total energy flux is however everywhere zero as the heat flux is compensated by an opposite coherent flux due to momentum transfer. Physically, the temperature bump can be interpreted as a sort of Joule effect: the transport of momentum involves dissipation, which in turn contributes to increasing the temperature, analogously to what happens when an electric wire is crossed by a flux of charges.

Similar effects are studied in [27] for the DNLS case, where the dependence of the Onsager matrix on temperature and chemical potential is considered. In this case the cross-coupling term (the equivalent of the Seebeck coefficient in the language of thermo-electricity) may change sign, leading to temperature- and mass-profiles with opposite slopes.

8.5.4 Integrable Models and Their Perturbations

The above results are mostly obtained in a strongly nonlinear regime or more generally far from any integrable limit. For the FPUT model, this means working with high enough energies/temperatures to avoid all the difficulties induced by quasi-integrability and the associated slow relaxation to equilibrium.

Integrable systems constitute *per se* a relevant case. In the framework of the present work the most important example is certainly the celebrated Toda chain, namely model (8.1) with $U = 0$ and

$$V(x) = e^{-x} + x - 1$$

As intuitively expected, heat transport is ballistic due to its integrability and the associated solitonic solution [70]. Mathematically, this is expressed by saying that there is a non-vanishing *Drude weight*, namely a zero-frequency component of the energy current power spectra [71, 72]. A lower bound of the Drude weight can be estimated making use of Mazur inequality [73] in terms of correlations between the currents themselves and the conserved quantities (see [71, 72] for the Toda case). However, the idea of solitons transporting energy as independent particles is somehow too simplistic. It has been recognized [74] that solitons experience a stochastic sequence of spatial shifts as they move through the lattice interacting with other excitations without momentum exchange [74]. At variance with the harmonic chain, which is also integrable, but whose proper modes are non-interacting phonons, the Toda chain, as proposed in [75], is an interacting integrable system. In particular, it is characterized by what has been termed a non–dissipative diffusion mechanism [74]. In fact, the calculation of the transport coefficients by the Green–Kubo formula indicates the presence of a finite Onsager coefficient, which corresponds to a diffusive process on top of the dominant ballistic one [72, 76, 77].

A natural question concerns the behavior when a generic perturbation is applied to an otherwise integrable system. For instance, adding a quadratic pinning potential $U(q) = q^2/2$ to the Toda chain is expected to restore standard diffusive transport, but numerical simulations show that long-range correlations are preserved over relatively long scales [77, 78]. Moreover, weak perturbations that conserve momentum (and are thus expected to display anomalous transport in the KPZ class) display instead significant differences [79] and even diffusive transport over the accessible simulation ranges [80]. Altogether a full unified picture of the problem is still lacking.

8.6 Overview and Open Problems

In spite of the noticeable progress that has been made over the last decades, the study of nonequilibrium processes in nonlinear systems remains a fascinating and challenging domain of research. On a methodological ground, it concerns mainly the scientific communities of mathematics and theoretical physics, but it is of primary interest also for optics, materials science and soft matter, just to mention a few among the related fields in experimental and applied research. Going through the reading of this chapter, one can easily realize that most achievements have been possible thanks to a fruitful combination of analytic approaches and numerical simulations and this can be reasonably expected to hold also in the future. For what concerns open problems, it is worth mentioning a few of them, that have already attracted some interest. The first one is the study of nonlinear models with long-range interactions. The main motivation stems from the observation that for this class of models the equivalence between statistical ensembles may not hold. This is expected to yield interesting consequences also for nonequilibrium phenomena. In fact, long-range systems are known to exhibit further peculiar properties, like long-living metastable states, anomalous energy diffusion, lack of thermalization when interacting with a single temperature reservoir, propagation of perturbations with infinite velocity, etc. (for a general review see [81]). The problem of heat transport in the long-range version of the rotor and FPUT chains has been recently tackled in a series of papers [82–86]. When the interaction is genuinely long-range, i.e. the long-range exponent α is smaller than 1, the heat transport process is dominated by parallel transport: a flat temperature profile sets in, simply because each oscillator, independently, takes a temperature value which is the average of those applied by the thermal baths. For α larger than 1, the long-range rotor model reproduces standard diffusion, i.e. normal heat conductivity, as in the short-range version. Conversely, the FPUT chain is characterized by an anomalous scaling exponent $\gamma(\alpha)$, which seems to recover the value of the short-range case only for large values of α. Anyway, a better understanding of the transport problem demands further refined investigations, that should take into account also a comparison with the unusual relaxation process to equilibrium characterizing long-range systems.

Let us conclude by mentioning one further open problem, which is related to energy localization induced by nonlinearity. In fact, localization processes may emerge in nonlinear systems even in the absence of disorder. The typical example is the spontaneous formation of breathers in the DNLS problem already discussed in Sect. 8.4. In the negative temperature region the phenomenon of condensation of DBs can be read as a process of ergodicity-breaking, because, in the microcanonical setup, energy equipartition is inhibited by its localization. It is well known that, when condensation phenomena are present, statistical ensemble equivalence is not granted [87]. This is the case of the DNLS Hamiltonian in the negative temperature region, where the only available statistical ensemble is the microcanonical one [25]. It is worth recalling that both total energy and mass are conserved quantities in the DNLS Hamiltonian: the breaking of ergodicity in the negative temperature phase indicates that a standard thermalization process for both quantities is suppressed. The analogy with the problem of the Eigenvalue Thermalization Hypothesis advanced for genuine quantum integrable systems (see the review paper [88]) suggests that ensemble equivalence should be reconsidered also in this context to properly formulate a thermalization hypothesis or, alternatively, the many-body localization phenomenon invoked in quantum integrable systems.

Acknowledgements We thank L. Chirondojan and G.-L. Oppo for useful discussions. We acknowledge support from the project MIUR-PRIN2017 *Coarse-grained description for non-equilibrium systems and transport phenomena (CO-NEST)* n. 201798CZL.

References

1. S. Lepri (ed.), *Thermal Transport in Low Dimensions: From Statistical Physics to Nanoscale Heat Transfer*, vol. 921, Lecture Notes in Physics (Springer, Heidelberg, 2016)
2. R. Livi, P. Politi, *Nonequilibrium Statistical Physics: A Modern Perspective* (Cambridge University Press, Cambridge, 2017)
3. M.J. Gillan, R.W. Holloway, J. Phys. C **18**, 5705 (1985)
4. B. Hu, B. Li, H. Zhao, Phys. Rev. E **57**, 2992 (1998)
5. K. Aoki, D. Kusnezov, Phys. Lett. A **265**, 250 (2000)
6. P.G. Kevrekidis, J. Cuevas-Maraver (eds.), *A Dynamical Perspective on the* ϕ^4 *Model: Past, Present and Future* (Springer, Cham, 2019)
7. G. Casati, J. Ford, F. Vivaldi, W.M. Visscher, Phys. Rev. Lett. **52**, 1861 (1984)
8. E. Fermi, J. Pasta, S. Ulam, Los Alamos Report LA-1940, p. 975 (1955)
9. D.N. Payton, M. Rich, W.M. Visscher, Phys. Rev. **160**, 706 (1967)
10. H. Nakazawa, Progr. Theor. Phys. Suppl. **45**, 231 (1970)
11. H. Kaburaki, M. Machida, Phys. Lett. A **181**, 85 (1993)
12. J.C. Eilbeck, P.S. Lomdahl, A.C. Scott, Phys. D **16**, 318 (1985)
13. J.C. Eilbeck, M. Johansson, in *Proceedings of the Conference on Localization and Energy Transfer in Nonlinear Systems*, eds. by L. Vazquez, R.S. MacKay, M.P. Zorzano. San Lorenzo del Escorial, June 2002 (World Scientific, Singapore, 2003), p. 44
14. P.G. Kevrekidis, *The Discrete Nonlinear Schrödinger Equation* (Springer, Berlin, 2009)
15. A. Scott, *Nonlinear Science. Emergence and Dynamics of Coherent Structures* (Oxford University Press, Oxford, 2003)
16. A.M. Kosevich, M.A. Mamalui, J. Exp. Theor. Phys. **95**, 777 (2002)
17. D. Hennig, G.P. Tsironis, Phys. Rep. **307**, 333 (1999)
18. R. Franzosi, R. Livi, G.L. Oppo, A. Politi, Nonlinearity **24**, R89 (2011)
19. S. Flach, A.V. Gorbach, Phys. Rep. **467**, 1 (2008)
20. M. Johansson, K.Ø. Rasmussen, Phys. Rev. E **70**, 066610 (2004)
21. S. Iubini, S. Lepri, R. Livi, A. Politi, J. Stat. Mech.: Theory Exp. **2013**, P08017 (2013)
22. H. Spohn, J. Stat. Phys. **154**, 1191 (2014)
23. K.Ø. Rasmussen, T. Cretegny, P.G. Kevrekidis, N. Grønbech-Jensen, Phys. Rev. Lett. **84**, 3740 (2000)
24. S. Iubini, R. Franzosi, R. Livi, G.L. Oppo, A. Politi, New J. Phys. **15**, 023032 (2013)
25. G. Gradenigo, S. Iubini, R. Livi, S.N. Majumdar, Localization in the Discrete Non-Linear Schrödinger equation: mechanism of a first-order transition in the microcanonical ensemble, arXiv:1910.07461
26. H.-H. Rugh, Phys. Rev. Lett. **78**, 772 (1997)
27. S. Iubini, S. Lepri, A. Politi, Phys. Rev. E **86**, 011108 (2012)
28. S. Lepri, R. Livi, A. Politi, Phys. Rep. **377**, 1 (2003)
29. H. Yoshida, Phys. Lett. A **150**, 262 (1990)
30. G. Benettin, A. Ponno, J. Stat. Phys. **144**, 793 (2011)
31. G. Benettin, H. Christodoulidi, A. Ponno, J. Stat. Phys. **152**, 195 (2013)
32. M. Onorato, L. Vozella, D. Proment, YuV Lvov, Proc. Natl. Acad. Sci. USA **112**, 4208 (2015)
33. G.P. Tsironis, S. Aubry, Phys. Rev. Lett. **77**, 5225 (1996)
34. F. Piazza, S. Lepri, R. Livi, J. Phys. A: Math. Gen. **34**, 9803 (2001)
35. F. Piazza, S. Lepri, R. Livi, Chaos **13**, 637 (2003)
36. M. Eleftheriou, S. Lepri, R. Livi, F. Piazza, Phys. D **204**, 230 (2005)

37. N. Cuneo, J.-P. Eckmann, Commun. Math. Phys. **345**, 185 (2016)
38. N. Cuneo, J.-P. Eckmann, C.E. Wayne, Nonlinearity **30**, R81 (2017)
39. J.-P. Eckmann, C.E. Wayne, Decay of Hamiltonian Breathers under Dissipation, arXiv:1907.12632
40. R. Livi, R. Franzosi, G.-L. Oppo, Phys. Rev. Lett. **97**, 060401 (2006)
41. G.S. Ng, H. Hennig, R. Fleischmann, T. Kottos, T. Geisel, New J. Phys. **11**, 073045 (2009)
42. S. Iubini, L. Chirondojan, G.-L. Oppo, A. Politi, P. Politi, Phys. Rev. Lett. **122**, 084102 (2019)
43. B. Rumpf, Phys. Rev. E **69**, 016618 (2004)
44. B. Rumpf, Phys. Rev. E **77**, 036606 (2008)
45. S. Iubini, S. Lepri, R. Livi, A. Politi, Phys. Rev. Lett. **112**, 134101 (2014)
46. S. Iubini, A. Politi, P. Politi, J. Stat. Mech.: Theory Exp. **2017**, 073201 (2017)
47. T. Mithun, Y. Kati, C. Danieli, S. Flach, Phys. Rev. Lett. **120**, 184101 (2018)
48. S. Iubini, S. Lepri, R. Livi, G.-L. Oppo, A. Politi, Entropy-Switz. **19**, 445 (2017)
49. A. Dhar, Adv. Phys. **57**, 457 (2008)
50. G. Basile, L. Delfini, S. Lepri, R. Livi, S. Olla, A. Politi, Eur. Phys J.-Spec. Top. **151**, 85 (2007)
51. S. Lepri, R. Livi, A. Politi, Phys. Rev. Lett. **78**, 1896 (1997)
52. S. Lepri, R. Livi, A. Politi, Europhys. Lett. **43**, 271 (1998)
53. S. Lepri, Eur. Phys J. B **18**, 441 (2000)
54. S. Denisov, J. Klafter, M. Urbakh, Phys. Rev. Lett. **91**, 194301 (2003)
55. P. Cipriani, S. Denisov, A. Politi, Phys. Rev. Lett. **94**, 244301 (2005)
56. S. Lepri, A. Politi, Phys. Rev. E **83**, 030107 (2011)
57. A. Kundu, C. Bernardin, K. Saito, A. Kundu, A. Dhar, J. Stat. Mech.: Theory Exp. **2019**, 013205 (2019)
58. H. van Beijeren, Phys. Rev. Lett. **108**, 180601 (2012)
59. A.L. Barabási, H.E. Stanley, *Fractal Concepts in Surface Growth* (Cambridge University Press, Cambridge, 1995)
60. S.G. Das, A. Dhar, K. Saito, C.B. Mendl, H. Spohn, Phys. Rev. E **90**, 012124 (2014)
61. P. Di Cintio, R. Livi, H. Bufferand, G. Ciraolo, S. Lepri, M.J. Straka, Phys. Rev. E **92**, 062108 (2015)
62. C.B. Mendl, H. Spohn, Phys. Rev. Lett. **111**, 230601 (2013)
63. S. Lepri, R. Livi, A. Politi, Phys. Rev. E **68**, 067102 (2003)
64. G.R. Lee-Dadswell, Phys. Rev. E **91**, 032102 (2015)
65. C. Mejia-Monasterio, H. Larralde, F. Leyvraz, Phys. Rev. Lett. **86**, 5417 (2001)
66. G. Casati, L. Wang, T. Prosen, J. Stat. Mech.: Theory Exp. **2009**, L03004 (2009)
67. G. Benenti, G. Casati, C. Mejia-Monasterio, New J. Phys. **16**, 015014 (2014)
68. S. Iubini, S. Lepri, R. Livi, A. Politi, New J. Phys. **18**, 083023 (2016)
69. A. Iacobucci, F. Legoll, S. Olla, G. Stoltz, Phys. Rev. E **84**, 061108 (2011)
70. M. Toda, Phys. Scr. **20**, 424 (1979)
71. X. Zotos, J. Low. Temp. Phys. **126**, 1185 (2002)
72. B. Sriram Shastry, A.P. Young, Phys. Rev. B **82**, 104306 (2010)
73. P. Mazur, Physica **43**, 533 (1969)
74. N. Theodorakopoulos, M. Peyrard, Phys. Rev. Lett. **83**, 2293 (1999)
75. H. Spohn, J. Math. Phys. **59**, 091402 (2018)
76. A. Kundu, A. Dhar, Phys. Rev. E **94**, 062130 (2016)
77. P. Di Cintio, S. Iubini, S. Lepri, R. Livi, Chaos Soliton Fract. **117**, 249 (2018)
78. A. Dhar, A. Kundu, J.L. Lebowitz, J.A. Scaramazza, J. Stat. Phys. **175**, 1298 (2019)
79. A. Iacobucci, F. Legoll, S. Olla, G. Stoltz, J. Stat. Phys. **140**, 336 (2010)
80. S. Chen, J. Wang, G. Casati, G. Benenti, Phys. Rev. E **90**, 032134 (2014)
81. A. Campa, T. Dauxois, D. Fanelli, S. Ruffo, *Physics of Long-Range Interacting Systems* (Oxford University Press, Oxford, 2014)
82. R.R. Avila, E. Pereira, D.L. Teixeira, Phys. A **423**, 51 (2015)
83. D. Bagchi, Phys. Rev. E **95**, 032102 (2017)
84. C. Olivares, C. Anteneodo, Phys. Rev. E **94**, 042117 (2016)
85. S. Iubini, P. Di Cintio, S. Lepri, R. Livi, L. Casetti, Phys. Rev. E **97**, 032102 (2018)
86. P. Di Cintio, S. Iubini, S. Lepri, R. Livi, J. Phys. A: Math. Theor. **52**, 274001 (2019)
87. D. Ruelle, *Statistical Mechanics: Rigorous Results* (World Scientific, Singapore, 1999)
88. L. Vidmar, M. Rigol, J. Stat. Mech.: Theory Exp. **2016**, 064007 (2016)

Chapter 9
Nonlinearity, Geometry and Field Theory Solitons

Nicholas S. Manton

Abstract Topological solitons occur in many types of nonlinear field theory. Their motion and interactions can be simulated classically, and can be well approximated by a finite-dimensional dynamics on a moduli space of collective coordinates. Interesting phenomena related to the curvature and topology of moduli spaces are illustrated here through the examples of vortices, sigma model lumps, and monopoles. Collective coordinate dynamics can be quantized, and it is shown how quantized Skyrmion dynamics is used to understand aspects of nuclear physics. A novel model for nuclear fusion, based on wormhole geometry, is also proposed.

9.1 Field Topology and Solitons

This chapter is about particles and multi-particle states modelled as solitons in quantum field theory, and mainly in field theories that are Lorentz invariant [1, 2]. Particles are usually regarded as pointlike and structureless. This makes sense for electrons, for which no internal structure has yet been discerned. But other particles, such as protons and neutrons, have a finite size. In field theory, such an extended particle can occur naturally as a soliton.

The usual treatment of quantum field theory is perturbation theory. Here, one first studies the linearised field equation and quantizes it as an infinite collection of harmonic oscillators. The linear classical field equation has oscillating, wavelike solutions—plane waves or dispersing wavepackets—that have no definite spatial structure. However, quantization of the wave amplitudes (note: not particle positions as in ordinary quantum mechanics) leads to states of definite momentum and energy, with the usual relativistic energy-momentum relation for a particle (in units where the speed of light is unity)

$$E^2 = m^2 + \mathbf{p} \cdot \mathbf{p} \,, \tag{9.1}$$

N. S. Manton (✉)
Department of Applied Mathematics and Theoretical Physics, University of Cambridge, Wilberforce Road, Cambridge CB3 0WA, UK
e-mail: N.S.Manton@damtp.cam.ac.uk

P. G. Kevrekidis et al. (eds.), *Emerging Frontiers in Nonlinear Science*, Nonlinear Systems and Complexity 32, https://doi.org/10.1007/978-3-030-44992-6_9

where the non-negative constant m is obtained from the field equation. This relation describes a point particle of mass m, momentum $\mathbf{p}$ and energy E, and the relativistic field theory allows for multi-particle states and antiparticles too. Each particle has definite momentum, but it has no obvious position coordinate, and no internal structure. Further terms in the classical field equation, beyond the linear terms, are treated as small, and in the perturbative approach to the quantum field theory, these terms generate interactions between the particles.

Here we treat the field theory differently. We consider the full field equation from the start. Its nonlinear structure can lead to isolated (solitary) wavelike solutions that have an obvious location and finite size, and do not disperse. In the principal examples the wave can be static in a certain rest frame, and we call it a soliton, by analogy with the localised nonlinear waves that occur in fluids and other condensed matter systems, and which are stationary or static in some (usually moving) frame of reference. The ending *-on* denotes a genuine particle.

So far the soliton is classical, but it has desirable features of a particle, like energy and momentum, and is localised. In the soliton's rest frame (in a relativistic theory) the energy is identified with the mass of the soliton. Quantization is much more tricky in the full nonlinear field theory than in perturbation theory, so some approximations are needed; this is the price to pay for the advantages of a nonlinear treatment of the classical theory. The most useful approximation is to restrict the dynamics of solitons to some finite number of collective coordinates. For a single soliton, these coordinates model the soliton's position and, if needed, its orientation in space and/or its orientation in internal field space. After quantization, the soliton acquires quantized spin, and sometimes an electric charge or another charge quantum number.

A fascinating feature of the collective coordinates is that they can in principle be used to deal with the dynamics of two or more solitons in interaction. The geometry of the collective coordinates is not just the product geometry for the two solitons separately. Two-soliton dynamics often combines a non-trivial geometry of interaction [3] with the more usual interaction potential energy. In practice, this means that for two solitons the kinetic energy coefficients vary with the soliton separation. This variation leads to velocity-dependent forces, supplementing the static forces derived from the gradient of the potential.

In special cases there is no potential, and the interaction is purely geometrical. We will discuss a few examples of the special solitons where this occurs. They are referred to as Bogomolny solitons [4], or BPS solitons, and they arise in field theories that have a remarkable extension as supersymmetric field theories. Bogomolny solitons are also solutions of a first order field equation, somewhat like a Cauchy–Riemann equation. We will hardly explore the supersymmetric extension here, because of limitations of space, but see [5].

Quantization of the collective coordinate dynamics is usually possible. For one soliton this is straightforward. For two or more, it gives a model of quantized interacting solitons, where the number of solitons doesn't change during the interaction, and no antisolitons are created. So collective coordinate quantization is essentially a non-relativistic approximation, appropriate for particles with relatively low energy. High-energy processes can generate soliton-antisoliton pairs, but this is hard to model. There are some new ideas how to introduce collective coordinates for solitons and antisolitons together, in the simplest field theories in one space dimension [6].

Let us introduce slightly more technical language. The collective coordinates are coordinates on a manifold, and are usually called moduli. The manifold as a whole is called the moduli space of the solitons, and there is a different one for each soliton number. In the Bogomolny set-up, where there is no interaction potential, and hence no static forces between solitons, the moduli space $\mathcal{M}_n$ is simply the manifold of static solutions of the field equation, representing n solitons. $\mathcal{M}_n$ is generally a smooth, connected Riemannian manifold of dimension proportional to n. Its metric is derived from the field theory kinetic energy, as we will explain further in Sect. 9.2. Defining $\mathcal{M}_n$ when there are static forces is more difficult. One way is to take a known moduli space where such forces vanish, and add a potential energy. The potential is adjusted to match the energy of known solutions, for example, merged solitons and infinitely separated solitons.

One may now ask—how is the number of solitons n defined? In many cases the answer lies in the topological character of the soliton field. We summarise this here, and refer to [1] for more details. Field theories always have a classical vacuum, a solution of minimal energy with spatially uniform fields. A localised wave is generally a continuous deformation of a unique vacuum, and hence topologically uninteresting. However, in some of the field theories of interest, the vacuum is not unique, because of symmetry breaking, combined possibly with gauge invariance. A soliton is then a field configuration that approaches a definite vacuum value along any radial line away from the soliton core in a fixed direction, but that value varies as the direction varies. There is a danger that such a soliton has infinite energy, because the angular gradients of the field do not fall off rapidly enough towards infinity. This problem is avoided in a gauge theory, because although the angular gradients fall off slowly, the gauge covariant angular gradients can be arranged to fall off much more rapidly, and these are what contribute to the energy. For (non-gauged) solitons with a possibly divergent energy, the divergence can often be regularised. Global vortices are like this. Two or more global vortices have a finite interaction energy, because the divergent part of the total energy is a constant, independent of the vortex positions, and can be discarded [7]. In practical situations, solitons are often in a finite region of space, and this also regularises the total energy.

For this type of soliton, the soliton number n is a conserved, topological feature of the field values at infinity. It is unchanged by any smooth evolution of the field, including high-energy collisions of solitons, and quantum fluctuations of the field are also believed not to affect its integer character or conservation.

To be more concrete, in several examples the classical field energy density has a contribution

$$V_{\text{pot}} = \lambda(1 - \boldsymbol{\phi} \cdot \boldsymbol{\phi})^2 , \tag{9.2}$$

where the number of components of the field $\boldsymbol{\phi}$ equals the dimension d of space, and λ is a positive constant. For the field energy density to be localised, and its integral finite after regularisation, $\boldsymbol{\phi}$ on the $(d-1)$-sphere at spatial infinity must be a continuous function of the angles there, and take its value in the unit $(d-1)$-sphere $\boldsymbol{\phi} \cdot \boldsymbol{\phi} = 1$, which is the sphere of possible vacua. The asymptotic field configuration therefore defines a mapping from the $(d-1)$-sphere to itself, which has a topological degree n. This degree is used in the context of vortices in two dimensions, and monopoles in three dimensions. The topological degree counts solitons, because if n basic solitons are well separated (radially, and in angles), then they each contribute one unit to the degree, as can be verified through a calculation. When the solitons approach each other, the degree does not change.

Both vortices and monopoles exist in gauge field theories, where the gradient terms in the energy density are gauge covariant. The total energy is finite, without need for regularisation. However, there is then always some curvature in the gauge field of the soliton, i.e. a non-vanishing magnetic field. For vortices in two dimensions this magnetic field decays rapidly towards infinity, but there is still a net magnetic flux localised around the vortex cores, whose total magnitude is proportional to the vortex number n. For monopoles in three dimensions the magnetic field has Coulomb-type decay, and there is a net magnetic flux through the sphere at infinity proportional to the monopole number n. So an n-monopole has a magnetic charge proportional to n. The magnetic Coulomb field does not have a singularity, because soliton fields are always smooth, and the energy is finite.

(Note that a 2-monopole is not a dipole, because a dipole consists of a monopole-antimonopole pair. Physically, true magnetic monopoles have never been seen, and they are forbidden by Maxwell's equations in standard electromagnetic theory, but they could exist in more exotic nonabelian gauge theories of particles, like Grand Unified Theories.)

An alternative topological characterisation of a soliton can occur when the field theory has a unique vacuum, but the field is everywhere subject to a nonlinear constraint. For example, a field $\boldsymbol{\psi}$ could be constrained to be everywhere a $(d+1)$-component unit vector, and also (for energetic reasons) take a unique vacuum value at spatial infinity. In this case one may topologically compactify space by identifying all points at infinity. This turns flat space $\mathbb{R}^d$ into a (large) d-sphere, and the field also takes values in a d-sphere. So the field (throughout space) defines a mapping from the d-sphere to itself, with a topological degree n, which is again identified with the soliton number. An example of this kind of field topology occurs for Skyrmions in three space dimensions [8, 9], where the field is constrained to a 3-sphere, not exactly because the field is a unit vector, but because it takes values in the group SU(2), which as a manifold is a 3-sphere. The unique vacuum is the identity group element, the unit matrix 1_2.

For the much-studied magnetic Skyrmions in two space dimensions [10, 11], the field is the magnetisation vector, which has values in a 2-sphere. The magnetisation can change its direction through the sample, but not easily its magnitude. The Skyrmion number n is the degree of the mapping from the physical surface to the target 2-sphere.

A topological degree can be defined more generally even when the field does not have a unique limiting value at spatial infinity, because it is the integral of a local density—the Jacobian of the mapping, suitably normalised. The integral then makes sense in any finite region, although it may not have integer values. Because of this, one can for example calculate the number of Skyrmions per unit volume, or unit area, in a periodic array of Skyrmions.

9.2 Moduli Spaces of Solitons

We consider here the moduli spaces of a few types of soliton, including one example where there isn't an underlying field theory. It is simplest to start with the mathematically precisely defined moduli spaces for Bogomolny solitons.

A key example is that of U(1) gauged vortices in a plane. Let x_1, x_2 be Cartesian coordinates, and introduce the complex coordinate $z = x_1 + ix_2$. The fields are a complex scalar field ϕ and a gauge potential with components a_1, a_2. $D_i\phi = \partial_i\phi - ia_i\phi$ denotes the gauge covariant derivative of ϕ, and $b = \partial_1 a_2 - \partial_2 a_1$ is the magnetic field. (In the fully dynamical theory, the fields also depend on time $t \equiv x_0$, so there is a time component a_0 of the gauge potential, and an electric field $e_i = \partial_0 a_i - \partial_i a_0$.) Static Bogomolny vortices are solutions of the first order coupled equations

$$(D_1 + iD_2)\phi = 0\,, \tag{9.3}$$

$$b - \frac{1}{2}(1 - |\phi|^2) = 0\,. \tag{9.4}$$

Taubes showed that these equations have a unique n-vortex solution for every choice of n (possibly coincident) points in the plane [12]. Let these n points have complex coordinates $Z_1, Z_2, \ldots, Z_n$, and let

$$p(z) = \prod_{k=1}^{n}(z - Z_k) = z^n + p_1 z^{n-1} + \cdots + p_n \tag{9.5}$$

be the polynomial with unit leading coefficient (i.e. the monic polynomial) having these points as zeros. There is then a convenient gauge choice where the vortex field ϕ, as a function of z and $\bar{z}$, is a real non-zero multiple of $p(z)$, and such that $|\phi| \to 1$ asymptotically. The points $Z_1, Z_2, \ldots, Z_n$ are therefore the zeros of ϕ, and are identified as the vortex locations. The winding number (degree) of ϕ at spatial infinity is n, because this is the winding of the phase of z^n, the leading term in $p(z)$.

Naively, the moduli are the points $Z_1, Z_2, \ldots, Z_n$, but these are unordered, so the moduli space $\mathcal{M}_n$ is not the obvious $\mathbb{C}^n$ for n labelled particles, but instead $\mathbb{C}^n/S_n$, where S_n is the permutation group. One might suspect that this quotient has conical singularities, but remarkably it is still a smooth manifold, and is again $\mathbb{C}^n$. Its coordinates are not the Z_k's but instead the coefficients $p_1, p_2, \ldots, p_n$ of the polynomial p. These coefficients are the elementary symmetric polynomials in the Z_k's, and have unconstrained values. (More generally, the symmetrised nth power of any Riemann surface is a smooth complex manifold.)

This vortex example illustrates a common feature of solitons. They have locations in space, but are indistinguishable at the classical level, because a permutation of their locations does not change the field. The moduli space therefore involves a quotient of what would occur for labelled particles. However, the moduli space remains smooth, and the dynamics of the solitons is smooth even if two or more solitons instantaneously coalesce at the same location. We will explain the consequence of this for two-soliton collisions shortly.

The dynamics of Bogomolny vortices only depends on the field kinetic energy (an integral involving $D_0\phi$ and e_i), and if this energy is relatively small, the field cannot escape far from the moduli space of static solutions. So the dynamics is effectively controlled by the kinetic energy restricted to the moduli space $\mathcal{M}_n$. The kinetic energy becomes a (hermitian) quadratic form in the collective coordinate velocities $\dot{p}_1, \dot{p}_2, \dots, \dot{p}_n$, and the coefficient matrix can be interpreted as a (hermitian) Riemannian metric on $\mathcal{M}_n$. So we need to know something about this metric on the moduli space. It is not flat. Samols calculated an implicit formula for it [13], showing that it was not just hermitian but Kähler—the details are not needed here.

With the dynamics controlled by a metric on the moduli space $\mathcal{M}_n$, the natural motion is along geodesics at constant speed [3]. Let us see what this means for two-vortex dynamics. Let the vortex locations be Z_1 and Z_2. As explained earlier, the good coordinates are $p_1 = -(Z_1 + Z_2)$ and $p_2 = Z_1 Z_2$. The centre of mass motion, i.e. the time-dependence of p_1, decouples, as shown by Samols, so let us consider centred vortex motion, where $Z_1 = Z$ and $Z_2 = -Z$. Then $p_2 = -Z^2$. The centred moduli space, with complex coordinate p_2, is a real surface of revolution having the shape of a cone, but smoothly rounded off near the vertex $p_2 = 0$ (see Fig. 9.1). The reasons are as follows. It is a surface of revolution because a rotation of z by angle α is a symmetry, causing a rotation of both Z and $-Z$, and a rotation of p_2 by 2α. The surface is a cone because for well separated vortices ($|p_2|$ large), Z is itself a good local coordinate and the metric is approximately Euclidean when expressed in terms of Z. So the centred moduli space is asymptotically a flat plane quotiented by the permutation group S_2 exchanging Z and $-Z$, i.e. a cone with semi-vertex angle $30°$. Finally, the metric is smooth at the vertex because vortex fields have no singularity as the two points Z and $-Z$ become coincident. Note that the smoothed cone has a smaller area than the corresponding flat cone with its sharp vertex.

One geodesic on the smoothed cone is through the vertex, down on one side and up on the other, at constant geometric speed. This corresponds to a straight line motion of p_2, passing through $p_2 = 0$, though not at constant speed in terms of this coordinate. Suppose the motion is along the real axis, from $-\infty$ to $+\infty$, and suppose that $p_2 = 0$ at time $t = 0$. Then, when $-t$ is replaced by t, the sign of p_2 is reversed. That means that the sign of Z^2 is reversed, so the locations Z and $-Z$ are replaced by iZ and $-iZ$. The incoming vortices are on the real z-axis, but the outgoing vortices are on the imaginary z-axis. Instantaneously, when the vortices are coincident, the field is circularly symmetric.

This $90°$ scattering of a vortex pair is very robust. It occurs during the short period of time while the vortices are close together even if there are other vortices present or the vortices are on a curved Riemann surface. It occurs because the vortices are

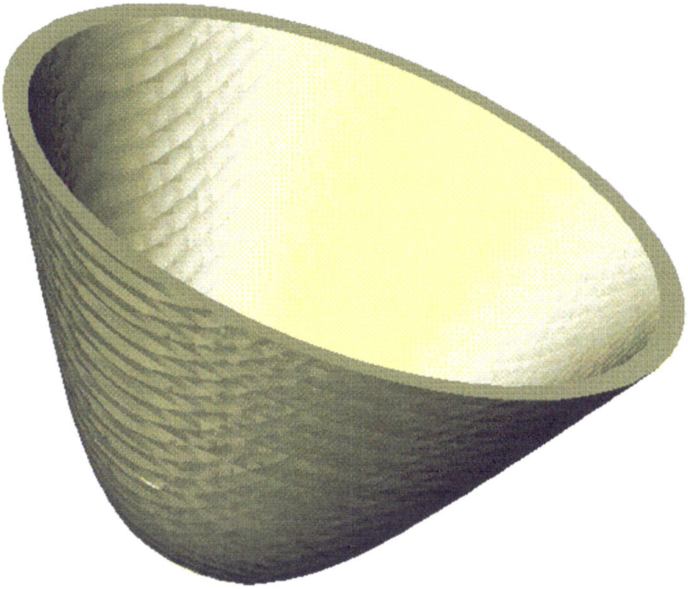

Fig. 9.1 Rounded cone—the moduli space for centred two-vortex dynamics. Reproduced from [1]

classically indistinguishable, and doesn't occur for point particle dynamics with any reasonable potential. Solitons behave as point particles while well separated, but not when they are close together.

An analogy for this scattering behaviour is what happens to the foci of the ellipse $x^2 + \beta y^2 = 1$ as β increases through 1. The ellipse evolves smoothly, passing through a circle, but the major and minor axes are exchanged, and the foci scatter through 90°. Instantaneously the foci coalesce and there is circular symmetry.

A few other soliton moduli spaces are known in some detail. Sigma model lumps are solitons in a nonlinear scalar field theory in two space dimensions. In the simplest case, the field is a mapping from a 2-sphere (the compactified spatial plane) to a target 2-sphere, and solutions of the relevant Bogomolny equation are meromorphic functions (using a complex, stereographic coordinate on both spheres) [14]. The soliton number n is the degree of the mapping. A meromorphic function of finite topological degree n has to be a rational function

$$R(z) = \frac{p(z)}{q(z)} \tag{9.6}$$

where (generically) p and q are polynomials of algebraic degree n, having no common zeros. The moduli space of such rational functions is understood, but is quite complicated because of the constraint of avoiding common zeros. Naively, the lumps are localised solitons with four real moduli each. These moduli specify the location, the size and a phase angle for each lump. Lumps whose separations are much larger than their sizes occur when the zeros of p and q are paired up close together. The metric on the moduli space is again Kähler [15]. Geodesic motion on moduli space can be studied for a single lump, when the lump moves either on a plane or on a sphere of finite radius. For the lump on a plane, the size and phase cannot vary, according to the geodesic dynamics, because the corresponding metric terms are divergent and the kinetic energy would be infinite. For a lump on a sphere, the metric is not divergent, but instead it is incomplete, and a lump can contract to zero size in finite time [16]. This contraction to a singular field is also observed in solutions of the full dynamical field equation on a plane, but the details are different [17].

For SU(2) Bogomolny (BPS) monopoles in three dimensions, the field equation is

$$B_i = D_i \Phi \tag{9.7}$$

where B_i is the SU(2) nonabelian magnetic field, and $D_i \Phi$ is the covariant gradient of an SU(2) triplet Higgs field. Φ is required to have unit magnitude at spatial infinity, and its angular variation there determines the topological degree, and hence the monopole number n. The metric on the moduli space $\mathcal{M}_n$ of n-monopole solutions is very interesting, but also complicated. The dimension of $\mathcal{M}_n$ is $4n$ and the metric is hyperKähler. Atiyah and Hitchin made a deep study of $\mathcal{M}_n$ and its metric, and explicitly constructed the metric on $\mathcal{M}_2$ [18]. From one point of view, $\mathcal{M}_n$ is again the space of rational maps of degree n (subject to a suitable base point condition), but the rational map does not have such a direct relation to the fields as it does for lumps.

A single monopole has a location in $\mathbb{R}^3$ and an internal phase. The monopole has a fixed magnetic charge -2π, and when its phase is time-dependent, it becomes electrically charged, with the electric charge proportional to the time-derivative of the phase. In classical geodesic motion, a single monopole has a conserved momentum and a conserved electric charge of arbitrary magnitude.

For two monopoles, the centre of mass and the overall phase decouple, so the system has a conserved total momentum and conserved total electric charge. The interesting part of the metric, known as the Atiyah–Hitchin metric, is on the 4-dimensional moduli space of the relative location and phase. Here, certain types of geodesic motion correspond to 90° scattering of monopoles similar to the 90° scattering of vortices. A separated monopole pair is not exactly axially symmetric around the line joining the centres, making 90° scattering possible. The initial condition for the relative phase angle determines the plane of scattering. Other types of scattering also occur; for example, if the incoming monopoles have no electric charges, then generically the outgoing monopoles have opposite non-zero charges. It is remarkable how angular momentum conservation is maintained in such processes.

Skyrmions also have interesting dynamics. Despite the absence of a good moduli space for more than one Skyrmion, it is still possible to numerically study the classical scattering of two Skyrmions. If the initial trajectories and orientations are carefully chosen, one observes the 90° scattering seen previously for vortices and monopoles

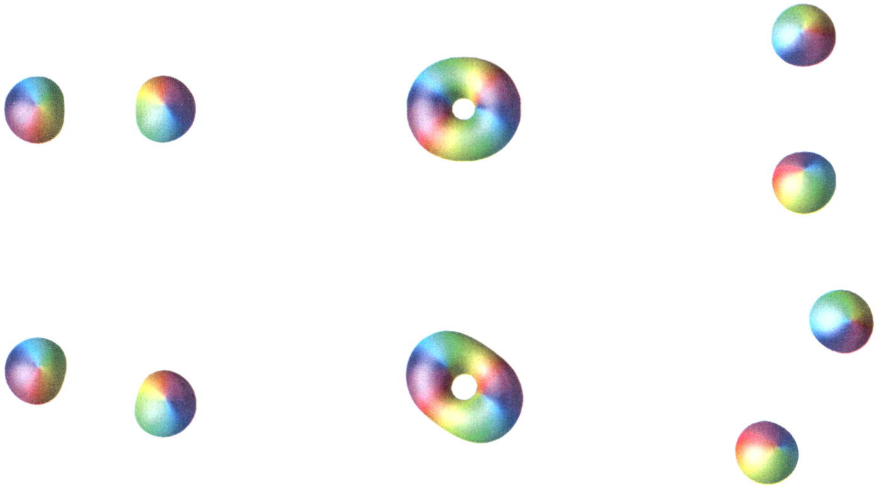

Fig. 9.2 Scattering of two Skyrmions. Top row: snapshots of 90° scattering in a head-on collision; bottom row: scattering with a small impact parameter. The colouring indicates field values. Reproduced from [19]. Creative Commons Attribution License (CC BY) https://creativecommons.org/licenses/by/4.0/

[19] (see Fig. 9.2). This phenomenon persists even if the Skyrmions are spinning [20].

9.2.1 *Half-Wormhole and Fusion*

There is a physical system involving the scattering or fusion of two objects with localised energy density, having somewhat analogous geometry to the rounded cone and 90° scattering we have just described. The system obeys dynamical equations, but there are no true fields or solitons here.

The system consists of a colliding Oxygen-16 nucleus and an alpha particle (a Helium-4 nucleus), modelled using the adiabatic self-consistent collective coordinate method [21]. The nuclear density is simply the matter density, and protons and neutrons are not distinguished. We do not need the details of the model, but only to know that the density distributions are spherical and localised, with smoothly decaying tails, when the two nuclei are initially well separated, and that the Oxygen-16 nucleus is larger. The overall density is therefore initially axially symmetric, with symmetry $C_{\infty v}$.

The nuclei may scatter elastically, or they may fuse into a Neon-20 nucleus, with excess energy being radiated away. The energies we consider here are not sufficient for the Oxygen-16 nucleus or alpha particle to break up.

It is assumed that the density remains axially symmetric during the collision. This is a simplification, because one model of Neon-20 is of five alpha particles arranged as a trigonal bipyramid, that is, a (vertical) equilateral triangle of three alpha particles with one additional alpha particle to the left and one to the right, arranged symmetrically. In the model of Neon-20 considered here, this structure is rotationally averaged about the C_3 axis to give a prolate density with $D_{\infty h}$ symmetry. The important property for us is its left/right symmetry.

The model predicts that a smooth one-parameter family of configurations and densities will occur during a head-on collision, where the Oxygen-16 starts on the right, and moves towards the alpha particle on the left. There is initially no left/right symmetry, but at some instant the nuclei merge into Neon-20 with left/right symmetry, before splitting again into an Oxygen-16 on the left and alpha particle on the right, which then separate. However, the outgoing alpha particle is not the incoming one—it is part of the incoming Oxygen-16, and the rest of this Oxygen-16 nucleus merges with the incoming alpha particle to produce the outgoing Oxygen-16. This dipole evolution is like Newton's cradle [22]: four touching particles come in and collide with one other, and four touching particles go out, leaving one behind. But there has been a permutation of the five particles.

There is a potential energy along this one-parameter family of densities, with a dip at the Neon-20, but also a Coulomb barrier to overcome to reach it. Calculations have been made of both this potential energy and the effective inertia, as a function of the separation, but there has been some uncertainty about what separation parameter to use. It is tempting to use the separation between the Oxygen-16 and alpha particle centres, but these centres fail to be well defined as the nuclei merge, and the separation vector may then appear to jump discontinuously. The permutation of the matter during the collision makes it unwise to use a separation distance at all, and one needs a different geometrical picture. We offer such a picture here.

The physics we have described appears to be essentially 1-dimensional, but it is better to allow for the separation between the Oxygen-16 and alpha particle to be along any line in space passing through the origin. The configurations are as before but with any orientation of the line, so the dimension of the configuration space is three, and there is an action of the rotation group SO(3). The generic SO(3) orbit is a 2-sphere because the Oxygen-16 and alpha particle are distinct. What were previously an Oxygen-16 on the right and an Oxygen-16 on the left are now distinct points on such an orbit. The Neon-20 cluster with its additional symmetry under a 180° end-over-end rotation has as its SO(3) orbit the real projective plane $\mathbb{RP}^2$, that is, a 2-sphere with antipodal points identified. The complete configuration space is a one-parameter family of 2-spheres, completed by this $\mathbb{RP}^2$. This is a standard smooth completion of such an orbit space. There is no problem defining good coordinates and a candidate for a smooth metric.

In fact, this manifold is a half-wormhole, or something very similar. A spatial three-dimensional wormhole is a Riemannian manifold with 'radial' coordinate r running from $-\infty$ to ∞, and spherical polar coordinates θ, ϕ, carrying the metric

$$ds^2 = dr^2 + (r^2 + a^2)(d\theta^2 + \sin^2\theta \, d\phi^2)\,, \qquad (9.8)$$

with a positive. This is asymptotic to two copies of $\mathbb{R}^3$, for $|r| \gg a$, smoothly joined by a throat of minimal radius a at $r = 0$. $|r|$ is the geodesic distance from the throat. The wormhole has an obvious action of SO(3) and all orbits are 2-spheres. There is also an antipodal map which combines the usual antipodal map on a 2-sphere $(\theta, \phi) \rightarrow (\pi - \theta, \phi + \pi)$ with $r \rightarrow -r$. This has no fixed points, so the quotient by this map is smooth and moreover still oriented. (The analogue quotient in two dimensions is a Möbius band and unoriented.) We call this quotient a half-wormhole. Its coordinates are r with range 0 to ∞, together with θ, ϕ. The generic SO(3) orbit is a 2-sphere but the orbit at $r = 0$ is $\mathbb{RP}^2$. The formula (9.8) is still the metric. A key property of this metric, required for the antipodal symmetry, is that the derivative of the radial scale factor $r^2 + a^2$ is zero at $r = 0$. Small or more substantial modifications of the scale factor do not change the orbit structure or geometry qualitatively, but the factor must remain symmetric under $r \rightarrow -r$ for the quotient to work, and for the quotient metric to be smooth, the factor's derivative must be zero at $r = 0$.

Note that the half-wormhole has some curvature. If one just takes flat $\mathbb{R}^3$, cuts out a ball and glues in an $\mathbb{RP}^2$, that does not give a smooth manifold (the radial scale factor does not have zero derivative anywhere). So curvature of the configuration space is inevitable in any smooth model with appropriate topology and orbit structure, though hard to see from a 1-dimensional perspective.

We propose here that a half-wormhole geometry, supplemented by a potential energy that depends only on r and has zero derivative at $r = 0$, is a useful model for the scattering or fusion of Oxygen-16 and an alpha particle, resolving earlier difficulties. Collisions can be head-on or have a non-zero impact parameter and non-zero angular momentum. However we have not constructed a detailed model for the potential, nor calculated the fusion cross section.

It is fairly easy to discuss quantum mechanics on the half-wormhole. The quantum Hamiltonian combines the negative of the Beltrami Laplacian on the half-wormhole with the potential. The Beltrami Laplacian is the natural generalisation of the flat-space Laplacian to a curved manifold. Stationary wavefunctions can be assumed to have definite angular momentum, so they are spherical harmonics $Y_{lm}(\theta, \phi)$ times a radial function $\chi(r)$. Crucially, χ must have zero derivative at $r = 0$ if l is even, and vanish at $r = 0$ if l is odd. These conditions follow if one starts with a smooth wavefunction on the complete wormhole and requires it to be invariant under the antipodal map. The fusion cross section is dominated by the S-wave with $l = 0$.

We can consider rotational bands of Neon-20, based on bound states in this model. The radial ground state allows for a $K^\pi = 0^+$ rotational band, built on the left/right symmetric configuration where the Oxygen-16 and alpha particle have merged. This band consists of states with spin/parity $0^+, 2^+, 4^+, \ldots$. The first radial excited state is dominated by a left/right asymmetric configuration of the Oxygen-16 and alpha particle, still axisymmetric around the line joining them. Building on this configuration one constructs a higher-energy $K^\pi = 0^-$ rotational band with $1^-, 3^-, 5^-, \ldots$ states. These rotational bands, and others, are observed in the Neon-20 spectrum [23–25].

9.3 Soliton Quantization

Let us consider first the moduli space $\mathcal{M}_1$ of a basic soliton with $n = 1$. The collective coordinates of the soliton are its position, and possibly further internal coordinates including its orientation. The potential energy is independent of all these, because $\mathcal{M}_1$ is the orbit of some symmetry group acting in the underlying field theory. Soliton dynamics is therefore determined just by the metric on $\mathcal{M}_1$, which encodes the kinetic energy of the soliton as it rigidly translates and rotates. The parameters of the metric are the soliton mass, and moment of inertia. There can also be further inertia parameters for internal motions, e.g. for a scale change in a theory with rescaling invariance, when the soliton mass is independent of size. We defer consideration of soliton vibrations until later.

The exact form of the metric sometimes needs to be calculated from the field theory by integration of the kinetic energy density of the moving soliton, but its general form is strongly constrained by symmetry. For example, for an SU(2) BPS monopole, the metric in the units of [1] is

$$ds^2 = d\mathbf{X} \cdot d\mathbf{X} + d\theta^2 . \tag{9.9}$$

Here $\mathbf{X}$ is the monopole position in $\mathbb{R}^3$ and θ is the phase angle with range 2π, and there are no spatial orientational moduli. The surprise here is that the inertia coefficients in the metric are all the same. This is because the energy associated with the monopole acquiring an electric charge is related to the standard kinetic energy of spatial motion.

The Lagrangian for the collective coordinate dynamics is

$$L = \pi(\dot{\mathbf{X}} \cdot \dot{\mathbf{X}} + \dot{\theta}^2) \tag{9.10}$$

where the prefactor π is half the mass of the monopole. The conjugate momenta are the spatial momentum $\mathbf{p}$ and electric charge Q,

$$\mathbf{p} = 2\pi\dot{\mathbf{X}} , \quad Q = 2\pi\dot{\theta} , \tag{9.11}$$

and the classical Hamiltonian (the Lagrangian expressed in terms of momenta) is

$$H = \frac{1}{4\pi}(\mathbf{p} \cdot \mathbf{p} + Q^2) . \tag{9.12}$$

Classical motion is with $\mathbf{p}$ and Q constant.

Canonical quantization in the Schrödinger picture is achieved by replacing the conjugate momenta by the operators

$$\mathbf{p} = -i\hbar\nabla , \quad Q = -i\hbar\frac{\partial}{\partial\theta} , \tag{9.13}$$

so the quantized Hamiltonian is

$$H = -\frac{\hbar^2}{4\pi}\left(\nabla^2 + \frac{\partial^2}{\partial\theta^2}\right) . \tag{9.14}$$

Stationary states with definite momentum and electric charge have the simple plane wave form

$$\psi(\mathbf{X}, \theta) = \exp\frac{i}{\hbar}(\mathbf{p}\cdot\mathbf{X} + Q\theta) , \tag{9.15}$$

where $\mathbf{p}$ and Q are now the eigenvalues of the corresponding operators.

The momentum $\mathbf{p}$ can have any value, but because θ is an angle, and ψ should be single-valued, Q is restricted in the quantum theory to be an integer multiple of $\hbar$. The magnetic charge of the classical monopole is -2π, and this is unchanged by collective coordinate quantization. The product of the magnetic and electric charges is therefore an integer multiple of $2\pi\hbar$ in the quantum theory. This is consistent with the quantization condition discovered by Dirac, using a quite different argument [26].

Some features of this quantization extend to various other types of soliton, in particular Skyrmions. The quantum Hamiltonian on moduli space is in general a multiple of the Beltrami Laplacian derived from the moduli space metric. The translational part is rather trivial and decouples, so when we discuss Skyrmions we shall concentrate on the rotational part, which is unaffected by the Skyrmion's spatial momentum.

It is a general principle of quantization that wavefunctions do not have to be single-valued on the classical configuration space C, but only on its simply-connected universal cover. Complex wavefunctions then transform under some 1-dimensional representation of the fundamental group $\Pi_1(C)$. Applying this to the monopole moduli space $\mathcal{M}_1$, where $\Pi_1(\mathcal{M}_1) = \mathbb{Z}$, we see that the electric charge could acquire a fractional shift, but if it does so than it can be shown that in the n-monopole sector, each monopole's electric charge acquires the same shift. There is a generalised Dirac quantization condition, due to Schwinger, to deal with this situation [27].

Care must be taken to ensure that the fundamental group of the field theory configuration space C is properly represented on any moduli space of solitons. Such a moduli space is always a finite-dimensional subspace of the much larger space C, and loops that are non-contractible in a moduli space may become contractible in C. This issue is important for Skyrmions. For U(1) gauged vortices, the moduli spaces for all n are topologically trivial.

The outline we have just given of quantization is far from complete. It assumes that the classical field Lagrangian has only quadratic terms in the time derivatives of the fields, but some field theories have Lagrangians with additional terms that are first order, and sometimes there are solely first order terms [28]. Such terms arise in Chern–Simons theories and in field theories where the classical field equation is more like a nonlinear Schrödinger equation than a nonlinear wave equation. Such theories are usually not Lorentz invariant, but they are still energy conserving. The classical dynamics restricted to moduli space becomes Hamiltonian, with the moduli space a symplectic rather than Riemannian manifold. This change of viewpoint is straightforward if the moduli space is Kähler, and has a consistent metric and symplectic structure. Quantization can be carried out using the ideas of geometric quantization, with a holomorphic polarisation. This has been applied to multivortex dynamics on a sphere [29], and on general compact Riemann surfaces [30].

(We are not considering here the solitons in condensed matter systems whose motion is non-relativistic and dissipative. Such solitons include vortices in superconductors and magnetic Skyrmions. These are observed as classical objects, and it appears that their motion should not be quantized.)

Quantization of n-soliton moduli spaces, for $n > 1$, is tricky, because the dimension of the moduli space becomes large as n increases, and the metric is often not known explicitly. Despite this, there has been considerable interest in quantizing multimonopoles. Sen understood that because multimonopole moduli spaces are hyperKähler (and there's no potential) it is natural to quantize in a supersymmetric way [31]. This means that wavefunctions are differential forms on moduli space rather than scalar functions. A non-zero degree of the form is related to the monopoles acquiring spin, as expected in a supersymmetric context. Sen could find a harmonic 2-form bound state on the Atiyah–Hitchin manifold—the moduli space for the relative motion of two monopoles. The Sen form represents a 2-monopole state of minimal energy (actually zero energy, because of supersymmetry). Its existence is related to the 2-sphere at the centre of the Atiyah–Hitchin manifold (not very different from the $\mathbb{R}P^2$ at the centre of the half-wormhole model of Sect. 9.2.1). Analogues of Sen's form on n-monopole moduli spaces with larger n have also been shown to exist [32]. The mathematics is challenging because of the non-compactness of these moduli spaces, but in the simpler case studied by Sen, the 2-form has been calculated explicitly, and it decays exponentially fast as the two monopoles separate.

Standard quantization on the Atiyah–Hitchin manifold, using scalar wavefunctions and a Hamiltonian proportional to the Beltrami Laplacian, is also possible [33]. Here most states are 2-monopole scattering states, but there are also some bound states of positive energy. The details are quite complicated, because of the possibility of electric charge exchange (in integer units of $\hbar$) during a collision, and bound state energies have to be calculated numerically.

9.4 Quantized Three-Dimensional Skyrmions as Models of Nuclei

Here, we discuss in a little more detail the three-dimensional Skyrmions and how they are quantized. For these Skyrmions, the degree n is identified with baryon number B, and this is the notation we use from now on.

Skyrme proposed this model for baryons (the particles now known to have three quark constituents) around 1960 [8, 9]. By then it was clear that proton and neutron interactions are mediated by the three pion fields and that the dynamics has an underlying approximate SO(4) symmetry, realised through a left and right action of SU(2). As pions are bosons, Skyrme imagined they could form a classical condensate obeying nonlinear equations. Having constructed a suitable Lagrangian and field equation for the pion fields, he realised that these equations have topologically stable solutions, characterised by the integer topological charge B, and that B can be identified with baryon number (atomic mass number A in nuclear physics). The basic solution with $B = 1$ could therefore be identified with a classical version of a proton or neutron. (Collectively, these particles are known as nucleons.) There is no need for explicit nucleon sources for the pion field—the nucleons emerge from the theory automatically as soliton solutions—and baryon number is conserved in all interactions, in agreement with experimental observations. To explain the distinction between proton and neutron, and also explain the spin of these particles, quantization is required. We discuss this below.

The fundamental Skyrme field is an SU(2) matrix $U(\mathbf{x})$ (also depending on time). Recall that as a manifold, the group SU(2) is a 3-sphere, S^3. The SO(4) symmetry is realised by left and right multiplication of U by ($\mathbf{x}$-independent) SU(2) matrices. The Skyrme Lagrangian is dominated by SO(4)-symmetric terms depending on the gradient and time-derivative of U. In addition there is a mild potential that fixes the vacuum to be $U = 1_2$. Finite-energy fields must approach the vacuum at spatial infinity. The field $U(\mathbf{x})$ then defines a map from compactified space S^3 to the target $SU(2) = S^3$, whose degree is the baryon number B.

The potential term breaks the SO(4) symmetry to an SO(3) subgroup, the isospin subgroup, and also gives the pions a small mass. (In the fully symmetric theory, the pions are massless Goldstone bosons, but this is known experimentally to be only approximately true.) There are many variants of the Skyrme Lagrangian, depending on how many powers of gradient terms are included, and also on whether fields representing the heavier ρ and ω mesons are added.

In all versions of the Skyrme model, fields close to the vacuum are described by the linearised Skyrme field

$$U(\mathbf{x}) \simeq 1_2 + i\boldsymbol{\pi}(\mathbf{x}) \cdot \boldsymbol{\tau} \tag{9.16}$$

where $\boldsymbol{\pi}$ is the (three-component) pion field and $\boldsymbol{\tau}$ are Pauli matrices. In the linearised theory, $\boldsymbol{\pi}$ obeys a massive scalar wave equation with plane wave solutions that can be quantized to give pion particles. By contrast, a $B = 1$ Skyrmion is a static solution of the nonlinear field equation for U, whose energy density and baryon density are spherically symmetric. The field has to be smooth at the centre, and this is achieved by having $U = -1_2$ there. The field U covers the whole of SU(2) once as $\mathbf{x}$ varies. Asymptotically, the Skyrmion has pion field components that are not individually spherically symmetric. Instead, they have a (Yukawa) dipole decay, with the three dipole moments mutually orthogonal. The frame of dipoles can be rotated, so a Skyrmion has an SO(3) of orientational degrees of freedom in addition to the translational degrees of freedom. For the $B = 1$ Skyrmion, the effect of a spatial rotation and of an isospin rotation are equivalent, but for Skyrmions with larger B, these SO(3) actions are generally independent. A reflection turns a Skyrmion into an antiSkyrmion.

The moduli space of the basic Skyrmion is $\mathbb{R}^3 \times SO(3)$. The field theory Lagrangian, restricted to motion in this moduli space, defines a mass and moment of inertia for the Skyrmion. The Skyrmion can then be quantized as a rigid body, a spherical rotor, leading to states with definite momentum and spin. Skyrme made the important observation that SO(3) is not simply connected, so wavefunctions need be single-valued only on its double cover SU(2). He suggested (on physical grounds) that one should choose the odd quantization, where wavefunctions change sign under a 2π rotation. This means that allowed quantum states have half-integer spin. The Skyrme model can then be calibrated by identifying the spin $\frac{1}{2}$ states with the proton and neutron, and the spin $\frac{3}{2}$ states with Delta resonances [34]. A rotor automatically has four spin $\frac{1}{2}$ states. In terms of Euler angles, these are the Wigner D-functions D^J_{sm}, with $J = \frac{1}{2}$. The label $s = \pm\frac{1}{2}$ distinguishes proton and neutron, and the label $m = \pm\frac{1}{2}$ the spin projection on the spatial 3-axis. The proton and neutron therefore form an isospin doublet with very similar masses, but different electric charges.

There are sophisticated arguments backing up Skyrme's choice of the half-integer spin quantization [35]. This choice is shown to be consistent with QCD, the theory of quarks and gluons, in the low-energy limit. In QCD the gauge group is SU(3), so baryons are made of three (an odd number) spin $\frac{1}{2}$ quarks. Topological features of the full gauge theory restrict how one should quantize the Skyrme field theory, regarded as a low-energy effective field theory approximating QCD. Baryons therefore have half-integer spin, in QCD and in Skyrme theory.

Skyrme began a study of Skyrmion interactions by calculating the small force between two well separated $B = 1$ Skyrmions, and how this force depends on the relative orientation. By quantizing the orientations, and projecting the interaction on to spin $\frac{1}{2}$ states, one finds features of nucleon-nucleon forces, in particular the tensor force [36].

As Skyrmions do not obey a Bogomolny-type equation, there is no easily defined moduli space for Skyrmions of higher baryon number. Some ideas for tackling this problem involve gradient flow [37, 38], and use of rational maps, but there is as yet no satisfactory solution. Two or more well separated $B = 1$ Skyrmions in certain orientations attract, so there are stable Skyrmion solutions with baryon number B, where the basic Skyrmions come together [39, 40] (see Fig. 9.3). These have the usual obvious moduli of translations and spatial rotations. In addition there are moduli of the SO(3) isorotations (isospin rotations). So in total there are usually nine moduli.

Although the $B = 1$ Skyrmions are spherical balls, the minimal energy solutions for larger B are not the clusters of balls familiar from popular depictions of nuclei. Neither do the $B = 1$ Skyrmions completely merge, because they are almost incompressible. Instead they partially merge, and the clusters have a fascinating, slightly enhanced symmetry, a point group symmetry that combines rotations with isorotations. For example, two Skyrmions merge into a torus, three into a tetrahedron, and four into a cube [39]. But the ball cluster structures are not far away.

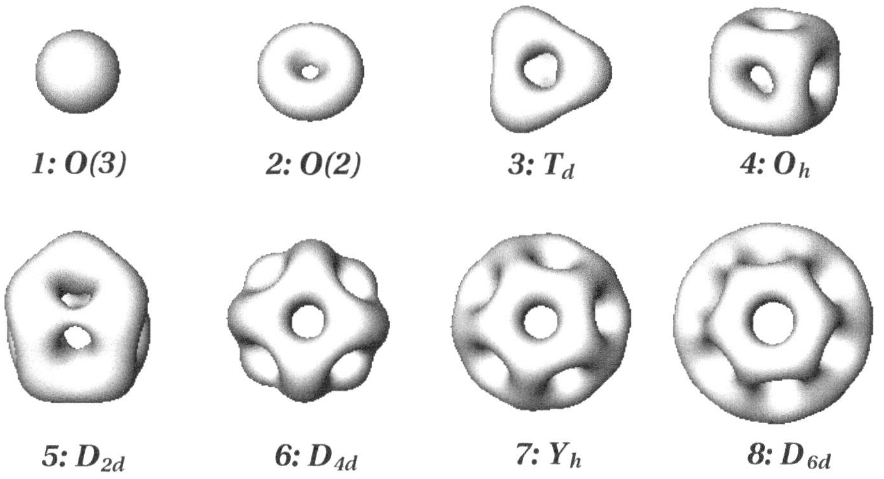

Fig. 9.3 Skyrmions with baryon numbers 1 to 8. A selected surface of constant energy density is shown, together with the baryon number and symmetry group. Reproduced from [1]

The classical energy is reduced by about 10% when B basic Skyrmions merge. This is much more than the binding energy of nucleons into nuclei, which is of order 1% of the rest energy. So classical Skyrmions are too tightly bound in the standard Skyrme model, and they are also too rigid, but quantum fluctuations can compensate for this. To model a nucleus, one should not just consider the rotational and isorotational degrees of freedom of the minimal energy Skyrmion. One should allow for vibrational degrees of freedom that deform the Skyrmion towards break-up into individual $B = 1$ Skyrmions.

The detailed, rigid-body quantization of a Skyrmion with baryon number $B > 1$ is quite tricky, but over many years, several examples have been tackled [41–46]. The orbit of rotations and isorotations is six-dimensional for $B > 2$; it is the product of two copies of SO(3) quotiented by the discrete symmetry group. (For $B = 2$ the orbit is five-dimensional as the symmetry group is $D_{\infty h}$.) There is a potentially complicated 6×6 inertia tensor, although this simplifies when there is symmetry. The rigid-body quantum states match a limited number of observed states quite well, but many states are missed, and the overall energy is too low because of the too large classical binding energy.

Much better nuclear spectra have been obtained by allowing shape changes of the Skyrmions. Examples are the deuteron [47] and the isospin doublet of Lithium-7/Beryllium-7 [48] (see Fig. 9.4). Another example is the quantized cubic $B = 4$ Skyrmion, whose ground state, modelling an alpha particle, has spin/parity 0^+ and whose lowest purely rotational excitation is an unobserved, high-energy 4^+ state. But the $B = 4$ Skyrmion has fairly low-lying vibrational modes transforming under the various irreducible representations of the cubic group. The quantized 1-phonon and 2-phonon states, coupled to rotations, give a realistic spectrum for more than ten

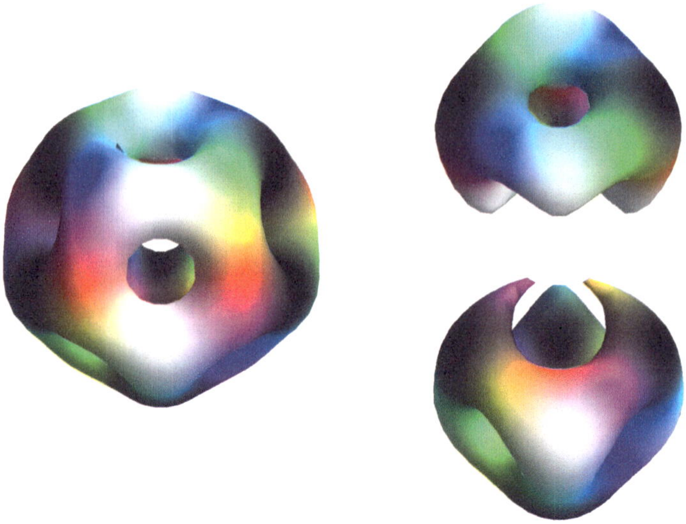

Fig. 9.4 Deformation of $B = 7$ Skyrmion into $B = 3$ and $B = 4$ clusters. Reproduced from [48].

excited states of an alpha particle, all with spins 0, 1 or 2 [49]. Degenerate vibrational modes have an internal angular momentum, and the Coriolis coupling of this to the rotational angular momentum has a significant effect on the energy levels.

For larger Skyrmions, it is tempting to consider $6B$ degrees of freedom in total (as each $B = 1$ Skyrmion has six degrees of freedom). It has also been suggested to consider $7B$ degrees of freedom, including a scale parameter for each Skyrmion, as the vibrational frequency spectrum appears to have a gap above this [50]. But quantizing so many degrees of freedom is impractical in the case of, say, the Oxygen-16 nucleus with $B = 16$. Instead one observes that the $B = 16$ Skyrmion looks like a tetrahedral arrangement of four fairly rigid alpha particles ($B = 4$ cubic Skyrmions) [51], so one may consider just the few vibrational degrees of freedom of this tetrahedron, as in [52] and earlier alpha-particle models. Those of lowest frequency form an E-mode doublet of the tetrahedral group T_d, but this mode should be treated nonlinearly, not as a pair of harmonic oscillators [53]. The E-mode deformations extend to a two-dimensional E-manifold of alpha particle configurations of fairly similar energy. Motion in the E-manifold allows a tetrahedron of alpha particles to transform into a square and then into the dual tetrahedron (see Fig. 9.5). $B = 16$ Skyrmions of nearly equal energies, with square and tetrahedral shapes, are known. The square has slightly higher energy and should be treated as a saddle point.

This E-manifold is topologically a 2-sphere, but six points on it represent configurations where two pairs of alpha particles separate to infinity. So metrically, the E-manifold is better described as a 6-punctured sphere with cubic symmetry, and in [53] we found it convenient to assume that its metric was of constant negative curvature, i.e. hyperbolic. In addition there is a potential on the E-manifold, matching the

Fig. 9.5 Dynamical path through tetrahedral and square $B = 16$ Skyrmions. Reproduced from [53]

harmonic oscillator potential for E-mode vibrations near to the tetrahedron. Quantization of the E-manifold gives a good selection of excited states of the Oxygen-16 nucleus, and an even better, almost complete spectrum of Oxygen-16 up to 20 MeV is obtained by coupling the E-manifold to harmonic F- and A-modes of vibration [54]. Again, Coriolis terms are important.

Considerable work has been done on the Carbon-12 nucleus, composed of three alpha particles, using similar ideas. So far, states of the Carbon-12 nucleus have been modelled by an equilateral triangle of alpha particles that can deform into a straight chain via isosceles triangles [55]. Many low-energy states—the ground state rotational band, the Hoyle state and its rotational excitations, and the lowest 1^- and 2^- states—are accounted for in this model. But these ideas could be taken further. The full shape space of triangles is a 3-punctured sphere, and one could use the hyperbolic metric on this. (Note, the hyperbolic 6-punctured sphere, with its eight ideal hyperbolic triangles glued together, is a 4-fold covering of the 3-punctured sphere, with its two ideal triangles.) The shape space also needs to be coupled to the A-mode of breather vibrations. This extended model includes scalene triangles of alpha particles, and these are needed to capture the lowest 1^+ state of Carbon-12, which has rather high energy.

These results, going beyond rigid-body quantization, show rather generally that Skyrmion quantization is a good way to study collective nuclear excitations. But larger nuclei are still to be explored. Many large nuclei exhibit spectra indicating an underlying intrinsic shape that is not spherical. Lead-208, a magic nucleus, has ground state 0^+ and first excited state 3^-. Above that are many states that can be modelled as 1-particle/1-hole states within the shell model, but the low-lying 3^- state and a few further states look collective. The latest thinking is that the 3^- state is not due to an octupole vibration, but is instead a rotational excitation of a tetrahedral, permanently deformed intrinsic structure. Deformations into icosahedrally symmetric shapes are also being considered [56]. It is a challenge to see if such shapes can be understood using Skyrmions.

9.5 Outstanding Issues

There is no doubt that topological soliton structures occur in condensed matter systems; images of them are widely available. The solitons can move, but their dynamics is classical, and usually dissipative. Much less clear is whether solitons are required in fundamental particle physics. This is surprising, as particle physics is currently understood using nonlinear quantum field theories, and such theories can have soliton solutions. The problem is that solitons often require scalar fields, and so far the only fundamental scalar field is the Standard Model Higgs field, coupled to gauge fields, and these fields do not support topologically stable solitons. The topology is wrong for this.

Some variant of the Skyrme model, as a low-energy limit of QCD, could provide a compelling understanding of how protons and neutrons interact to form nuclei. Work on this is continuing. Curiously, one of the most popular current approaches to nuclear forces is via an effective field theory of pions and heavier mesons [57]. This is a perturbative approach to a field theory that could support non-perturbative Skyrmions as solitons, replacing the fundamental proton and neutron. However, the soliton connection has not yet been made.

When solitons are well separated they can be accurately treated as point particles interacting through a potential and conventional forces. The field theory gives a prediction for the strength of the long-range interactions, e.g. the strength of the Yukawa dipoles making up a Skyrmion. However, we have seen that the short-range interaction of two or more solitons is often of a very surprising kind, having a clear geometrical description; it is not due solely to a potential. The classical soliton dynamics is then modelled by geodesic motion on a smooth, curved moduli space, possibly modified by an additional potential. The quantum dynamics is derived from the classical dynamics, and an ingredient of the quantum Hamiltonian is the Beltrami Laplacian on the moduli space, which takes curvature into account. We explained that nontrivial, curved geometry is an essential consequence of the permutation symmetry acting on classical solitons. Solitons exhibit features of "identical particles" that one usually attributes only to quantized bosons and fermions.

It would be desirable to find compelling evidence for this geometrical picture of the interactions. Unfortunately it is hard to scatter the solitons that occur in condensed matter systems, although this might be achievable. It is also hard to calculate the scattering cross section of quantized Skyrmions, as there is no convincing moduli space for two Skyrmions that can be quantized. We have suggested here a new, geometrical model for the fusion of Oxygen-16 with an alpha particle. This half-wormhole model has an analogy with Newton's cradle, and the dynamics is smooth. If a detailed version of this model is successful, then one should try to generalise it, and apply it elsewhere.

The best evidence for the usefulness of Skyrmions with moderate baryon numbers is the occurrence of intrinsic structures similar to what has been assumed in simple alpha-particle models, but allowing for a more complete analysis of rotational and vibrational excitations, and even isospin excitations. Good understandings of the nuclear spectra of Helium-4, Lithium-7/Beryllium-7, Beryllium-8, Carbon-12 and Oxygen-16 have now been obtained. More should be done to model Coulomb effects for larger Skyrmions, responsible for the neutron excess observed in most large nuclei. There could then be a reconsideration of the Bethe–Weizsäcker nuclear mass formula from a Skyrmion perspective. It might be possible to determine some of the parameters in this formula.

There is currently more interest in magnetic Skyrmions than in Skyrme's original Skyrmions of nuclear physics. Magnetic Skyrmions also have geometrical features, and can be studied on curved surfaces like spheres and cylinders, as well as on planar domains. Some exact solutions have recently been constructed, but it is not yet clear if these are physically important [58, 59].

Acknowledgements The half-wormhole model for Oxygen-16 interacting with an alpha particle was developed following a talk by Takashi Nakatsukasa and a related question concerning Newton's cradle by Panagiota Papakonstantinou, at the 1st APCPT-TRIUMF Joint Workshop, Pohang, Korea, 2018. I also thank J. Martin Speight and Maciej Dunajski for helpful discussions. This work has been partially supported by STFC consolidated grant ST/P000681/1.

References

1. N. Manton, P. Sutcliffe, *Topological Solitons* (Cambridge University Press, Cambridge, 2004)
2. E.J. Weinberg, *Classical Solutions in Quantum Field Theory* (Cambridge University Press, Cambridge, 2012)
3. N.S. Manton, Phys. Lett. B **110**, 54 (1982)
4. E.B. Bogomolny, Sov. J. Nucl. Phys. **24**, 449 (1976)
5. M. Shifman, A. Yung, *Supersymmetric Solitons* (Cambridge University Press, Cambridge, 2009)
6. N.S. Manton, K. Oleś, A. Wereszczyński, JHEP **2019**, 86 (2019)
7. Y.N. Ovchinnikov, I.M. Sigal, Nonlinearity **11**, 1277 (1998)
8. T.H.R. Skyrme, Proc. Roy. Soc. Lond. A **260**, 127 (1961)
9. T.H.R. Skyrme, Nucl. Phys. **31**, 556 (1962)
10. S. Mühlbauer, B. Binz, F. Jonietz, C. Pfleiderer, A. Rosch, A. Neubauer, R. Georgii, P. Böni, Science **323**, 915 (2009)
11. X.Z. Yu, Y. Onose, N. Kanazawa, J.H. Park, J.H. Han, Y. Matsui, N. Nagaosa, Y. Tokura, Nature **465**, 901 (2010)
12. C.H. Taubes, Commun. Math. Phys. **72**, 277 (1980)
13. T.M. Samols, Commun. Math. Phys. **145**, 149 (1992)
14. A.A. Belavin, A.M. Polyakov, JETP Lett. **22**, 245 (1975)
15. A.M. Din, W.J. Zakrzewski, Nucl. Phys. B **259**, 667 (1985)
16. L.A. Sadun, J.M. Speight, Lett. Math. Phys. **43**, 329 (1998)
17. B. Piette, W.J. Zakrzewski, Nonlinearity **9**, 897 (1996)
18. M.F. Atiyah, N.J. Hitchin, *The Geometry and Dynamics of Magnetic Monopoles* (Princeton University Press, 1988)
19. D. Foster, S. Krusch, Nucl. Phys. B **897**, 697 (2015)
20. D. Foster, N.S. Manton, Nucl. Phys. B **899**, 513 (2015)
21. K. Wen, T. Nakatsukasa, Phys. Rev. C **96**, 014610 (2017)
22. P. Papakonstantinou, A.I.P. Conf, Proc. **1947**, 020023 (2018)
23. A.E. Litherland, J.A. Kuehner, H.E. Gove, M.A. Clark, E. Almqvist, Phys. Rev. Lett. **7**, 98 (1961)

24. M. Bouten, Nuovo Cim. **26**, 63 (1962)
25. Y. Fujiwara, H. Horiuchi, K. Ikeda, M. Kamimura, K. Katō, Y. Suzuki, E. Uegaki, Prog. Theor. Phys. Suppl. **68**, 29 (1980)
26. P.A.M. Dirac, Proc. R. Soc. Lond. A **133**, 60 (1931)
27. J. Schwinger, Science **165**, 757 (1969)
28. N.S. Manton, Ann. Phys. **256**, 114 (1997)
29. N.M. Romão, J. Math. Phys. **42**, 3445 (2001)
30. D. Eriksson, N.M. Romão, Kähler quantization of vortex moduli. arXiv:1612.08505
31. A. Sen, Phys. Lett. B **329**, 217 (1994)
32. G. Segal, A. Selby, Commun. Math. Phys. **177**, 775 (1996)
33. B.J. Schroers, Nucl. Phys. B **367**, 177 (1991)
34. G.S. Adkins, C.R. Nappi, E. Witten, Nucl. Phys. B **228**, 552 (1983)
35. D.S. Freed, J. Diff. Geom. **80**, 45 (2008)
36. A. Jackson, A.D. Jackson, V. Pasquier, Nucl. Phys. A **432**, 567 (1985)
37. N.S. Manton, Phys. Rev. Lett. **60**, 1916 (1988)
38. M.F. Atiyah, N.S. Manton, Commun. Math. Phys. **153**, 391 (1993)
39. E. Braaten, S. Townsend, L. Carson, Phys. Lett. B **235**, 147 (1990)
40. R.A. Battye, P.M. Sutcliffe, Phys. Rev. Lett. **79**, 363 (1997)
41. E. Braaten, L. Carson, Phys. Rev. D **38**, 3525 (1988)
42. L. Carson, Phys. Rev. Lett. **66**, 1406 (1991)
43. T.S. Walhout, Nucl. Phys. A **547**, 423 (1992)
44. P. Irwin, Phys. Rev. D **61**, 114024 (2000)
45. R.A. Battye, N.S. Manton, P.M. Sutcliffe, S.W. Wood, Phys. Rev. C **80**, 034323 (2009)
46. P.H.C. Lau, N.S. Manton, Phys. Rev. Lett. **113**, 232503 (2014)
47. R.A. Leese, N.S. Manton, B.J. Schroers, Nucl. Phys. B **442**, 228 (1995)
48. C.J. Halcrow, Nucl. Phys. B **904**, 106 (2016)
49. J.I. Rawlinson, Nucl. Phys. B **949**, 114800 (2019)
50. S.B. Gudnason, C. Halcrow, Phys. Rev. D **98**, 125010 (2018)
51. R.A. Battye, N.S. Manton, P.M. Sutcliffe, Proc. Roy. Soc. Lond. A **463**, 261 (2007)
52. R. Bijker, F. Iachello, Phys. Rev. Lett. **112**, 152501 (2014)
53. C.J. Halcrow, C. King, N.S. Manton, Phys. Rev. C **95**, 031303(R) (2017)
54. C.J. Halcrow, C. King, N.S. Manton, Int. J. Mod. Phys. E **28**, 1950026 (2019)
55. J.I. Rawlinson, Nucl. Phys. A **975**, 122 (2018)
56. A. Heusler, Eur. Phys. J. A **53**, 215 (2017)
57. E. Epelbaum, PoS **CD15**, 014 (2016)
58. B. Barton-Singer, C. Ross, B.J. Schroers, Commun. Math. Phys. (2020). https://doi.org/10.1007/s00220-019-03676-1
59. E. Walton, Some exact Skyrmion solutions on curved thin films. arXiv:1908.08428

Chapter 10
Nonlinear and Novel Phenomena in Non-Hermitian Photonics

Li Ge and Wenjie Wan

Abstract Photonic systems provide a convenient and versatile platform to study and test novel phenomena in non-Hermitian physics, an emerging field that has benefitted from a series of quantum-inspired symmetries without the Hermiticity restriction in canonical quantum mechanics. In this chapter we will highlight a few recent achievements in non-Hermitian photonics, with a focus on nonlinear behaviors and systems. We will first discuss the realizations of parity-time symmetry and antisymmetric parity-time symmetry using nonlinear optics, before turning our attention to phase modulated nonlinear behaviors in non-Hermitian photonics. Nonlinear modal interaction in lasers with parity-time symmetry, non-Hermitian particle-hole symmetry and supersymmetry will also be reviewed.

10.1 Introduction

Quantum mechanics is arguably the most important foundation of modern physics and technology. The entire semiconductor industry would not be possible without the establishment of quantum mechanism, not to mention the current pursuit of quantum computing and quantum communication. Canonical quantum mechanics, in its Hamiltonian form, deals with closed and hence lossless systems. As a result, the Hamiltonian is Hermitian and gives rise to real-valued energy levels.

L. Ge (✉)
Department of Physics and Astronomy, College of Staten Island, CUNY, Staten Island, NY 10314, USA
e-mail: li.ge@csi.cuny.edu

The Graduate Center, CUNY, New York City, NY 10016, USA

W. Wan
The State Key Laboratory of Advanced Optical Communication Systems and Networks, University of Michigan-Shanghai Jiao Tong University Joint Institute, Shanghai Jiao Tong University, Shanghai 200240, China
e-mail: wenjie.wan@sjtu.edu.cn

MOE Key Laboratory for Laser Plasmas and Collaborative Innovation Center of IFSA, Department of Physics and Astronomy, Shanghai Jiao Tong University, Shanghai 200240, China

P. G. Kevrekidis et al. (eds.), *Emerging Frontiers in Nonlinear Science*, Nonlinear Systems and Complexity 32, https://doi.org/10.1007/978-3-030-44992-6_10

However, most physical systems we encounter are open by nature; they exchange energy and particles with their environment constantly. One effective way to describe the openness of the system is to use a non-Hermitian formalism, which is a common practice, for example, in optics and photonics: to describe the material loss for light propagation, we include an imaginary part in the refractive index, and the wave equation immediately becomes non-Hermitian. Even when the material loss is negligible, the finite size of the system usually implies the existence of incoming and outgoing optical fluxes at its boundary, which also cause the system to be non-Hermitian similar to nuclear physics [1].

A consequence of non-Hermiticity in these cases is that energy or frequency associated with an eigenstate of the system becomes complex, and its imaginary part, when negative in sign, gives the decay rate or inverse life time. To achieve a stationary state with sustained oscillation, obviously we require an energy input to balance out the loss or dissipation. The laser, widely used in optics and photonics, is such an example. The energy input here takes the form of either another light source (a flashtube, LED, another laser, etc.) or electrical current, and when its power is increased to a critical value, the energy input and output is exactly balanced, and the resonant frequency associated with the optical state of lasing, becomes real. This critical value is known as the laser threshold, and upon further increase of the input power, the intensity of the laser gradually increases from zero, while the resonant frequency stays real due to optical saturation nonlinearity [2].

Whether it is possible to impose this explicit energy balance in quantum mechanism, on the other hand, was less clear. Therefore, it stirred up quite an excitement when Carl Bender and collaborators identified unbroken parity-time symmetry in a non-Hermitian potential [3–5], which possesses an entirely real energy spectrum. While the physical ramifications of a non-Hermitian quantum theory based on parity-time symmetry are still under debate, the introduction of parity-time symmetry to non-Hermitian photonics has led to many inspiring and fruitful explorations [6–9].

Unsurprisingly, the laser again played an important role in these endeavors. Its narrow spectral linewidth (i.e. color purity) enabled the isolation of individual states, which can be coupled conveniently across spatially separated elements, including optical fibers, waveguides, and resonators. The loss of different optical modes, instead of being a nuisance, is now an important tuning parameter to reach various regimes, including both unbroken and broken parity-time symmetry.

In this chapter we will highlight a few recent achievements in non-Hermitian photonics, with a focus on nonlinear behaviors and systems. An earlier survey on this subject [10] has already covered many different approaches, and we have tried to minimize the overlap for the benefit of the readers. Nonlinear optics describes a broad range of phenomena and techniques that rely on nonlinear responses of the optical medium. These nonlinear responses are typically very weak in normal materials, and it requires a strong coherent light source to reveal their effects, i.e. a laser. The first nonlinear optical phenomenon, i.e. second harmonic generation, was observed one year after the first demonstration of the laser. We note that while photonics deals with open systems and is hence non-Hermitian in nature, here we limit the scope of our discussion to the subset that is based on quantum-inspired symmetries, including parity-time (PT) symmetry, antisymmetric parity-time (anti-PT) symmetry [11], non-Hermitian particle-hole (NHPH) symmetry [12], and supersymmetry (SUSY) [13, 14].

We will start by discussing in Sect. 10.2 the realizations of these novel symmetries using nonlinear optics. In Sect. 10.3 we turn our attention to phase modulated nonlinear behaviors in non-Hermitian photonics, including asymmetric interferometric control near an exceptional point (Sect. 10.3.1), time-reversed second harmonic generation and optical parametric amplification (Sect. 10.3.2), and anomalous parity-time transition away from an exceptional point (Sect. 10.3.3). In Sect. 10.4 the focus is shifted to modal interaction in lasers with novel non-Hermitian symmetries, including parity-time symmetry (Sect. 10.4.1), non-Hermitian particle-hole symmetry (Sect. 10.4.2) and supersymmetry (Sect. 10.4.3). We will conclude by a brief outlook on emerging quests related to these topics.

10.2 Realizing PT, Anti-PT and Other Novel Symmetries Using Nonlinear Optics

Parity-time (PT) symmetry, in its original context, requires the system to be invariant under a combined parity and time-reversal operation. Therefore, its optical realization imposes the following constraint: $n(-x) = n^*(x)$. Here $n(x)$ is the complex-valued refractive index, and we have assumed a one-dimensional system along the x-direction for simplicity. Two most important properties of a PT-symmetric system are: (1) the possibility of having a real-valued spectrum; and (2) the transition of two real energy levels to a pair of complex conjugated ones when they meet at an exceptional point [15], where the two corresponding eigenstates also become the same.

It should be noted that these two properties survive when parity is replaced by another linear operator $\boldsymbol{L}$, satisfying $\boldsymbol{L}(a\phi_1 + b\phi_2) = a\boldsymbol{L}\phi_1 + b\boldsymbol{L}\phi_2$. Here $\phi_{a,b}$ are two arbitrary states and a, b are their complex superposition coefficients. This observation opened the possibility of studying a broader range of systems under the PT-symmetric framework [16], for example, those with rotation-time and inversion-time symmetries [17]. In fact, this linear operator should not be limited to perform a spatial operation; it can also act on the frequency domain, which does not necessarily exchange the frequencies of two modes or flip the frequency axis.

Utilizing such a "synthetic parity operator" offers a more versatile approach to PT symmetry in synthetic photonics [18], beyond spatially balanced gain and loss. This additional, synthetic dimension is enabled by coupling modes with different frequencies, achieved by imposing nonlinear wavelength conversion in resonant whispering gallery mode optical microcavities. Different optical nonlinearities can be employed for this purpose, including stimulated Raman scattering [19], Brillouin scattering [20], four-wave mixing [21] and optomechanical oscillations [22]. This nonlinear

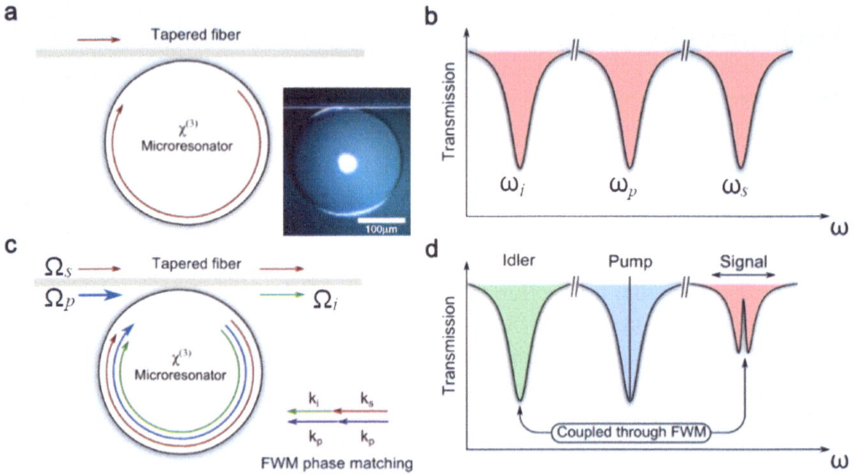

Fig. 10.1 Synthetic PT symmetry via degenerate FWM. **a** Weak light coupling into a micro-cavity through a tapered fiber at near-critical coupling condition. Inset: Experimental image of the coupled system. **b** Multiple resonances in the transmission spectrum observed by varying the frequency of the input light. **c** Schematics of the FWM process, fulfilling the phase matching condition in linear momentum (inset). **d** A transparent window at the signal frequency indicates nonlinear coupling to the idler wave through the FWM process. Adapted from [21]

coupling approach has led to the observation of several novel photonic dynamics, e.g. optically induced transparency and slow light [23, 24].

One proposal to realize PT-symmetric photonics in the synthetic dimension relies on cavity-enhanced degenerate four wave mixing (FWM) [25]. This process converts two photons of the same frequency (denoted as "pump") into a pair of photons of different frequencies (denoted as "signal" and "idler"). A nonlinear coupling is achieved between the signal and the idler, and it can be verified by observing a transparency window in the otherwise opaque resonance dip of the signal, which is the manifestation of the interference between two resonant transitions, i.e. direct absorption and the FWM process. This phenomenon is known as optically induced transparency and similar to electromagnetically induced transparency in atomic physics. It is highly tunable via varying a strong pump beam, and a small frequency-detuning of the pump can lead to a Fano-like asymmetric resonance.

Under this framework of nonlinear FWM, let us consider a nonlinear microcavity with passive resonances ω_i, ω_p, and ω_s (see Fig. 10.1). A pump beam is injected into the nonlinear medium near ω_p, and a weak signal wave of frequency $\Omega_s \sim \omega_s$ and amplitude A_s^{in} is used to probe the response of the system. Under appropriate phase matching conditions (both in energy and linear momentum), the idler wave, which is another sideband at frequency $\Omega_i \sim \omega_i$, is generated. Such a degenerate FWM process can be described by a coupled mode theory:

$$i\frac{\partial A_s}{\partial t} = (-\Delta_s - i\kappa_s)A_s - pA_s^{in} - gA_i^*, \quad i\frac{\partial A_i}{\partial t} = (-\Delta_i - i\kappa_i)A_i - gA_s^*. \tag{10.1}$$

Here $A_{s,i}$ are the slowly-varying amplitudes of the signal and idler waves inside the cavity, and their decay rates are denoted by $\kappa_{s,i}$ (taken to be equal for simplicity). $\Delta_{s,i}$ are the frequency detunings from the corresponding passive frequencies, i.e. $\Delta_{s,i} = \omega_{s,i} - \Omega_{s,i} + G_{s,i}$ [21]. Note that $G_{s,i}$ are the nonlinear frequency shifts due to the coupling to the pump beam (i.e. "cross phase modulation"). Here g is the nonlinear coupling between signal and idler induced by FWM (see Fig. 10.1d), and p is the coupling rate of the injected signal wave into the system.

Clearly the coupled mode equation above describes a driven-dissipative system. To understand its inherent dynamics, we extract its effective Hamiltonian H for the amplitudes $(\tilde{A}_s, \tilde{A}_i^*)^T \equiv [A_s e^{(-i\delta+\kappa)t}, A_i^* e^{(-i\delta+\kappa)t}]^T$:

$$i\frac{\partial}{\partial t}\begin{pmatrix}\tilde{A}_s \\ \tilde{A}_i^*\end{pmatrix} = H\begin{pmatrix}\tilde{A}_s \\ \tilde{A}_i^*\end{pmatrix}, \quad H \equiv \begin{pmatrix}-\Delta & -g \\ g^* & \Delta\end{pmatrix}. \tag{10.2}$$

Here $\delta \equiv (\Delta_i - \Delta_s)/2$ and $\Delta \equiv (\Delta_i + \Delta_s)/2$. Unlike a typical PT-symmetric Hamiltonian, here the couplings g, g^* also include implicitly the nonlinear gain and loss. Therefore, its synthetic parity operator must take a different form [26]:

$$P_s = \frac{ig_r\sigma_y + \Delta\sigma_z}{\sqrt{\Delta^2 - g_r^2}}. \tag{10.3}$$

It can be easily checked that the synthetic PT symmetry $[H, P_sT] = 0$ holds, where $\sigma_{y,z}$ are two Pauli matrices, $g_r = \text{Re}[g]$, and the denominator warrants $(P_sT)^2 = 1$.

The effective Hamiltonian given by (10.2) in fact has a series of other symmetries. First of all, if it is PT symmetric, then it must be pseudo-Hermitian [27] as well, i.e. $H^\dagger = \eta H \eta^{-1} \neq H$, where η is an arbitrary matrix operator, η^{-1} is its inverse, and "$\dagger$" denotes the Hermitian conjugation. In this case we find that η is given conveniently by σ_z.

In addition, H also satisfies antisymmetric PT symmetry [11], also anti-PT for short. Different from PT symmetry, anti-PT symmetry was first constructed in non-Hermitian photonics using $n(-x) = -n^*(x)$ and the anti-commutation relation $\{n(x), PT\} = 0$. The real part of the refractive index now needs to be anti-symmetric about the mirror plane at $x = 0$. Therefore, negative index materials are necessary and another relation $\mu(-x) = \mu(x) \in \mathbb{R}$ was initially imposed. In terms of the imaginary part of the refractive index, it only needs to be symmetric and does not necessarily require both optical gain and loss. An extension of this original idea was applied to the whole effective Hamiltonian, which is required to anti-commute with the combined PT operation [28]. Anti-PT symmetry is responsible for many intriguing properties, including a continuous lasing spectrum and a flat total transmission band [11]. It has been demonstrated in hot mobile atoms [29], electric RLC resonators [30], nonlinear optics [26, 31], coupled waveguides [32], and diffusive heat transfer [33].

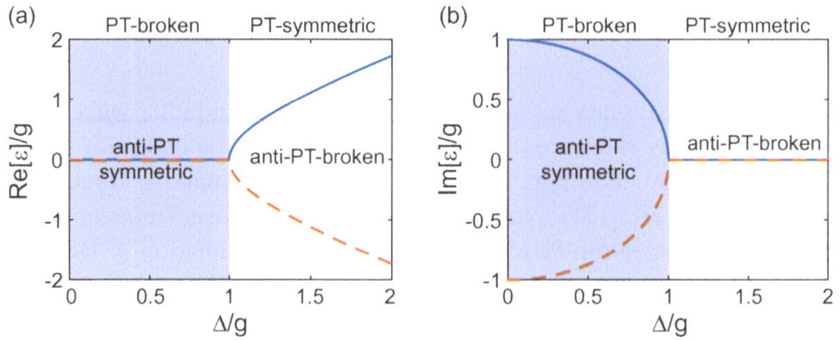

Fig. 10.2 Opposite phases of PT and anti-PT symmetry. Adapted from [25]

For H given by (10.2), the "parity" operator in anti-PT symmetry is given by the Pauli matrix σ_x. It should be noted that this "parity" operator also lacks the meaning of a mirror reflection in the spatial domain. It can however, be interpreted as a permutation operator in the synthetic (frequency) dimension, which exchanges the spectral order of the signal wave and the idler wave.

In Sect. 10.4.2 we will discuss non-Hermitian particle-hole (NHPH) symmetry, which is defined in the same form as anti-PT symmetry, i.e. with H anti-commutating with an anti-linear operator. Therefore, they also share similar manifestations, including the existence of non-Hermitian zero modes. We will discuss these consequences in Sect. 10.4.2 in the general case (and in the absence of PT symmetry of any form). Here we want to point out that the two eigenvalues of H in (10.2) are given by $\varepsilon_{\pm} = \pm\sqrt{\Delta^2 - |g|^2}$, which can be real or complex conjugates as a result of synthetic PT symmetry or pseudo-Hermiticity. They are also imaginary or form a negative complex conjugate pair (i.e. $\varepsilon_+ = -\varepsilon_-^*$) as a consequence of anti-PT symmetry. Since there are only two eigenvalues in this coupled mode theory, the exceptional points at $\Delta = \pm|g|$ are shared by PT and anti-PT symmetry: when the system is in its PT-symmetric (PT-broken) phase, it must be in the anti-PT-broken (anti-PT-symmetric) phase (see Fig. 10.2). We also mention in passing that a similar Hamiltonian can be achieved using stimulated Brillouin scattering in a microcavity [26].

10.3 Phase Modulated Nonlinear Behavior in Non-Hermitian Photonics

The phase angle of the electromagnetic field is fundamental to its wave properties, including refraction, diffraction, and interference. In this section we showcase a few examples where by controlling or engineering the phase between two optical beams or modes, one can switch on and off an intensive beam using a weak control near an exceptional point [34], perfectly absorb light in one frequency and convert it into light of a different frequency [35], or induce a transition between broken and unbroken phases of PT symmetry in a fashion that is otherwise impossible [36].

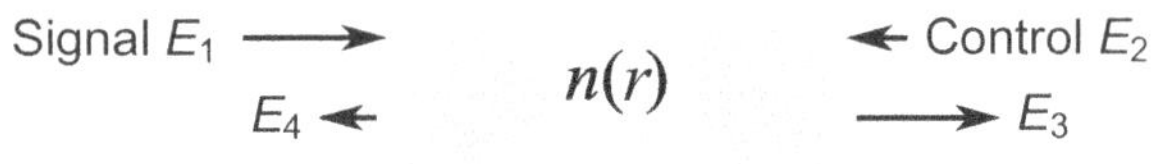

Fig. 10.3 Schematic of an asymmetric interferometric light-light switch

10.3.1 Asymmetric Interferometric Control Near an Exceptional Point

The interaction between two photons in vacuum, i.e. the constituent particles of light, is negligible under normal conditions. To achieve light-light interaction in a tabletop experiment, one hence relies on the nonlinear response of the optical medium in which light propagates. Even so, to switch a beam of light on and off using another beam typically requires intense laser fields, which impose a high power consumption and limit practical applications of all-optical devices.

Interferometric control, on the other hand, does not have this drawback. Its basic principle is straightforward: let $E_{1,2}$ denote the complex amplitudes of a signal beam and a control beam, respectively. When the relative phase θ between them is zero, the presence of the control beam enhances the overall amplitude of the optical field: $|E_1 + E_2| = |E_1| + |E_2| > |E_1|$. When θ is 180°, the control reduces the overall amplitude instead: $|E_1 + E_2| = ||E_1| - |E_2|| < |E_1|$. Using this principle and that of time-reversed lasing [37], it was shown that coherent perfect absorption (CPA) can be achieved by tuning the phase of the control beam [38, 39], which enters a symmetric optical cavity from the opposite side of the signal beam.

This setup requires the same magnitude of the two optical beams (i.e. $|E_1| = |E_2|$) and hence cannot be used to switch a strong signal beam using a weak control. To achieve this goal, one can operate the optical cavity near an exceptional point of its scattering matrix [40–42]. This scattering matrix, denoted by S below, connects the incoming channels (e.g. the signal E_1 from the left and the control E_2 from the right; see Fig. 10.3), to the scattered channels with amplitudes $E_{3,4}$:

$$\begin{pmatrix} E_3 \\ E_4 \end{pmatrix} = S \begin{pmatrix} E_1 \\ E_2 \end{pmatrix} = \begin{pmatrix} t & r_r \\ r_l & t \end{pmatrix} \begin{pmatrix} E_1 \\ E_2 \end{pmatrix}. \tag{10.4}$$

Here t is the reciprocal transmission coeffient and $r_{l,r}$ are the reflection coefficients from the two sides. An exceptional point of S is reached when either r_l or r_r becomes zero, and S takes the Jordan normal form [15]. This condition leads to an anisotropic transmission resonance in a PT-symmetric structure [41], where the transmission from both sides is unity (i.e. $|t| = 1$) while the reflection vanishes only from one side. Assuming $r_l = 0$, the two coalesced eigenstates of S at this exceptional point are given by $[E_1\ E_2]^T \propto [1\ 0]^T$.

To understand why operating near this exceptional point can achieve asymmetric interferometric control and switching, we note that the reflected amplitude E_3 is a superposition of the reflection from the signal (i.e. $r_l E_1$) and the transmission from the control (i.e. tE_2). Therefore, to turn off the scattering light E_3 with $|E_2| \ll |E_1|$,

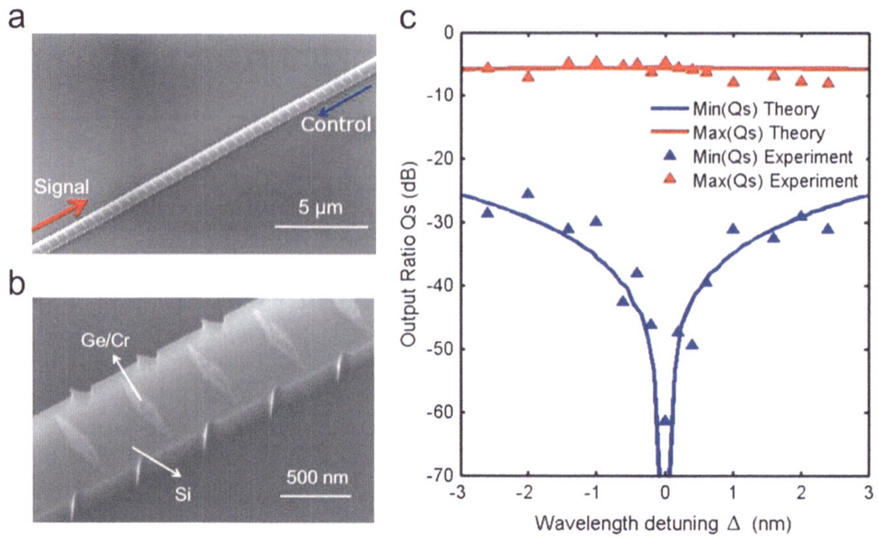

Fig. 10.4 An asymmetric interferometric light-light switch using a PT-symmetric grating. Adapted from [34]

we require $|r_l| \ll |t|$. Similarly, we find $|r_l| \gg |t|$ by considering the vanishing of E_4 in CPA. These two conditions then imply $|r_l| \ll |r_r|$, which is the case near the aforementioned exceptional point.

The experiment was performed using a silicon waveguide with a PT-symmetric grating on top (see Fig. 10.4a, b). CPA was detected at the designed wavelength of 1550 nm (see Fig. 10.4c), where the output to input ratio $Q_s \equiv 10\log_{10}\frac{|E_3|^2+|E_4|^2}{|E_1|^2+|E_2|^2} + C$ is below 60 dB, providing the "off" state of the signal. Here C is a constant taking into consideration the overall detection loss. By changing the phase of the control beam by 180°, a dramatically larger Q_s was observed, which provides the "on" state of the signal.

10.3.2 Time-Reversed Second Harmonic Generation and Optical Parametric Amplification

Even though CPA mentioned in the last section can be interpreted as time-reversed lasing, it should be stressed that the latter is at its lasing threshold and the saturation nonlinearity is absent. To explore nonlinear optical effects in similar time-reversed processes where the phase modulation is crucial, theoretical and experimental work have investigated time-reversed second harmonic generation (SHG) and optical parametric amplification (OPA) [35, 43]. Unlike the case of CPA (in the linear regime) where incident light is perfectly absorbed and converted into heat, the disappearance of certain incident light in such nonlinear time-reversed processes leads to the generation of light at a different frequency. This mechanism offers new techniques for flexible controls in nonlinear optics and has potential applications in all-optical computing.

Here we briefly review the SHG process. It converts a fundamental wave of frequency ω_1 to its second harmonic of frequency $\omega_2 = 2\omega_1$ in a nonlinear crystal. Under the slowly-varying amplitude approximation, the amplitude of the second harmonic ("A_2") along the propagation (z) direction is given by

$$\partial_z A_2 = \frac{i\omega_2^2 d_{\text{eff}} A_1^2}{k_2 c^2} e^{i\Delta k z}, \tag{10.5}$$

where $\Delta k = 2k_1 - k_2$ and $k_{1,2}$ are the wave vectors of the two waves in the nonlinear crystal. d_{eff} is the effective nonlinearity and c is the speed of light.

Time reversal of SHG should manifest as the generation of the fundamental wave from the second harmonic. Since z plays the role of time in (10.5), it would require a backward propagation that complicates the experimental setup. A more convenient scheme keeps the propagation direction unchanged, which requires a second crystal with the opposite phase match condition ($\Delta k' = -\Delta k$) *and* a phase shift of π between the fundamental wave and the second harmonic [35]. The process is illustrated in Fig. 10.5a, where the two nonlinear crystals (denoted by BBO1 and BBO2) are cut into the same length and configured with controllable phase-mismatching vectors.

Figure 10.5b shows the amplitude of the second harmonic in the two crystals under phase matched ($\Delta k = 0$) and unmatched ($\Delta k \neq 0$) conditions. Without the π phase shift between the two nonlinear crystals, A_2 would keep increasing monotonically in the phase matched case in the second crystal (red dashed line). The unmatched case in this example will lead to an initial increase of A_2 (blue dashed line) in the second crystal, and eventually the energy flows back into the fundamental wave. With the π phase shift between the two crystals, this time-reversed process of SHG is triggered immediately once light enters the second crystal, producing a symmetric intensity profile in the two crystals (blue solid curve). Two different sets of experimental data are shown in Fig. 10.5c, for $\Delta kL/2 = 0.35\pi$ and 0.6π. With the π phase shift implemented using a thin quartz plate, it is clear that the amplitude of the second harmonic at the end of the second nonlinear crystal is reduced drastically when $\Delta k' = -\Delta k$. Here Δk and $\Delta k'$ can be tuned independently by rotating the crystals accordingly.

One might wonder why OPA cannot be considered as the time-reversed SHG. After all, one photon is split into two of lower frequencies in OPA, which is exactly the opposite of SHG. The reason lies in the low conversion efficiency of both processes, and hence an intensity beam at the low frequency (fundamental wave) is required for SHG, while an intensity beam at the higher frequency (second harmonic) is required for OPA. Nevertheless, time reversal of OPA can be achieved in a similar scheme as that in SHG, and we refer the reader to Refs. [35, 43] for details.

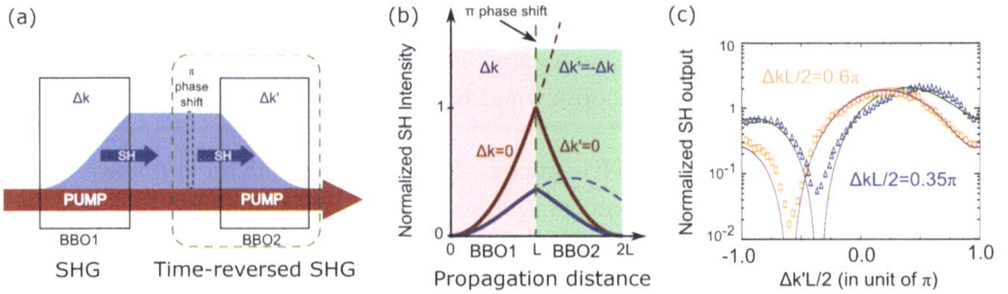

Fig. 10.5 Phase engineering in time-reversed SHG. **a** Schematic showing the two nonlinear crystals. **b** Intensity of the second harmonic as a function of propagation distance. Its value at $z = 2L$ is plotted against $\Delta k'$ in **c** for two different values of Δk's. Adapted from [35]

10.3.3 Anomalous PT Transition Away from an Exceptional Point

As we have mentioned previously, a PT-symmetric Hamiltonian has two distinct phases: one has real energy levels and the other has complex conjugate energy eigenvalues. It was believed that the transition between these two phases can only occur at an exceptional point, but an exception to this rule was observed and explained in [36], where the couplings between two resonances change nonlinearly with the field amplitudes, including their phases.

To understand this behavior, let us consider the following effective Hamiltonian that couples two harmonic oscillators and depends nonlinearly on their complex oscillation amplitudes $c_{a,b}$:

$$H = \begin{bmatrix} E_0 + 2\varepsilon|c_a|^2 + i\kappa_0 & g_0 + \varepsilon\beta c_a^* c_b + \varepsilon\gamma|c_a|^2 \\ g_0 + \varepsilon\beta c_b^* c_a + \varepsilon\gamma|c_b|^2 & E_0 + 2\varepsilon|c_b|^2 - i\kappa_0 \end{bmatrix} \equiv \begin{bmatrix} E_a(\varepsilon) & g_a(\varepsilon) \\ g_a(\varepsilon) & E_b(\varepsilon) \end{bmatrix}. \quad (10.6)$$

Even though the nonlinearity in H looks to be artificially constructed, it is extracted and simplified from an actual system with the Kerr nonlinearity [36].

H reduces to the well-studied linear PT-symmetric Hamiltonian H_0 when the nonlinearity ε is zero, and here we analyze the properties of H_0 first. H_0 is in its PT-symmetric phase when $|g_0| > \kappa_0$, where the two eigenvalues of H_0 given by $E^{(1,2)} = E_0 \pm \sqrt{g_0^2 - \kappa_0^2}$ are real. The system is in the PT-broken phase when $|g_0| < \kappa_0$, where $E^{(1,2)}$ become complex conjugates. Therefore, for the linear system to go from one phase to the other, it must pass through the exceptional point at $|g_0| = \kappa_0$. We emphasize that $|c_a| = |c_b|$ in the PT-symmetric phase, while $|c_a| \neq |c_b|$ in the broken PT phase. These properties play an important role in our analysis below.

For simplicity, we take the gain and loss strength κ_0 to be independent of the nonlinearity ε in (10.6). Here γ and β are two dimensionless real constants: γ specifies the influence of the nonlinearity on the couplings relative to that of the energy shifts, and β indicates how strongly the phases of the couplings depend on

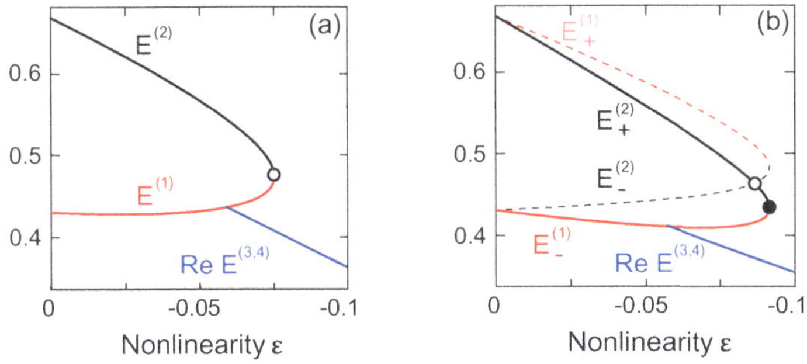

Fig. 10.6 Conventional (**a**) and anomalous (**b**) PT transition in nonlinear systems. The parameters used are: $E_0 = 0.55$, $\kappa_0 = 0.22$, $g_0 = -0.25$, and $\gamma = -0.8$. $\beta = 0$, 0.6 in **a** and **b**, respectively. Adapted from [36]

the nonlinearity, which is crucial for an anomalous PT transition: if $\beta = 0$, the two original eigenvalues of H meet each other at an exceptional point as the nonlinearity increases (see the red and black curves in Fig. 10.6a), which is the same scenario as in the linear case. Afterward they disappear and the system is left with only a broken PT phase, given by two additional eigenvalues of H that satisfy $E^{(3)} = [E^{(4)}]^*$ (blue line in Fig. 10.6). However, if $\beta \neq 0$, it is no longer an exceptional point where the two original eigenvalues of H "annihilate" each other (filled dot in Fig. 10.6b). Afterwards, the system again has only two additional eigenvalues satisfying $E^{(3)} = [E^{(4)}]^*$.

Below we only focus on the two original nonlinear eigenstates of H. The $\beta = 0$ case is easy to understand. As mentioned previously, $|c_a| = |c_b|$ holds for both eigenvalues of the effective Hamiltonian when $\varepsilon = 0$. As we increase $|\varepsilon|$, $c_{a,b}$ change but only in their relative phase. As a result, the two nonlinear oscillators still have the same frequency (i.e. Re E_a = Re E_b) and the couplings are still real and equal (i.e. $g_a = g_b$). Consequently, H itself remains PT symmetric, and its two nonlinear eigenvalues annihilate each other at the exceptional point $\varepsilon = -2(|g_0| - |\kappa_0|)/|\gamma| = -0.075$ in Fig. 10.6a. We have used the normalization $|c_a|^2 + |c_b|^2 = 1$ as usual.

The $\beta \neq 0$ case is more intriguing. We start with the ansatz $|c_a| = |c_b|$ when $|\varepsilon|$ is small. We again find Re E_a = Re E_b, but this time $g_{a,b}$ are no longer real and we have $g_a = g_b^*$ instead. It is important to note that the values of $g_{a,b}$ depend on which nonlinear eigenvalue of H we follow, due to the different relative phases between c_a and c_b in the two corresponding nonlinear eigenstates. For the jth nonlinear eigenvalue ($j = 1, 2$), we linearize H at the corresponding $c_{a,b}^{(j)}$. The resulting linearized Hamiltonian $H^{(j)}$ is also PT symmetric; it has two real eigenvalues $E_\pm^{(j)}$ when the nonlinearity is small, and one of them equals the jth nonlinear eigenvalue (solid black and red lines in Fig. 10.6b):

$$E_\pm^{(j)} = E_0 + \varepsilon \pm \sqrt{|g_a^{(j)}(\varepsilon)|^2 - \kappa_0^2}. \tag{10.7}$$

The corresponding eigenvectors of $H^{(j)}$ still satisfy $|c_a| = |c_b|$, which is consistent with our ansatz.

Equation (10.7) shows clearly that each $H^{(j)}$ can have an exceptional point at $|g_a^{(j)}(\varepsilon)| = \kappa_0$, which in principle could cause two PT transitions. However, these points are not where the two original nonlinear eigenvalues of H annihilate each other. One of them materializes in this case and is shown as the open dot in Fig. 10.6b. It can be shown that $H^{(j)}$ actually bounces right back into its PT-symmetric phase at this exceptional point, instead of going into its broken PT phase [36]. This is the reason that $E_\pm^{(1)}$ stay real between the values of ε specified by the open and closed dots in Fig. 10.6b.

10.4 Modal Interaction in Lasers with Novel Non-Hermitian Symmetries

The intense and coherent light provided by the laser largely enabled the development of nonlinear optics. At the same time, the laser itself is also a nonlinear system in disguise. Simple analyses based on the rate equation or the semiclassical theory show that the laser intensity increases linearly with the pump power [2, 44]. However, this seemingly linear behavior is impossible without the saturation nonlinearity of the gain medium, which provides a large number of photons under stimulation. In this section we discuss the effects of this saturation nonlinearity in lasers with several non-Hermitian symmetries, operated under steady-state conditions.

10.4.1 Photonic Molecule Lasers with PT Symmetry

PT-symmetric lasers have attracted increasing amount of attention due to their intriguing properties and promising applications, such as single-mode lasing and enhanced sensitivity near an exceptional point. While linear threshold analysis explained some features of PT-symmetric lasers, it is important to consider nonlinear modal interactions in the analysis of such novel lasers.

In the analysis below, we consider a photonic molecule laser that consists of two laser cavities (e.g. waveguides, microdisks, and microrings) coupled via evanescent waves [45–48]. Under steady-state conditions, several scenarios of gain clamping can take place, for lasing operation in both the PT-symmetric and PT-broken phases. In particular, their behaviors in the nonlinear regime fall into two distinct categories [49]: in one the system remains "frozen" in the PT phase space (defined by the PT parameter τ below) as the external pump strength increases, while in the other the system approaches its exceptional point.

We illustrate these two contrasting behaviors using a coupled mode theory:

$$H^{(\mu)} = \begin{bmatrix} \omega_0^{(\mu)} + i(\gamma_a - \kappa_a) & g \\ g & \omega_0^{(\mu)} + i(\gamma_b - \kappa_b) \end{bmatrix}. \tag{10.8}$$

This effective Hamiltonian acts on the wave functions $\varphi^{(\mu)} = [\psi_a^{(\mu)} \ \psi_b^{(\mu)}]^T$ in the two cavities a and b. Here $\omega_0^{(\mu)}$ is the identical frequency of two cavity modes close to the peak of the gain spectrum, one in each cavity in the absence of the coupling g (taken to be real and positive). Note that $\kappa_{a,b}$ and $r_{a,b}$ are the loss and applied gain in the two cavities respectively, and we model gain saturation by $\gamma_{a,b} = r_{a,b}/(1 + \sum_\mu I_{a,b}^{(\mu)})$, where $I_{a,b}^{(\mu)} \equiv |\psi_{a,b}^{(\mu)}|^2$ are the intensities of mode μ in the two cavities, scaled by their natural units and hence dimensionless [50–54]. It is this (homogeneously broadened) saturation nonlinearity that couples modes of different frequencies and their respective effective Hamiltonians $H^{(\mu)}$ in our model. Saturable absorption can be modeled similarly if needed [55]. It should be noted that $H^{(\mu)}$ is PT-symmetric without requiring physically balanced gain and loss, i.e. $\gamma_a - \kappa_a = -(\gamma_b - \kappa_b)$. Instead, its PT symmetry holds with respect to the average gain and loss [46].

A lasing mode μ in steady-state operation is given by $\varphi^{(\mu)}(t) = \varphi^{(\mu)}(0) \exp(-i\varepsilon^{(\mu)}t)$, where $\varepsilon^{(\mu)}$ is the corresponding *real* eigenvalue of $H^{(\mu)}$:

$$\varepsilon^{(\mu)} = \omega_0^{(\mu)} + i(\overline{\gamma} - \overline{\kappa}) \pm i\sqrt{(\Delta - \delta)^2 - g^2}. \tag{10.9}$$

Here $\overline{\kappa}, \overline{\gamma}$ denote the average of the losses and saturated gains in the two cavities respectively, and Δ, δ are their half differences, i.e. $\Delta = (\kappa_b - \kappa_a)/2$, $\delta = (\gamma_b - \gamma_a)/2$. The conditions that warrant a real-valued $\varepsilon^{(\mu)}$ depend on whether the laser is in its PT-symmetric or PT-broken phase, determined in turn by the PT parameter $\tau \equiv (\Delta - \delta)^2 - g^2$. In the PT-symmetric phase, the last term in (10.9) is real and we find $\overline{\gamma} = \overline{\kappa}$, i.e. the average gain and loss in the two cavities need to be the same. These PT-symmetric modes are stable in the nonlinear regime, even when the underlying linear Hamiltonian (i.e. setting $\gamma_{a,b} = r_{a,b}$) enters its PT-broken phase when δ is large [56].

Below we will focus on the more interesting case of a laser in its PT-broken phase, where the last term in (10.9) is now imaginary. In this case we find $\overline{\gamma} = \overline{\kappa} \pm \sqrt{(\Delta - \delta)^2 - g^2}$, which also gives the thresholds of the two lasing modes when $I_{a,b}^{(\mu)} = 0$. More specifically, the lasing threshold $r_{\text{TH}}^{(\mu)}$ of mode μ can be defined using the value of the applied gain at which the corresponding ε becomes real. We refer to $\text{Im}\,\varepsilon$ as the modal gain, which is negative for a mode below its threshold and equals zero at or above its threshold.

Here we consider two configurations: in the first configuration, the two cavities have the same loss ($\kappa_a = \kappa_b$) and the gain is only applied to the first cavity ($\gamma_b = r_b = 0$). In the second configuration, the same amount of gain is applied to both cavities ($r_a = r_b$), and we assume one cavity is lossier than the other ($\kappa_a < \kappa_b$). In both cases, the lasing mode at threshold is localized in cavity a, which warrants a

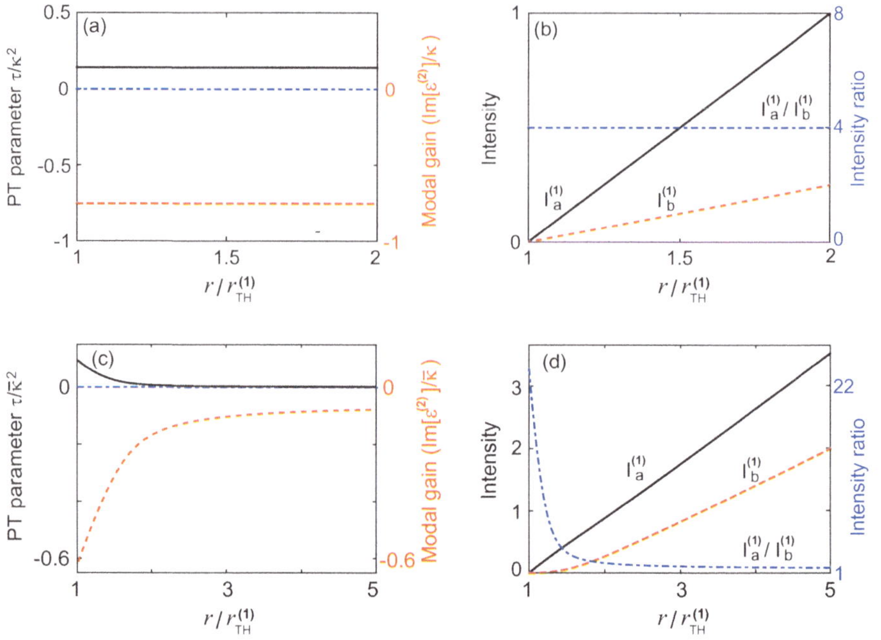

Fig. 10.7 Lasing in the PT-broken phase in configuration 1 (**a**, **b**) and 2 (**c**, **d**). **a** The frozen PT parameter $\tau = 0.14\kappa^2$ (solid line) and the modal gain of the second mode $\text{Im}\varepsilon^{(2)} = -0.75\kappa$ (dashed line) as a function of the applied gain r. The exceptional point ($\tau = 0$) and the lasing condition ($\text{Im}\varepsilon = 0$) are marked by the dash-dotted line. **b** Intensity of the first mode in both cavity a (solid line) and b (dashed line). Their ratio is shown by the dash-dotted line. $\omega_0/\kappa = 10^4$ and $g/\kappa = 0.5$. **c**, **d** Same as **a**, **b** but for configuration 2. Here $\omega_0/\kappa_a = 10^4$, $\kappa_b/\kappa_a = 2$ and $g/\kappa_a = 0.2$. Adapted from [49]

lower threshold due to the stronger overlap with the pump in configuration 1 and a lower loss in configuration 2.

In configuration 1, we have $\Delta = 0$ ($\kappa_a = \kappa_b \equiv \kappa$) and $\delta = -\gamma_a/2 = -\overline{\gamma}$. It requires $\kappa > g$ for the photonic molecule to start lasing in the PT-broken phase, and the first threshold is given by $r_{\text{TH}}^{(1)} = \kappa(1 + g^2/\kappa^2)$. For ε to remain real above threshold, $\gamma_a = r_{\text{TH}}^{(1)}$ must hold, i.e. the saturated gain (in cavity a) is clamped at its threshold value. Consequently, the system is frozen in the PT-broken phase, with a constant $\tau = \kappa^2(1 - g^2/\kappa^2)^2/4 > 0$ (see Fig. 10.7a) and a constant intensity ratio $I_a^{(1)}/I_b^{(1)} = \kappa^2/g^2 > 1$ (see Fig. 10.7b). The intensity of mode 1 can be calculated directly from the clamped gain: $I_a^{(1)} = \kappa r(\kappa^2 + g^2)^{-1} - 1$, and the modal gain of mode 2 stays negative (i.e. suppressed) and clamped at $\text{Im}\varepsilon^{(2)} = -2\sqrt{\tau} < 0$.

In configuration 2, even though the gain is applied equally to both cavities, it is not necessarily true that $\delta = (\gamma_b - \gamma_a)/2 = 0$ in the nonlinear regime. The laser starts lasing at $r_{\text{TH}}^{(1)} = \overline{\kappa} - \sqrt{\Delta^2 - g^2}$ in the PT-broken phase if $\Delta > g$. The intensity of this mode is higher in cavity a than in cavity b above threshold: $I_a^{(1)}/I_b^{(1)} =$

$(\sqrt{\tau}+\sqrt{\tau+g^2})^2/g^2 > 1$ (see Fig. 10.7d). Therefore, as the applied gain increases above threshold, it is saturated more in cavity a, leading to an increasing $\delta > 0$. Therefore, the PT parameter $\tau = (\Delta-\delta)^2 - g^2$ decreases (see Fig. 10.7c).

This asymmetric saturation hence acts back on the lasing mode itself, and the system is pulled towards its exceptional point where $\tau = 0$. The intensity ratio $I_a^{(1)}/I_b^{(1)}$ also reduces as a result (see Fig. 10.7d):

$$\frac{I_a^{(1)}}{I_b^{(1)}} \to \left[\frac{\overline{\kappa}+\Delta}{g+\sqrt{\overline{\kappa}^2+g^2-\Delta^2}}\right]^2. \tag{10.10}$$

This limit is approximately 1 when $\Delta \approx g$. Meanwhile, the saturated gains in the two cavities also approach their clamped values in the large γ limit, leading to an asymptotic value of the PT parameter:

$$\tau = \frac{g^2}{4}\left[\frac{g+\sqrt{\overline{\kappa}^2+g^2-\Delta^2}}{\overline{\kappa}+\Delta} - \frac{\overline{\kappa}+\Delta}{g+\sqrt{\overline{\kappa}^2+g^2-\Delta^2}}\right]^2. \tag{10.11}$$

Similar to configuration 1, one can show that the modal gain of the next mode is negative (given by $\mathrm{Im}\varepsilon^{(2)} = -2\sqrt{\tau}$; see Fig. 10.7c) and hence suppressed, also giving rise to a single-mode behavior.

Above we briefly discussed the nonlinear behavior of PT-symmetric lasers based on photonic molecules, lasing in their PT-broken phase. To confirm whether the single-mode behavior is robust, one also needs to consider other cavity modes of different frequencies. A semiclassical approach is best suited for this purpose because of spatial hole burning [49], and the extended range of single-mode operation was indeed verified in [48]. We also mention in passing that a PT-symmetric laser can also be realized in a single microcavity, which itself has separate gain and loss regions. Two typical examples are a multilayer waveguide with balanced gain and loss [40, 41] and a microring cavity with a PT-symmetric grating on top [57–59]. The stability analysis of the latter in the nonlinear regime can be found in [60].

10.4.2 Laser Arrays with NHPH Symmetry

We have discussed anti-PT symmetry and briefly mentioned its connection to NHPH symmetry in Sect. 10.2. To gain some understanding of NHPH symmetry, we first resort to the original particle-hole symmetry in Hermitian systems, first identified in particle physics and playing an important role in condensed matter systems as well. It is satisfied when the Hamiltonian anti-commutes with an anti-linear operator CT: $\{H, CT\} = 0$. T here is the time reversal operator in the form of the complex conjugation, and C is the charge conjugation operator that is a linear operator by itself. Different from a linear operator, an anti-linear operator $\boldsymbol{A}$ is defined by

$A(a\phi_1 + b\phi_2) = a^* A\phi_1 + b^* A\phi_2$, where $\phi_{a,b}$ are two arbitrary states and a, b are their complex superposition coefficients.

One important property of Hermitian systems with particle-hole symmetry is the possible existence of zero modes, such as the majoranas [61–65]. Their topological protection and non-Abelian nature have stimulated wide research interests, especially in robust topological quantum computation. In a Hermitian system, its energy spectrum is symmetric about the energy level that divides the energy of particles and holes, i.e. $E_\mu = -E_\nu$, as a direct consequence of $\{H, CT\} = 0$.

Remarkably, a non-Hermitian version of particle-hole symmetry hides in plain sight in many previously investigated PT-symmetric systems [12, 66, 67]. If we restrict parity to a mirror reflection in one dimension or an inversion in three dimensions, NHPH symmetry then has a significant difference from anti-PT symmetry: a system can have PT symmetry and NHPH symmetry simultaneously, which cannot be true for PT and anti-PT symmetries. This NHPH symmetry imposes no restriction on the spatial profile of gain and loss landscape. All it requires are that (1) the system consists of coupled optical elements that form a lattice; and that (2) it has sublattice symmetry (also known as chiral symmetry in condensed matter systems) in the Hermitian limit when the overall gain and loss strength approaches zero. The resulting non-Hermitian Hamiltonian can be written as $H = H_0 + i\gamma_p\delta_{pq}$, where γ_p on the main matrix diagonal describes the gain and loss landscape. H_0 is the Hermitian part with chiral symmetry, and in the block matrix form, it has two zero diagonal blocks and two off-diagonal blocks $G_{1,2}$ satisfying $G_1 = G_2^*$.

This NHPH symmetry warrants an energy spectrum satisfying $E_\mu = -E_\nu^*$, where the mode indices μ and ν are not necessarily the same. When they indeed are the same, the system is in its NHPH symmetric phase and we find $E_\nu = -E_\nu^*$ which defines a non-Hermitian zero mode, on the imaginary axis of the complex energy plane. The phase of its wave function has a unique distribution: it can only take four discrete values, with those on the two sublattices differ by $\pi/2$ [12]. In other words, one can adopt the convention of making the wave function real on one sublattice and imaginary on the other. We also note that this "zero" energy or frequency does not represent infinite wavelength in this context. Rather, it refers to a well-defined frequency in the system, from which all revelent frequencies ("detunings") are measured.

We have in fact already encountered examples of non-Hermitian zero modes in the previous section: the photonic molecule lasers in their PT-broken phase are exactly lasing in their zero modes. These modes have the same frequency as the individual cavity modes, i.e. $\varepsilon^{(\mu)} = \omega_0^{(\mu)}$ in (10.9), from which the detunings are measured.

To exemplify the NHPH symmetry and its non-Hermitian zero modes in a more general case (e.g. in the absence of PT symmetry of any form), we discuss a laser array arranged into a square lattice (Fig. 10.8c). Each of the two sublattices, marked by filled and open dots, has eight lattice sites that represent coupled optical cavities. Here $g_{x,y}$ denote the horizontal and vertical couplings, and we assume that each cavity has a loss equal to $g_x/2$. The system does not have a non-Hermitian zero mode initially (Fig. 10.8b) when the two lattice "A" sites in Fig. 10.8d are pumped equally. If the pump strength is above a certain threshold, however, a pair of modes

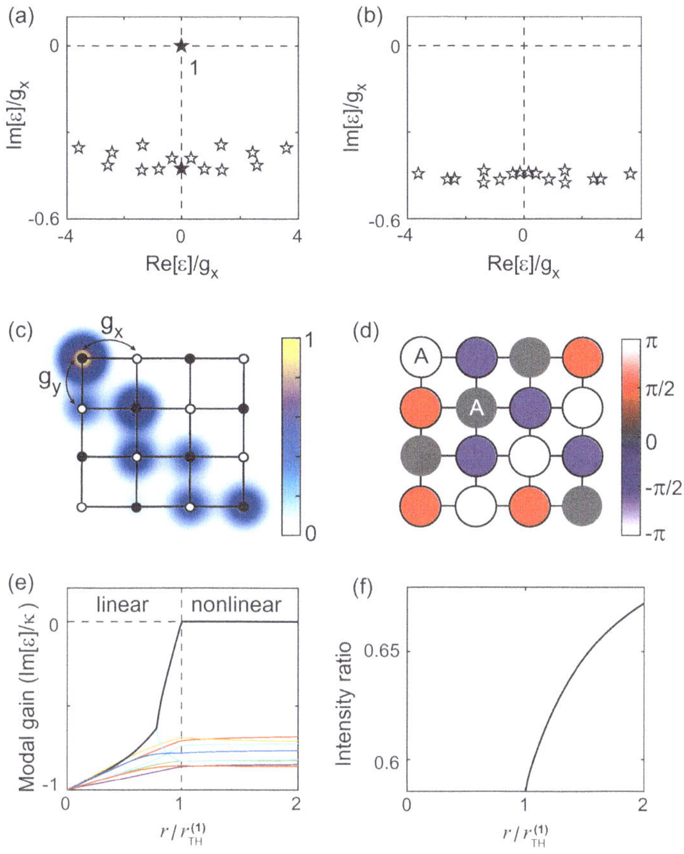

Fig. 10.8 Non-Hermitian zero mode laser. **a** Complex energy spectrum of a rectangular lattice with NHPH symmetry at the laser threshold $r_{\mathrm{TH}}^{(1)} = 1.07g_x$. The couplings satisfy $g_y = 1.25g_x \in \mathbb{R}$. The cavity decay rate is given by $\kappa = 0.5g_x$. **b** Same as **a** but with a lower pump strength ($r = 0.38g_x$) and no zero modes. **c, d** Schematics showing the spatial profile and phase distribution of the lasing mode at threshold. **e** Modal gain for all modes below and above the lasing threshold. **f** Intensity ratio of the lasing mode in the two pumped cavities

first meet and then move along the imaginary axis in the complex energy plane (filled stars in Fig. 10.8a), which are the non-Hermitian zero modes. Once one of them reaches the real axis and starts lasing, the rest of the modes are suppressed due to saturation nonlinearity when the pump is further increased (Fig. 10.8e).

The non-Hermitian zero modes depend on the gain (and loss) landscape across the lattice: with a different pump configuration, one may find the lasing mode is a zero mode with a completely different spatial profile. Nevertheless, they share the same frequency since they are all zero modes, which provides a symmetry-protected scheme to realize a laser with a fixed frequency but a tunable spatial profile, and

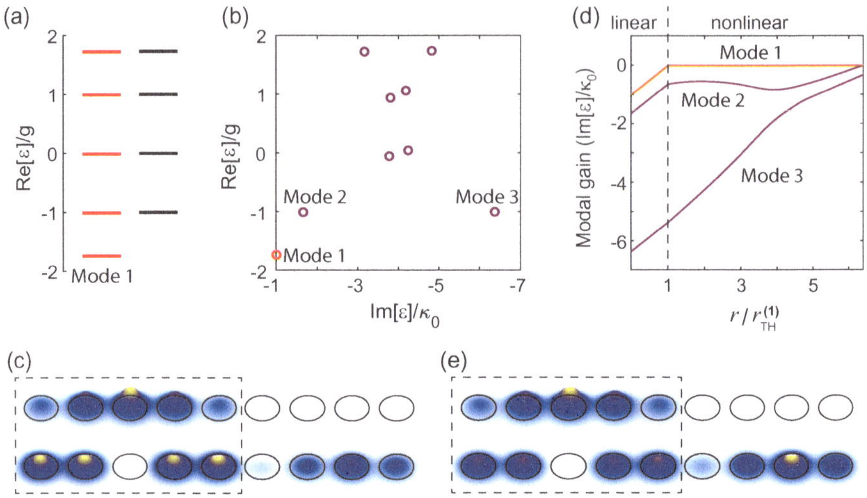

Fig. 10.9 A supersymmetric laser array. **a** Mode spectra of the original array (left) and the partner array (right). **b** Complex energy spectrum of the whole system. Here g and κ_0 are the uniform coupling and cavity loss in the original array. $g' = 0.2g$ is the coupling between the two arrays, and $\kappa = 7\kappa_0$ is the uniform cavity loss in the partner array. **c** Intensity profiles of mode 1 and 2. Dashed box marks the original array. **d** Modal gain versus pump strength. Mode 2 reaches its lasing threshold at $r = 6.4r_{\mathrm{TH}}^{(1)}$. **e** Same as **c** but at $r = 6.4r_{\mathrm{TH}}^{(1)}$

this additional degree of freedom may find applications in telecommunication and encoding.

10.4.3 Laser Arrays with Supersymmetry

Supersymmetry, first introduced in particle physics, has also found its way into non-Hermitian photonics. Instead of treating bosons and fermions on equal footing, here two optical systems are constructed using a supersymmetry transformation, which warrants identical frequency spectra except for an extra state in one of them, usually chosen as its ground state. This principle can be applied, for example, to engineering band structures in a quantum cascade laser [68], mode converters in coupled waveguides [69], and single-mode laser arrays [70–73]. In this section we will focus on the last example.

A single-mode laser array based on supersymmetry relies on the principle of Q-spoiling. The basic idea is illustrated in Fig. 10.9a: even though the cavities in the original array are identical, their couplings spread out the spectrum of the array. By applying a supersymmetry transformation, one constructs another array with one less cavity and one less mode [70]. The alignment of the two spectra then provides an interesting scheme to single-mode lasing: by making the partner array lossier than the original array, the former acts as the "sink" and spoils the quality (Q) factor of the frequency-matched states in the latter. The unmatched mode in the original array, however, is barely affected (see Fig. 10.9b) and becomes the only lasing mode. If it is chosen to be the ground state of the original array, the envelope of the field distribution has a single intensity lobe (see Fig. 10.9c) and the same phase across all cavities in the original array. This choice hence makes it ideal to combine the laser radiation from all cavities into a strong beam in the farfield [72, 73].

We should caution that each individual cavity itself in fact has multiple modes, and they lead to multiple copies of the frequency ladder shown in Fig. 10.9a. Therefore, unless each individual cavity is already optimized to have a single lasing mode by its own, the unmatched modes from all the copies of the frequency ladder will lead to multiple peaks in the lasing spectrum.

Previous studies of supersymmetry laser arrays in the nonlinear regime considered uniform pumping in both the original array and the partner array. Because the unmatched mode (1 in Fig. 10.9b) in the original array has little intensity in the partner array (see Fig. 10.9c), the gain in the latter is barely saturated and can in principle support other modes to lase, once the pump is strong enough. However, the coupling of the aligned modes between the two arrays are similar to that in the PT-symmetric photonic molecule, and the resulting mode with the lowest loss (still higher than that of the unmatched mode; see mode 2 in Fig. 10.9b) is necessarily in the "PT broken phase." Here the PT symmetry is not exact: the two arrays are not symmetric, and two matched modes also couple weakly to other modes of different frequencies.

Nevertheless, this approximate PT symmetry indicates that the next mode to lase is also strongly localized in the original array (see Fig. 10.9c), and hence the unsaturated gain in the partner array is ineffective in bringing this mode to its own lasing threshold (see Fig. 10.9d). Interestingly, as the pump increases, this mode evolves to have almost equal intensities in the original array and the partner array (see Fig. 10.9e), i.e. approaching its "PT-symmetric" phase similar to what we have shown in Fig. 10.7d. This more or less balanced intensity distribution has a much stronger overlap with the gain, and eventually, this mode reaches its threshold as shown in Fig. 10.9d. The range of single-mode operation can be extended by pumping just the original array.

One inconvenience of supersymmetric transformation is that even though all the cavities in the original array are identical and equally spaced, the ones in the partner array are not. Their sizes and spacings vary from cavity to cavity, which makes the fabrication extremely challenging. One workaround is to apply the supersymmetry transformation twice [71], with which the cavities in the partner array are of the same size as in the original array, even though their spacings are still nonuniform. The partner array now lacks both the ground state and the highest frequency mode in one copy of the frequency ladder, and another "sink" in the form of a single cavity is required to spoil the Q-factor of the highest frequency mode.

10.5 Outlook

Nonlinearity covers a broad range of topics in both physics and mathematics. In this section we have only briefly touched on some interesting aspects in non-Hermitian photonics, not all of which fall into standard nonlinear optics. Nevertheless, we hope that these discussions may stimulate further investigations of nonlinear and novel phenomena in non-Hermitian photonics.

One particular direction is the role of non-Hermitian symmetries on the dynamics of non-Hermitian systems. So far the majority of studies have focused on the steady-state operations or stationary-state solutions, and the exploration of dynamical behaviors including limit cycles [74] and chaos [75] is rather limited. We do want to stress that the selection of a physically sound model is crucial for these studies to be meaningful. For example, modeling optical cavities without the linear cavity loss will lead to erroneous dynamical behaviors, and the results obtained from a two-mode coupled theory in a PT-symmetric photonic molecule should be taken with a grain of salt far above the laser threshold.

Due to the similarity of anti-PT symmetry and NHPH symmetry, we suggest considering them in the same category that may be referred to as "fermionic antilinear symmetries." Accordingly, PT symmetry and those with a similar form may be referred to as "bosonic antilinear symmetries," since not all synthetic operators can be regarded comfortably as an extension of parity. Here "bosonic" and "fermionic" refer to the commutation and anti-commutation relation in their respective definitions, and we remind the reader that $T^2 = -1$ (instead of $T^2 = 1$) holds in some of these systems [76, 77]. Beyond these two categories, there are also recent findings that include the transpose or Hermitian conjugate of the Hamiltonian [78], similar to the definition of pseudo-Hermiticity. Further categorizing them according to their topological properties is an interesting development [79], and we expect more work to emerge that explore their unique nonlinear behaviors.

The intertwining of nonlinear optics and non-Hermitian photonics can undoubtedly enrich both fields simultaneously. So far however, it seems that the favor has been heavily tilted towards the former. For example, new nonlinear behaviors have been identified for both PT-symmetric nonlinear Schrödinger equation [80] and Dirac equation [81], and even a new scheme to achieve OPA without phase matching has been proposed [82]. We believe that there are emerging possibilities to expand the boundaries of non-Hermitian photonics using nonlinear optics. The introduction of the synthetic dimension enabled by nonlinear optical processes looks to be particularly promising, especially when the nonlinear coupling can be extended to form a synthetic lattice [18]. This approach may finally allows us to investigate PT symmetry experimentally in three and higher dimensions, where more exciting discoveries await.

Acknowledgements LG acknowledges the National Science Foundation under Grant No. PHY-1847240

References

1. J. Okołowicz, M. Płoszajczak, I. Rotter, Phys. Rep. **374**, 271 (2003)
2. H. Haken, *Light*, vol. 2 (North-Holland, Amsterdam, 1985)
3. C.M. Bender, S. Boettcher, Phys. Rev. Lett. **80**, 5243 (1998)
4. C.M. Bender, S. Boettcher, P.N. Meisinger, J. Math. Phys. **40**, 2201 (1999)
5. C.M. Bender, D.C. Brody, H.F. Jones, Phys. Rev. Lett. **89**, 270401 (2002)
6. L. Feng, R. El-Ganainy, L. Ge, Nat. Photonics **11**, 752 (2017)
7. R. El-Ganainy, K.G. Makris, M. Khajavikhan, Z.H. Musslimani, S. Rotter, D.N. Christodoulides, Nat. Phys. **14**, 11 (2018)
8. Ş.K. Özdemir, S. Rotter, F. Nori, L. Yang, Nat. Mater. **18**, 783 (2019)
9. B. Qi, H.-Z. Chen, L. Ge, P. Berini, R.M. Ma, Adv. Opt. Mater. **7**, 1900694 (2019)
10. V.V. Konotop, J. Yang, D.A. Zezyulin, Rev. Mod. Phys. **88**, 035002 (2016)
11. L. Ge, H.E. Türeci, Phys. Rev. A **88**, 053810 (2013)
12. L. Ge, Phys. Rev. A **95**, 023812 (2017)
13. G. Lévai, M. Znojil, J. Phys. A: Math. Gen. **35**, 8793 (2002)
14. M. Znojil, J. Phys. A: Math. Gen. **35**, 2341 (2002)
15. W.D. Heiss, J. Phys. A: Math. Gen. **37**, 2455 (2004)
16. C.M. Bender, P.N. Meisinger, Q. Wang, J. Phys. A: Math. Gen. **36**, 1029 (2003)
17. L. Ge, A.D. Stone, Phys. Rev. X **4**, 031011 (2014)
18. L. Yuan, Q. Lin, M. Xiao, S. Fan, Optica **5**, 1396 (2018)
19. S.M. Spillane, T.J. Kippenberg, K.J. Vahala, Nature **415**, 621 (2002)
20. G. Bahl, J. Zehnpfennig, M. Tomes, T. Carmon, Nat. Commun. **2**, 403 (2011)
21. Y. Zheng, J. Yang, Z. Shen, J. Cao, X. Chen, X. Liang, W. Wan, Light Sci. Appl. **5**, e16072 (2016)
22. T.J. Kippenberg, K.J. Vahala, Opt. Express **15**, 17172 (2007)
23. J. Kim, M.C. Kuzyk, K. Han, H. Wang, G. Bahl, Nat. Phys. **11**, 275 (2015)
24. C.H. Dong, Z. Shen, C.L. Zou, Y.L. Zhang, W. Fu, G.C. Guo, Nat. Commun. **6**, 6193 (2015)
25. L. Ge, W. Wan, Pseudo-Hermitian Transition in Degenerate Nonlinear Four-Wave Mixing (2016), arXiv:1603.05624
26. F. Zhang, Y. Feng, X. Chen, L. Ge, W. Wan, Synthetic anti-PT symmetry in a single microcavity. Phys. Rev. Lett. **124**, 053901 (2020)
27. A. Mostafazadeh, Int. J. Geom. Methods Mod. Phys. **7**, 1191 (2010)
28. J.H. Wu, M. Artoni, G.C. La Rocca, Phys. Rev. A **91**, 33811 (2015)
29. P. Peng, W. Cao, C. Shen, W. Qu, J. Wen, L. Jiang, Y. Xiao, Nat. Phys. **12**, 1139 (2016)
30. Y. Choi, C. Hahn, J.W. Yoon, S.H. Song, Nat. Commun. **9**, 2182 (2018)
31. Y. Jiang, Y. Mei, Y. Zuo, Y. Zhai, J. Li, J. Wen, S. Du, Phys. Rev. Lett. **123**, 193604 (2019)
32. X.-L. Zhang, T. Jiang, C.T. Chan, Light Sci. Appl. **8**, 1 (2019)
33. Y. Li, Y. Peng, L. Han, M. Miri, W. Li, M. Xiao, X. Zhu, J. Zhao, A. Alu, S. Fan, C. Qiu, Science **364**, 170 (2019)
34. H. Zhao, W.S. Fegadolli, J. Yu, Z. Zhang, L. Ge, A. Scherer, L. Feng, Phys. Rev. Lett. **117**, 193901 (2016)
35. Y. Zheng, H. Ren, W. Wan, X. Chen, Sci. Rep. **3**, 3245 (2013)
36. L. Ge, Phys. Rev. A **94**, 013837 (2016)
37. Y.D. Chong, L. Ge, H. Cao, A.D. Stone, Phys. Rev. Lett. **105**, 53901 (2010)
38. W. Wan, Y. Chong, L. Ge, H. Noh, A.D. Stone, H. Cao, Science **331**, 889–892 (2011)
39. J. Zhang, K.F. MacDonald, N.I. Zheludev, Light Sci. Appl. **1**, e18 (2012)
40. Y.D. Chong, L. Ge, A.D. Stone, Phys. Rev. Lett. **106**, 93902 (2011)
41. L. Ge, Y.D. Chong, A.D. Stone, Phys. Rev. A **85**, 023802 (2012)
42. Z. Lin, H. Ramezani, T. Eichelkraut, T. Kottos, H. Cao, D.N. Christodoulides, Phys. Rev. Lett. **106**, 213901 (2011)
43. S. Longhi, Phys. Rev. Lett. **107**, 033901 (2011)
44. M. Sargent III, M.O. Scully, W.E. Lamb, Jr., *Laser Physics* (Addison-Wesley, Reading, 1974)

45. M. Liertzer, L. Ge, A. Cerjan, A.D. Stone, H.E. Türeci, S. Rotter, Phys. Rev. Lett. **108**, 173901 (2012)
46. R. El-Ganainy, M. Khajavikhan, L. Ge, Phys. Rev. A **90**, 013802 (2014)
47. P. Hamel, S. Haddadi, F. Raineri, P. Monnier, G. Beaudoin, I. Sagnes, A. Levenson, A.M. Yacomotti, Nat. Photon. **9**, 311 (2015)
48. H. Hodaei, M.A. Miri, M. Heinrich, D.N. Christodoulides, M. Khajavikhan, Science **346**, 975 (2014)
49. L. Ge, R. El-Ganainy, Sci. Rep. **6**, 24889 (2016)
50. H.E. Türeci, L. Ge, S. Rotter, A.D. Stone, Science **320**, 643 (2008)
51. H.E. Türeci, A.D. Stone, B. Collier, Phys. Rev. A **74**, 043822 (2006)
52. H.E. Türeci, A.D. Stone, L. Ge, Phys. Rev. A **76**, 013813 (2007)
53. L. Ge, Y.D. Chong, A.D. Stone, Phys. Rev. A **82**, 063824 (2010)
54. A. Cerjan, Y.D. Chong, A.D. Stone, Opt. Express **23**, 6455 (2015)
55. A.U. Hassan, H. Hodaei, M.-A. Miri, M. Khajavikhan, D.N. Christodoulides, Phys. Rev. A **92**, 063807 (2015)
56. L. Ge, to be submitted to Phys. Rev. A
57. L. Feng, Z.J. Wong, R.-M. Ma, Y. Wang, X. Zhang, Science **346**, 972 (2014)
58. P. Miao, Z. Zhang, J. Sun, W. Walasik, S. Longhi, N.M. Litchinitser, L. Feng, Science **353**, 464 (2016)
59. H.-Z. Chen, T. Liu, H.-Y. Luan, R.-J. Liu, X.-Y. Wang, X.-F. Zhu, Y.-B. Li, Z.-M. Gu, S.-J. Liang, H. Gao, L. Lu, L. Ge, S. Zhang, J. Zhu, R.-M. Ma, Nat. Phys. (2020). https://doi.org/10.1038/s41567-020-0807-y
60. Z. Zhang, P. Miao, J. Sun, S. Longhi, N.M. Litchinitser, L. Feng, ACS Photon. **5**, 3016 (2018)
61. M.Z. Hasan, C.L. Kane, Rev. Mod. Phys. **82**, 3045 (2010)
62. X.-L. Qi, S.-C. Zhang, Rev. Mod. Phys. **83**, 1057 (2011)
63. J. Alicea, Rep. Prog. Phys. **75**, 076501 (2012)
64. C.W.J. Beenakker, Rev. Mod. Phys. **87**, 1037 (2015)
65. C. Nayak, S.H. Simon, A. Stern, M. Freedman, S. Das Sarma, Rev. Mod. Phys. **80**, 1083 (2008)
66. B. Qi, L. Zhang, L. Ge, Phys. Rev. Lett. **120**, 093901 (2018)
67. L. Ge, Photon. Res. **6**, A10 (2018)
68. S. Weinberg, *The Quantum Theory of Fields*, vol. III (Cambridge University Press, Cambridge, 2005)
69. M. Heinrich, M.A. Miri, S. Stützer, R. El-Ganainy, S. Nolte, A. Szameit, D.N. Christodoulides, Nat. Commun. **5**, 3698 (2014)
70. R. El-Ganainy, L. Ge, M. Khajavikhan, D.N. Christodoulides, Phys. Rev. A **92**, 033818 (2015)
71. M.H. Teimourpour, L. Ge, D.N. Christodoulides, R. El-Ganainy, Sci. Rep. **6**, 33253 (2016)
72. M.P. Hokmabadi, N.S. Nye, R. El-Ganainy, D.N. Christodoulides, M. Khajavikhan, Science **363**, 623 (2019)
73. B. Midya, H. Zhao, X. Qiao, P. Miao, W. Walasik, Z. Zhang, N.M. Litchinitser, L. Feng, Photon. Res. **7**, 363 (2019)
74. S. Yu, X. Piao, N. Park, Adv. Sci. **6**, 1900771 (2019)
75. J.P. Deka, A.K. Sarma, Nonlinear Dynam. **96**, 565–571 (2019)
76. K. Jones-Smith, H. Mathur, Phys. Rev. A **82**, 042101 (2010)
77. V.V. Konotop, D.A. Zezyulin, Phys. Rev. Lett. **120**, 123902 (2018)
78. J.D.H. Rivero, L. Ge, Chiral symmetry in non-Hermitian systems: product rule, Clifford algebra and pseudo-chirality (2019), arXiv:1903.02231
79. K. Kawabata, S. Higashikawa, Z. Gong, Y. Ashida, M. Ueda, Nat. Commun. **10**, 297 (2019)
80. A.K. Sarma, M.-A. Miri, Z.H. Musslimani, D.N. Christodoulides, Phys. Rev. E **89**, 052918 (2014)
81. J. Cuevas-Maraver, P.G. Kevrekidis, A. Saxena, F. Cooper, A. Khare, A. Comech, C.M. Bender, IEEE, J. Sel. Top. Q. Electron. **22**, 67 (2016)
82. R. El-Ganainy, J.I. Dadap, R.M. Osgood, Opt. Lett. **40**, 5086 (2015)

Chapter 11
Computational Challenges of Nonlinear Systems

Laurette S. Tuckerman

Abstract We survey some of the major types of dynamical-systems computations that can be carried out for two or three-dimensional systems of partial differential equations. In order of increasing complexity, we describe methods for calculating steady states and bifurcation diagrams, linear stability and Floquet analysis, and heteroclinic orbits. These are illustrated by computations for Rayleigh–Bénard convection in a cylindrical geometry, the Faraday instability of a fluid layer, the flow past a cylinder and over a square cavity, flow in a cylindrical container with counter-rotating lids, and Bose–Einstein condensation. We discuss some mathematical questions raised by these computations and the need for improved numerical tools.

11.1 Time Integration

Many dynamical systems of interest in the sciences and engineering can be written as

$$\partial_t U = \mathcal{F}(U) = \mathcal{L}U + \mathcal{N}(U) \tag{11.1}$$

where U consists of one or more time-dependent functions of spatial variables x, (x, y) or (x, y, z), and $\mathcal{F}$ combines algebraic and differential operators. In (11.1), the evolution operator $\mathcal{F}$ has been decomposed into a linear operator $\mathcal{L}$, while $\mathcal{N}$ contains any other terms. Examples of such dynamical systems are the Navier–Stokes equations, the Boussinesq equations, the Swift–Hohenberg equation, and the Nonlinear Schrödinger equation. $\mathcal{L}$ is often the Laplacian ∇^2 but may also be a higher-order differential operator, as in the case of the Swift–Hohenberg equation, and it may contain other types of terms, such as buoyancy forces.

L. S. Tuckerman (✉)
Laboratoire de Physique et Mécanique des Milieux Hétérogènes (PMMH), CNRS, ESPCI Paris, PSL Research University, Sorbonne Université, Université Paris Diderot, 75005 Paris, France
e-mail: laurette@pmmh.espci.fr

P. G. Kevrekidis et al. (eds.), *Emerging Frontiers in Nonlinear Science*, Nonlinear Systems and Complexity 32, https://doi.org/10.1007/978-3-030-44992-6_11

The most well-known and fundamental role of computation in nonlinear dynamics is that of a perfect experiment: following the evolution of $U(t)$ in time from an initial condition U_0. Geometries and forces can be implemented perfectly, without noise, and parameters can be scanned, even taking on values that are not physically realizable. In addition, numerical solutions provide all of the details of the system, e.g. the velocity or temperature at every point in space. Research communities in computational physics, chemistry, and engineering all have their own preferred methods for discretizing partial differential equations in space and time.

Numerical integration in time can be seen as replacing the temporally continuous dynamical system (11.1) by a temporally discrete system. The best choice of method is dictated by the nature of $\mathcal{F}$. If (11.1) is a Hamiltonian system, then methods that conserve volume in phase space are essential. If $\mathcal{L}$ represents a highly dissipative operator, then implicit methods are required for its integration. This is a vast topic and here we will consider only the simple first-order Euler implicit-explicit scheme:

$$U(t+\Delta t) = U(t) + \Delta t\,(\mathcal{L}U(t+\Delta t) + \mathcal{N}(U(t)))$$

leading to

$$U(t+\Delta t) = \mathcal{B}_{\Delta t}U(t) \equiv (I - \Delta t\mathcal{L})^{-1}(I + \Delta t\mathcal{N})U(t) \tag{11.2}$$

Scheme (11.2) will be used as a building block for other algorithms. When the goal is time integration, other schemes that are more accurate than (11.2) can be used.

11.2 Steady States and Bifurcation Diagrams: Cylindrical Convection

Dynamical systems can be organized around objects of increasing complexity: steady states, periodic orbits, and then tori, heteroclinic orbits, and so on. The study of the steady-state problem might seem prosaic, but in fact it is anything but that. For motivation, we present in Figs. 11.1, 11.2, 11.3, 11.4, 11.5, 11.6, 11.7, and 11.8 the results of experimental and numerical investigations of Rayleigh–Bénard convection in a fluid-filled cylinder whose radius is twice the height and whose upper and lower boundaries are maintained at different temperatures.

The governing equations of this system are the Navier–Stokes and Boussinesq equations:

$$\partial_t H + (\mathbf{U}\cdot\nabla)\,H = Ra\;U_z + \nabla^2 H \tag{11.3a}$$

$$Pr^{-1}\,(\partial_t\mathbf{U} + (\mathbf{U}\cdot\nabla)\,\mathbf{U}) = -\nabla P + \nabla^2\mathbf{U} + H\mathbf{e}_z \tag{11.3b}$$

$$\nabla\cdot\mathbf{U} = 0 \tag{11.3c}$$

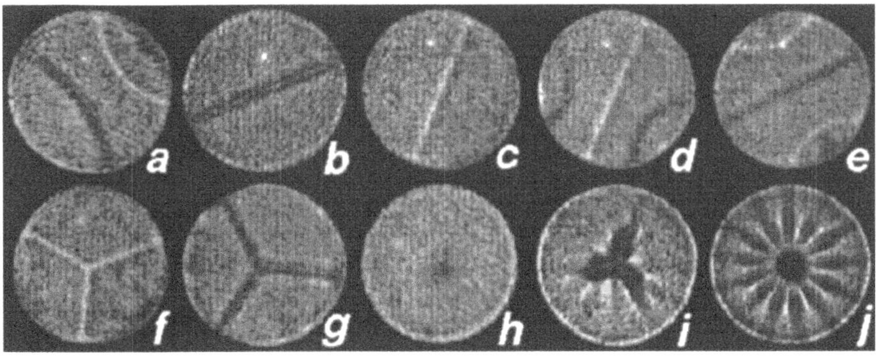

Fig. 11.1 Patterns observed in the experiment of Hof et al. **a**–**h** are all observed at $Ra = 14\,200$. **a** three rolls, **b** two rolls, **c** inverted two rolls, **d** four rolls, **e** inverted four rolls, **f** mercedes, **g** inverted mercedes, **h** axisymmetric pattern. **i** is a rotating pattern at $Ra = 26\,600$ while **j** is a pulsed spoked pattern at $Ra = 33\,000$. Dark areas correspond to hot (rising) and bright to cold (descending) fluid. Reproduced from [1]. ©1999 by the American Institute of Physics

The conductive temperature profiles is linear in the vertical direction; H is the deviation of the temperature from this profile while $\mathbf{U}$ is the velocity and P is the pressure. Ra is the Rayleigh number, which is a nondimensional measure of the temperature difference imposed between the upper and lower plates, and Pr is the Prandtl number, which is the ratio of the kinematic viscosity to the thermal diffusivity. The boundary conditions for the velocity correspond to no-penetration and no-slip on the horizontal plates and sidewalls:

$$\mathbf{U} = 0 \quad \text{for} \quad z = \pm 1/2 \quad \text{and for} \quad r = 2. \tag{11.4a}$$

The boundary conditions on the temperature correspond to perfectly conducting horizontal bounding plates and perfectly insulating sidewalls:

$$H = 0 \quad \text{for} \quad z = \pm 1/2 \quad \text{and} \quad \partial_r H = 0 \quad \text{for} \quad r = 2 \tag{11.4b}$$

In their experiments, Hof et al. [1] observed a multiplicity of distinct patterns—tori, dipoles, roll patterns squeezed into a baseball-like shape, and a three-fold pattern that they named *mercedes*—at the same Rayleigh and Prandtl numbers, as shown in Fig. 11.1. Boronska & Tuckerman were able to reproduce these states [2] and found additional states as well by numerical time integration via, e.g. (11.2), of the spatially discretized equations (11.3a)–(11.4b) in a domain with resolution $N_r \times N_\theta \times N_z = 40 \times 120 \times 20 \approx 10^5$, i.e. state vectors of size $M \approx 4 \times 10^5$. These time-dependent simulations are summarized in Fig. 11.2.

An understanding of the origin of these states and their connections and ranges of existence and stability can only be obtained from a full bifurcation-theoretic study. This was done by Boronska & Tuckerman [3] in a large-scale calculation via methods that we will describe below. The bifurcation diagram that we have computed is shown

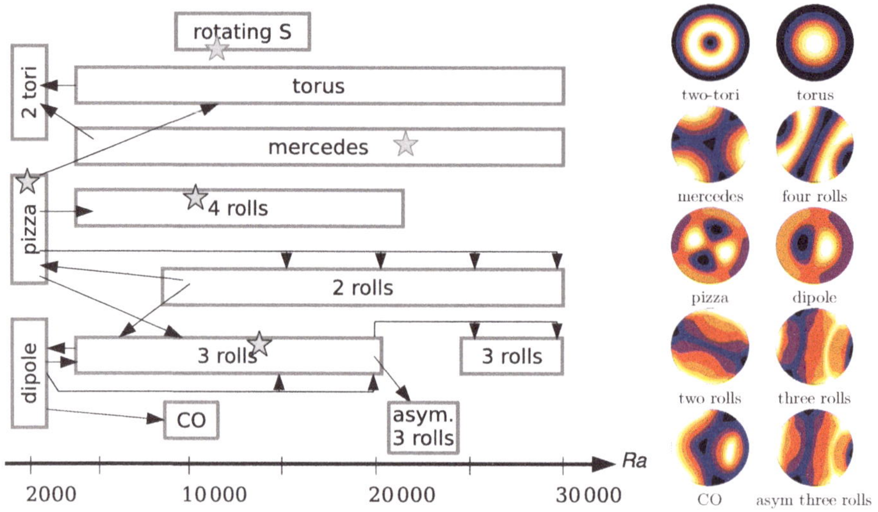

Fig. 11.2 Overview of the patterns observed in time-dependent simulations of convection in a cylinder with insulating radial wall. Initial simulations used a small perturbation of the conductive state at the Rayleigh numbers indicated by stars and lead to the patterns listed on the boxes. The Rayleigh number was then raised or lowered, either gradually, or abruptly as indicated by the arrows. The panels on the right show the temperature field in the midplane, with rising warm fluid indicated by light colors and descending cold fluid by dark colors. Reproduced from [3]

in Fig. 11.3. As complicated as it appears, it is still incomplete: many other branches exist that have not been followed and so have been omitted.

Below the convective threshold, the system has symmetry $O(2) \times Z_2$, derived from the azimuthal rotation and reflection symmetry of the cylinder, and the Boussinesq reflection symmetry. Basic principles of bifurcation theory then dictate that branches bifurcating from the conductive state must have a trigonometric dependence on θ with some azimuthal wavenumber m, and Fig. 11.4 shows that this is indeed the case.

In Figs. 11.5 and 11.6, we follow the branches emanating from the $m = 0$ (axisymmetric) and $m = 3$ bifurcations. Figure 11.5 shows surprising and specific features: the branch that bifurcates from the conductive state disappears again after undergoing a saddle-node bifurcation, and a second disconnected set of axisymmetric states exists. Figure 11.6 shows the tortuous path taken by the $m = 3$ branch, from its birth at a bifurcation from the conductive state (breaking the $O(2)$ symmetry) followed by a second pitchfork bifurcation (breaking the remaining Z_2 symmetry) and then two saddle-node bifurcations, finally leading to the mercedes states.

We now explain the computational method by which the bifurcation diagrams in Figs. 11.3, 11.4, 11.5 and 11.6 have been calculated. Steady states are solutions to

$$0 = \mathcal{F}(U) \tag{11.5}$$

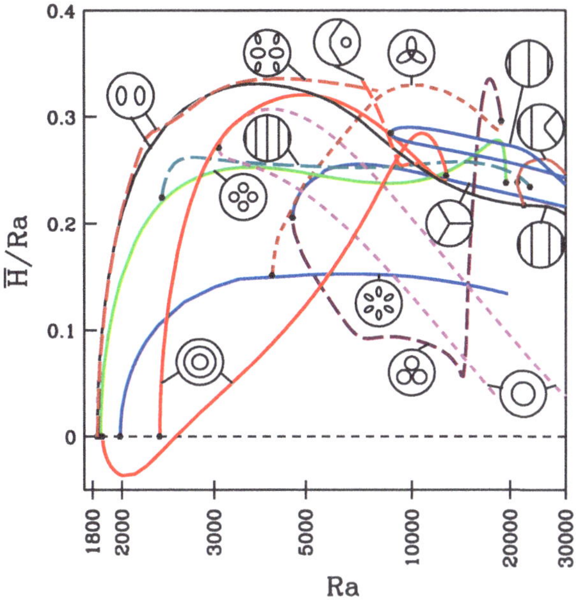

Fig. 11.3 Bifurcation diagram for cylindrical convection. There are 17 branches of steady states, as well as the conductive branch (shown as the short-dashed horizontal line). We call the branches pizza (solid green), four-roll (long-dashed turquoise), two-tori (solid red; 2), torus (long-dashed magenta; 2), marigold (solid blue), mitsubishi (short-dashed purple), cloverleaf (long-dashed purple) and mercedes (solid blue), three-roll (solid black), tiger (long-dashed brick), asymmetric three-roll (solid brick; 2), two-roll (solid blue; 2), and CO (long-dashed red), where (·; 2) indicates related pairs of branches. Dots indicate turning points or pitchfork bifurcations. No information about stability is provided in this diagram. Reproduced from [3]

Fig. 11.4 Primary bifurcations from conductive state. The first four critical wavenumber and Rayleigh numbers are ($m = 1$, $Ra = 1828$; black), ($m = 2$, $Ra = 1849$; green), ($m = 0$, $Ra = 1861$; red), and ($m = 3$, $Ra = 1985$; blue) for the radius-to-height aspect ratio or 2 with insulating radial walls. Reproduced from [3]

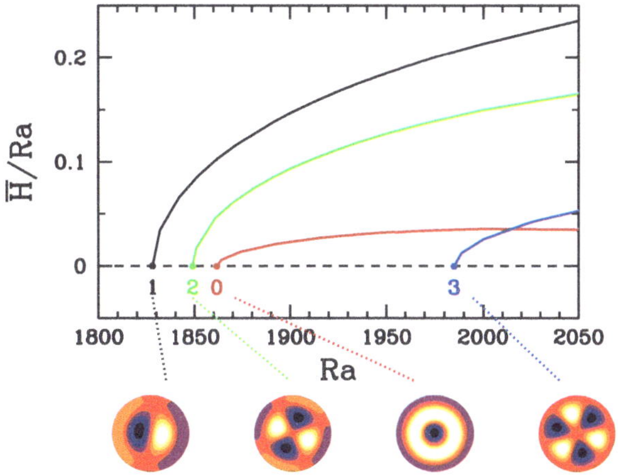

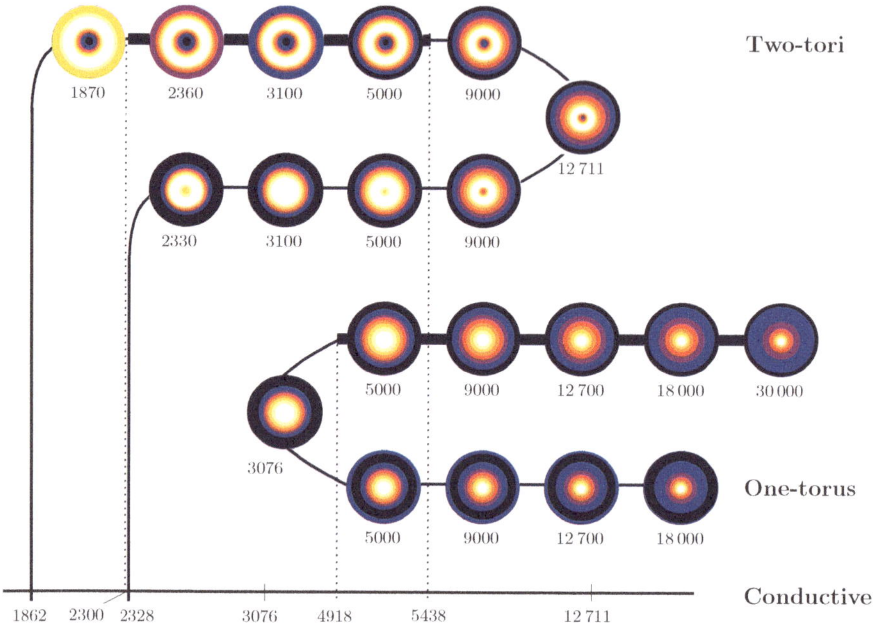

Fig. 11.5 Axisymmetric branches displayed via a schematic partial bifurcation diagram. Pitchfork bifurcations at $Ra = 1862$ and 2328 from the conductive state lead to the *two-tori* branches, which are themselves connected via a saddle-node bifurcation at $Ra = 12\,711$; the upper one is stable for $2300 \leq Ra \leq 5438$. A turning point at $Ra = 3076$ leads to the *one-torus* branches; the upper one is stable for $Ra \geq 4918$. Reproduced from [3]

The main way to find steady states, i.e. to solve (11.5), is via Newton's method

$$\begin{aligned} \mathcal{F}_U u &= \mathcal{F}(U) \\ U &\leftarrow U - u \end{aligned} \tag{11.6}$$

where the subscript U designates, here and elsewhere, the linearization of an operator about U. The main challenge is the solution of the linear system (11.6). Indeed, calculating the solutions of linear systems and the eigenvectors of large matrices is the main bottleneck in numerical computations for studying dynamical systems.

Combining (11.1) and (11.6), Newton's method calls for solving the linear system

$$(\mathcal{L} + \mathcal{N}_U)\, u = (\mathcal{L} + \mathcal{N})U \tag{11.7}$$

where $\mathcal{N}_U$ is the linearization of $\mathcal{N}$ about the steady state U. Linear systems can be solved directly or iteratively. For the calculations shown in Figs. 11.5 and 11.6, the domain is represented by nearly 10^5 points and the state vector is of size $M = 4 \times 10^5$. Inverting a matrix of this size directly is usually impossible for a discretized

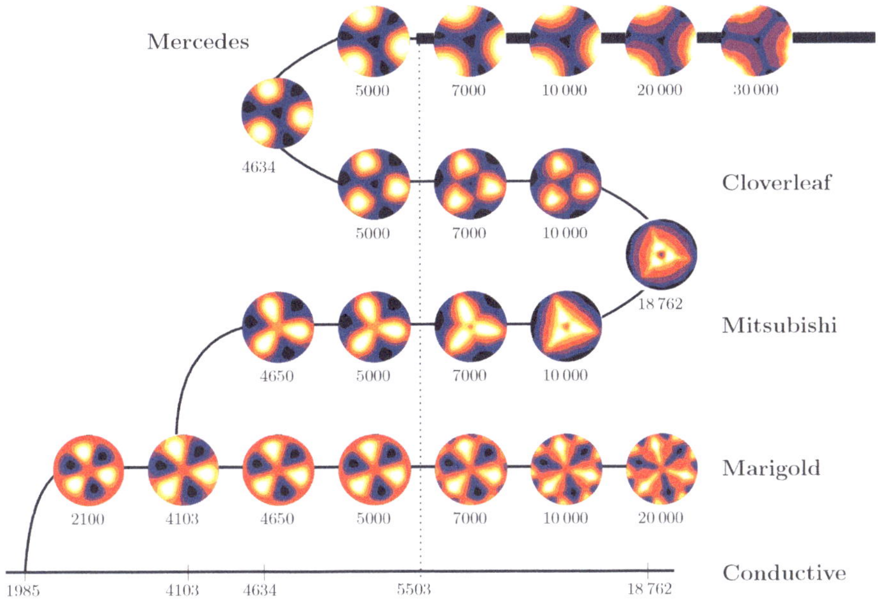

Fig. 11.6 Branches originating from the $m = 3$ bifurcation displayed via a schematic partial bifurcation diagram. The *Marigold* branch is created via an $m = 3$ circle pitchfork bifurcation from the Conductive branch at $Ra = 1985$. A pitchfork bifurcation at $Ra = 4103$, creates the *Mitsubishi* branch. A saddle-node at $Ra = 18\,762$ produces the *Cloverleaf* branch, and another saddle-node at $Ra = 4634$ leads to the *Mercedes* branch. The thick line for $Ra > 5503$ indicates the only stable regime, a portion of the Mercedes branch. Reproduced from [3]

PDE in two or three spatial dimensions, since inversion takes a time proportional to M^3.

The class of iterative methods called Krylov methods generalize the conjugate gradient (CG) method to matrices that are not symmetric definite and include the algorithms GMRES [4], BiCGSTAB [5] and IDR [6]; interested readers should consult these references. These methods construct a solution from the Krylov vectors, formed by taking successive products of the matrix and the right-hand-side vector. In order to simplify the notation, we write the problem we wish to solve as

$$Ax = b \tag{11.8}$$

The general idea is to act repeatedly with A in order to form vectors

$$b, Ab, A^2b, \ldots A^{K-1}b \tag{11.9}$$

These vectors—or orthonormalized versions of them—are called Krylov vectors and their linear combinations comprise the Krylov space. Krylov methods approximate the solution x within the Krylov space:

$$x \approx \sum_{k=0}^{K-1} c_k A^k b \tag{11.10}$$

In the worst-case analysis, if the number K of Krylov vectors attains the size M of the vectors, then generically the Krylov space spans $\mathcal{R}^M$ and x can always be expressed via (11.10). Since a general matrix-vector product requires $O(M^2)$ operations, the operation count for constructing M Krylov vectors is $O(M^3)$, the same as for a direct method.

The challenge is thus to make the Krylov method less time-consuming. A first savings is realized from the fact that matrices derived from spatially discretizating PDEs are not arbitrary: the time required to act with them (rather than inverting them) is typically closer to M than to M^2. A second economy is realized by reducing $K \ll M$. The number K of Krylov vectors required to approximate the solution depends on the condition number of the matrix, basically the ratio of its largest to its smallest eigenvalue. By *preconditioning*, i.e. by acting on both sides of (11.8) with an operator that approximates the inverse of A, the ratio between maximum and minimum eigenvalues is reduced. The matrix in (11.7) is badly conditioned, primarily (in the cases of interest to us) because $\mathcal{L}$ is. We multiply (11.7) by $(I - T\mathcal{L})^{-1}\, T$, where T is a large timestep. This acts to counterbalance $\mathcal{L}$ and leads to

$$\begin{aligned}
(I - T\mathcal{L})^{-1}\, T\, (\mathcal{L} + \mathcal{N}_U)u &= (I - T\mathcal{L})^{-1}\, T\, (\mathcal{L} + \mathcal{N})U \\
(I - T\mathcal{L})^{-1}(I + T\mathcal{N}_U - I + T\mathcal{L})u &= (I - T\mathcal{L})^{-1}(I + T\mathcal{N} - I + T\mathcal{L})U \\
\left[(I - T\mathcal{L})^{-1}(I + T\mathcal{N}_U) - I\right] u &= \left[(I - T\mathcal{L})^{-1}(I + T\mathcal{N}) - I\right] U \\
\left[\mathcal{B}_{U,T} - I\right] u &= \left[\mathcal{B}_T - I\right] U
\end{aligned} \tag{11.11}$$

where $\mathcal{B}_{U,T} \equiv (I - T\mathcal{L})^{-1}(I + T\mathcal{N}_U)$. (We will omit subscripts when they seem unnecessary) Equation (11.11) shows that steady states, which are the roots of $(\mathcal{L} + \mathcal{N})$, are also roots of the operator $(\mathcal{B}_T - I)$, where $\mathcal{B}_T$ is obtained from a single Euler timestep (11.2), for *any* value of T. A large value of T, on the order of $T = 100$ or 1000, serves to insure that $(I - T\mathcal{L})^{-1}\, T \approx -\mathcal{L}^{-1}$. Solving (11.11) using method BiCGSTAB for the cylindrical convection problem above then requires a much smaller (30–800) number of Krylov vectors K than the system size $M = 4 \times 10^5$. This method, called Stokes preconditioning (because $\mathcal{L}$ is associated with the Stokes problem in hydrodynamics), has been presented and used in, e.g. [7–15].

Another method for finding steady states also relies on time-integration, using the operator $\mathcal{B}_{\Delta t}^{T/\Delta t} - I$, meaning that $T/\Delta t$ timesteps are taken with a timestep of conventional size, e.g. $\Delta t \approx 0.01$ for a typical non-dimensionalized hydrodynamic stability problem. Since for $\Delta t \ll 1$,

$$\begin{aligned}
U(\Delta t) \approx B_{\Delta t} U(0) &\Longrightarrow U(T) \approx (B_{\Delta t})^{T/\Delta t}\, U(0) \\
U(T) - U(0) &\approx \left((B_{\Delta t})^{T/\Delta t} - I\right) U(0)
\end{aligned}$$

steady states are also roots of $\left((B_{\Delta t})^{T/\Delta t} - I\right)$ (in the limit $\Delta t \to 0$), and the corresponding Newton step is

$$\left((B_{U,\Delta t})^{T/\Delta t} - I\right) u = \left((B_{\Delta t})^{T/\Delta t} - I\right) U \tag{11.12}$$

We call this the integration method. Each operator action of (11.12) requires $T/\Delta t$ timesteps, rather than the single timestep required by the action of $\mathcal{B}_T - I$. However $\left((B_{\Delta t})^{T/\Delta t} - I\right)$ is better conditioned than $\mathcal{B}_T - I$ and so fewer Krylov vectors are required to represent the solution.

Theories depicting transitional turbulence as chaotic trajectories visiting unstable steady states and traveling waves [19] have inspired the calculation of a large number of such states, mostly by method (11.12), e.g. [16]. (In this context, unstable steady

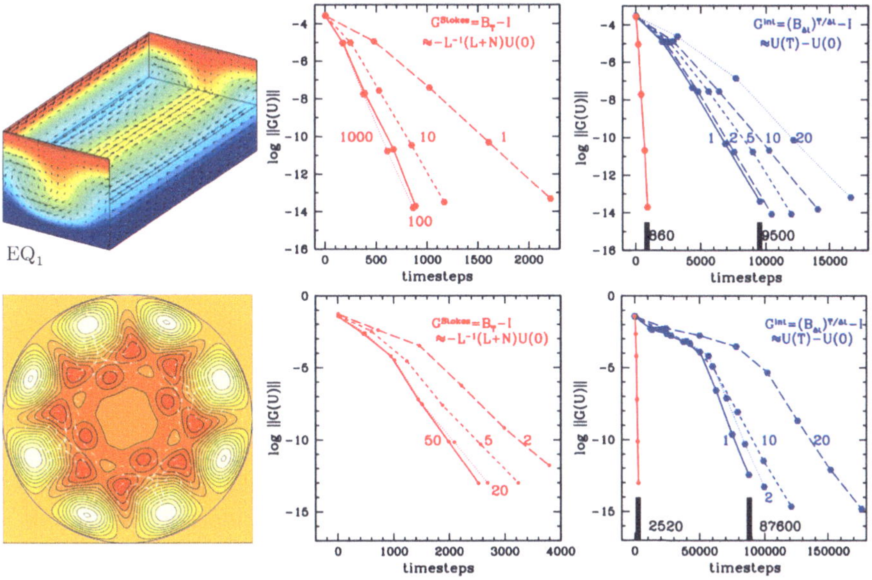

Fig. 11.7 Upper row: Steady state in plane Couette flow. Lower row: Traveling wave in pipe flow. Left column: depiction of flow. Middle and right columns: Performance of Newton's method, with timings measured by the number of timesteps taken. U is considered to be a steady state when $||G(U)|| < 10^{-13}$. Dots show the number of timesteps at the end of each Newton step, while numbers show the time T used in either a single large timestep (Stokes method) or for multiple timesteps (integration method). Middle column: performance of Newton's method using the Stokes method (11.11). Convergence is much faster when T is increased, showing the effectiveness of Stokes preconditioning. Right column: performance of Newton's method using the integration method (11.12). The Stokes method is 11 times faster than the integration method for the plane Couette flow calculation and 50 times faster for the pipe flow calculation. Top left panel is reproduced from [16] (©2009 by Cambridge University Press), bottom left panel is reproduced from [17] (©2008 by the Royal Society) and the rest of panels are reproduced from [18] (©2019 by Springer International Publishing)

states are often called Exact Coherent Structures.) In [18] a comparison of methods (11.12) and (11.11) has been carried out for the computation of unstable steady states for plane Couette flow via Channelflow [20] and traveling waves for pipe flow via OpenPipeFlow [21]. This comparison is summarized in Fig. 11.7. We find that the tradeoff between fewer operator actions for $(B_{\Delta t})^{T/\Delta t} - I$ vs. fewer timesteps for each operator action for $\mathcal{B}_T - I$ is such that the $\mathcal{B}_T - I$ method is 10 to 50 times faster in the Reynolds-number range that we have studied. However, some time-integration codes rely on Δt being small (in particular, to impose incompressibility), making them difficult to incorporate into a single-timestep Newton's method.

Traveling waves can be calculated using the same methods as steady states. Writing $U = U(x - Ct)$, U is a solution to

$$\partial_t U = -CU = \mathcal{F}(U) \tag{11.13}$$

i.e. (U, C) is a solution to

$$0 = CU + \mathcal{F}(U) \tag{11.14}$$

An additional condition on the phase must be imposed to compensate for the additional variable C. Although traveling waves can be computed using either (11.11) or (11.12), general periodic orbits cannot use (11.11), since the entire periodic orbit must be integrated. That is, one must find roots of $U(T) - U(0)$ and not of $\mathcal{F}(U)$ or $CU + \mathcal{F}(U)$.

11.3 Linear Stability Analysis

In the bifurcation diagrams presented in Figs. 11.5 and 11.6, thick lines denote sections of branches in which the steady states are linearly stable. Unlike states computed via time-integration, which are necessarily stable, those computed by Newton's method can also be unstable. How does one then determine their stability? The linear stability of a steady state is controlled by the eigenvalues λ_j of $\mathcal{F}_U$. If none of the eigenvalues have positive real parts, then the steady state is stable. Otherwise it is unstable to perturbations of the form of the eigenvectors corresponding to the eigenvalues whose real parts are positive. In Fig. 11.8a, we show the leading eigenvalues corresponding to the axisymmetric states of Fig. 11.5. For the same reasons as for the conductive state, bifurcation-theoretic principles require that the corresponding eigenvectors each be trigonometric in the azimuthal direction, with a single wavenumber m. Another bifurcation-theoretic principle states that a new bifurcating branch (with azimuthal wavenumber m) is generated via a circle pitchfork at each zero-crossing of these eigenvalue curves. This is one of the reasons that we know that many other branches exist in addition to those shown in the already quite complex Fig. 11.3. Figure 11.8b shows the single leading eigenvalue (that with largest real part) corresponding to the states in Fig. 11.6. States for which this eigenvalue

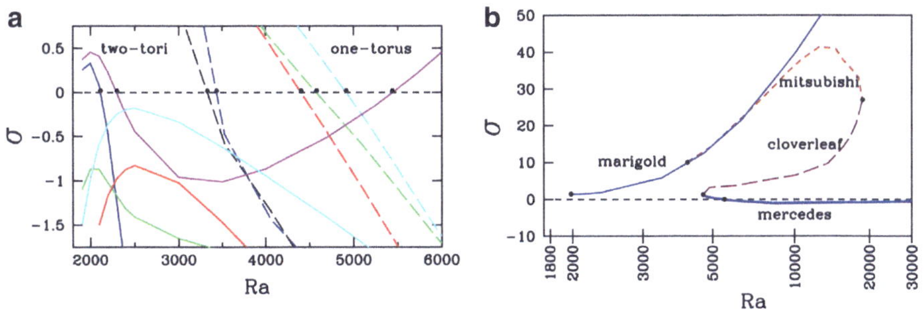

Fig. 11.8 **a** Leading eigenvalues of upper two-tori (solid) and one-torus (dashed) branches. The upper two-tori branch has two positive eigenvalues at onset, which subsequently cross zero, leading to stability over the interval $2300 \leq Ra \leq 5438$. The upper one-torus branch has five positive eigenvalues at onset. These subsequently cross zero, leading to stability for $Re > 4918$, above which the branch is stable. Each zero-crossing is associated with the formation of a new branch. **b** Leading eigenvalue for each of the three-fold-symmetric branches: Marigold, Mitsubishi, Cloverleaf, Mercedes (from highest to lowest). Dots indicate bifurcations between the branches and final stabilization of Mercedes branch for $Ra > 5503$. Reproduced from [3]

is positive are unstable. From this figure it can be seen that all of the $m = 3$ states are unstable until the leading eigenvalue of the Mercedes branch finally becomes negative, as also seen in Fig. 11.6.

We now discuss the method by which these eigenvalues are computed. The most straightforward way to conduct linear stability analysis is to form the matrix $\mathcal{F}_U$ and to diagonalize it, i.e. to compute all of its eigenvalues. If, however, (11.1) consists of one or more partial differential equations in two or three spatial dimensions, then U and u are on the order of size $100^2 = 10^4$ or $100^3 = 10^6$, and so the matrix $\mathcal{F}_U$ is too large to diagonalize completely. The solution is to calculate only the leading eigenvalues, i.e. those whose real part is the largest and their eigenvectors. To do so, we turn again to Krylov methods.

Like the Krylov methods for solving linear equations, those for finding selected eigenvalues rely on repeated operator action to build up a Krylov space. The eigenvalues that are easiest to find are the dominant ones, those of largest magnitude. The simplest such method is the power method, in which only one vector is retained, which converges to the dominant eigenvector. Generalizations, e.g. the block power method, converge to the dominant eigenvectors which, for dissipative systems like the Navier–Stokes equations, correspond to highly damped modes of $\mathcal{F}_U$ rather than the bifurcating modes that are of interest. We can apply a mapping to $\mathcal{F}_U$, using the fact that this maps the eigenvalues of $\mathcal{F}_U$ to their images under this mapping, while leaving the eigenvectors unchanged. Thus, we continue to use the power method or the block power method, but on a different matrix whose dominant eigenvalues are those we seek.

Two such operator mappings are illustrated in Fig. 11.9. Consider the two operators

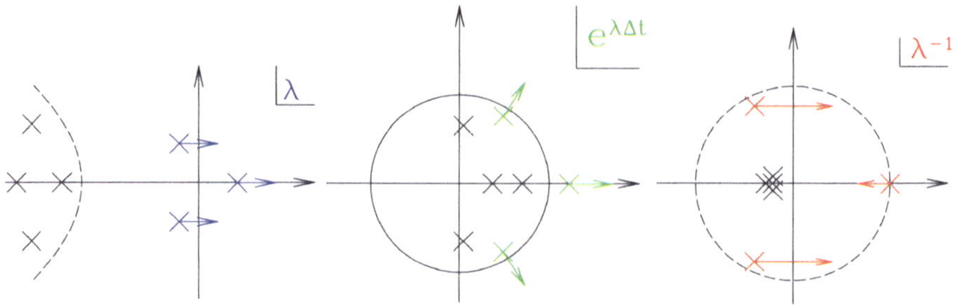

Fig. 11.9 Spectrum of a matrix A, along with that of $e^{A\Delta t}$ and A^{-1}. We seek the leading eigenvalues of A (blue crosses) but these are difficult to obtain via the power or block power methods because they are not the dominant ones. When the exponential or inverse is applied to A, the leading eigenvalues of A become the dominant ones of $e^{A\Delta t}$ (green crosses) and of A^{-1} (red crosses) and are hence much easier to obtain

$$\exp(\Delta t(\mathcal{L} + \mathcal{N}_U)) \quad \text{and} \quad (\mathcal{L} + \mathcal{N}_U)^{-1} \tag{11.15}$$

We approximate the action of these operators on a vector $u^{(n)}$ by using variants of the Euler time-stepping operator (11.2). The exponential power method uses

$$u^{(n+1)} = \mathcal{B}_{U,\Delta t} u^{(n)} \equiv (I - \Delta t \mathcal{L})^{-1}(I + \Delta t \mathcal{N}_U) u^{(n)} \approx \exp(\Delta t(\mathcal{L} + \mathcal{N}_U)) u^{(n)} \tag{11.16}$$

for small Δt. The inverse power method uses

$$(\mathcal{L} + \mathcal{N}_U) u^{(n+1)} = u^{(n)} \tag{11.17}$$

which, as seen in (11.11), is equivalent to

$$\left[(I - T\mathcal{L})^{-1}(I + T\mathcal{N}_U) - I\right] u^{(n+1)} = (I - T\mathcal{L})^{-1} u^{(n)} \tag{11.18}$$

Equations (11.16) and (11.18) are analogous to the two methods described in (11.12) and (11.11) for computing steady states.

For simplicity, we have presented the power method, which computes a single eigenvalue-eigenvector pair. Both the exponential power method or the inverse power method can be generalized to compute a number of eigenpairs, by keeping a number K of iterates of the matrix as follows. We renumber the first retained iterate to be u_0. The vectors $\{u_0, \ldots u_{K-1}\}$ are orthonormalized to yield a set $\{v_0, \ldots v_{K-1}\}$, whose span is the Krylov space. A $K \times K$ matrix is defined via $H_{i,j} \equiv \langle v_i, Av_j \rangle$. The small matrix H captures the action of the full matrix A on the Krylov space if Av_{K+1} does not venture too far outside the Krylov space, i.e. if there is some choice of coefficients $\{c_j\}$ for which $||Av_{K-1} - \sum_{j=0}^{K-1} c_j v_j||$ is sufficiently small. In this case the K eigenvalues of H are candidates for approximate eigenvalues of A. To capture complex eigenvalues, K must be chosen such as to include both members of the pair.

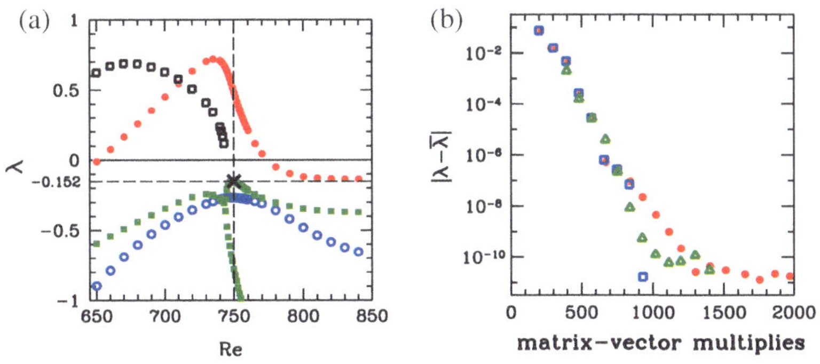

Fig. 11.10 **a** Four leading eigenvalues for spherical Couette flow. Two eigenvalues form a complex conjugate pair for $Re < 744$; their real part is shown as solid green squares and the imaginary part as hollow black squares. The cross at $(Re = 750, \lambda = -0.152)$ shows the targeted eigenvalue. **b** Convergence of error $|\lambda - \bar{\lambda}|$ for various Krylov space dimensions and spatial resolutions. Solid red circles: $K = 2$ and spatial resolution 16×128. Hollow green triangles: $K = 4$ and spatial resolution 16×128. Hollow blue squares: $K = 2$ and 32×256. Reproduced from [22]

The eigenvalues in Fig. 11.8 were obtained using the exponential power method (11.16), i.e. timestepping the linearized equations. This method is reliable and easy but, because the approximation in (11.16) requires a small timestep Δt, the operator resembles the identity and so the iteration proceeds slowly. In contrast, iteration using (11.18) is extremely rapid, converging to the eigenvector corresponding to the eigenvalue closest to zero, typically in 10 or fewer actions of the inverse. Each inverse action requires the solution of (11.18), typically via GMRES, BiCGSTAB or IDR. A shift is easily included in (11.18) by subtracting sI from the explicit term $\mathcal{N}_U$, thus finding eigenvalues close to s instead of 0. Figure 11.10a shows eigenvalues computed for spherical Couette flow (flow between differentially rotating concentric spheres) using different shifts via the inverse power method (11.18); see [22]. Figure 11.10b, like Fig. 11.7, shows the speed of convergence in terms of number of actions of $\mathcal{B}_T$, and establishes that it is unchanged when the spatial resolution is increased, thus demonstrating the effectiveness of Stokes preconditioning.

A complex shift can be used to target eigenvalues on or close to the imaginary axis, which are those that lead to Hopf bifurcations. However, it is not clear how to scan large portions of the imaginary axis, if the frequencies being sought are unknown.

11.4 Floquet Analysis

We now discuss the stability of periodic orbits $U(t)$, i.e. solutions such that $U(t + T) = U(t)$. Infinitesimal perturbations u to a periodic orbit $U(t)$ are governed by the equation

$$\partial_t u = \mathcal{F}_{U(t)} u \tag{11.19}$$

which is a Floquet problem, i.e. a linear differential equation whose coefficients are periodic in the independent variable, here time. Solutions to (11.19) are of the form

$$u(x,t) = e^{\lambda t}\hat{v}(x, t \bmod T) \tag{11.20}$$

where x is used to designate any possible spatial dependence, λ is a Floquet exponent, and $\hat{v}$ is a Floquet function. As is the case for ordinary linear stability problems, there are as many pairs $(\lambda, \hat{v})$ as there are degrees of freedom in U. $U(t)$ is stable if none of the Floquet exponents λ have positive real part. Otherwise, $U(t)$ is unstable to perturbations of the form of the Floquet function $\hat{v}(x, t \bmod T)$ corresponding to λ.

Floquet problems cannot generally be solved analytically, even when U is a scalar and (11.19) is an ordinary differential equation. Floquet problems can be solved numerically in a variety of ways. One way is to integrate (11.19) starting from an initial condition $u(t = 0)$ that is a unit vector, simultaneously with the determination of the limit cycle $U(t)$ by integrating the original problem (11.1). Assembling all the resulting vectors $u(t = T)$ leads to the monodromy matrix, whose eigenvalues are the Floquet multipliers $e^{\lambda T}$.

As is the case for ordinary stability problems, for problems with two or three spatial dimensions, it is usually too onerous to assemble and diagonalize the full monodromy matrix. The usual procedure is then to use Krylov methods, which compute a subset of the leading eigenvalues via a number K of matrix actions where K is much smaller than the total size of the system. The matrix action in this case is the integration of (11.19) over a time T.

We now present a sample of some large-scale Floquet problems that have been solved computationally. When a fluid layer is subjected to vertical oscillations of a sufficient amplitude, a standing-wave pattern forms on the surface, as was first studied by Faraday in 1831. The flat surface is a solution; the imposed vertical oscillation means that its destabilization is described by a Floquet problem. In an infinite horizontally homogeneous fluid layer, a solution is the product of a plane wave, an exponential, and a Floquet function

$$\zeta(\mathbf{x},t) = e^{i\mathbf{k}\cdot\mathbf{x}}\zeta(t) = e^{i\mathbf{k}\cdot\mathbf{x}}e^{\lambda t}\sum_n \zeta_n e^{in\omega t} \tag{11.21}$$

where ζ is the interface height, $\omega = 2\pi/T$ is the frequency of the imposed vertical oscillation and $\mathbf{k}$ is the wavevector of the plane wave that is a candidate for instability. The sum in (11.21) is the temporal Fourier series of the unknown T-periodic Floquet function. When viscosity is absent, the problem reduces to the Mathieu equation [23]:

$$\ddot{\zeta} + \left(gk + \frac{\sigma}{\rho}k^3\right)\zeta = a\cos(\omega t)\zeta \tag{11.22}$$

where a is the acceleration associated with the imposed oscillation, g is the gravitational acceleration, σ is the surface tension, and ρ is the density.

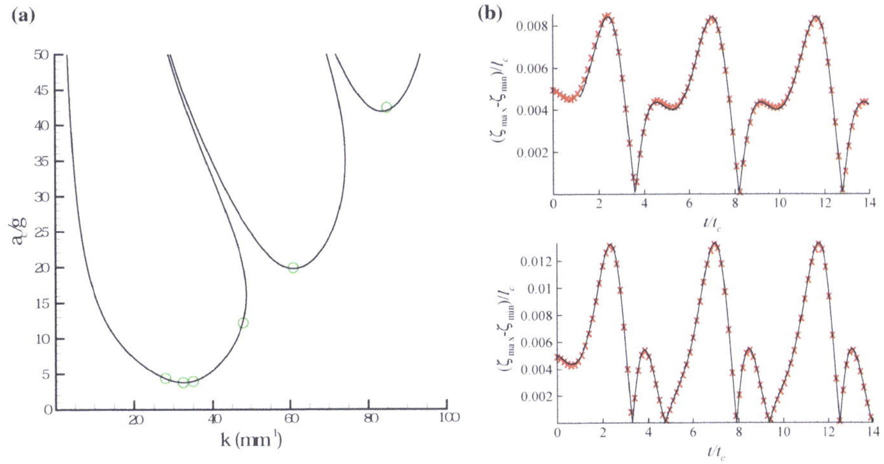

Fig. 11.11 **a** Critical acceleration for the onset of Faraday waves as a function of the wavenumber. The solid curves represent the neutral curves obtained by Floquet analysis, while the values found via full nonlinear simulation are indicated by the circles. **b** Linear evolution of the surface height of Faraday waves, in the first two instability tongues. Floquet functions are plotted with the solid line, while results from full nonlinear simulation near threshold are plotted with symbols. Reproduced from [25]

This Floquet problem has been solved by an original method [24]: rather than fixing the oscillation amplitude a and calculating the resulting Floquet exponents, the Floquet exponent λ is set to 0 (the harmonic case) or to $i\omega/2$ (the subharmonic case) thus determining the threshold a_c directly as follows. In terms of the Fourier coefficients ζ_n, (11.22) becomes

$$\left((\lambda + in\omega)^2 + \left(gk + \frac{\sigma}{\rho}k^3\right)\right)\zeta_n = a\,[\cos(\omega t)\zeta\,]_n \tag{11.23}$$

where $[\cdot]_n$ means the nth component of the expression within. Equation (11.23) is an eigenvalue problem with the vector of temporal Fourier coefficients $\{\zeta_n\}$ as the eigenvector, a as the eigenvalue, a diagonal matrix on the left-hand-side, and a bi-diagonal matrix on the right-hand-side. The inclusion of viscosity complicates the diagonal matrix on the left-hand-side but the basic idea of the method remains the same. Results obtained using this method are shown in Fig. 11.11. These are compared with the threshold calculated from a full three-dimensional free-surface code [25].

One of the most classic hydrodynamic configurations is the wake of a circular cylinder in a uniform flow. Above a critical Reynolds number, a temporally periodic two-dimensional flow develops, in which vortices of alternating sign are generated close to the cylinder and advected downstream, a flow called the Bénard–von Kármán vortex street. Above a second critical Reynolds number, this spanwise-independent

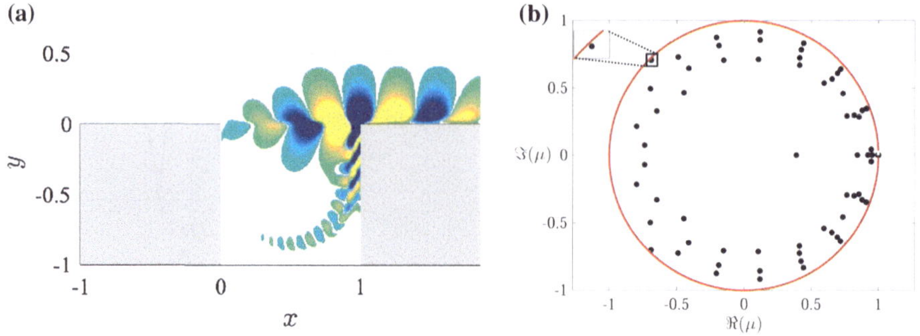

Fig. 11.12 **a** Instantaneous visualization of a temporally periodic state in an open shear-driven cavity at $Re = 4500$. The vertical velocity fluctuations travel rightwards over the cavity. **b** Floquet multipliers of the periodic state in panel **a**, emphasizing the multiplier that traverses the unit circle at a slightly higher Reynolds number, thus destabilizing this periodic state. Reproduced from [26]

temporally periodic flow $\mathbf{U}_{2D}(x, y, t \bmod T)$ becomes unstable to a spanwise periodic mode. This instability is governed by a Floquet problem, whose leading mode can be written as

$$\mathbf{u}_{3D}(x, y, z, t) \sim e^{i\beta z} e^{\lambda t} \mathbf{v}(x, y, t \bmod T) \tag{11.24}$$

Barkley and Henderson [27] showed that the Bénard–von Kárman vortex street becomes unstable at $Re = 188$ to a Floquet mode with spanwise wavelength of about four cylinder diameters and a Floquet exponent $\lambda = 0$, meaning that no new time dependence is introduced into the resulting spanwise-dependent flow.

Another classic configuration is that generated by a uniform flow over a square cavity. Above a critical Reynolds number, a new flow appears that is temporally periodic and approximately spatially periodic in the streamwise direction, shown via its vertical velocity fluctuations in Fig. 11.12a. A Floquet analysis [26] shows that above a second critical Reynolds number, this flow is in turn destabilized via a Floquet mode with a complex Floquet exponent, as shown in Fig. 11.12b.

11.5 Heteroclinic Orbits

Among the more exotic phenomena displayed by dynamical systems are heteroclinic orbits: limit cycles whose period is infinite. Although the theory and understanding of heteroclinic orbits is well-grounded, they cannot be seen experimentally, since they are broken by any perturbation, and in fact can barely be seen computationally. Heteroclinic orbits can be divided into two broad categories. In one case, they occur at a single parameter value, at one end of the parameter range of existence of a limit cycle. Such a heteroclinic cycle is the signature of a global bifurcation that creates or destroys a limit cycle that, at neighboring parameter values, has a long but finite period. These cycles evade experimental observation because they occur only at isolated parameter values. Their existence is inferred from the divergence of the nearby periodic orbits. The other case is that of heteroclinic cycles that are robust, meaning that they exist over a range of parameters. These heteroclinic cycles result from symmetry properties. They evade experimental observation because symmetries in nature are rarely exact.

11.5.1 SNIPER Bifurcations

One mechanism by which heteroclinic cycles appear is via a SNIPER (Saddle-Node In a PERiodic orbit, or Saddle-Node-Infinite-PERiod bifurcation) bifurcation [29, 30]. On one side of the bifurcation, there exists a closed curve consisting of a chain of alternating stable and unstable fixed points connected by trajectories of the system. (In the simplest case, there is a single pair of stable and unstable fixed points; in a symmetric system, additional pairs exist.) The stable and unstable points meet and annihilate one another in saddle-node bifurcations, leaving in their wake a closed curve that forms one continuous closed trajectory, i.e. a limit cycle. At a bifurcation point ϵ_c, the period of the limit cycle is infinite; slightly beyond ϵ_c, the period behaves like $(\epsilon - \epsilon_c)^{-1/2}$.

A heteroclinic cycle was first observed for a full hydrodynamic configuration in a simulation of Rayleigh–Bénard convection in a cylinder [28]. In this case, the configuration is axisymmetric and the radius is five times the height. Two successive pitchfork bifurcations lead to four steady states, each consisting of a set of four or five concentric convection rolls, as shown in Fig. 11.13. When the Rayleigh number is increased past the SNIPER bifurcation point, the rolls begin to travel radially inwards. The innermost roll shrinks and disappears, while a new roll appears at the outer boundary.

SNIPER bifurcations have been computed in other systems. In another axisymmetric cylindrical convective configuration, the top and bottom disks bounding the domain are rotated at equal and opposite speeds (the von Kármán flow described below in Sect. 11.5.2), in addition to having fixed and different temperatures [31]. The height and radius are equal, with one large concentric roll filling almost the entire domain. During the cycle a small roll appears alternatively at the outer radial boundary, alternating between the top and the bottom of the cylinder. The limit cycle exists in a region of the (Reynolds number, Rayleigh number) parameter plane delimited by a SNIPER bifurcation curve and a Hopf bifurcation curve. SNIPER bifurcations have also been observed in simulations of other variants of cylindrical Rayleigh–Bénard convection [32, 33], and have been hypothesized to play a role in the reversals of the earth's magnetic field [34].

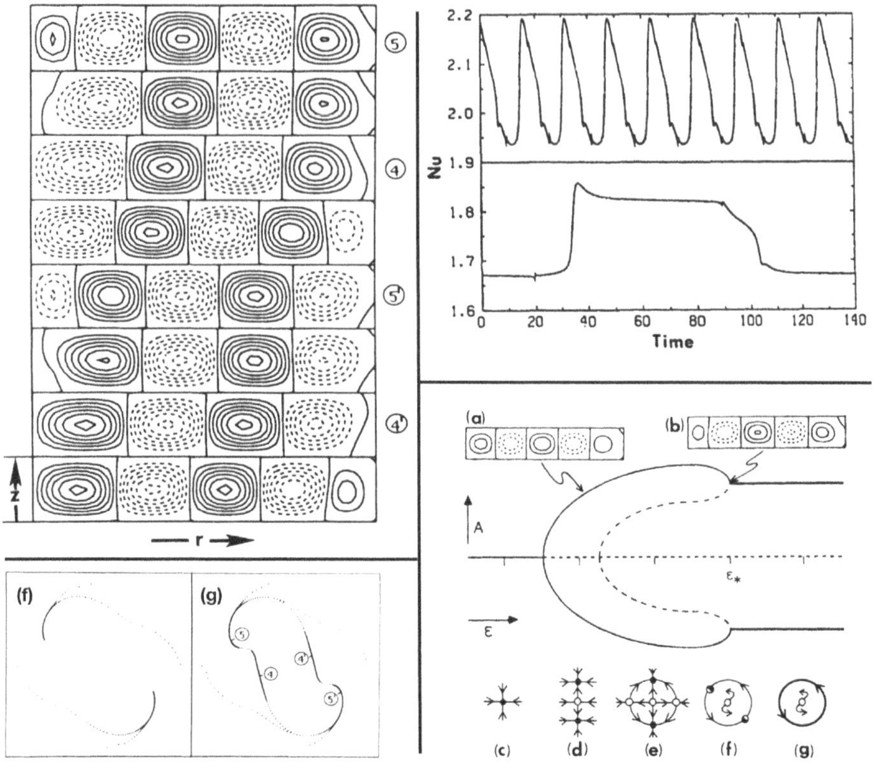

Fig. 11.13 SNIPER bifurcation in axisymmetric cylindrical convection. Upper left: instantaneous streamfunction during the limit cycle at reduced Rayleigh number $\epsilon \equiv (Ra - Ra_c) = 1.39$ at successive times. The numbers label slowly varying quasi five-roll and quasi four-roll states. Upper right: Nusselt number time series for $\epsilon = 2.60$ (above) and for $\epsilon = 1.39$ (below). The period diverges at the SNIPER bifurcation. Lower right: Schematic bifurcation diagram using reduced Rayleigh number ϵ as the control parameter and a projection A of the states as order parameter. Above (**a**, **b**) are numerically calculated streamfunction contours of representative five-roll steady states. Below (**c**–**g**) are phase portraits at the values of ϵ denoted by tick marks. Solid lines and filled circles denote stable states, while dashed lines and hollow circles denote unstable states. The straight bold curve represents the limit cycle formed in the SNIPER bifurcation. Lower left: Phase portraits using as coordinates the projections onto the two most unstable eigenvectors of the conductive state. Letters correspond to the schematic phase portraits shown below the schematic bifurcation diagram on the right. **f** is computed at $\epsilon = 1.38$, for which all trajectories terminate at one of two steady states. **g** is computed at $\epsilon = 1.39$, for which the SNIPER bifurcation has led to a limit cycle. The numbers (4, 5, 4′, 5′) describe the instantaneous states corresponding to those in the streamfunction visualizations above. Adapted from [28]

11.5.2 1:2 Mode Interaction

Turning now to robust heteroclinic cycles, i.e. those that exist over a range of parameters, one important mechanism is provided by the 1:2 mode interaction. This scenario occurs in a configuration with $O(2)$ symmetry, when pitchfork bifurcations to modes with azimuthal wavenumbers $m = 1$ and $m = 2$ modes occur at close parameter values. In this case, Armbruster et al. [35] predict the existence of $m = 1$ and $m = 2$ steady states, traveling waves, modulated traveling waves, and robust heteroclinic cycles. These cycles are trajectories between two $m = 2$ states that differ by a change in phase of $\pi/2$. Several full hydrodynamic computations have displayed realizations of this 1:2 mode interaction scenario. Nore et al. [13] investigated the flow in a cylinder whose height is twice the radius and whose two bounding disks cylinder rotate in equal and opposite directions, sometimes called von Kármàn flow. The basic state is axisymmetric and consists of an axial shear of the azimuthal velocity combined with two superposed toroidal cells driven by Ekman pumping and separated at the midplane between the top and bottom disks. The flow undergoes circle pitchfork bifurcations that break the axisymmetry, leading to an undulation in the separating surface and to radially oriented vortices.

Figure 11.14a shows results of linear stability analysis, in particular, the $m = 1$ and $m = 2$ modes (called mixed M and pure P, respectively, because the higher harmonics of $m = 1$ contain both odd and even wavenumbers, while those of $m = 2$ contain only even wavenumbers). The eigenmodes are represented visually via the vertical velocity at the midplane. Figure 11.14b shows the sequence of bifurcations in this case, adapted from those predicted in [35]. The mixed branch engenders traveling waves (via a drift pitchfork bifurcation) and then modulated traveling waves (via

Fig. 11.14 The 1:2 mode interaction in counter-rotating von Kármán flow. **a** Eigenvalues corresponding to azimuthal wavenumbers $m = 1$ and 2 of the basic state for $Re < Re_P$, the threshold of bifurcation to the pure mode state, and of the pure mode state for $Re > Re_P$. Contours of vertical velocity of each eigenvector are shown. **b** Schematic bifurcation diagram showing bifurcations: circle pitchforks from the basic branch to mixed (M) and pure (P) branches, drift to traveling waves (TW), Hopf to modulated waves (MWH), pitchforks (MP, M'P) connecting mixed and pure branches. The heteroclinic orbit is stable between thresholds Het and M'P. Timeseries of traveling waves and the heteroclinic cycle are shown. Adapted from [13]

a Hopf bifurcation). The branch of pure states is unstable at onset, but engenders a heteroclinic cycle that consists of alternation between states that are $\pi/2$ out of azimuthal phase from one another.

Returning to Fig. 11.14a, many of these transitions coincide with the crossings of eigenvalues, as explained in the Armbruster et al. [35] scenario. The left part ($Re \leq 401$) shows the result of linearization about the basic state; the mixed and pure branches are created via pitchfork bifurcations when the $m = 1$ and $m = 2$ eigenvalues cross zero. The right part ($Re \geq 401$) shows the result of linearization about the pure states. Azimuthal phase becomes significant and so the eigenmodes of the basic state each split into those symmetric (S) and antisymmetric (A) with respect to the pure mode. The $m = 2$ eigenmodes split into the classic negative amplitude and neutral phase modes. The $m = 1$ eigenvalues decrease; the mixed branch is annihilated when the eigenvalue associated with 1 S becomes negative, and when its magnitude surpasses that of the still-positive eigenvalue associated with 1A, the heteroclinic cycle becomes stable. Excursions consist of growth along the 1 A direction and decay along the 1 S direction. When the eigenvalue associated with 1 A becomes negative, the pure mode becomes stable. This case shows the power of combining full nonlinear time-dependent simulations with computational linear stability analysis.

Heteroclinic cycles associated with the 1:2 mode interaction have been computed in other hydrodynamic systems. Mercader et al. [37] simulated non-Boussinesq two-dimensional Rayleigh–Bénard convection in a rectangle of width π (between the favored wavelengths of approximately 2 and 4). Bengana and Tuckerman [36] simulated Taylor–Couette flow between counter-rotating cylinders and discovered two very different cycles, one whose excursions resemble the non-axisymmetric azimuthally traveling ribbon state, and another whose excursions remain axisymmetric, shown via the phase portraits in Fig. 11.15. Both cycles are based at the same saddles and the crossover between them corresponds to the crossover between two leading eigenvalues, that corresponding to 1 A and a complex eigenvalue correspond-

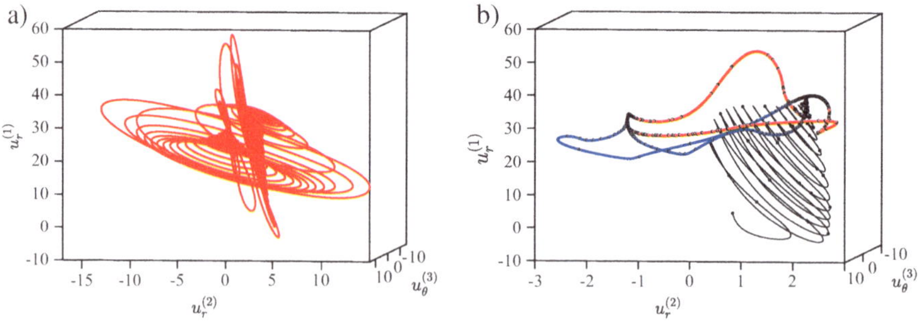

Fig. 11.15 Phase portraits of two heteroclinic orbits in counter-rotating Taylor–Couette flow. **a** Excursions are non-axisymmetric and spiral in and out of the saddles. **b** After a spiralling initial transient, the excursions are non-oscillatory and chosen quasi-randomly from several possible trajectories. Both cycles are anchored by the same saddles. Reproduced from [36]

ing to a non-axisymmetric eigenmode. In all of these computations, the cycles may behave in a regular or slightly irregular way. Their period is not infinite, so strictly speaking, they should be called near-heteroclinic cycles.

All of the cycles described above were calculated using only time-integration. However, there exist more specialized numerical methods aimed specifically at calculating heteroclinic cycles [38–40].

11.6 Hamiltonian Systems: Bose–Einstein Condensation

A Bose–Einstein condensate can be represented by a complex wave function $\Psi(\mathbf{x}, t)$ that obeys the Gross–Pitaevskii equation

$$-i\partial_t\Psi = \left[\frac{1}{2}\nabla^2 - V(x) - a|\Psi|^2 + \mu\right]\Psi \tag{11.25}$$

where a is the nondimensionalized scattering length and

$$V(x) \equiv \tfrac{1}{2}|\boldsymbol{\omega}\cdot\mathbf{x}|^2 \tag{11.26}$$

is a confining harmonic potential. The Lagrange multiplier μ is associated with the constraint that the particle number $\mathcal{N} \equiv \int|\Psi|^2 d^3x$ be kept constant. (In this section, we use $\mathcal{N}$ to denote particle number, in contrast to Sects. 11.1–11.3, in which it denotes the nonlinear terms).

We have considered a potential with cylindrical symmetry, i.e. with inverse length scales $\omega_x = \omega_y = \omega_r$ and ω_z, in either the pancake ($\omega_r < \omega_z$) or cigar ($\omega_r > \omega_z$) configuration. Equation (11.25) is distinguished from the equations considered in the previous sections by the presence of the imaginary i. As a Hamiltonian system, its bifurcations are different from those of dissipative systems. With the cylindrically symmetric potential, we find that (11.25) exhibits a Hamiltonian saddle-node bifurcation. A saddle and a center coexist and annihilate one another at a critical value of μ or, equivalently, $\mathcal{N}$.

The search for steady states is not affected by the presence of i; we merely solve

$$0 = \left[\frac{1}{2}\nabla^2 - V(\mathbf{x}) - a|\Psi|^2 + \mu\right]\Psi \tag{11.27}$$

This is done via the same technique as in Sect. 11.2, with $\mathcal{L}\Psi = \frac{1}{2}\nabla^2\Psi$ and $\mathcal{N} = -(V(x) + a|\Psi|^2 - \mu)\Psi$. These steady states are shown for the pancake case via their energy

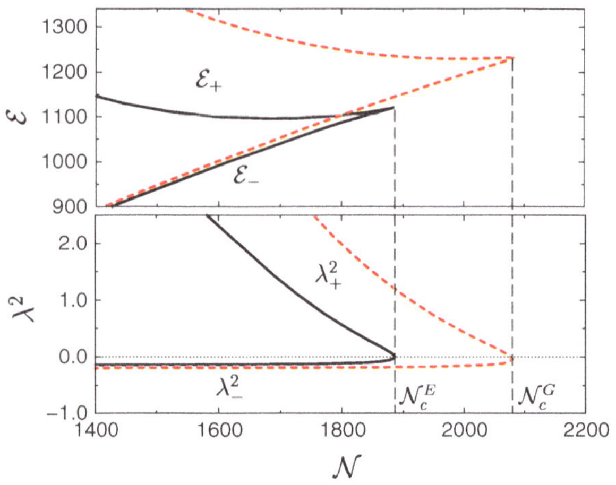

Fig. 11.16 Stationary solutions of the GP equation as a function of the particle number $\mathcal{N}$ for a potential (11.26) of pancake form with $\omega_z = 5\omega_r$. Energy functional (above) and square of the bifurcating eigenvalue $\lambda^2_\pm$ (below). Exact numerical solution (black solid lines) is compared with Gaussian approximation (red dashed lines). Reproduced from [41]

$$\mathcal{E} \equiv \int d^3x \left[\frac{1}{2}|\nabla\Psi|^2 + V(\mathbf{x})|\Psi|^2 + \frac{a}{2}|\Psi|^4 \right] \tag{11.28}$$

in the upper part of Fig. 11.16. We mention some other methods that have been used to compute such states. For a spherically symmetric potential $\omega_r = \omega_z$, the wave-function is one-dimensional, so that linear systems arising in Newton's method are much smaller and can easily be solved directly. One method approximates the functional dependence of $\Psi(\mathbf{x})$ as Gaussian; this yields a problem without any spatial dependence in which the few scalar parameters of this approximation are computed via Newton's method. Figure 11.16 includes the resulting branches, which are qualitatively, but not quantitatively, the same as the exact solutions. Another method, called relaxation or imaginary time, integrates

$$\partial_t \Psi = \left[\frac{1}{2}\nabla^2 - V(x) - a|\Psi|^2 + \mu \right] \Psi \tag{11.29}$$

turning the system into a dissipative one. However, this method does not find the branches that are unstable under the evolution of (11.29), and so cannot obtain the full diagram of Fig. 11.16.

We now turn to the stability of the real steady states Ψ of (11.25). The eigenmodes $(\lambda, \psi_R, \psi_I)$ satisfy

$$\lambda \begin{pmatrix} \psi_R \\ \psi_I \end{pmatrix} = \begin{bmatrix} 0 & -(\mathcal{L} + DW_I) \\ \mathcal{L} + DW_R & 0 \end{bmatrix} \begin{pmatrix} \psi_R \\ \psi_I \end{pmatrix} \tag{11.30}$$

where

$$\mathcal{L}\Psi \equiv \frac{1}{2}\nabla^2\Psi, \qquad DW_R \equiv \mu - V(\mathbf{x}) - 3a\Psi^2, \qquad DW_I \equiv \mu - V(\mathbf{x}) - a\Psi^2. \tag{11.31}$$

The eigenvalues occur in equal and opposite pairs and the pair closest to zero determines the nature of each branch. For states on the stable elliptic branch $\mathcal{E}_-$, this pair is imaginary, so λ^2 is negative; $|\lambda_-|$ is the energy of excitations around this branch. Along the unstable hyperbolic branch $\mathcal{E}_+$, this pair is real, so that λ_+^2 is positive. At the saddle-node bifurcation, these eigenvalues meet at zero. All of the other eigenvalues are imaginary pairs.

In order to calculate the critical eigenvalue pair, we have used the inverse power method (see Sect. 11.3). Moreover, we square the matrix, which leads to a block diagonal matrix whose blocks both have the same eigenvalues λ^2:

$$\lambda^2 \begin{pmatrix} \psi_R \\ \psi_I \end{pmatrix} = \begin{bmatrix} -(\mathcal{L} + DW_I)(\mathcal{L} + DW_R) & 0 \\ 0 & -(\mathcal{L} + DW_R)(\mathcal{L} + DW_I) \end{bmatrix} \begin{pmatrix} \psi_R \\ \psi_I \end{pmatrix} \tag{11.32}$$

Preconditioning by $\mathcal{L}^{-2}$, we carry out the inverse square iteration via:

$$-\mathcal{L}^{-2}(\mathcal{L} + DW_I)(\mathcal{L} + DW_R)\psi_R^{(n+1)} = \mathcal{L}^{-2}\psi_R^{(n)} \tag{11.33}$$

These methods and results are described in detail in [11, 41].

In a separate investigation, Ψ is prolonged to include a second complex component and a periodic lattice component is added to the confining potential

$$V(r, z) = \frac{1}{4}|\boldsymbol{\omega} \cdot \mathbf{x}|^2 + A\left[\sin^2(2\omega_r x) + \sin^2(2\omega_r y)\right] \tag{11.34}$$

leading to a problem that is not axisymmetric. Using the steady-state methods described above, we have computed a non-axisymmetric solution consisting of stable symbiotic vortex-bright solitons shown in Fig. 11.17; see [42].

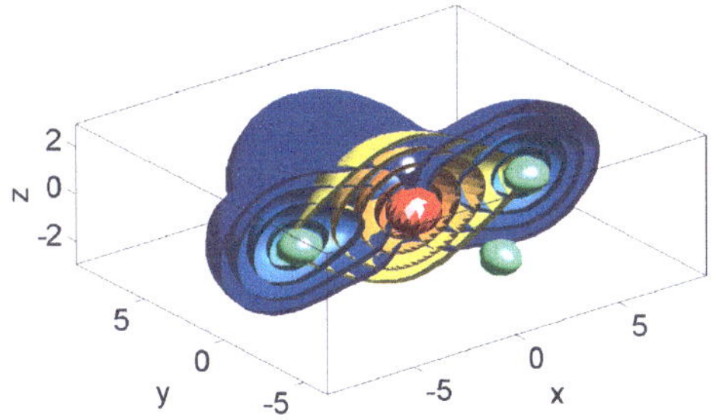

Fig. 11.17 Stable symbiotic vortex-bright structure in the presence of an optical lattice with potential (11.34) with $\omega_z = 5\omega_r$. Surfaces of constant density are shown in blue for the vortex and in yellow for the soliton. This 3D stationary state is stabilized by the second component, which displaces the vortex component at its core. Reproduced from [42]

11.7 Looking Ahead

In this chapter, we have described a number of dynamical-systems phenomena, namely steady states and limit cycles; bifurcations, instability and Floquet modes; and heteroclinic cycles, that occur in many nonlinear physical systems. Many of these phenomena, such as unstable steady states or heteroclinic orbits, are accessible only to computation.

The numerical simulations described in the previous sections have uncovered several new dynamical-systems scenarios. Faraday waves provide an example for which full nonlinear three-dimensional simulations have led to the discovery of new phenomena. First, in simulations of Faraday waves in a minimal hexagonal domain [43, 44], the hexagons give way after many subharmonic oscillation periods to a pattern we have called beaded stripes. These are succeeded in turn by quasi hexagons, and then by regular alternation between asymmetric beaded stripes and quasi hexagons, as shown in Fig. 11.18. The bifurcation-theoretic genesis of this complicated scenario is unknown. Secondly [45], a large square domain containing Faraday waves spontaneously divides into a two-by-two grid in which the square waves are in phase with those diagonally opposite, as shown in Fig. 11.19. Finally, Faraday waves on the surface of a drop induced by a radially directed oscillatory force display patterns [46], many of which resemble Platonic solids, as expected [47]. However, some of the patterns we observe undergo a slow and long-lasting drift in orientation, such as the axisymmetric pattern of Fig. 11.20. This feature

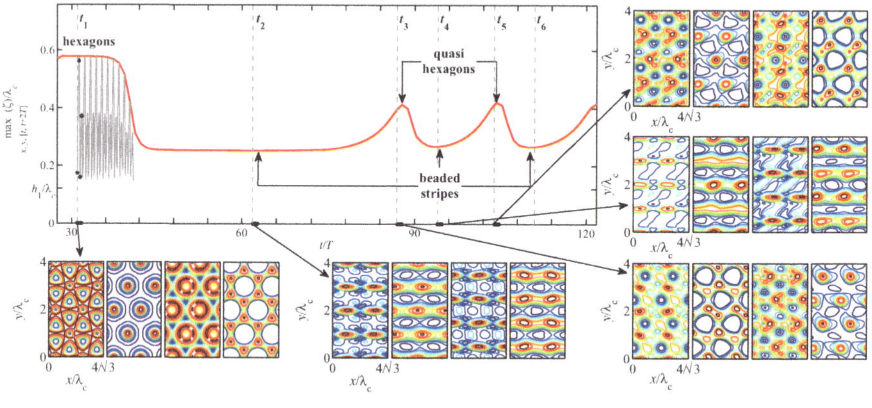

Fig. 11.18 Complex dynamics beginning with hexagons at time t_1, beaded stripes at time t_2, quasi-hexagons at t_3 and t_5, and asymmetric beaded stripes at t_4 and t_6. Curves show maxima of interface height, with the rapidly oscillating curve showing $\max_{x,y} \zeta(x, y, t)$ for initial times and the smooth curve showing $\max_{x,y,[t,t+T]} \zeta(x, y, t)$. Surrounding visualizations show instantaneous contour plots of $\zeta(x, y, t)$ at four temporal phases within one oscillation period. The size of the box has been doubled in each dimension. Over the large white areas, the interface is very close to the bottom and almost flat. Reproduced from [43].

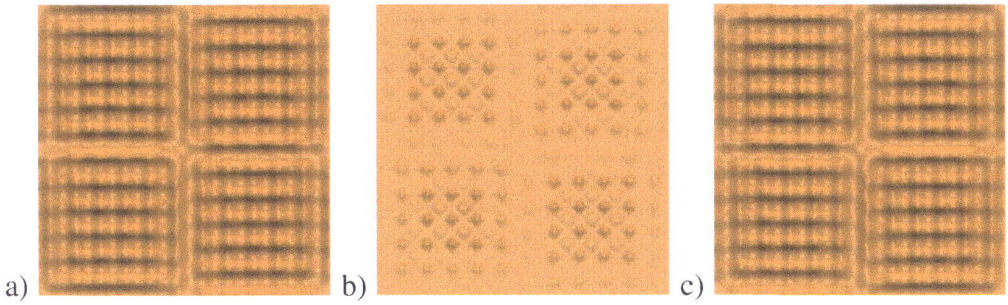

Fig. 11.19 Supersquare pattern at instants separated by intervals of $T/2$, where T is the forcing period and $2T$ is the subharmonic response period. In **a** and **c**, wells are surrounded by ridges with peaks at each corner, while in **b**, the squares are centered around low peaks. The large squares on the bottom left and top right are in phase, as are those on the bottom right and top left. Reproduced from [45]

Fig. 11.20 Axisymmetric $\ell = 4$ pattern seen in capillary waves during one subharmonic response period $2T$. The interface sometimes resembles a top and sometimes a finite cylinder. Each fluid drop is surrounded by its spherical domain. The magnitude of the velocity is indicated by the colors. Reproduced from [46]

remains unexplained. The configurations in Figs. 11.19 and 11.20 were computed with a massively parallel multiphase time-integration code [48]. Although bifurcation analysis has been carried out for thin films [49], most tools for numerical bifurcation analysis have not yet been applied to full three-dimensional nonlinear free-surface problems.

In most quantitative sciences, a surprisingly important bottleneck is that of numerical linear algebra, and this is certainly the case for nonlinear dynamics. The main focus of computational science is time-integration, which has been tamed to yield algorithms whose timing is approximately linear in the size M of the system. Over the years, many monographs, e.g. [50–53], dedicated volumes, e.g. [54–56], review articles, e.g. [15, 57, 58], and software packages such as AUTO [59], DSTOOL [60], PDECont [61], MatCont [62], LOCA [63], JuliaDynamics [64], pde2path [65] have been dedicated to targeting simple or complicated bifurcation-theoretic objects for general dynamical systems. However, such algorithms usually require the solution of linear systems or matrix diagonalization. Straightforward algorithms for these tasks scale like M^3, and if the matrix is derived from a two or three-dimensional partial differential equation then $M = M_x M_y M_z$ may be on the order of 10^6 or more. Krylov space methods can be used, in which the solution is obtained as a superposition of K vectors obtained by matrix actions. The timings for these scale like KM or K^2M, so the crucial challenge is to keep K low.

A number of strategies for speeding up the iterative solution of linear systems have been proposed in this chapter, more specifically transforming the matrix or the nonlinear problem to one that is more tractable. These include using the exponential or the inverse of the operator and preconditioning by the Laplacian. An extension of this approach is as follows. Consider a steady-state problem containing a linear operator $\mathcal{L}$ and a bilinear operator $\mathcal{N}$.

$$0 = \mathcal{F}(U) \equiv \mathcal{L}U + \mathcal{N}(U, U) \tag{11.35}$$

In order to compute u, an improvement to the estimate U, Newton's method requires us to solve

$$\begin{aligned} \mathcal{F}_U u &= F(U) \\ \mathcal{L}u + \mathcal{N}(U, u) + \mathcal{N}(u, U) &= \mathcal{L}U + \mathcal{N}(U, U) \end{aligned} \tag{11.36}$$

where u is a decrement to U that should decrease $||\mathcal{L}U + \mathcal{N}(U)||$. The Stokes method described in Sect. 11.2 preconditions (11.36) with $\mathcal{L}^{-1}$

$$\mathcal{L}^{-1}\left[\mathcal{L}u + \mathcal{N}(U, u) + \mathcal{N}(u, U)\right] = \mathcal{L}^{-1}\left[\mathcal{L}U + \mathcal{N}(U, U)\right] \tag{11.37}$$

It is easy to generate other decompositions of (11.35) that lead to other preconditioners. For any fixed field U_{base}, we may write

$$U = U_{\text{base}} + \tilde{U} \tag{11.38}$$

The operator whose roots are sought is unchanged, but it now formally acts on the unknown $\tilde{U}$

$$F(\tilde{U}) = \mathcal{L}U_{\text{base}} + \mathcal{L}\tilde{U} + \mathcal{N}(U_{\text{base}}, U_{\text{base}}) + \mathcal{N}(U_{\text{base}}, \tilde{U}) + \mathcal{N}(\tilde{U}, U_{\text{base}}) + \mathcal{N}(\tilde{U}, \tilde{U}) \tag{11.39}$$

and has Jacobian

$$DF_{\tilde{U}}\tilde{u} = \mathcal{L}\tilde{u} + \mathcal{N}(U_{\text{base}}, \tilde{u}) + \mathcal{N}(\tilde{u}, U_{\text{base}}) + \mathcal{N}(\tilde{U}, \tilde{u}) + \mathcal{N}(\tilde{u}, \tilde{U}) \tag{11.40}$$

where $\tilde{u}$ is a decrement to $\tilde{U}$ that should decrease $F(\tilde{U})$. If U_{base} is sufficiently simple, it may be possible to invert

$$\mathcal{L} + \mathcal{N}(U_{\text{base}}, \cdot) + \mathcal{N}(\cdot, U_{\text{base}}) \tag{11.41}$$

The inverse of (11.41) may be a much better preconditioner for F than $\mathcal{L}^{-1}$ since it includes parts of $\mathcal{N}$ that were previously left out of the preconditioning. As an example of this approach, [18] discusses the computation of steady states and traveling waves in wall-bounded shear flows. For such flows, there is a known laminar base flow $\mathbf{U}_{\text{base}} = U(y)\mathbf{e}_x$ for which

$$(\mathbf{U}_{\text{base}} \cdot \nabla)\,\mathbf{u} + (\mathbf{u} \cdot \nabla)\,\mathbf{U}_{\text{base}} = U(y)\partial_x \mathbf{u} + u_y \partial_y U(y)\mathbf{e}_x \tag{11.42}$$

Incorporating one or both of these terms into the Stokes preconditioner would almost surely greatly lower the number K of Krylov vectors needed.

Another way in which the search for steady states may be accelerated is by applying an operator or functional $\mathcal{G}$ to $\mathcal{F}$, transforming the steady-state problem to

$$0 = \mathcal{G}(0) - \mathcal{G}(\mathcal{F}(U)) \tag{11.43}$$

Again, this is advantageous if the Jacobian of $\mathcal{G}(\mathcal{F})$, i.e. $\mathcal{G}_{\mathcal{F}}\mathcal{F}_U$ is better conditioned than $\mathcal{F}_U$, or if a better preconditioner for it is available. In fact the two approaches (11.11) and (11.12) constitute a special case of such a transformation, where $\mathcal{G}$ is taken to be the time integral of $\mathcal{F}$

$$\mathcal{G}(\mathcal{F}(U)) \equiv \int_t^{t+T} \mathcal{F}(U(\tau))d\tau \tag{11.44}$$

Similarly, if $\mathcal{F}$ is decomposed into $\mathcal{L} + \mathcal{N}$, as in (11.35), then applying $\mathcal{G}$ leads to the problem

$$0 = \mathcal{G}(\mathcal{L}(U)) - \mathcal{G}(-\mathcal{N}(U)) \tag{11.45}$$

that could be preconditioned by $[\mathcal{G}(\mathcal{L})]^{-1}$ if $\mathcal{G}$ is linear. It would be desirable to popularize such strategies and to augment these by finding new ones, or, even better, to automate the search for an optimal transformation.

References

1. B. Hof, P.G. Lucas, T. Mullin, Phys. Fluids **11**, 2815 (1999)
2. K. Borońska, L.S. Tuckerman, Phys. Rev. E **81**, 036320 (2010)
3. K. Borońska, L.S. Tuckerman, Phys. Rev. E **81**, 036321 (2010)
4. Y. Saad, M.H. Schultz, SIAM J. Sci. Stat. Comput. **7**, 856 (1986)
5. H.A. Van der Vorst, SIAM J. Sci. Stat. Comput. **13**, 631 (1992)
6. P. Sonneveld, M.B. Van Gijzen, SIAM J. Sci. Comput. **31**, 1035 (2008)
7. L.S. Tuckerman, in *11th International Conference on Numerical Methods in Fluid Dynamics*, ed. by D.L. Dwoyer, M.Y. Hussaini, R.G. Voight. Lecture Notes in Physics, vol. 323 (Springer, Berlin, 1989), p. 573
8. C.K. Mamun, L.S. Tuckerman, Phys. Fluids **7**, 80 (1995)
9. S. Xin, P. Le Quéré, L.S. Tuckerman, Phys. Fluids **10**, 850 (1998)
10. A. Bergeon, D. Henry, H. Benhadid, L.S. Tuckerman, J. Fluid Mech. **375**, 143 (1998)
11. C. Huepe, S. Metens, G. Dewel, P. Borckmans, M.E. Brachet, Phys. Rev. Lett. **82**, 1616 (1999)
12. L.S. Tuckerman, D. Barkley, in *Numerical Methods for Bifurcation Problems and Large-Scale Dynamical Systems*, ed. by E. Doedel, L.S. Tuckerman. The IMA Volumes in Mathematics and its Applications, vol. 119 (Springer, New York, 2000), p. 453

13. C. Nore, L.S. Tuckerman, O. Daube, S. Xin, J. Fluid Mech. **477**, 51 (2003)
14. I. Mercader, A. Alonso, O. Batiste, Eur. Phys. J. E **15**, 311 (2004)
15. H.A. Dijkstra, F.W. Wubs, A.K. Cliffe, E. Doedel, I.F. Dragomirescu, B. Eckhardt, A.Y. Gelfgat, A.L. Hazel, V. Lucarini, A.G. Salinger, E.T. Phipps, J. Sanchez-Umbria, H. Schuttelaars, L.S. Tuckerman, U. Thiele, Commun. Comput. Phys. **15**, 1 (2014)
16. J.F. Gibson, J. Halcrow, P. Cvitanović, J. Fluid Mech. **638**, 243 (2009)
17. C.C. Pringle, Y. Duguet, R.R. Kerswell, Philos. Trans. R. Soc. A **367**, 457 (2009)
18. L.S. Tuckerman, J. Langham, A. Willis, in *Computational Modelling of Bifurcations and Instabilities in Fluid Dynamics*, ed. by A. Gelfgat (Springer, Cham, 2019), p. 3
19. P. Cvitanović, B. Eckhardt, J. Phys. A **24**, L237 (1991)
20. J.F. Gibson, Channelflow: a spectral Navier-Stokes simulator in C++ (2014), http://channelflow.org. Accessed 27 Jan 2020
21. A.P. Willis, SoftwareX **6**, 124 (2017)
22. L.S. Tuckerman, Commun. Comput. Phys. **18**, 1336 (2015)
23. T.B. Benjamin, F.J. Ursell, Proc. R. Soc. A **225**, 505 (1954)
24. K. Kumar, L.S. Tuckerman, J. Fluid Mech. **279**, 49 (1994)
25. N. Perinet, D. Juric, L.S. Tuckerman, J. Fluid Mech. **635**, 1 (2009)
26. Y. Bengana, J.C. Loiseau, J.C. Robinet, L.S. Tuckerman, J. Fluid Mech. **875**, 725 (2019)
27. D. Barkley, R.D. Henderson, J. Fluid Mech. **322**, 215 (1996)
28. L.S. Tuckerman, D. Barkley, Phys. Rev. Lett. **61**, 408 (1988)
29. A. Andronov, E. Leontovich, Mat. Sb. **48**, 335 (1959)
30. Y.A. Kuznetsov, *Elements of Applied Bifurcation Theory* (Springer, New York, 2004)
31. L. Bordja, L.S. Tuckerman, L.M. Witkowski, M.C. Navarro, D. Barkley, R. Bessaih, Phys. Rev. E **81**, 036322 (2010)
32. J.M. Lopez, A. Rubio, F. Marques, J. Fluid Mech. **569**, 331 (2006)
33. J.H. Siggers, J. Fluid Mech. **475**, 357 (2003)
34. F. Pétrélis, S. Fauve, E. Dormy, J.-P. Valet, Phys. Rev. Lett. **102**, 144503 (2009)
35. D. Armbruster, J. Guckenheimer, P. Holmes, Physica D **29**, 257 (1988)
36. Y. Bengana, L.S. Tuckerman, Phys. Rev. Fluids **4**, 044402 (2019)
37. I. Mercader, J. Prat, E. Knobloch, Int. J. Bifurc. Chaos **12**, 2501 (2002)
38. J. Halcrow, J.F. Gibson, P. Cvitanović, D. Viswanath, J. Fluid Mech. **621**, 365 (2009)
39. L. Van Veen, G. Kawahara, M. Atsushi, SIAM J. Sci. Comput. **33**, 25 (2011)
40. M. Farano, S. Cherubini, J.C. Robinet, P. De Palma, T.M. Schneider, J. Fluid Mech. **858**, (2019)
41. C. Huepe, L.S. Tuckerman, S. Métens, M.E. Brachet, Phys. Rev. A **68**, 023609 (2003)
42. K.J. Law, P.G. Kevrekidis, L.S. Tuckerman, Phys. Rev. Lett. **105**, 160405 (2010)
43. N. Périnet, D. Juric, L.S. Tuckerman, Rev. Cuba. Fís. **29**, 1E6 (2012)
44. N. Périnet, D. Juric, L.S. Tuckerman, Phys. Rev. Lett. **109**, 164501 (2012)
45. L. Kahouadji, N. Périnet, L.S. Tuckerman, S. Shin, J. Chergui, D. Juric, J. Fluid Mech. **772**, (2015)
46. A.H. Ebo-Adou, L.S. Tuckerman, S. Shin, J. Chergui, D. Juric, J. Fluid Mech. **870**, 433 (2019)
47. P. Chossat, R. Lauterbach, I. Melbourne, Arch. Ration. Mech. Anal. **113**, 313 (1991)
48. S. Shin, J. Chergui, D. Juric, J. Mech. Sci. Technol. **31**, 1739 (2017)
49. P. Beltrame, U. Thiele, SIAM J. Appl. Dyn. Syst. **9**, 484 (2010)
50. M. Kubicek, M. Marek, *Computational Methods in Bifurcation Theory and Dissipative Structures* (Springer, Berlin, 1983)
51. H.B. Keller, *Lectures on Numerical Methods in Bifurcation Problems*. Lectures on Mathematics and Physics Mathematics, vol. 79 (Springer, Berlin, 1988)
52. R. Seydel, *From Equilibrium to Chaos: Practical Bifurcation and Stability Analysis* (Elsevier, Amsterdam, 1988)
53. W.J.F. Govaerts, *Numerical Methods for Bifurcations of Dynamical Equilibria* (SIAM, Philadelphia, 2000)
54. E. Doedel, L.S. Tuckerman (eds.), *Numerical Methods for Bifurcation Problems and Large-Scale Dynamical Systems*. The IMA Volumes in Mathematics and Its Applications, vol. 119 (Springer, New York, 2000)

55. B. Krauskopf, H.M. Osinga, J. Galán-Vioque (eds.), *Numerical Continuation Methods for Dynamical Systems* (Springer Netherlands, Dordrecht, 2007)
56. A.S.D. Roose, B. De Dier, A. Spence (eds.), *Continuation and Bifurcations: Numerical Techniques and Applications*. NATO Science Series C, vol. 313 (Springer Netherlands, Dordrecht, 1990)
57. K.A. Cliffe, A. Spence, S.J. Tavener, Acta Numer. **9**, 39 (2000)
58. E.J. Doedel, in *Numerical Continuation Methods for Dynamical Systems*, ed. by B. Krauskopf, H.M. Osinga, J. Galán-Vioque (Springer Netherlands, Dordrecht, 2007), p. 1
59. E.J. Doedel, Congr. Numer. **30**, 265 (1981)
60. A. Back, J. Guckenheimer, M. Myers, F. Wicklin, P. Worfolk, Not. Am. Math. Soc. **39**, 303 (1992)
61. K. Lust, D. Roose, SIAM J. Sci. Comput. **19**, 1188 (1998)
62. A. Dhooge, W. Govaerts, Y.A. Kuznetsov, A.C.M. Trans, ACM Trans. Math. Softw. **29**, 141 (2003)
63. A.G. Salinger, N.M. Bou-Rabee, R.P. Pawlowski, E.D. Wilkes, E.A. Burroughs, R.B. Lehoucq, L.A. Romero, LOCA 1.0 library of continuation algorithms: theory and implementation manual. Tech. Rep. SAND2002-0396 (Sandia National Laboratories, 2002)
64. G. Datseris, J. Open Source Softw. **3**, 598 (2018)
65. H. Uecker, D. Wetzel, J.D.M. Rademacher, Numer. Math.: Theory Methods Appl. **7**, 58 (2014)

Chapter 12
Dissipative Systems

Edgar Knobloch

Abstract This chapter focuses on pattern formation, both extended and localized, in driven dissipative systems. The origin of localized states and their organization into a snakes-and-ladders structure is explained and the results illustrated with several examples drawn from the physical sciences. The corresponding results for systems with a conserved quantity are described, focusing on crystallization from a supercooled melt via a propagating crystallization front. Unsolved questions are highlighted throughout.

12.1 Introduction

Dissipative dynamical systems are fundamental to the physical sciences. These arise in macroscale descriptions of physical systems with dissipation modeling the transfer of energy, momentum, etc. from the macroscale to the microscale. Because of dissipation, phase space volumes contract, and the long time behavior of the system can be described in terms of attractors. For nontrivial attractors, energy, momentum, etc. have to be supplied at the macroscale, i.e. the systems have to be driven. This chapter is about pattern formation in infinite-dimensional pattern-forming systems, i.e. in continuum systems, an area where many advances have recently taken place. Discrete dynamical systems are not considered. For background reading, the books [1–4] are recommended, as is the review [5].

E. Knobloch (✉)
Department of Physics, University of California at Berkeley, Berkeley, CA 94720, USA
e-mail: knobloch@berkeley.edu

P. G. Kevrekidis et al. (eds.), *Emerging Frontiers in Nonlinear Science*, Nonlinear Systems and Complexity 32, https://doi.org/10.1007/978-3-030-44992-6_12

12.2 Nonconserved Systems

We first discuss the generic case, focusing on pattern-forming systems of the form

$$\mathbf{u}_t = \mathcal{L}\mathbf{u} + N[\mathbf{u}], \tag{12.1}$$

where $\mathbf{u}(\mathbf{x}, t) \in \mathbb{R}^n$ is a vector of state variables, $\mathcal{L}$ is a linear (possibly nonlocal) operator depending on a parameter r and N represents nonlinear terms. Many systems can be written in this form, including reaction-diffusion systems and the equations of fluid mechanics. We assume that the spatial variable $\mathbf{x} \in \mathbb{R}^d$ is continuous.

In the special but important case where $\mathcal{L}$ is an autonomous operator and (12.1) has a trivial solution $\mathbf{u} = 0$ the properties of the latter are described by a dispersion relation

$$\sigma = \sigma(\mathbf{k}; r) \in \mathbb{C} \tag{12.2}$$

that determines the growth rate σ_r and frequency $\sigma_i \equiv \omega$ of wavenumber $\mathbf{k}$ at a given value of the parameter r. Here the subscripts r, i indicate real and imaginary parts. In one spatial dimension $\omega(k)$ determines the phase speed $c_p \equiv \omega/k$ and group speed $c_g \equiv d\omega/dk$ of infinitesimal perturbations of $\mathbf{u} = 0$ with wavenumber k. In systems that are invariant with respect to spatial reflection, $x \to -x$, the waves travel in both directions and an initial disturbance will spread in both positive and negative directions. In cases where this symmetry is broken the waves will have a preferred direction and the disturbance will spread asymmetrically. In both cases the long time behavior of the disturbance depends on whether the domain is bounded so that the disturbance ultimately interacts with the domain boundary, or infinite, in which case it continues to propagate, evolving as it does so. When $\mathcal{L}$ is a dissipative operator the waves ultimately die unless the dissipation is compensated for by forcing that injects energy into the system. Two types of forcing are frequently considered, boundary forcing in which energy is injected into the system at a single location (the boundary) and spatial forcing where energy is injected at every location. An example of the former might be flow driven by a wall oscillating at frequency Ω; such a flow is confined to the vicinity of the wall and decays away from it in a distance that can be computed by solving $i\Omega = \sigma(k)$ in the relevant domain and locating the wavenumber k with the smallest imaginary part k_i. An example of the latter is provided by buoyancy-driven flow above a uniformly heated horizontal plate.

In this review we are interested in describing the effects of the nonlinear term $N[\mathbf{u}]$ in (12.1). In view of the above it will come as no surprise that nonlinear dissipative systems exhibit both spatially localized and spatially extended states. Nonlinearities can generate strongly localized structures (hereafter LSs) via a self-pinning process which steepens the fronts connecting the state to the background and allows locking or pinning of these fronts to heterogeneities within the structure. In the case of extended states, nonlinearities select among the possible spatially periodic structures whose growth is triggered by the loss of stability of $\mathbf{u} = 0$ with respect to symmetry-breaking ($\mathbf{k} \neq 0$) perturbations. The resulting stable periodic structures typically form the core of the possible LS. In generic systems this occurs in the presence of bistability between an extended pattern state and a homogeneous state such as $\mathbf{u} = 0$. For these reasons we consider first spatially periodic patterns and then the properties of fronts that may connect them to the background homogeneous state.

12.2.1 Pattern Formation

In the vicinity of a pattern-forming instability at $r = 0$ the system (12.1) is described by a low-dimensional system of equations for the amplitudes of the unstable modes [6]. Periodic solutions can be found by restricting the dynamics to a periodic lattice in one, two or three dimensions corresponding to the critical wavenumber k_c at $r = 0$. Center manifold reduction can then be employed to reduce (12.1) to a finite-dimensional system of ordinary differential equations for the amplitudes of the (slowly) growing modes. The form of these equations is entirely determined by the symmetries of the problem when posed on a lattice. Consequently, group-theoretic techniques can be used to find all primary solutions with maximal isotropy and to determine their stability in terms of the coefficients in these equations [7]. When truncated the amplitude equations are of polynomial type and may be high-dimensional. Specialized tools have therefore been developed that guarantee that one has found all the solutions of the resulting (truncated) equations [8]. Software enabling the computation of the coefficients for periodic domains or domains equipped with Neumann boundary conditions has also been developed [9] but in more complex cases the coefficients have to be found by solving a series of linear problems numerically following weakly nonlinear theory. These techniques apply to both stationary and oscillating patterns.

In unbounded domains the situation is understood less well. Here, there is usually no spectral gap and stable eigenvalues accumulate on zero from below. These situations are usually studied using formal multiscale modulation equations which are partial differential equations for the slowly varying amplitudes (in space and time) of the critical modes. These equations are often local but in cases with strong advection ($c_g = O(1)$) they may be nonlocal [10]. In one spatial dimension the most common of these equations, the real Ginzburg–Landau equation, has been justified rigorously [11] but in two or more dimensions additional complications arise due to orientational degeneracy, a property that is usually lost when the modulation equations are truncated. The resulting equations reveal that there is a range of wavenumbers for which a pattern may be stable, bounded in one dimension by the Eckhaus stability boundary, and in two dimensions by additional instabilities such as the zig-zag, skewed varicose or oscillatory instabilities [1, 5, 12]. The resulting stability region in the (k, r) plane is called the Busse balloon after the work of F. H. Busse on the stability of convection rolls in two-dimensional Rayleigh–Bénard convection [13]. Similar stability regions are present for traveling and standing waves. The issue of wavenumber selection as r increases represents a major unsolved problem in the theory of pattern formation, and it now appears that in unbounded systems there is no unique answer. This is not so in bounded domains in which the presence of distant boundaries has in general a profound effect on the observed wavenumber in the domain interior [5]. We mention that it may happen that as r increases the system reaches and crosses the Eckhaus boundary. In this case a phase slip is initiated that ultimately restores a stable wavenumber. The process may then repeat, leading to repeated loss or gain of phase, i.e. wavelengths of the pattern, as r varies [14, 15].

Systems with advection in one direction behave differently. Here a localized, slowly growing perturbation near a pattern-forming instability is advected downstream with the result that the perturbation eventually dies out at any fixed location. Such systems are called convectively unstable. If the growth rate is increased sufficiently that the localized perturbation can expand upstream, against the flow, we call the instability absolute. This transition can be computed from the dispersion relation and occurs at $r = r_a$ for which the dispersion relation $\sigma = \sigma(k)$ has a double root in the complex k plane subject to appropriate pinching conditions [16, 17]. At this point the (complex) group velocity c_g vanishes. The real and imaginary parts of k determine the wavenumber of the instability selected by the advection and its growth rate in space. Convectively unstable systems are of interest because they are sensitive to the injection of small amplitude noise at the upstream inlet. The noise is spatially amplified in the downstream direction and may lead to complex noise-sustained structures downstream of the inlet that propagate out of the system when the noise is turned off [18]. The distinction between convective and absolute instability persists in finite domains, with the threshold for the transition to absolute instability shifted by $O(L^{-2})$ when the domain size L is large compared to $2\pi/k_r$. In this case the convective regime is associated with the presence of spatially amplifying transients (when $\mathcal{L}$ is non-normal) that ultimately decay. Thus instability is only present for $r > r_g = r_a + O(L^{-2})$ for which the linear problem possesses a global eigenfunction. In this regime the upstream boundary serves as a pacemaker that selects the frequency ω; the dispersion relation in turn selects the complex wavenumber k that determines the spatial growth rate k_i in the downstream direction and the wavenumber k_r of the downstream pattern. The selected wavenumber may fall outside of the Busse stability region, however, triggering a secondary instability whose evolution once again depends on whether the instability is convective or absolute [19]. In particular, the far-field breakup of target patterns or spirals requires the presence of an absolute secondary instability [20]. In the convectively unstable regime $r < r_g$ the system is highly noise-sensitive, and noise-sustained structures are present, just as in the unbounded case [21, 22]. In contrast when advection is absent, the frequency (if any) is determined by the dominant wavenumber, and noise-sensitivity is absent.

12.2.2 *Localized Patterns*

The above analysis has identified numerous systems where extended patterns coexist with a homogeneous state in both one, two and three dimensions. This bistability region is typically created as a result of a subcritical bifurcation, and we consider here the case where this instability is to a 1D pattern of stripes, or in the context of thermal convection, 2D rolls, since the flow is then confined between a pair of horizontal plates the lower of which is maintained at a higher temperature than the upper. The essence of what happens in this case is captured by the Swift–Hohenberg equation

$$u_t = r\,u - (k_c^2 + \nabla^2)^2\,u + f(u) \tag{12.3}$$

here written in d dimensions. The function $f(u)$ represents a bistable nonlinearity of the form $f(u) = b_2u^2 - u^3$, $b_2 > 0$, and we refer to the resulting equation as SH23. The parameters r and k_c represent the bifurcation parameter and a characteristic wavenumber (inverse lengthscale). In unbounded domains the wavenumber k_c can be set equal to $k_c = 1$ but this is not the case on finite domains. In addition, (12.3) has the minimum number of spatial derivatives for the presence of robust heteroclinic cycles. As a result we shall use it as a "normal form" for understanding systems exhibiting spatially localized structures.

Equation (12.3) is equivariant with respect to the reflection $R_1 : x \to -x, u \to u$ and possesses a trivial state $u = 0$, hereafter referred to as O. In addition it has variational structure, i.e. it possesses a Lyapunov functional $F[u(x,t)]$, such that

$$u_t = -\frac{\delta F}{\delta u}, \tag{12.4}$$

where F is given by

$$F = \int_{-\infty}^{\infty} \mathrm{d}^d x \left[-\frac{1}{2}\,r\,u^2 + \frac{1}{2}\left[(k_c^2 + \nabla^2)\,u\right]^2 - \int_0^u f(v)\,\mathrm{d}v \right]. \tag{12.5}$$

It follows that

$$\frac{\mathrm{d}F}{\mathrm{d}t} = -\int_{-\infty}^{\infty} \mathrm{d}^d x \left(\frac{\partial u}{\partial t}\right)^2 \le 0, \tag{12.6}$$

and hence that $\mathrm{d}F/\mathrm{d}t < 0$ provided $\partial u/\partial t \neq 0$ somewhere in the domain. Thus on a finite domain all solutions evolve towards stationary states; on unbounded or periodic domains solutions in the form of steadily moving fronts are possible. In the following we will think of the functional $F[u]$ as the (free) energy of the system. Stable (unstable) solutions correspond to local minima (saddle points) of this energy. As emphasized below in appropriate parameter regimes the energy landscape described by (12.5) can be exceedingly complex.

Figure 12.1 shows the L^2 norm, $||u|| = \{\int_{-\infty}^{\infty} u^2(x)\,\mathrm{d}x\}^{1/2}$, of the localized states $L_{0,\pi}$ in SH23 in one spatial dimension as a function of the bifurcation parameter r. The L^2 norm (per unit length) of the periodic state, labeled γ, is shown for comparison. The figure shows that the two branches of even parity LS that bifurcate subcritically from O at $r = 0$ enter a shaded region, hereafter the snaking or pinning region, in which they undergo repeated saddle-node bifurcations as they snake

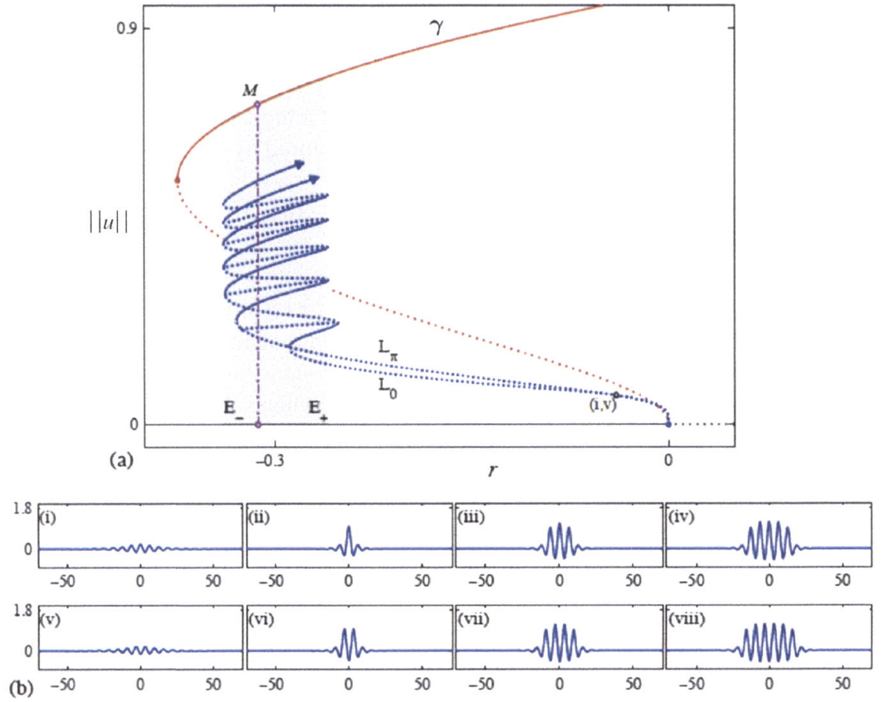

Fig. 12.1 **a** Bifurcation diagram showing the snakes-and-ladders structure of LS in SH23. Away from the origin the snaking branches L_0 and L_π are contained within a snaking region (shaded) between E_- and E_+, where $r(E_-) \approx -0.3390$ and $r(E_+) \approx -0.2593$. Solid (dotted) lines indicate stable (unstable) states. **b** Sample LS profiles $u_L(x)$: (i)–(iv) lie on L_0, near onset and at the 1st, 3rd, and 5th saddle-nodes from the bottom, respectively; (v)–(viii) lie on L_π, near onset and at the 1st, 3rd, and 5th saddle-nodes, respectively. Parameters: $b_2 = 1.8$, $k_c = 1$. Reproduced from [24, 25]

back and forth across the region. These saddle-nodes converge exponentially rapidly to a pair of r values, hereafter $r(E_-)$ and $r(E_+)$, representing the boundaries of the shaded region. The convergence is monotonic and from the right in both cases. The lower panels show a series of profiles of $u_L(x)$ along $L_{0,\pi}$ (Fig. 12.1), starting near $r = 0$ and followed by the profiles at successive saddle-nodes at the right of the pinning region. These reveal that states L_0 are characterized by a peak in the center while L_π have a dip in the center. As one proceeds up either branch each LS nucleates a pair of peaks or cells, one on either side, in the vicinity of $r = r(E_-)$. These grow to the amplitude of the coexisting periodic state γ by the time one reaches the next fold on the right, at $r = r(E_+)$, and the branch turns around to repeat the process. Thus as one proceeds up the intertwined $L_{0,\pi}$ branches the localized states repeatedly add cells on either side while preserving their parity, each back-and-forth oscillation increasing the width of the state by two wavelengths $2\pi/k_c$. On the real

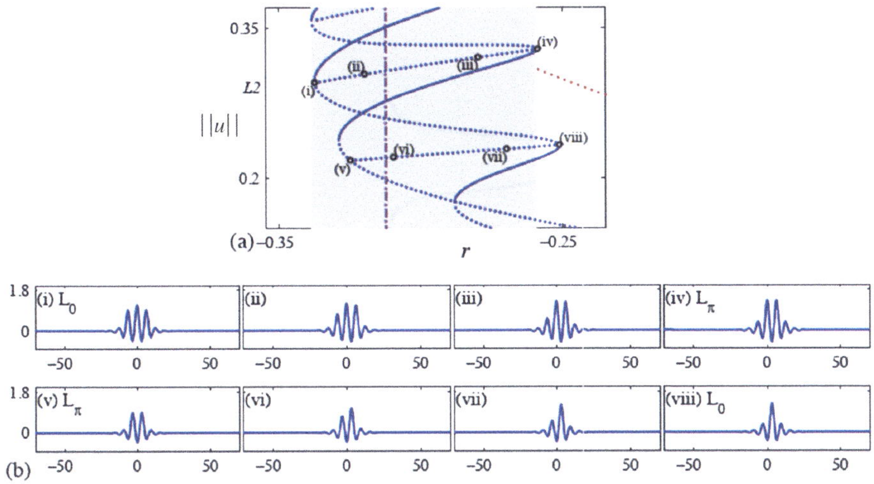

Fig. 12.2 **a** Close-up view of Fig. 12.1a showing two rungs connecting the snaking branches L_0 and L_π. Solid (dotted) lines indicate stable (unstable) states. **b** The profiles (i) and (viii) lie on L_0 while (iv) and (v) lie on L_π. Reproduced from [24, 25]

line this process continues indefinitely as both branches approach the periodic state γ. These results are further explored in the review [23].

Figure 12.2a is a close-up view of Fig. 12.1, focusing on the rung states which connect the $L_{0,\pi}$ snaking branches. These states are asymmetric (Fig. 12.2b), unstable but stationary and are created in pitchfork bifurcations which break the R_1 symmetry of the $L_{0,\pi}$ states. Consequently each rung in the figure corresponds to two states related by R_1 and hence of identical L^2 norm. We refer to the resulting structure as the snakes-and-ladders structure of the pinning region [24, 25].

In addition to these single-pulse states the pinning region also contains a variety of multipulse states [26]. These are generally found on nested isolas (not shown) instead of intertwined branches.

12.2.2.1 Mathematical Explanation of the Pinning Region

For applications it is important to understand what determines the width of the pinning or snaking region. For this purpose it is helpful to observe that equilibria of (12.3) satisfy a fourth order ordinary differential equation in space that defines a (nonintegrable) autonomous Hamiltonian system with Hamiltonian

$$H = -\frac{1}{2}\left(r - k_c^4\right)u^2 + k_c^2 u_x^2 - \frac{1}{2}u_{xx}^2 + u_x u_{xxx} - \int_0^u f(v)\mathrm{d}v. \qquad (12.7)$$

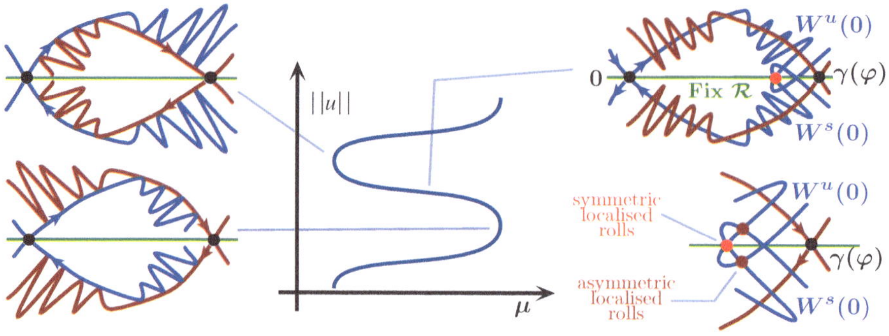

Fig. 12.3 A cartoon showing the stable and unstable manifolds for the fixed points O and $\gamma(\phi)$ in the level set $H = 0$, at different values of $\mu \equiv -r$ in the pinning region. Reproduced from [29]. ©2009 by the Society for Industrial and Applied Mathematics

Thus $\mathrm{d}H/\mathrm{d}x = 0$ and any homoclinic orbit $O \to O$ must lie in the level set $H = 0$, i.e. in a three-dimensional surface in four dimensions, and this is so for the heteroclinic cycle $O \to \gamma \to O$ as well.

Because of translation invariance periodic orbits of (12.3) are not isolated – for each H there is a continuous family of such orbits. In the following we pick $H = 0$ and select one representative from this family, for example by assigning the origin $x = 0$ to the maximum value of u along the orbit. We call the resulting orbit γ. A point on this orbit with phase ϕ relative to $x = 0$, $\gamma(\phi)$, will be a fixed point of a "time-T" map, where T is the (spatial) period of the orbit [27]. By construction the "time-T" map is two-dimensional and has two fixed points, O and $\gamma(\phi)$. The result of repeated application of the "time-T" map can therefore be represented in a plane, as shown in Fig. 12.3. The figure shows the two fixed points as solid black points; these lie on the green line, $\mathrm{Fix}(R_1)$, representing solutions with the symmetry R_1. The figure shows the intersections of the stable and unstable manifolds of O, labeled $W^{s,u}(O)$, with the surface $H = 0$. In the cartoon these are one-dimensional (blue curves), and consist of points that approach O after an infinite number of backward and forward applications of the map. The intersection of the corresponding (three-dimensional) center-stable and center-unstable manifolds of γ with $H = 0$ at phase $\phi = 0$ is shown in brown and is also one-dimensional. Since we are dealing with a discrete map these manifolds consist of discrete sequences of points obtained by applying the map to different points in the stable and unstable manifolds of these fixed points. Because of the discrete nature of the resulting two-dimensional dynamics we expect the unstable manifold $W^u(O)$ to intersect transversally with the center-stable manifold $W^s(\gamma)$ (top right panel in the figure). The point of intersection is simultaneously on both manifolds implying that forward iterations take it to $\gamma(\phi)$ while backward iterations take it to O, i.e. such a point is a heteroclinic point. Each image of this point, forward or backward, will also be a heteroclinic point since it must again lie on an intersection of these manifolds. Since the forward iterates accumulate on γ the unstable manifold $W^u(O)$ must execute increasingly wild gyrations near $\gamma(\phi)$ as indicated in the figure. This is a consequence of the Hamiltonian nature of (12.3) which implies that the "time-T" map is area-preserving. Thus the areas of the (primary) lobes are all the same and since the foot of the lobes shrinks towards $\gamma(\phi)$ their length must grow in proportion. Spatial reversibility (i.e. the symmetry R_1) implies that $W^s(O)$ undergoes identical behavior and hence that $W^u(O)$ also intersects $W^s(O)$. The primary intersections must lie on the green curve (corresponding to the fixed point subspace Fix(R_1)) and hence correspond to solutions with $u(-x) = u(x)$ that lie simultaneously in $W^u(O)$ and $W^s(O)$ (large red dot). Such solutions represent symmetric homoclinic solutions of (12.3). Observe that since the primary intersections accumulate on $\gamma(\phi)$ there will in fact be an infinite number of such homoclinic solutions corresponding to symmetric LS of ever larger length. The figure also indicates that associated with each primary intersection there is a pair of secondary intersections (small red dots, bottom right panel in the figure). These do not lie in the green line and hence correspond to asymmetric homoclinic points, i.e. the rung states.

Figure 12.3 shows that the *heteroclinic tangle* described above is created, as the bifurcation parameter r increases, at the point of first tangency between $W^u(O)$ and $W^s(\gamma)$ (top left panel) and destroyed at the point of last tangency (bottom left panel). Thus the snaking region is bounded on either side by the tangencies at $E_\pm$ and no (long) LS are present outside of the parameter interval between these tangencies [27–29].

An essentially identical picture applies to reversible but nonvariational systems since the fundamental properties of the heteroclinic tangle depend only on the presence of a transversal intersection between $W^u(O)$ and $W^s(\gamma)$ together with spatial reversibility. For this reason the geometrical picture sketched here has a far greater applicability than one may imagine at first sight. This is a consequence of the fact that a transversal intersection between manifolds cannot be destroyed by small perturbations such as a change in the parameter r, or the addition of an R_1-preserving but nonvariational term to the equation itself, i.e. it is a consequence of *structural stability*.

12.2.2.2 Physical Explanation of the Pinning Region

Consider now the energetics of the system. The free energy F allows us to compare the energy of the homogeneous state O with that of the periodic state γ. The point where these are equal is called the "Maxwell point" by analogy with first order phase transitions between two homogeneous phases such as a liquid and a gas. At the Maxwell point the two phases coexist; away from it one or other is energetically favored and a front separating the two will move so as to lower the energy of the system. In the present case $F(O) < F(\gamma)$ when $r < r_M$ implying that O is energetically preferred and vice versa when $r > r_M$. However, the phase γ is structured and small changes in r will not result in front motion as the front is held back by a "pinning potential" due to the structured state behind it [30]. This self-pinning allows stationary fronts over a range of r straddling r_M, and r must be changed by a finite amount to overcome the pinning potential and allow the fronts to move. Within this interval many steady states coexist since it costs little to insert fronts between the two competing phases. The pinning region can thus be thought of as an "unfolding" of the Maxwell point due to the heterogeneity of one of the states.

The wavelength $\ell(r)$ of the pattern within LS depends on the value of r in the pinning region. This wavelength is determined by the requirement $H = 0$. One finds that for $r < r_M$ the wavelength is compressed relative to that at r_M but that it is stretched for $r > r_M$ [24].

Note that the presence of the fronts at either end of the structure leads to a unique wavenumber between them, however far apart they are. Thus the fronts collapse the Busse balloon [1, 5].

In nonvariational systems F does not exist, and neither does H. In this case the role of the Maxwell point is played by a stationary front between the two competing states, and the location of such fronts must now be determined numerically. As mentioned, the snakes-and-ladders bifurcation diagram is expected in an interval straddling this location but in the absence of a spatial Hamiltonian H the wavelength selection problem within this interval remains unsolved.

12.2.3 Depinning

If r is moved sufficiently far from r_M the energy difference between O and γ exceeds the pinning potential and the fronts depin [28] . Direct integration of (12.3) reveals time-dependent growth of the structure via sequential nucleation of new cells (Fig. 12.4a). The nucleation time depends on the distance from the edge of the pinning region, as indicated in Fig. 12.4b. The time diverges at the edge of the pinning region (where it takes an infinite amount of time to nucleate a new cell) and decreases as the distance from the pinning region increases. The speed of the front, which is a 'pushed front' because it propagates into a stable state [17], can be calculated from

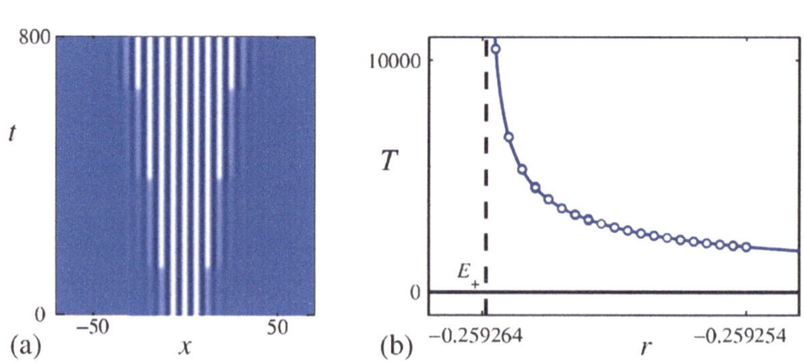

Fig. 12.4 **a** Space-time plot of the evolution of LS in SH23 at $r \approx -0.2587$, and **b** the nucleation time T as a function of r. Parameters: $b_2 = 1.8$, $k_c = 1$. Reproduced from [25]

the time it takes to nucleate cells at the front [24, 31]. This time can be computed by projecting SH23 onto the three near-marginal eigenfunctions present at $E_{\pm}$, the amplitude and phase eigenfunctions and the translation mode [24]. To the left of the pinning region (where the solution evolves towards the lower energy state O) the fronts move inwards via sequential annihilation of cells with the same qualitative dependence on the distance from E_{-} as in the E_{+} case.

12.2.4 Generalizations

- *Finite size effects*. The multiple bifurcation at $r = 0$ (Fig. 12.1a) is unfolded when $L < \infty$, i.e. it breaks up into a primary bifurcation to a periodic wavetrain, together with a secondary bifurcation from this state to $L_{0,\pi}$. Thus the LS bifurcate from O only in the limit $L \to \infty$. Moreover, on a finite periodic domain snaking does not continue forever. Once the LS has grown to fill the domain no additional growth is possible and the LS branch exits the snaking region and terminates near the fold on a branch of periodic states. The details of this transition are in general complex since they depend on exactly how much space is left, i.e. on L mod $\ell(r)$, where $\ell(r)$ is the wavelength within the pinning region [32]. Moreover as L increases the termination points must "jump" between different periodic states, a process studied in [32, 33]. Near the fold the LS resemble holes in an otherwise periodic wavetrain. This bifurcation is present even on the real line and gives rise to two branches of holes that snake once the hole deepens enough to come close to $u = 0$ and starts to broaden. On the real line the LS branches bifurcating from $u = 0$ at $r = 0$ and the branches of holes remain distinct, but on a periodic domain with $L < \infty$ they connect up pairwise: a broad LS can, after all, be viewed as a hole in a periodic state. Moreover, on smaller domains a single pulse state bifurcates from the periodic states at a larger amplitude, and the resulting branch terminates farther from the fold.
  If the boundary conditions are changed from periodic to Robin (or mixed) boundary conditions, the effect is dramatic. Since periodic states are now absent LS bifurcate directly from $u = 0$ in a *primary* bifurcation. Thereafter they snake normally since LS are insensitive to the details of the boundary conditions but when the domain is almost full the snaking branch evolves continuously into an extended large amplitude (almost) periodic state with defects at the boundaries [34, 35].
- *Broken reversibility*. The symmetry R_1 can be broken, for example, by adding dispersion to SH23. In this case all states generically drift and numerical computations [36] show that dispersion destroys the snakes-and-ladders structure of the snaking region with the drifting LS now located on a stack of figure-eight isolas. As the dispersion increases the isolas shrink and eventually disappear. Thus drifting LS are absent for large dispersion.
- *Broken translation invariance*. Broken translation invariance, either through the imposition of Robin boundary conditions at the boundaries of a finite domain [34] or through space-dependent forcing [37, 38], has also been studied. The details are

complex and not well understood. Either type of forcing selects preferred locations for the LS [39], and may result in incomplete snaking depending on whether local maxima are in phase with the LS maxima or out of phase. Large scale heterogeneity may cause an overall slant to the snaking structure, as discussed in a different context below. Moreover spatially periodic forcing can generate stable LS even in systems where snaking would not otherwise be present [40]. In fact, this is always the case in the vicinity of a Maxwell point between two homogeneous states [41, 42] and is a consequence of the pinning introduced into the system by the forcing. Discrete problems in which the spatial Laplacian is replaced, for example, by nearest neighbor coupling, also lack translation invariance and constitute an important class of systems with multiple coexisting LS that snake [43–45].

- *Additional symmetry*. The Swift–Hohenberg equation with $f(u) = b_3u^3 - u^5$ (hereafter SH35) possesses an extra symmetry $R_2 : x \to x, u \to -u$ that is analogous to the so-called Boussinesq symmetry of Rayleigh–Bénard convection with identical boundary conditions at the top and bottom [6]. For this reason predictions based on SH35 find a number of applications, particularly in fluid mechanics. The most important include: (i) the periodic states with $k = k_c$ bifurcate subcritically if $b_3 > 0$. (ii) Exponentially small terms select four phases ϕ: $0, \pi/2, \pi, 3\pi/2$ [46, 47]. The $\phi = 0, \pi$ states are related by R_2 and likewise for the $\phi = \pi/2, 3\pi/2$ states. A bifurcation diagram showing the L^2 norm of u_L thus shows two branches, a branch of even parity $\phi = 0, \pi$ states, and a branch of odd parity states $\phi = \pi/2, 3\pi/2$. The former are invariant under R_1 as in SH23; the latter are invariant under $R_2 \circ R_1$. Both bifurcate subcritically from $u = 0$ at $r = 0$. (iii) The odd and even parity branches are organized within a similar snakes-and-ladders structure as found in SH23, and possess the same stability properties [46].
- *Noise*. It is natural to consider the effect of small amplitude additive noise on LS in the pinning region. Each local minimum will have a lowest barrier in energy across which escape is most probable. The type of evolution that results is expected to depend on whether $r < r_M$ or $r > r_M$. In the former case the LS should gradually shrink; in the latter case it should gradually grow. Despite some numerical studies of this process [48, 49] the details of the resulting evolution are by no means clear. The effects of multiplicative noise have not been studied.
- *Higher dimensions*. Similar snaking behavior accompanies both two-dimensional [50, 51] and three-dimensional localized patterns [52]. In addition, such systems exhibit a variety of spot structures and localized target patterns that exhibit collapsed snaking, an effect that arises from decreasing curvature as the structure grows in size following the solution branch [53].

12.2.5 Examples

Gradient systems such as (12.3) are nongeneric. Generic systems no longer possess an energy functional but the snakes-and-ladders structure of the snaking or pinning region persists when a gradient system is perturbed by nongradient terms. However, while solutions with the symmetry R_1 remain stationary, the asymmetric rung states now drift; the drift direction is determined by the asymmetry. In addition, secondary Hopf bifurcations may appear on the stable segments of the primary snaking branches leading to breathing [54]. We now briefly discuss four systems of this type.

1. *Bright and dark optical solitons in the Lugiato–Lefever equation*. The Lugiato–Lefever equation [55]

$$\partial_t A = -(1 + i\theta)A + i\nu\partial_x^2 A + i|A|^2 A + \rho \tag{12.8}$$

   describes the dynamics of the electric field (proportional to $A(x, t) \in \mathbb{C}$) in a wide area Fabry–Perot interferometer partially filled with a nonlinear medium or temporal dynamics in an optical ring cavity. In the latter case the variable x represents a slow time [56]. Thus the equation describes both spatial optical solitons and temporal solitons. The equation is dissipative but $\rho \in \mathbb{R}$ represents injected power, while $\theta \in \mathbb{R}$ represents cavity detuning. In temporal systems bright and dark solitons can be found. Taking into account only second order dispersion two regimes can be identified, characterized by either normal ($\nu = -1$) or anomalous ($\nu = 1$) chromatic dispersion. In the latter case the only type of dissipative solitons that exist are bright solitons and these are present in both monostable [57] and bistable regimes [58–60]. In contrast, in the normal dispersion case the main type of dissipative solitons that appear are dark solitons [60–63].
   For $\theta > \sqrt{3}$ (the bistable regime) the spatially uniform states form a S-shaped curve as a function of ρ, and LS bifurcate from both the left and the right folds. Of particular interest is the situation $\theta > 2$ in which the lower branch becomes unstable to a pattern-forming bifurcation referred to in the present context as a modulational instability. The periodic pattern that results bifurcates subcritically and hence a pair of branches of localized states bifurcate from modulational instability in addition to the pattern branch. In this case no LS bifurcate from the right fold. The different scenarios that enable the two LS branches that bifurcate at modulational instability to terminate at the upper left fold on the S curve are complex and are discussed in [64, 65].
2. *Reaction-diffusion models*. Many two-species reaction-diffusion models behave in a similar way to that just described. The Gierer–Meinhardt model serves as example [66]

$$\frac{\partial u}{\partial t} = \frac{u^2 v^{-1}}{1 + u^2} - u + D\frac{\partial^2 u}{\partial x^2}, \qquad \frac{\partial v}{\partial t} = Gu^2 - Ev + S + \frac{\partial^2 v}{\partial x^2}. \tag{12.9}$$

   Here the nonzero uniform state is also folded and for suitable parameter values the upper state can lose stability at a pattern-forming Turing bifurcation. The pattern state that results is initially subcritical before turning around, generating bistability between the uniform state and the pattern. In this case LS are also found and these snake as described by the Swift–Hohenberg equation [66]. Other equations with related behavior include the Brusselator model [67] and the Schnakenberg model [68].

3. *Vegetation models*. A number of model equations have recently been proposed to describe vegetation cover in arid ecosystems [4]. The simplest of these are of Gray-Scott type and consist of two equations, one for the biomass and the other for water. Both the biomass and available water are assumed to diffuse, the former more slowly than the latter. The models capture tristability, that is, coexistence of uniform vegetation cover, patchy cover and bare soil, and have the desired property that decreasing precipitation results in (hysteretic) transitions from uniform cover to patchy cover and eventually to vegetation collapse. Because of folds in the uniform state, and the presence of a subcritical Turing bifurcation from this state one finds two classes of LS, holes in the uniform vegetation cover, and patches of vegetation on bare background [69, 70]. In general these have different origins but can interact in the nonlinear regime. These properties appear to be a general feature of models of this type [71].
4. *Binary fluid convection*. This system provides a more complex example of LS because the system is confined between two horizontal plates. Thus in a 2D system only the horizontal direction is spatially extended. Simulations show that this system can form stable LS (hereafter convectons) via a focusing instability of a complex state called dispersive chaos [72]. These convectons resemble those in SH35 but are embedded in an unstable background state provided the latter is only convectively unstable [73, 74]. The convectons are interesting because they channel all the heat flux through the layer. Presumably as the domain size increases the single convecton state transfers stability to multiconvecton states in order to cope with the increasing heat flux although this has not been demonstrated. Odd parity convectons will travel if the reflection symmetry in the layer midplane (analogous to R_2) is broken and undergo strongly inelastic collisions [75] resembling those in SH35 with an extra term that simultaneously breaks R_2 and its gradient structure [76]. No stable convectons in 3D binary convection have been found [77] although LS in 3D porous media [78] and in 3D vertical cavities [79] have been computed. A classical fluid system exhibiting snaking LS is provided by plane Couette flow [80, 81].

12.3 Conserved Systems

Physical systems arising in nature frequently possess a conserved quantity and in such systems the order parameter field has a fixed mean value [82]. Systems of this type are distinguished from the standard snaking scenario described above by the following properties: (i) the snaking becomes slanted (sometimes referred to as "sidewinding"), (ii) LS may be present outside of the region of coexistence of the homogeneous and periodic states, (iii) LS are present even when the periodic states bifurcate supercritically, i.e. when the coexistence region is absent entirely. The slanting [83, 84] of the snakes-and-ladders structure is a finite size effect: in a finite domain expulsion of the conserved quantity from an LS implies its increase outside, a fact that progressively delays (to stronger forcing) the nucleation events whereby LS grow in length. The net effect is that LS are found in a much broader region of parameter space than in nonconserved systems.

12.3.1 The Conserved Swift–Hohenberg Equation

The conserved Swift–Hohenberg (cSH) equation provides arguably the simplest illustration of the above results. Although this equation appears in earlier work in different fields [85, 86] it arises naturally in the phase field crystal (PFC) model of soft matter [87]. This model is in turn derived [87–89] from dynamical density functional theory (DDFT) and represents the simplest microscopic model for the freezing transition. In this model the transition from a homogeneous state to a periodic state corresponds to the transition from a uniform density liquid to a periodic crystalline solid. The LS of interest in this model then correspond to states in which a finite size portion of the periodic crystalline phase coexists with the uniform density liquid phase, and these are expected to be present in the coexistence region between the two phases. In fact, it turns out that LS of this type are also present at state points outside of the coexistence region [90]. Some rather striking examples of LS in large 2D systems with conserved mass include snow-flake-like and dendritic structures [87, 91–93].

We write the cSH (or PFC) equation in the form

$$\partial_t \phi(\mathbf{x}, t) = \alpha \nabla^2 \frac{\delta F[\phi]}{\delta \phi(\mathbf{x}, t)}, \tag{12.10}$$

where $F[\phi] \equiv \int \mathrm{d}^d x \left[\frac{\phi}{2}[r + (k_c^2 + \nabla^2)^2]\phi + \frac{\phi^4}{4} \right]$ and $\phi(\mathbf{x}, t)$ is an order parameter field that corresponds in the PFC context to a scaled density profile. Here r plays the role of temperature ($r < 0$ for a supercooled liquid) and α is a (constant) mobility coefficient. It follows that the system evolves according to the cSH equation

$$\partial_t \phi = \alpha \nabla^2 \left[r\phi + (k_c^2 + \nabla^2)^2 \phi + \phi^3 \right]. \tag{12.11}$$

Stationary states thus obey the equation

$$r\phi + (k_c^2 + \nabla^2)^2 \phi + \phi^3 = \mu, \tag{12.12}$$

where μ is the chemical potential. This quantity controls the fraction of the mass that is in the liquid and solid phases at any given temperature r, and hence determines ϕ_0, the conserved average value of the order parameter $\phi(\mathbf{x})$. Stability analysis of the liquid phase in 1D shows it first becomes linearly unstable when $r = 0$, $\phi_0 = 0$. With decreasing r the system passes as tricritical point below which one finds thermodynamic coexistence between the liquid and stripe phases. Figure 12.5 shows the resulting bifurcation diagrams in terms of the L^2 norm of $\delta\phi(x) \equiv \phi(x) - \phi_0$ revealing the presence of a supercritical bifurcation to a spatially periodic state as ϕ_0 increases, as well as a pair of branches of symmetric spatially localized states (LS_{odd} or LS_{even}, according to whether the number of peaks is odd or even) exhibiting slanted snaking, complete with rung states representing asymmetric LS, provided r is sufficiently below the thermodynamic tricritical point. The LS described by these diagrams tend to have lower energy than the periodic crystal [90], and when replotted as a function of the chemical potential μ standard snaking is recovered (Fig. 12.6). This is because μ enforces the conservation of ϕ_0 and so is the proper thermodynamic variable for systems of this type. It would be of interest to identify a quantity that plays the role of the chemical potential in the fluid systems with conserved dynamics described below. Direct numerical simulation of the cSH equation in 2D and 3D [90] reveals a similar thermodynamic preference for LS in particular intervals of ϕ_0 (Figs. 12.7 and 12.8).

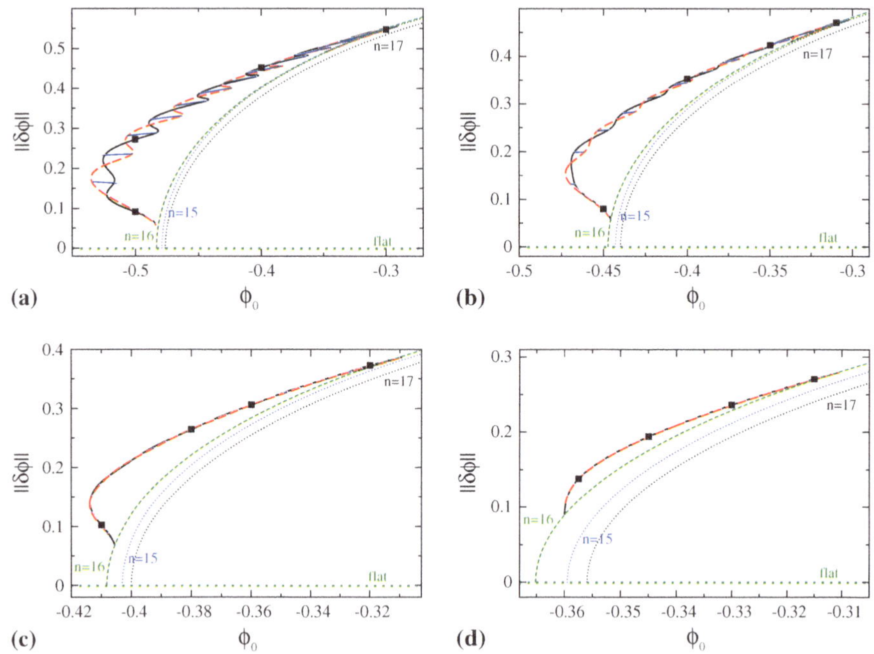

Fig. 12.5 The L^2 norm of the homogeneous, periodic and LS solutions of the cSH equation (12.11) with $k_c = 1$ as a function of the mean order parameter ϕ_0, for a fixed domain size $L = 100$ and **a** $r = -0.7$, **b** $r = -0.6$, **c** $r = -0.5$ and **d** $r = -0.4$. The $n = 16$ periodic state (16 wavelengths $2\pi/k_c$ in the domain) is the first to set in as $|\phi_0|$ decreases and the homogeneous state loses stability. The solid black (dashed red) lines correspond to LS_{odd} (LS_{even}). Reproduced from [90]

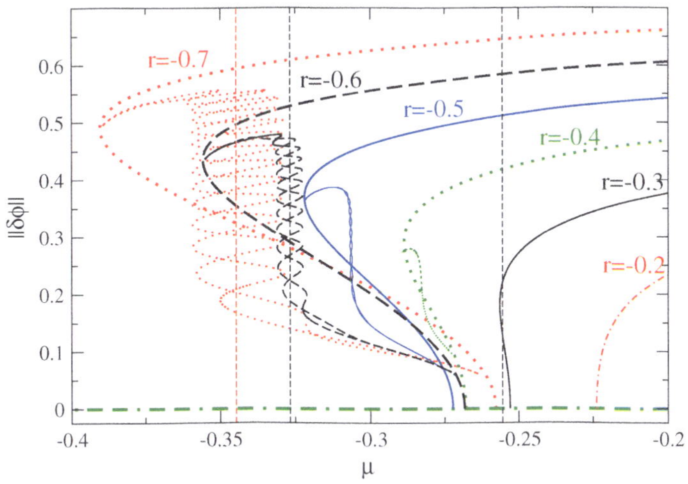

Fig. 12.6 The L^2 norm $||\delta\phi||$ of the homogeneous ($\phi(x) = \phi_0$, dot-dashed green line), periodic ($n = 16$) and localized solutions of the cSH equation (12.11) with $k_c = 1$ as a function of the chemical potential μ, for a fixed domain size of $L = 100$ and various values of r, showing that standard snaking is recovered when slanted snaking is replotted against the correct thermodynamic parameter. The dashed vertical lines indicate the coexistence values of the chemical potential for $r = -0.7, -0.6$ and -0.3. The snaking region occupied by the two types of symmetric LS shrinks as $|r|$ decreases. Reproduced from [90]

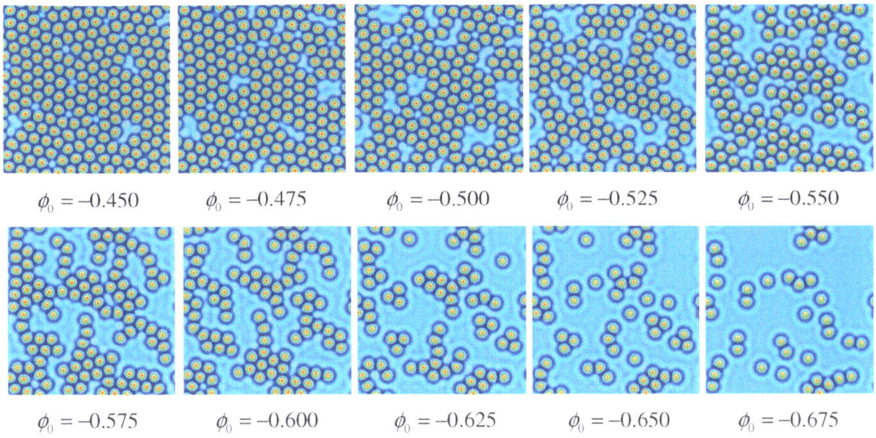

Fig. 12.7 Steady-state solutions of the cSH equation (12.11) in 2D for $r = -0.9$, $k_c = 1$ and different values of ϕ_0 in the range $-0.675 < \phi_0 < -0.45$, where LS are present. The domain size is 100×100. Reproduced from [90]

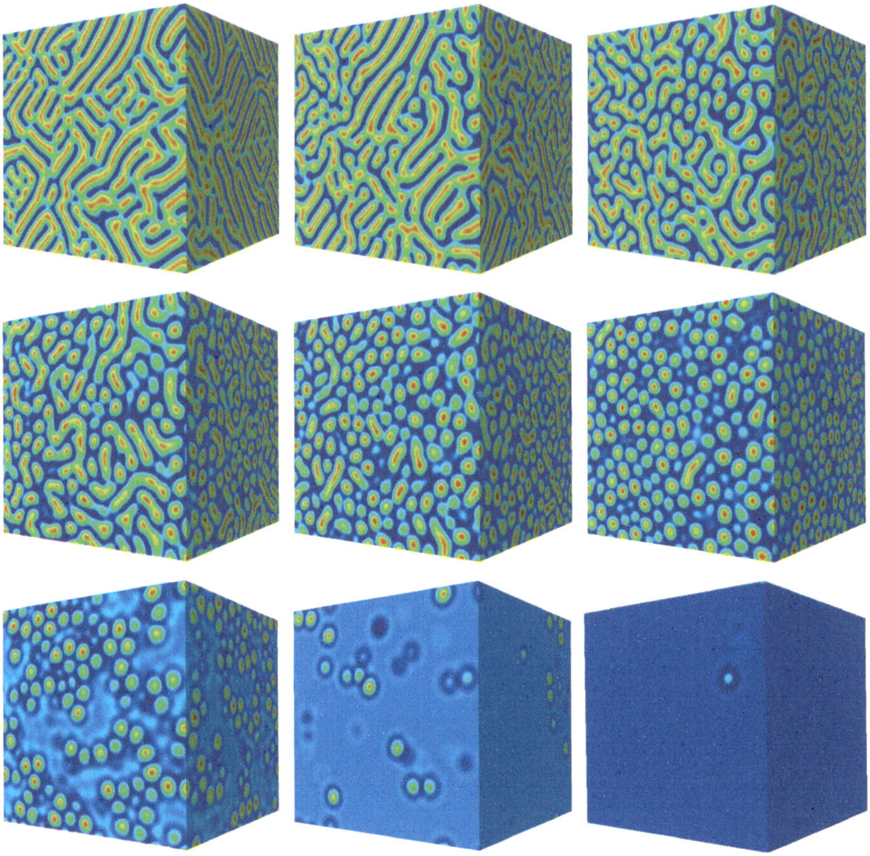

Fig. 12.8 Steady-state solutions of the cSH equation (12.11) in 3D for $r = -0.9$, $k_c = 1$ and different values of ϕ_0: from top left to bottom right $|\phi_0| = 0.025, 0.125, 0.225, 0.325, 0.425, 0.525, 0.625, 0.725$ and 0.750. The domain size is $100 \times 100 \times 100$. Reproduced from [90]

Figure 12.6 does not indicate the stability properties of the LS. Focusing on the thermodynamically stable state at each ϕ_0 one obtains Fig. 12.9a, showing that the state corresponding to the global energy minimum changes with increasing ϕ_0, and that the corresponding chemical potential becomes concentrated in the vicinity of the Maxwell point μ_M. Indeed one finds that the spread $\Delta\mu$ around μ_M obeys $\Delta\mu \sim 1.027L^{-0.998}$ [94] showing that in the thermodynamic limit one recovers the Maxwell construction for a first order phase transition. Similar results hold in 2D as well [94].

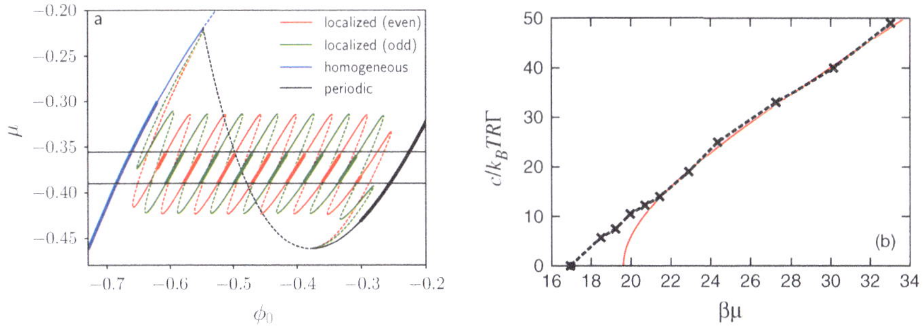

Fig. 12.9 **a** Chemical potential μ for the liquid (homogeneous) state (blue line) and the crystalline (periodic) state (black line) in the 1D PFC system on a domain of size $L = 16L_c$, together with the crystallites (spatially localized states) with even (red line) and odd (green line) numbers of bumps, all as a function of ϕ_0. Linearly stable and unstable states are indicated by solid and dashed lines, respectively. Global energy minima are indicated by thick line segments. When connected up, the global minima corresponding to localized states lie on a curve with piecewise constant slope: a zigzag curve. The width $\Delta\mu$ of this zigzag region is indicated by a pair of horizontal lines. Reproduced from [94]. **b** The front speed c as a function of the chemical potential μ for a GEM-4 fluid with temperature $k_B T = \epsilon$, where ϵ is the energy scale for particle-particle repulsion. The red solid line shows the result from (12.13) while the black dashed line results from numerical tracking of the hexagonal front. The black circle denotes the coexistence value $\beta\mu \approx 17.0$, where $\beta = (k_B T)^{-1}$. Reproduced from [98]

12.3.2 Crystallization

Within the PFC model the dynamics of a crystallization front invading a supercooled liquid can be determined from the dispersion relation $\sigma = \sigma(k)$ for infinitesimal density perturbations of the liquid state. Owing to mass conservation this dispersion relation has a neutral mode at $k = 0$, in addition to unstable modes near $k = k_c$ when $r < 0$. Since the front is pulled (stable crystalline state in $x < 0$ invades an unstable liquid phase in $x > 0$) the marginal stability hypothesis [95] leads to the predictions

$$ic + \frac{\mathrm{d}\omega(k)}{\mathrm{d}k} = 0, \qquad \mathrm{Re}[ikc + \omega(k)] = 0. \tag{12.13}$$

These three equations are to be solved for the speed c of the front and $k = k_r + ik_i$, i.e. the wavenumber k_r selected by the moving front and the spatial decay rate $k_i > 0$ of the front profile: $\delta\rho(x,t) \sim \exp(-k_i x)\sin(k_r(x - ct) + \mathrm{Im}[\omega(k)]t)$. If no phase slips take place, then the wavenumber of the density modulations left behind by the front is [7, 8, 16, 24]:

$$k^* = k_r + \frac{1}{c}\mathrm{Im}[\omega(k)]. \tag{12.14}$$

It follows that the wavelength $2\pi/k^*$ of the density modulation generated behind the advancing front differs in general from the equilibrium crystal lattice spacing, requiring substantial subsequent rearrangement in order to form a defect-free crystal without strain. In some cases this rearrangement can lead to quasicrystalline structure [96] and when these states are subcritical they are accompanied by LS with a quasicrystalline motif [97].

Figure 12.9b shows the speed of a 2D front separating a supercooled liquid and the growing crystalline phase in a more realistic soft matter model known as a GEM-4 fluid [98]. The front advances via a stripe phase precursor that subsequently breaks up into a hexagonal phase. Figure 12.9b compares the results of numerical tracking of the front (crosses) with the predictions of the marginal stability hypothesis [17, 95] determined from the linear dispersion relation for the growth of infinitesimal disturbances of the liquid phase (red curve). This 1D theory describes the invasion of the liquid state by the stripe state and thus ignores the subsequent instability of the stripe state; we see that the prediction works very well under strongly supercooled conditions when the front speed is large but fails dramatically for slow speeds.

12.3.3 Example: Magnetoconvection

Convection in an imposed magnetic field [99] provides a simple example of a system with a conserved quantity, the magnetic flux imposed across the layer. Thus the flow within the layer can at most redistribute flux. Recent computations [100] confirm that all three departures (i)–(iii) from standard snaking behavior mentioned above take place in this system and that these are a consequence of the coupling to a large-scale magnetic mode.

Two-dimensional convection in an imposed vertical magnetic field is described by the dimensionless equations [99]

$$\sigma^{-1}\left[\nabla^2\psi_t + J(\psi, \nabla^2\psi)\right] = Ra\theta_x + \zeta Q J(x + A, \nabla^2 A) + \nabla^4\psi, \quad (12.15)$$

$$\theta_t + J(\psi, \theta) = \psi_x + \nabla^2\theta, \quad (12.16)$$

$$A_t + J(\psi, A) = \psi_z + \zeta\nabla^2 A, \quad (12.17)$$

with ψ, θ and A defined such that the velocity $\mathbf{u} = \nabla \times \psi\hat{\mathbf{y}}$, the temperature $\Theta = 1 - z + \theta$ and the magnetic field perturbation $\mathbf{b} = \nabla \times A\hat{\mathbf{y}}$. In addition to the Rayleigh number Ra and the Prandtl number $\sigma = \nu/\kappa$ the system is characterized by two additional dimensionless parameters, the Chandrasekhar number Q measuring the strength of the imposed magnetic field, and the diffusivity ratio $\zeta = \eta/\kappa$. Here κ and η are the thermal and ohmic diffusivities and ν is the kinematic viscosity.

With stress-free, fixed temperature, force-free boundary conditions [99]

$$\psi = \psi_{zz} = \theta = A_z = 0 \quad \text{on} \quad z = 0, 1, \quad (12.18)$$

and periodic boundary conditions in the horizontal with dimensionless period L one finds that the perturbation flux $\bar{A} \equiv L^{-1}\int_0^L \int_0^1 A(x, z, t)\, dx\, dz$ across the layer

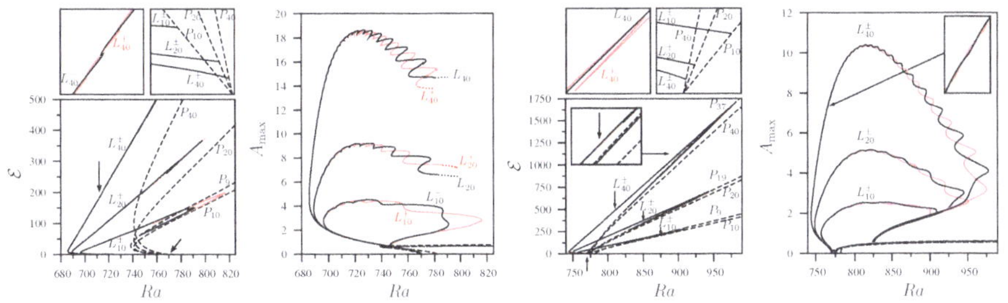

Fig. 12.10 The convecton branches L^+ (red) and L^- (black) in magnetoconvection together with the periodic states P_{10} (dashed) when $L = 10\lambda_c, 20\lambda_c, 40\lambda_c$ and (a) $\zeta = 0.1$, (b) $\zeta = 0.5$, shown in terms of the total kinetic energy $\mathcal{E}$ in the domain (left panels) and the maximum $A_{\max}$ of the potential $A(x, z)$ in the domain (right panels), both as functions of the Rayleigh number Ra. The insets at top left show enlargements at the locations indicated by arrows, and reveal that with increasing L the convecton branches bifurcate from P_{10} at smaller and smaller amplitude and that they develop pairs of saddle-node bifurcations but remain intertwined. In all cases the convectons on the upper branch between the far left and right saddle-node bifurcations are stable, except for the short intervals between the intervening pairs of saddle-nodes. Parameters: $Q = 4, \sigma = 1$. Reproduced from [100]

remains constant in time, $\bar{A} = 0$. Figure 12.10 shows the solutions for (a) $\zeta = 0.1$ (subcritical bifurcation to periodic states) and (b) $\zeta = 0.5$ (supercritical bifurcation to periodic states) obtained by numerical continuation [100]. Two-dimensional rotating convection with stress-free boundary conditions at top and bottom is very similar [101]. Here the conserved mode is the zonal velocity and the convectons that form expel shear from regions of strong convection thereby reducing the local shear. Thus the convectons create the conditions required for their existence just as the convectons in the magnetic problem expell magnetic field to facilitate their presence. While this behavior is characteristic of subcritical instabilities, the presence of a conserved quantity extends it into the supercritical regime, just as in the cSH equation.

12.4 Future

The field of pattern formation and in particular the study of spatially localized structures is evolving rapidly. The following areas appear to be promising.

1. Fronts and defects: Although much is known about the dynamics of fronts between homogeneous phases this is not the case when one or both phases are structured. Existing studies of the invasion of a liquid phase by a crystalline phase have employed amplitude equations [102, 103]. Such descriptions suffer from two shortcomings—they assume that the crystalline state is weakly nonlinear and they ignore pinning of the front to the crystal phase deposited behind the front. Defects in wavetrains, particularly nontopological defects, also pin to heterogeneities, in general leading to a hierarchy of potentially stable defects undergoing snaking [68, 104]. These structures cannot be identified or stud-

ied using amplitude-phase descriptions based on multiple scale approaches and new approaches have to be developed. Multiple applications of these ideas in materials science may be expected.
2. Swarming: Agent-based models of swarming and schooling lead naturally to localized structures or groups. These may arise kinematically or dynamically through mutual interactions. While much work is based on model simulations continuum models have been successful in describing the properties of such structures [105]. These are typically nonlocal, zero-temperature models. As a result the density distributions are often compactly distributed. These systems have much in common with studies of the dynamics of LS in activator-inhibitor systems in the semi-strong regime. In these systems the LS are described via geometric singular perturbation techniques and interact via a large scale inhibitor field, as described by ordinary differential equations for their location [106]. In the presence of overcrowding some LS may collapse while in low density situations LS may split and multiply in number leading to new types of dynamics absent from agent-based models.
3. Elasticity: Many systems of interest in biological sciences are elastic, leading to the interaction between fluid flow and elastic structures and chemical reactions and elastic structures. These systems lead to new types of dynamics in which the natural scale of the system has to compete with intrinsic scales imposed by the Gaussian curvature of the surface [107, 108]. Other examples include compression of floating elastica [109, 110], delamination phenomena [111], as well as buckling and crack propagation.
4. Vortices: Vortices are quintessentially localized structures. In fluid mechanics vortices are usually studied in the inviscid context although the equations of fluid mechanics are dissipative. As a result vortices have to be forced, for example by buoyancy forcing at small scales in rotating convection. In such flows the vorticity profile in a vortex is nonmonotonic dramatically reducing the circulation and hence vortex-vortex interaction [112]. The resulting system can then be described in terms of a "gas" of weakly interacting vortices and so may be amenable to techniques from statistical physics. Other dramatic examples of vortex generation occur in geostrophic turbulence [113–115] and other strongly anisotropic systems [116, 117]. Here vortices develop spontaneously on top of a turbulent state via the transfer of energy from small convective scales directly into the boxscale mode in a process that resembles spectral condensation [118]. In other cases jet-like structures may form by the same process [119, 120]. Similar structures form in 2D active turbulence [121]. Much remains to be learned about the development and properties of these remarkable coherent structures.

Acknowledgements This chapter is dedicated to the memory of Patrick Woods whose work [27] started it all; its preparation was supported in part by NSF grant DMS-1908891.

References

1. R.B. Hoyle, *Pattern Formation: An Introduction to Methods* (Cambridge University Press, Cambridge, 2006)
2. L.M. Pismen, *Patterns and Interfaces in Dissipative Dynamics* (Springer, Berlin, 2006)
3. M. Cross, H. Greenside, *Pattern Formation and Dynamics in Nonequilibrium Systems* (Cambridge University Press, Cambridge, 2009)
4. E. Meron, *Nonlinear Physics of Ecosystems* (CRC Press, Boca Raton, 2015)
5. M.C. Cross, P.C. Hohenberg, Rev. Mod. Phys. **65**, 851 (1993)
6. J.D. Crawford, E. Knobloch, Annu. Rev. Fluid Mech. **23**, 341 (1991)
7. T. Callahan, E. Knobloch, Nonlinearity **10**, 1179 (1997)
8. D.J. Bates, J.D. Hauenstein, A.J. Sommese, C.W. Wampler, BERTINI™ software (2013)
9. H. Uecker, D. Wetzel, J.D.M. Rademacher, Numer. Math. Theory Me. **7**, 58 (2014)
10. E. Knobloch, J. DeLuca, Nonlinearity **3**, 975 (1990)
11. I. Melbourne, J. Nonlinear Sci. **8**, 1 (1998)
12. A. Doelman, J.D.M. Rademacher, S. van der Stelt, Discret. Contin. Dyn. Syst. **5**, 61 (2012)
13. F.H. Busse, Phys. Rep. **41**, 1929 (1978)
14. L. Kramer, W. Zimmermann, Phys. D **16**, 221 (1985)
15. K. Siteur, E. Siero, M.B. Eppinga, J.D.M. Rademacher, A. Doelman, M. Rietkerk, Ecol. Complex. **20**, 81 (2014)
16. A. Bers, R.N. Sudan (eds.), *Space-time Evolution of Plasma Instabilities - Absolute and Convective* (North-Holland, Amsterdam, 1983)
17. W. van Saarloos, Phys. Rep. **386**, 29 (2003)
18. R.J. Deissler, Phys. D **25**, 233 (1987)
19. S.M. Tobias, M.R.E. Proctor, E. Knobloch, Phys. D **113**, 43 (1998)
20. S.M. Tobias, E. Knobloch, Phys. Rev. Lett. **80**, 4811 (1998)
21. M.R.E. Proctor, S.M. Tobias, E. Knobloch, Phys. D **145**, 191 (2000)
22. D. Avitabile, M. Desroches, E. Knobloch, M. Krupa, Proc. R. Soc. A **473**, 20170018 (2017)
23. E. Knobloch, Annu. Rev. Cond. Matter Phys. **6**, 325 (2015)
24. J. Burke, E. Knobloch, Phys. Rev. E **73**, 056211 (2006)
25. J.R. Burke, Localized states in driven dissipative systems. Ph.D. thesis, University of California, Berkeley, 2008
26. J. Burke, E. Knobloch, Discr. Cont. Dyn. Syst. Suppl. **2009**, 109 (2009)
27. P.D. Woods, A.R. Champneys, Phys. D **129**, 147 (1999)
28. P. Coullet, C. Riera, C. Tresser, Phys. Rev. Lett. **84**, 3069 (2000)
29. M. Beck, J. Knobloch, D.J.B. Lloyd, B. Sandstede, T. Wagenknecht, SIAM J. Math. Anal. **41**, 936 (2009)
30. Y. Pomeau, Phys. D **23**, 3 (1986)
31. I.S. Aranson, B.A. Malomed, L.M. Pismen, L.S. Tsimring, Phys. Rev. E **62**, R5 (2000)
32. A. Bergeon, J. Burke, E. Knobloch, I. Mercader, Phys. Rev. E **78**, 046201 (2008)
33. J.H.P. Dawes, SIAM J. Appl. Dyn. Syst. **8**, 909 (2009)
34. S.M. Houghton, E. Knobloch, Phys. Rev. E **80**, 026210 (2009)
35. I. Mercader, O. Batiste, A. Alonso, E. Knobloch, Phys. Rev. E **80**, 025201(R) (2009)
36. J. Burke, S.M. Houghton, E. Knobloch, Phys. Rev. E **80**, 036202 (2009)
37. H. Uecker, J. Nonlinear Sci. **11**, 89 (2001)
38. H.-C. Kao, C. Beaume, E. Knobloch, Phys. Rev. E **89**, 012903 (2014)
39. G. Kozyreff, P. Assemat, S.J. Chapman, Phys. Rev. Lett. **103**, 164501 (2009)
40. H. Schmidt, A. Hutt, L. Schimansky-Geier, Phys. D **238**, 1101 (2009)
41. M. Clerc, C. Falcon, Phys. A **356**, 48 (2005)
42. B.C. Ponedel, E. Knobloch, Eur. Phys. J. Spec. Top. **225**, 2549 (2016)
43. C. Chong, R. Carretero-González, B.A. Malomed, P.G. Kevrekidis, Phys. D **238**, 126 (2009)
44. A.V. Yulin, A.R. Champneys, SIAM J. Appl. Dyn. Syst. **9**, 391 (2010)
45. J.J. Bramburger, B. Sandstede, J. Nonlinear Sci. (2019). https://doi.org/10.1007/s00332-019-09584-x

46. J. Burke, E. Knobloch, Phys. Lett. A **360**, 681 (2007)
47. A.D. Dean, P.C. Matthews, S.M. Cox, J.R. King, Nonlinearity **24**, 3323 (2011)
48. H. Sakaguchi, H.R. Brand, Phys. D **97**, 274 (1996)
49. M.G. Clerc, C. Falcon, E. Tirapegui, Phys. Rev. Lett. **94**, 148302 (2005)
50. D.J.B. Lloyd, B. Sandstede, D. Avitabile, A.R. Champneys, SIAM J. Appl. Dyn. Syst. **7**, 1049 (2008)
51. D. Avitabile, D.J.B. Lloyd, J. Burke, E. Knobloch, B. Sandstede, SIAM J. Appl. Dyn. Syst. **9**, 704 (2010)
52. H. Uecker, D. Wetzel, Snaking branches of planar BCC fronts in the 3D Brusselator, arXiv:1906.10446
53. S. McCalla, B. Sandstede, Phys. D **239**, 1581 (2010)
54. J. Burke, J.H.P. Dawes, SIAM J. Appl. Dyn. Syst. **11**, 261 (2012)
55. L.A. Lugiato, R. Lefever, Phys. Rev. Lett. **58**, 2209 (1987)
56. F. Leo, S. Coen, P. Kockaert, S.P. Gorza, P. Emplit, M. Haelterman, Nat. Photonics **4**, 471 (2010)
57. D. Gomila, A.J. Scroggie, W.J. Firth, Phys. D **227**, 70 (2007)
58. F. Leo, L. Gelens, P. Emplit, M. Haelterman, S. Coen, Opt. Express **21**, 9180 (2013)
59. P. Parra-Rivas, D. Gomila, M.A. Matías, S. Coen, L. Gelens, Phys. Rev. A **89**, 043813 (2014)
60. C. Godey, I.V. Balakireva, A. Coillet, Y.K. Chembo, Phys. Rev. A **89**, 063814 (2014)
61. X. Xue, Y. Xuan, Y. Liu, P.-H. Wang, S. Chen, J. Wang, D.E. Leaird, M. Qi, A.M. Weiner, Nat. Photonics **9**, 594 (2015)
62. V.E. Lobanov, G. Lihachev, T.J. Kippenberg, M.L. Gorodetsky, Opt. Express **23**, 7713 (2015)
63. P. Parra-Rivas, D. Gomila, E. Knobloch, S. Coen, L. Gelens, Opt. Lett. **41**, 2402 (2016)
64. P. Parra-Rivas, E. Knobloch, D. Gomila, L. Gelens, Phys. Rev. A **93**, 063839 (2016)
65. P. Parra-Rivas, D. Gomila, L. Gelens, E. Knobloch, Phys. Rev. E **97**, 042204 (2018)
66. A. Yochelis, Y. Tintut, L.L. Demer, A. Garfinkel, New J. Phys. **10**, 055002 (2008)
67. J.C. Tzou, Y.-P. Ma, A. Bayliss, B.J. Matkowsky, V.A. Volpert, Phys. Rev. E **87**, 022908 (2013)
68. H. Uecker, D. Wetzel, SIAM J. Appl. Dyn. Syst. **13**, 94 (2014)
69. Y.R. Zelnik, P. Gandhi, E. Knobloch, E. Meron, Chaos **28**, 033609 (2018)
70. P. Gandhi, Y.R. Zelnik, E. Knobloch, Phil. Trans. Royal Soc. A **376**, 20170375 (2018)
71. D. Ruiz-Reynés, L. Martin, E. Hernández-García, E. Knobloch, D. Gomila, Patterns, localized structures and fronts in a reduced model of clonal plant growth, arXiv:2001.00224
72. C.S. Bretherton, E.A. Spiegel, Phys. Lett. A **96**, 152 (1983)
73. O. Batiste, E. Knobloch, A. Alonso, I. Mercader, J. Fluid Mech. **560**, 149 (2006)
74. T. Watanabe, M. Iima, Y. Nishiura, J. Fluid Mech. **712**, 219 (2012)
75. I. Mercader, O. Batiste, A. Alonso, E. Knobloch, J. Fluid Mech. **722**, 240 (2013)
76. S.M. Houghton, E. Knobloch, Phys. Rev. E **84**, 016204 (2011)
77. I. Mercader, A. Alonso, O. Batiste, Phys. Rev. E **77**, 036313 (2008)
78. D. Lo Jacono, A. Bergeon, E. Knobloch, J. Fluid Mech. **730**, R2 (2013)
79. C. Beaume, A. Bergeon, E. Knobloch, J. Fluid Mech. **840**, 74 (2018)
80. T.M. Schneider, J.F. Gibson, J. Burke, Phys. Rev. Lett. **104**, 104501 (2010)
81. J.F. Gibson, T.M. Schneider, J. Fluid Mech. **794**, 530 (2016)
82. E. Knobloch, IMA J. Appl. Math. **81**, 457 (2016)
83. W.J. Firth, L. Columbo, A.J. Scroggie, Phys. Rev. Lett. **99**, 104503 (2007)
84. J.H.P. Dawes, SIAM J. Appl. Dyn. Syst. **7**, 186 (2008)
85. E. Knobloch, Phys. Rev. A **40**, 1549 (1989)
86. P.C. Matthews, S.M. Cox, Nonlinearity **13**, 1293 (2000)
87. H. Emmerich, H. Löwen, R. Wittkowski, T. Gruhn, G. Tóth, G. Tegze, L. Gránásy, Adv. Phys. **61**, 665 (2012)
88. S. van Teeffelen, R. Backofen, A. Voigt, H. Löwen, Phys. Rev. E **79**, 051404 (2009)
89. A.J. Archer, M.J. Robbins, U. Thiele, E. Knobloch, Phys. Rev. E **86**, 031603 (2012)
90. U. Thiele, A.J. Archer, M.J. Robbins, H. Gomez, E. Knobloch, Phys. Rev. E **87**, 042915 (2013)

91. T. Pusztai, G. Tegze, G.I. Tóth, L. Kornyei, G. Bansel, Z.Y. Fan, L. Gránásy, J. Phys. Condens. Matter **20**, 404205 (2008)
92. G. Tegze, L. Gránásy, G.I. Tóth, F. Podmaniczky, A. Jaatinen, T. Ala-Nissila, T. Pusztai, Phys. Rev. Lett. **103**, 035702 (2009)
93. G. Tegze, L. Gránásy, G.I. Tóth, J.F. Douglas, T. Pusztai, Soft Matter **7**, 1789 (2011)
94. U. Thiele, T. Frohoff-Hülsmann, S. Engelnkemper, E. Knobloch, A.J. Archer, New J. Phys. **21**, 123021 (2019)
95. G. Dee, J.S. Langer, Phys. Rev. Lett. **50**, 383 (1983)
96. A.J. Archer, A.M. Rucklidge, E. Knobloch, Phys. Rev. Lett. **111**, 165501 (2013)
97. P. Subramanian, A.J. Archer, E. Knobloch, A.M. Rucklidge, New J. Phys. **20**, 122002 (2018)
98. A.J. Archer, M.C. Walters, U. Thiele, E. Knobloch, Phys. Rev. E **90**, 042404 (2014)
99. N.O. Weiss, M.R.E. Proctor, *Magnetoconvection* (Cambridge University Press, Cambridge, 2014)
100. D. Lo Jacono, A. Bergeon, E. Knobloch, J. Fluid Mech. **687**, 595 (2011)
101. C. Beaume, A. Bergeon, H.-C. Kao, E. Knobloch, J. Fluid Mech. **717**, 417 (2013)
102. A. Hari, A.A. Nepomnyashchy, Phys. Rev. E **61**, 4835 (2000)
103. A. Doelman, B. Sandstede, A. Scheel, G. Schneider, Eur. J. Appl. Math. **14**, 85 (2003)
104. E. Knobloch, H. Uecker, D. Wetzel, Phys. Rev. E **100**, 012204 (2019)
105. A.J. Bernoff, C.M. Topaz, SIAM Rev. **55**, 709 (2013)
106. W. Chen, M.J. Ward, SIAM J. Appl. Dyn. Syst. **10**, 582 (2011)
107. Y. Klein, E. Efrati, E. Sharon, Science **315**, 1116 (2007)
108. N. Stoop, R. Lagrange, D. Terwagne, P.M. Reis, J. Dunkel, Nat. Mater. **14**, 337 (2015)
109. L. Pocivavsek, R. Dellsy, A. Kern, S. Johnson, B. Lin, K.Y.C. Lee, E. Cerda, Science **320**, 912 (2008)
110. M. Piñeirua, N. Tanaka, B. Roman, J. Bico, Soft Matter **9**, 10985 (2013)
111. A. Juel, D. Pihler-Puzović, M. Heel, Annu. Rev. Fluid Mech. **50**, 691 (2018)
112. I. Grooms, K. Julien, J.B. Weiss, E. Knobloch, Phys. Rev. Lett. **104**, 224501 (2010)
113. K. Julien, A.M. Rubio, I. Grooms, E. Knobloch, Geophys. Astro. Fluid Dyn. **106**, 392 (2012)
114. B. Favier, L.J. Silvers, M.R.E. Proctor, Phys. Fluids **26**, 096605 (2014)
115. C. Guervilly, D.W. Hughes, C.A. Jones, J. Fluid Mech. **758**, 407 (2014)
116. H. Xia, D. Byrne, G. Falkovich, M. Shats, Nat. Phys. **7**, 321 (2011)
117. A. Alexakis, L. Biferale, Phys. Rep. **767–769**, 1 (2019)
118. M. Chertkov, C. Connaughton, I. Kolokolov, V. Lebedev, Phys. Rev. Lett. **99**, 084501 (2007)
119. C. Guervilly, D.W. Hughes, Phys. Rev. Fluids **2**, 113503 (2017)
120. K. Julien, E. Knobloch, M. Plumley, J. Fluid Mech. **837**, R4 (2018)
121. M. Linkmann, G. Boffetta, M.C. Marchetti, B. Eckhardt, Phys. Rev. Lett. **122**, 214503 (2019)

Chapter 13
Synchronization in Discrete Models

Alexandre Rosas, Daniel Escaff, and Katja Lindenberg

Abstract In this chapter we discuss the emergence of a collective behavior in systems of discrete units with all to all interactions. Our focus is two fold: first we discuss the differences between systems with an infinite number of units (corresponding to a mean field approximation) and systems with finite populations; we also discuss conditions for systems with a finite number of states to be able to describe the continuous Kuramoto model.

13.1 Introduction

The Merriam-Webster Thesaurus [1] lists the following definition for the word "synchronize": *to occur or exist at the same time.* It lists the following synonyms: *accompany, attend, co-occur, coexist, coincide, concur.* None of these convey the full breadth of synchronization phenomena, nor do they exclude behaviors that occur due to an external force, such as electrons flowing in the same direction due to an applied field. The latter are not usually thought of as synchronization. From Wikipedia [2], *Tropical fireflies, in particular, in Southeast Asia, routinely synchronise their flashes among large groups. This phenomenon is explained as phase synchronization and spontaneous order.* There is a plethora of examples of synchronization phenomena of discrete units, each most likely responding to a different detailed mechanism: *heart*

A. Rosas
Departamento de Física, CCEN, Universidade Federal da Paraíba, Caixa Postal 5008, 58059-900 João Pessoa, Brazil
e-mail: arosas@fisica.ufpb.br

D. Escaff
Complex Systems Group, Facultad de Ingeniería y Ciencias Aplicadas, Universidad de los Andes, Av. Monseñor Alvaro del Portillo, 12455 Las Condes, Santiago, Chile
e-mail: descaff@miuandes.cl

K. Lindenberg (✉)
Department of Chemistry and Biochemistry and BioCircuits Institute, University of California San Diego, La Jolla, CA 92093-0340, USA
e-mail: klindenberg@ucsd.edu

P. G. Kevrekidis et al. (eds.), *Emerging Frontiers in Nonlinear Science*, Nonlinear Systems and Complexity 32, https://doi.org/10.1007/978-3-030-44992-6_13

tissues of different origins can 'beat' in sync, dancers coordinating their movements, synchronization in neural networks. The examples are endless in the world of computer science. One can think of two units or many units or an infinite number of units that synchronize.

Synchronization usually refers to emergent macroscopic behavior in systems consisting of microscopic or mesoscopic nonlinearly interacting units. Usually synchronization refers to temporal coincidence in a time dependent situation, for example, a moving pattern of most units being in the same state (e.g. an oscillation). The moving or oscillating units may move in continuous time as do dancers, or in discrete steps as do fireflies. The interactions among units may be short range or long range or anything in between. Furthermore, synchronization may also refer to a static coincidence in the state space of units, that is, to the formation of stationary patterns or agglomerations in a particular state. The topic is vast and can and does fill many books [3, 4].

In this Chapter we must narrow our discussion a great deal, and we focus on the emergence of collective behavior in systems of identical discrete units stepping from one state to another in discrete jumps. We assume all-unit to all-unit interactions. The interactions may all be of the same strength (which we assume), in which case geometry plays no role. In the absence of any disturbances, if the number of units is infinite, the system behaves as described by mean field theory. The final state of the system is then sharply determined.

The synchronization may not be perfect, for example, if there is noise in the system. Imperfect synchronization leads to a distribution of behaviors around a maximum that usually represents the behavior if the synchronization were perfect. This distribution may be stationary or may move in time, depending on the details of the model. This could happen, for instance, if the units are not identical, a case that we do not address here. We do address an important source of noise: when the number of units is finite rather than infinite. Although the synchronization is not perfect in this case, we will loosely use the terminology of dynamical systems to describe the stochastic counterpart. In any case, our goal is and has been to understand synchronization models that are simple enough for entire or partial analytic study. In all cases we choose units that can be described by the smallest possible number of variables that still allow for synchronization.

We introduce two-state models ("on-off") as well as three-state models. This is accomplished in Sect. 13.2. We take the interactions among units to be nonlinear. Nonlinear interactions are essential to achieve synchronization, and we work with polynomial interactions in the two-state case and exponential interactions in the three-state model. In the two-state models a Markovian transition rate of each unit between the two states leads to a patterned stationary distribution, that is, one of the states turns out to be more populated than the other. To obtain time dependent patterning in the two-state case it is necessary to introduce a memory whereby one of the two transitions of each unit is non-Markovian. In the three-state case we take the transitions to be unidirectional and obtain time-dependent effects such as waves of the majority of units being in one of the three states followed by them being in another state, in turn followed by the third state and then back to the first state.

When the number of units in the system is infinite, the final behaviors, be they time dependent or time independent, can often be handled analytically using mean field theory, and the long-time distributions are infinitely sharp. However, if the number of units is finite the problem becomes much more difficult because the evolution equations now acquire a noise term (i.e. they are now Langevin equations) and need to be handled at least in part numerically. We address the mean field theory cases as well as the cases of finite numbers of units in Sect. 13.2.

Finally, it is of course well known that Kuramoto studied one of the first mathematical models of synchronization [5]. His original model describes a continuous phase continuous time array. We asked ourselves this question: can we take Kuramoto's model and coarse grain it to arrive at discrete models? We discuss this in Sect. 13.3 and arrive at a result that is difficult to "guess" a priori. We will leave this suspense until the reader arrives at that section. Finally, in Sect. 13.4 we end with some conclusions and perspectives.

An additional final note: the various models that we discuss, namely arrays of two-state units, arrays of three-state units, and the coarse graining of Kuramoto's model are separate in the sense that the interactions among units are different, albeit all nonlinear. This Chapter is thus meant as a presentation of various discrete unit models without necessarily a comparison between them.

13.2 Finite Versus Infinite Population Models

In this section we will discuss the role of the number of units on the synchronization of two-state units (Sect. 13.2.1) and three state units (Sect. 13.2.2) with global coupling. In particular, whenever the steady state ($t \to \infty$ limit) for infinite populations presents bistability, the large population limit for finite population destroys this bistability. That is, the order of the limits of large times and large number of units does matter in determining the fate steady state.

13.2.1 Two-State Models

The problem of synchronization of arrays of globally coupled two-state units was discussed in [6, 7]. Here the term synchronization is used loosely to indicate that a transition to an ordered state with more units in one state than the other is achieved.

13.2.1.1 Infinite Population

An infinite ensemble of two-state (states 1 and 2) stochastic units is governed by the mean field equation

$$\dot{n}_1(t) = \gamma_2 n_2(t) - \gamma_1 n_1(t) = \gamma_2 - (\gamma_1 + \gamma_2) n_1(t), \tag{13.1}$$

where $n_1(t)$ and $n_2(t)$ are the densities of units in state 1 or 2 at time t, respectively. Here, we used density normalization $n_1(t) + n_2(t) = 1$ to write the last equality. Despite the apparent simplicity of this mean field equation, there is a wealth of possibilities hidden in the transition rates. For instance, the units may be coupled or uncoupled, explicitly time-dependent or not, etc. Here, we are concerned with Markovian globally coupled units, that is, the transition rates at any time t depend on the densities of units in the states 1 and 2 at that time. Again, using the density normalization, we may write the transition rates as $\gamma_1(n_1)$ and $\gamma_2(n_1)$.

It is worth noting that no fluctuations appear because the population is infinite. Therefore, for infinite populations we have a deterministic evolution that is completely determined by the initial conditions and, obviously, the transition rates. Hence, determining the steady state for infinite populations is a matter of finding the steady state of unidimensional dynamical systems, and phase transitions for those systems correspond to the bifurcations of the dynamical systems. Bearing that in mind, we try to map the mean field equation (13.1), onto one of the well-known normal forms presented, for example, in [8]. In order to do that, we write the transition rates as polynomials,

$$\gamma_1(n_1) = \sum_{k=0}^{\infty} \gamma_1^{(k)} n_1^k, \tag{13.2}$$

$$\gamma_2(n_1) = \sum_{k=0}^{\infty} \gamma_2^{(k)} n_1^k. \tag{13.3}$$

Different relations between these series lead to different normal forms. For more general transition rates, the normal forms can still be seen as approximations near the bifurcations (phase transitions). Therefore, we can write the mean field equation as

$$\dot{n}_1 = \sum_{k=0}^{\infty} a_k n_1^k, \tag{13.4}$$

where

$$\begin{aligned} a_0 &= \gamma_2^{(0)}, \\ a_k &= -\gamma_1^{(k-1)} - \gamma_2^{(k-1)} + \gamma_2^{(k)}. \end{aligned} \tag{13.5}$$

13.2.1.2 Finite Populations

For finite populations, we need to take into account fluctuations due to the finite number or units. Mathematically, we need to consider a Langevin equation instead of the deterministic mean field equation of the previous section. Hence, we start our analysis writing the time-evolution of the number of units in state 1, $N_1(t)$,

$$N_1(t+\mathrm{d}t) = N_1(t) - \sum_{k=1}^{N_1} \theta\left(\gamma_1(n_1)\mathrm{d}t - \zeta_k\right) + \sum_{k=N_1+1}^{N} \theta\left(\gamma_2(n_1)\mathrm{d}t - \zeta_k\right), \quad (13.6)$$

where $\theta(\cdot)$ is the Heaviside step function and ζ_k is a random variable uniformly distributed in the interval [0, 1]. Furthermore, $\mathrm{d}t$ is an infinitesimal time increment, so that $\gamma_k(n_1)\mathrm{d}t$ is the probability of jumping for a unit in state k. Thus, the first sum represents the number of units that jump out of state 1 and the second sum represents the number of units jumping into state 1. Equation (13.6) leads to the Langevin equation [6]

$$\dot{n}_1 = \gamma_2(n_1) - \left[\gamma_1(n_1) + \gamma_2(n_1)\right] n_1 + \sqrt{(1-n_1)\,\gamma_2(n_1) + n_1\gamma_1(n_1)}\,\frac{\xi(t)}{\sqrt{N}}, \quad (13.7)$$

where $\xi(t)$ is a Gaussian zero-centered white noise. Comparing Langevin equation (13.7) and the mean field equation (13.1) we notice that they differ by the fluctuation term (last term in the Langevin equation). In the limit of large populations $N \to \infty$, the fluctuations vanish and we recover the mean field equation. However, it should be noticed that there are two limits to be taken in order to get the steady state for large populations: time and number of units must both go to infinity. While the mean field equation approach takes the limit $N \to \infty$ first, the Langevin equation approach takes the limit $t \to \infty$ first. Therefore, when considering the Langevin approach we should not take the limit $N \to \infty$ at this point. Instead, we must find the steady state before taking the infinite population limit.

Using the Itô interpretation [9] for this Langevin equation, we obtain a Fokker–Planck equation for the probability of finding a fraction of n_1 of units in state 1 at time t,

$$\frac{\partial P(n_1,t)}{\partial t} = -\frac{\partial}{\partial n_1}\left[\mu(n_1)P(n_1,t)\right] + \frac{\partial^2}{\partial n_1^2}\left[D(n_1,N)P(n_1,t)\right], \quad (13.8)$$

where

$$\mu(n_1) = \gamma_2(n_1) - \left[\gamma_1(n_1) + \gamma_2(n_1)\right] n_1 \quad (13.9)$$

is the drift and

$$D(n_1,N) = \frac{\gamma_2(n_1) + \left[\gamma_1(n_1) - \gamma_2(n_1)\right] n_1}{2N} \quad (13.10)$$

is the diffusion coefficient. The Fokker–Planck equation (13.8) has the stationary solution

$$P_{st}(n_1) = c_N \frac{\exp\left[\int_0^{n_1} \frac{\mu(n)}{D(n,N)} dn\right]}{D(n_1,N)}, \quad (13.11)$$

where c_N is an N-dependent normalization constant for the probability density $P_{st}(n_1)$. At this point, a few comments are relevant. First, since the drift of the Fokker–Planck equation equals the right hand side of the mean field equation, we may expect that for large populations (small fluctuations) the two approaches should give similar results. However, contrary to the infinite population case, the steady state probability density for finite N does not hold any memory about the initial conditions. Therefore, they must differ in case of coexistence of two or more stable solutions for the mean field equation. Here, we will discuss this point in more detail for the saddle-node bifurcation. The reader interested in other types of bifurcations should refer to [7].

Typically, the saddle-node bifurcation creates two fixed points (one stable and one unstable) out of nothing. Hence, there is no coexistence of stable solutions and we should not see the behavior described above. However, the normal form for the saddle-node bifurcation $\dot{x} = r + x^2$, poses a problem for our model. For $r < 0$, there are two fixed points: the stable $x^* = \sqrt{-r}$ and the unstable $x^* = -\sqrt{-r}$ fixed points. Any initial condition greater than the unstable fixed point will grow unboundedly, while our variable n_1 must lie in the interval [0,1]. Therefore, we must add another fixed point above the unstable fixed point. A possible way to add this fixed point without perturbing the bifurcation is to consider the following dynamical system

$$\dot{n}_1 = \left[r + (n_1 - n_B)^2\right]\left\{1 - A\left[r + (n_1 - n_B)^2\right]\right\}, \tag{13.12}$$

where A is a positive constant and n_B is a positive constant in the interval [0, 1] chosen so that the two fixed points arising from the saddle node bifurcation are both positive and also lie in this interval. For certain values of A near the bifurcation point, this modified formula introduces two new fixed points—a stable fixed point for large values of n_1 (but still smaller than 1) and an unstable negative fixed point. Therefore, there is only one new fixed point in the interval of interest. Moreover, this new fixed point introduces a bistability region (see Fig. 13.1). For $r < 0$ there are two stable fixed points: the one from the saddle-node bifurcation and the new fixed point introduced by the $\left\{1 - A\left[r + (n_1 - n_B)^2\right]\right\}$ term. Consequently, in this bistability region for the infinite population model, any initial condition below the dashed line in the figure (unstable fixed point) will end up at the bottom fixed point, while initial conditions above the dashed line evolve to the upper fixed point. As the bifurcation parameter r increases and crosses zero, there is a saddle-node bifurcation and the bottom stable fixed point collides with the unstable fixed point and they disappear, leaving only one stable fixed point to which all the initial conditions evolve.

For finite population, however, comparing (13.4) and (13.12), and using (13.5), we have

$$\begin{aligned}
\gamma_2^{(0)} &= -An_B^4 - 2An_B^2 r - Ar^2 + n_B^2 + r, \\
\gamma_2^{(1)} - \gamma_1^{(0)} &= 4An_B^3 + 4An_B r - 2n_B + \gamma_2^{(0)}, \\
\gamma_2^{(2)} - \gamma_1^{(1)} &= -6An_B^2 - 2Ar + 1 + \gamma_2^{(1)},
\end{aligned}$$

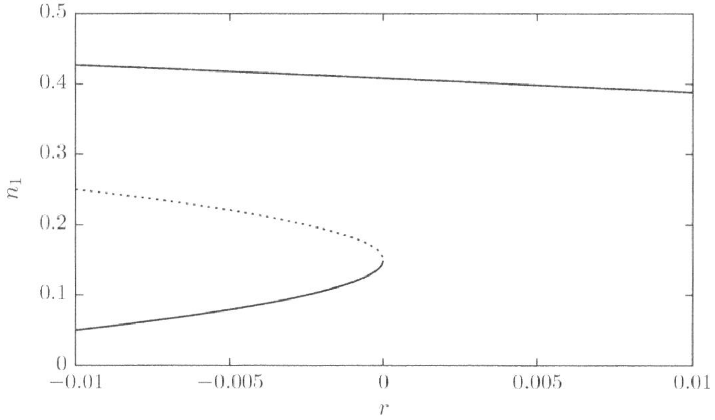

Fig. 13.1 Bifurcation diagram for the model given by (13.12) with $A = 15$ and $n_B = 0.15$. The full lines represent stable fixed points while the dashed line represents unstable fixed points

$$\begin{aligned}\gamma_2^{(3)} - \gamma_1^{(2)} &= 4An_B + \gamma_2^{(2)},\\ \gamma_2^{(4)} - \gamma_1^{(3)} &= -A + \gamma_2^{(3)},\\ \gamma_2^{(k+1)} - \gamma_1^{(k)} &= \gamma_2^{(k)} \qquad \text{for } k > 3.\end{aligned}$$

Hence, the definition of the bifurcation model does not completely define the transition rates. As a matter of fact, different choices of the transition rates $\gamma_1(n_1)$ and $\gamma_2(n_1)$ lead to the same mean field equation. Therefore, different finite population models with different steady states may lead to the same infinite population model. Moreover, the steady state probability density $P_{st}(n_1)$ favors one state. The finite population fluctuations thus destroy the coexistence, as shown in Fig. 13.2. In this figure, we can clearly see that as the number of units increases, the predominance of one peak becomes stronger. That is, as the number of units goes to infinity, one state becomes more and more probable thus confirming the destruction of the coexistence.

We end this section with an important comment: while arrays of Markovian two-state units can only lead to stationary ordering, the inclusion of a memory, e.g. a refractory period that forces units arriving in state 2 to wait a certain amount of time before returning to state 1, yield time-dependent oscillations [10].

13.2.2 Three-State Model

While two-state Markovian models can only provide synchronization as an asymmetric steady state for which one state is more populated than the other, three-state Markovian models may lead to the more striking form of synchronization in which an aggregate of units move together from one state to the next, to the next, and so on. We consider a set of states (in this case states 1, 2 and 3) and transition rates that

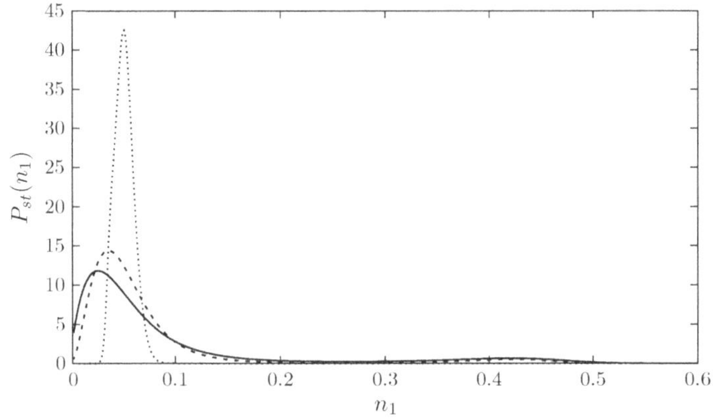

Fig. 13.2 Steady state probability density as a function of the density of units in state 1. For this figure, we used the following values for the parameters: $A = 15, n_B = 0.15, r = -0.01, \gamma_2^{(1)} = 3/2, \gamma_2^{(2)} = 1, \gamma_2^{(3)} = 1/5, \gamma_2^{(4)} = 1/4$ and $\gamma_2^{(k)} = 0$ for $k > 4$. The different curves correspond to different numbers of units: $N = 300$ full line, $N = 500$ dashed line and $N = 5000$ dotted line

dictate the dynamics. Here, we use Wood's model [11], for which the transitions are unidirectional (units in a given state can only stay there or move to the next state in a cyclic way: $1 \to 2 \to 3 \to 1$).

13.2.2.1 Infinite Population

An infinite array is thus governed by the mean field equations

$$\begin{aligned} \dot{n}_1 &= \gamma_{31} - (\gamma_{12} + \gamma_{31})\, n_1 - \gamma_{31} n_2, \\ \dot{n}_2 &= \gamma_{12} n_1 - \gamma_{23} n_2. \end{aligned} \tag{13.13}$$

Once again, we used the density normalization, which in this case reads $n_1 + n_2 + n_3 = 1$, to eliminate the density of one of the states (n_3). Moreover, in Wood's model [11], the transition rates are given by

$$\gamma_{i,i+1} = \gamma \exp\left[a\,(U n_{i+1} + V n_{i-1} + W n_i)\right], \tag{13.14}$$

where the indices are cyclical as noted above.

The symmetry of the model implies that the point $n_1 = n_2 = n_3 = 1/3$ is always a fixed point. A linear analysis [11] shows that this fixed point is stable for $a < a_c = 3/(U - W)$. Further, for $U \neq V$, there is a Hopf bifurcation at $a = a_c$. The type of Hopf bifurcation (subcritical or supercritical) is determined by the sign of the first Lyapunov coefficient l_1, which is found to be

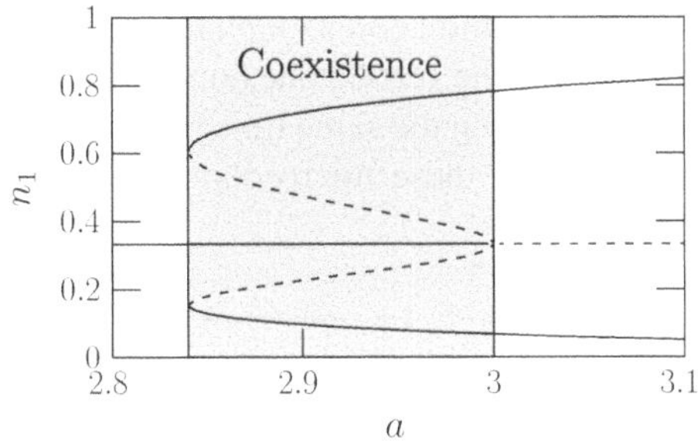

Fig. 13.3 Bifurcation diagram for the three-state model. For the transition rates we used $U = 1$, $V = 4$, and $W = 0$. The horizontal line represents the fixed point and the curves indicate the maximum and minimum values of n_1 in the limit cycle. In both cases the solid lines represent the stable attractors and the dashed lines the unstable ones

$$l_1 = -\frac{9\sqrt{3}\,(U + V - 2W)}{3\,(U - W)}. \tag{13.15}$$

For $l_1 > 0$, the Hopf bifurcation is supercritical, that is, we have a continuous transition and no coexistence region. More interesting for our discussion, for $l_1 < 0$, the Hopf bifurcation is subcritical and presents a coexistence region (see Fig. 13.3). In this case, for $a \equiv a_{lc}$ a pair of limit cycles (one stable and one unstable) is created. The stable limit cycle coexists with the symmetry-dictated fixed point. As a increases further and approaches a_c, the unstable limit cycle shrinks while the stable one grows. At $a = a_c$ the unstable limit cycle radius vanishes whilst it collides with the fixed point $n_1 = n_2 = n_3 = 1/3$. For $a < a_{lc}$ and $a > a_c$ there is only one attractor, the fixed point or the limit cycle, respectively. Between the two, $a_{lc} < a < a_c$, however, there are two stable attractors and there is coexistence.

13.2.2.2 Finite Populations

Next we move to the finite population case. As in the two-state model, we start by writing the evolution equation for the number of units in each state. However, for the three-state case, we need to follow the number of units in two states, say 1 and 2, the third one being determined from the condition $N = N_1 + N_2 + N_3$. A simple counting protocol leads to

$$\begin{aligned} N_1(t + \mathrm{d}t) &= N_1(t) - \sum_{k=1}^{N_1} \theta\left(\gamma_{12}(n_1, n_2)\,\mathrm{d}t - \zeta_k\right) + \sum_{k=N_1+N_2+1}^{N} \theta\left(\gamma_{31}(n_1, n_2)\,\mathrm{d}t - \zeta_k\right), \\ N_2(t + \mathrm{d}t) &= N_2(t) - \sum_{k=N_1+1}^{N_1+N_2} \theta\left(\gamma_{23}(n_1, n_2)\,\mathrm{d}t - \zeta_k\right) + \sum_{k=1}^{N_1} \theta\left(\gamma_{12}(n_1, n_2)\,\mathrm{d}t - \zeta_k\right). \end{aligned} \tag{13.16}$$

where the first sum in the N_1 equation counts the number of units leaving state 1 to state 2 between t and $t + \mathrm{d}t$, and the second one counts the number of units arriving at state 1 coming from state 3 during the same time interval. A similar interpretation holds for the N_2 equation. From these microscopic equations, one arrives at the Langevin equations

$$\dot{n}_1 = \gamma_{31}(1 - n_1 - n_2) - \gamma_{12}n_1 + \sqrt{\gamma_{31}(1 - n_1 - n_2)}\,\frac{\xi_1(t)}{\sqrt{N}} - \sqrt{\gamma_{12}n_1}\,\frac{\xi_2(t)}{\sqrt{N}},$$
$$\dot{n}_2 = \gamma_{12}n_1 - \gamma_{23}n_2 + \sqrt{\gamma_{12}n_1}\,\frac{\xi_2(t)}{\sqrt{N}} - \sqrt{\gamma_{23}n_2}\,\frac{\xi_3(t)}{\sqrt{N}}. \tag{13.17}$$

From there using the Itô interpretation we obtain the Fokker–Planck equation

$$\frac{\partial P(n_1, n_2, t)}{\partial t} = \frac{\partial \Phi_1}{\partial n_1} + \frac{\partial \Phi_2}{\partial n_2} \tag{13.18}$$

where

$$\Phi_i = -\mu_i P + \sum_{j=1}^{2} \frac{\partial (D_{ij} P)}{\partial n_j}, \tag{13.19}$$

with

$$\mu = \begin{pmatrix} \gamma_{31}(1 - n_1 - n_2) - \gamma_{12}n_1 \\ \gamma_{12}n_1 - \gamma_{23}n_2 \end{pmatrix}, \tag{13.20}$$

being the drift vector, and

$$D = \frac{1}{2N}\begin{pmatrix} \gamma_{12}n_1 + \gamma_{31}(1 - n_1 - n_2) & -\gamma_{12}n_1 \\ -\gamma_{12}n_1 & \gamma_{12}n_1 + \gamma_{23}n_2 \end{pmatrix}, \tag{13.21}$$

the diffusion matrix.

The steady state solution $P_{ss}(n_1, n_2)$ of the Fokker–Planck equation, (13.18), can be obtained numerically. We illustrate the qualitative behavior of $P_{ss}(n_1, n_2)$ for two values of the control parameter a (both in the infinite population coexistence region) and two different numbers of units. In Fig. 13.4, for the smaller value of a ($a = 2.85$), we can see that for small N (left panel) there is a coexistence of the symmetric fixed point (center of the triangle) and the limit cycle, characterized by the brighter triangular region. For larger populations (right panel) there is only one bright spot in the middle of the triangle, indicating that the steady state only presents small fluctuations around the state $n_1 = n_2 = n_3 = 1/3$ and there is no limit cycle and hence no coexistence. This result is thus similar to that of the two-state model and is again due to the fact that the $t \to \infty$ and $N \to \infty$ limits do not commute.

When we increase the value of a and approach a_c (Fig. 13.5) the situation for small populations (left panel) barely changes—we can see that the limit cycle now is slightly favored in comparison to the fixed point, but the coexistence is still there. However, for the larger population (right panel) the limit cycle clearly dominates

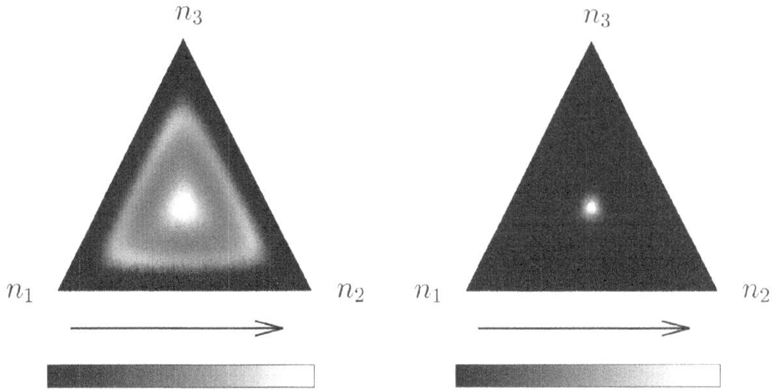

Fig. 13.4 Steady state probability density for the three-state model with $U = 1$, $V = -4$ and $W = 0$. For both panels, $a = 2.85$, while $N = 500$ for the left panel and $N = 5000$ for the right panel. The gray shaded horizontal bar indicates the code for the values of P_{ss} and the arrows indicate the direction of increase (small values of P_{ss} are darker and larger values are brighter)

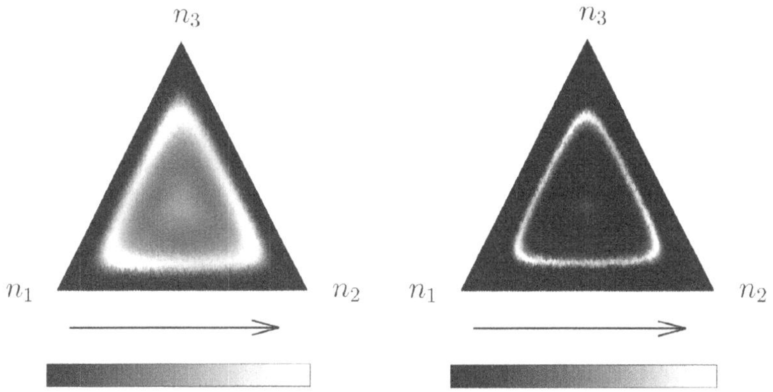

Fig. 13.5 Steady state probability density for the three state model with $U = 1$, $V = -4$ and $W = 0$. For both panels, $a = 2.87$, while $N = 500$ for the left panel and $N = 5000$ for the right panel. The gray shaded horizontal bar indicates the code for the values of P_{ss} and the arrows indicate the direction of increase (small values of P_{ss} are darker and larger values are brighter)

(there is only a very weak bright spot in the center of the triangle that completely fades out for even larger populations). Therefore, once again, the finite population fluctuations destroy the coexistence in the limit of large populations.

An order parameter that is a discrete version of one used by Kuramoto for continuous phases is

$$r(n_1, n_2) = \left| \frac{1}{N} \sum_{k=1}^{N} \exp(i\phi_k) \right| = |n_1 + n_2 \exp(i2\pi/3) + (1 - n_1 - n_2) \exp(i4\pi/3)| , \tag{13.22}$$

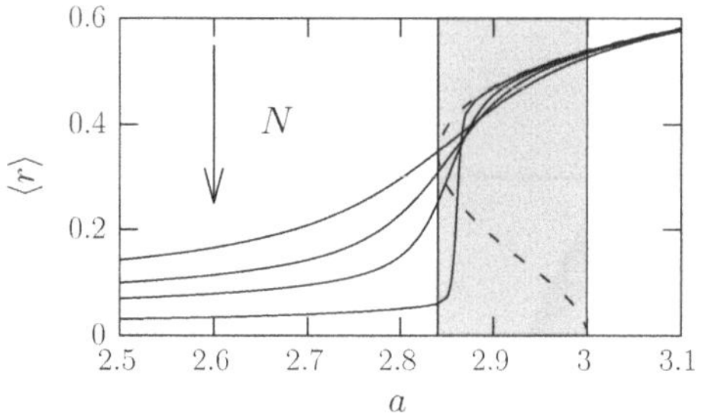

Fig. 13.6 Kuramoto order parameter as a function of the control parameter a for $U = 1$, $V = -4$ and $W = 0$. The full lines represent different numbers of units ($N = 250$, 500, 1000, 5000 and 50000) and the dashed line corresponds to the case of an infinite number of units ($N \to \infty$ and then $t \to \infty$). The gray region, indicates the coexistence region for the infinite number of units case, for which the order parameter is double valued

where the phase ϕ_k of state k is defined as $\phi_k = \frac{2\pi}{3}(k-1)$. The average of the order parameter over the steady state is then given by

$$\langle r \rangle = \int \int r(n_1, n_2) P_{ss}(n_1, n_2) dn_1 dn_2. \tag{13.23}$$

In Fig. 13.6 we show the average order parameter as a function of a for several numbers of units and also for infinite N. As N increases, the order parameter curve becomes stiffer and stiffer, indicating a first-order transition and confirming our assertion that the coexistence is destroyed.

13.3 Coarse Graining Kuramoto's Model

So far we have only discussed discrete-state models. In this section we will present a formal connection between continuous phase and discrete phase stochastic dynamics to explore whether coarse graining a continuous phase model can lead to, say, a three-state model. The results, as we will see, are somewhat unexpected. For this purpose we will use the normal form formalism. Full details of this approach can be found in [12].

We start with a globally coupled array of N continuous phase oscillators. The state of each oscillator can be described by a d-dimensional vector $\mathbf{X}$. That is, the entire array is described by the variables $\{\mathbf{X}_s\}_{s=1}^N$, which obey the equations of motion

$$\dot{\mathbf{X}}_s = \mathcal{F}(\mathbf{X}_s) + \mathcal{I}(\mathbf{X}_1, \ldots, \mathbf{X}_N) + \chi_s(t). \tag{13.24}$$

Here $\mathcal{F}$ accounts for the internal dynamics of each unit. These dynamics may be schematically represented by Fig. 13.7a, which is meant to show an arbitrary limit cycle. The function $\mathcal{I}$ accounts for the interactions among the members of the ensemble, which, in the mean field approach, takes the form

$$\mathcal{I}\left(\mathbf{X}_1, \ldots, \mathbf{X}_N\right) = \mathcal{I}\left(\mathcal{R}\right) \quad \text{where} \quad \mathcal{R} = \frac{1}{N}\sum_{s=1}^{N} A_s. \tag{13.25}$$

The functions $\mathcal{F}$ and $\mathcal{I}$ are identical for all the oscillators and hence do not carry an index. The last term of (13.24), $\chi_s(t)$, represents the inherent fluctuations of each oscillator.

In the vicinity of a Hopf bifurcation that gives rise to the formation of a limit cycle, the dynamics of each oscillator can be reduced to a two-dimensional complex amplitude (central manifold theorem) that obeys the normal form

$$\dot{A}_s = J\left(1 - |A_s|^2\right)A_s + Kf\left(|\mathcal{R}|^2\right)\mathcal{R} + \sqrt{\eta}\zeta_s(t). \tag{13.26}$$

Here the real constant parameter J governs the internal dynamics of each unit, and we have scaled out irrelevant constants. In particular, this equation is in a moving frame: we have removed the natural frequency ω of the oscillators. In these amplitude variables, the internal dynamics lead to a perfectly circular limit cycle of the array, as illustrated in Fig. 13.7b. The parameter K is a measure of the strength of the interactions which is written in a way that respects the phase invariance. Moreover, the generic function f is positive definite so as to model an attractive interaction between oscillators. In the original Kuramoto model [5], $f(|\mathcal{R}|^2) = 1$ and the interaction is then linear in (R). For the fluctuations we choose δ-correlated complex Gaussian noises:

$$\zeta_s(t) = \zeta_R^s(t) + i\zeta_I^s(t), \tag{13.27}$$

where $\zeta_R^s(t)$ and $\zeta_I^s(t)$ are independent real Gaussian white noises of zero mean and correlation functions

$$\left\langle\zeta_R^s(t)\,\zeta_R^{s'}(t')\right\rangle = \left\langle\zeta_I^s(t)\,\zeta_I^{s'}(t')\right\rangle = \delta_{ss'}\delta\left(t - t'\right), \quad \text{and} \quad \left\langle\zeta_R^s(t)\,\zeta_I^{s'}(t')\right\rangle = 0. \tag{13.28}$$

If the internal dynamics dominates over the interaction and fluctuations ($J \gg K$ and $J \gg \eta$), after a short transient $|A_s| \sim 1$, and the system can be described by the phase equations

$$\dot{\phi}_s = KF(r)\sin\left(\psi - \phi_s\right) + \sqrt{\eta}\xi_s(t) \quad \text{where} \quad R = \frac{1}{N}\sum_{s=1}^{N} e^{i\phi_s} \equiv re^{i\psi}, \tag{13.29}$$

with $F(r) = rf(r^2)$. The phase evolution is represented by Fig. 13.7c, where only the phase of oscillation is relevant.

At the mean field level, this set of stochastic differential equations can be described by a nonlinear Fokker–Planck equation for the one-particle probability density

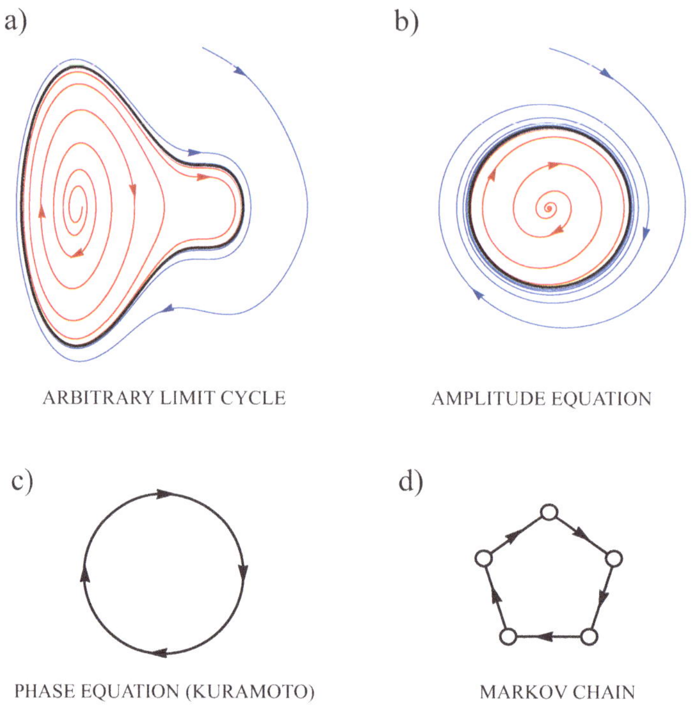

Fig. 13.7 Schematic picture of the coarse-graining of the phase variable. **a** Schematic of an arbitrary limit cycle. **b** Amplitude equation near the point where the limit cycle first develops. **c** Reduction to phase dynamics. **d** Markov chain model for coarse-grained phases

$\rho(\phi, t)$, that takes the form

$$\frac{\partial \rho}{\partial t} = \frac{\eta}{2} \frac{\partial^2 \rho}{\partial \phi^2} - K \frac{\partial}{\partial \phi} \{\rho \Omega [\rho, \phi]\}, \tag{13.30}$$

where the second derivative term on the right is the diffusion term, and where the drift contains

$$\Omega [\rho, \phi] = F(r [\rho]) \sin (\psi [\rho] - \phi), \tag{13.31}$$

with

$$R = r [\rho] e^{i\psi[\rho]} \equiv \int_0^{2\pi} \rho(\phi, t) e^{i\phi} d\phi. \tag{13.32}$$

The asynchronous state corresponds to a uniform distribution of phases,

$$\rho(\phi) = \frac{1}{2\pi}, \tag{13.33}$$

which destabilizes at the critical point

$$K_c = \frac{\eta}{f(0)}. \tag{13.34}$$

Near the onset of synchronization, the dynamics can be described by the normal form

$$\dot{R} = \left(\alpha - \beta\,|R|^2\right) R \text{ with } \alpha = \frac{f(0)}{2}\,(K - K_c)\,, \text{ and } \beta = \frac{K_c}{2}\left(\frac{1}{2} f(0) - f'(0)\right), \tag{13.35}$$

and $f'(0) \equiv (df(r)/dr)|_{r=0}$. The bifurcation is supercritical if $\beta > 0$ and subcritical if $\beta < 0$. Furthermore, the synchronization is tighter if $f(r)$ is an increasing function, representing stronger interactions with increasing r. Since we are interested in a coarse-graining of the phase, for which a coarser synchronization is more propitious, we will focus on a decreasing function $f(r)$. Then β is always positive and the bifurcation will always be supercritical.

We are now ready to perform the coarse graining of the phase. Consider M discrete phases,

$$\phi \in [0, 2\pi] \;\rightarrow\; \phi \in \{j\Delta\phi\}_{j=0}^{M-1}, \quad \text{where} \quad \Delta\phi = \frac{2\pi}{M}.$$

Discretizing the nonlinear Fokker–Planck (13.30), we obtain

$$\dot{P}_j = -\left(w_{j\to j+1} + w_{j\to j-1}\right) P_j + w_{j+1\to j} P_{j+1} + w_{j-1\to j} P_{j-1}, \tag{13.36}$$

where $P_j(t)$ is the probability to be in the jth phase, and

$$w_{j\to j\pm1} = \frac{\eta}{2(\Delta\phi)^2} \mp \frac{K}{2\Delta\phi}\Omega_j, \quad \text{with} \quad \Omega_j = F(r)\sin(\psi - j\Delta\phi)\,. \tag{13.37}$$

For this to be an acceptable physical description, the transition rates (13.37) must be positive. A bound to insure this is (see [12])

$$K < K_{\max} = \frac{\eta}{F_{\max}\Delta\phi}, \tag{13.38}$$

where $F_{\max}$ is the maximum of the function $F(r)$ in the interval $r \in [0, 1]$. The coarse-grained dynamics can be interpreted as a Markov chain, as illustrated in Fig. 13.7d.

This discrete formalism reproduces the bifurcation structure of the continuous phase model for $M \geq 4$. Figure 13.8 displays our numerical results for both dynamics, continuous and discrete phases. Figure 13.8a shows numerical simulations of the amplitude equations (13.26), $|A_s| \sim 1$ with a small dispersion but with an agglomeration of the phases that indicates that the system has crossed the critical point where phase synchronization first occurs. The seven-bar histogram of Fig. 13.8b has used the same data as Fig. 13.8a, while the continuous curve is a numerical solution of

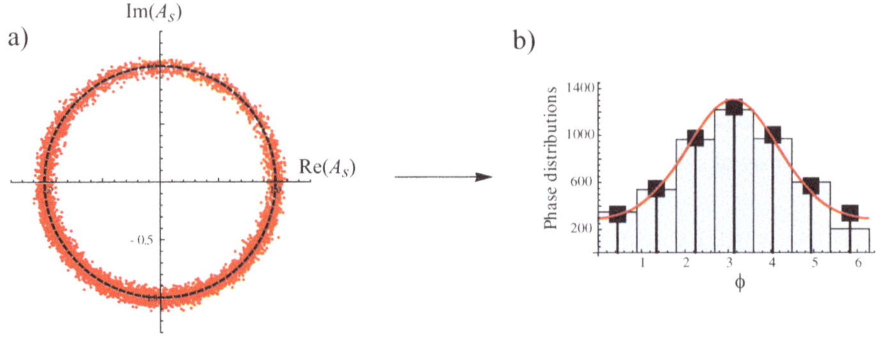

Fig. 13.8 Numerical computations for $J = 200$, $N = 5000$, $K = 1.5708 > K_c = 0.98696$, $\eta = 0.98696$, and $f(r) = \exp(-r/a)$ with $a = 0.3$. **a** Amplitude equations (13.26), showing an agglomeration of phases along the limit cycle. **b** Comparison between phase distribution from the solution of the nonlinear Fokker–Planck equation (13.30), and a seven-node Markov chain

the nonlinear Fokker–Planck equation (13.30). The black squares in Fig. 13.8b are the result of a seven-node Markov chain following the prescription described above, showing good agreement between both approaches.

The case of $M = 3$, however, turns out to be pathological, which we did not anticipate. The bifurcation turns transcritical, and the system displays a very different dynamical behavior than in the continuous phase model. A full analysis of these sorts of three-state units can be found in [12].

13.4 Conclusions and Perspectives

The vast majority of the enormous literature on synchronization phenomena assumes continuous time and often continuous state space for the evolution of the interacting units of interest. However, a number of years ago we concluded that discrete time and discrete state space would make these problems more approachable. In this Chapter we discussed two aspects of synchronization related to discrete state models, namely, the role of the number of units and of the number of states in the synchronization.

We presented arrays of two-state Markovian units and arrays of three-state Markovian units. In both cases, when the arrays of an infinite number of units present coexistence of stable states, the fluctuations created by a finite number of units destroy the coexistence and the system "chooses" one of the stable states in the limit of a large number N of units. We traced the apparent contradiction of the limit of a large number of units unable to recover the coexistence to an ergodicity breaking—the order of the limits $t \to \infty$ and $N \to \infty$ matters.

For arrays of globally coupled two-state Markovian units, when there is an infinite number of these units in the array the problem becomes deterministic (described by mean field equations) unless there is some external source of noise, which we have not included in our description. We were particularly interested in the steady state of the system, which is necessarily time independent—two-state Markovian units do not lead to time dependent configurations as $t \to \infty$. We assumed a (modified) nonlinear (polynomial) normal form for the interactions among the units which leads the system to a saddle node bifurcation. For values of the control parameter below the bifurcation value, there are two stable states, one arising from the ordinary normal form for this type of bifurcation, and the other from the modification that keeps the fractions of systems in each of the two states in the physical regime 0 to 1. (There is also an unstable state in the bifurcation regime.) Hence, there is a coexistence of two stable states in this region, the steady state being entirely settled by the deterministic equation and the initial condition.

We then considered a globally coupled array of a finite number of two-state units. The finite number introduces a noise term that entirely changes the behavior of the system. In particular, as $t \to \infty$ one state becomes more populated than the other, and any memory of the initial state is forgotten. This imbalance in the population of the two states describes a steady state synchronization. If we now take the limit $N \to \infty$, the system does not arrive at the description in the previous paragraph. The initial state is forgotten in this case. In other words, as noted above, the limits $t \to \infty$ and $N \to \infty$ do not commute. As an aside, we noted that the inclusion of a memory, for instance a refractory period that forces any unit that arrives at say state 2 to wait for some time before returning to state 1, yields time-dependent oscillations [10].

We next considered a globally coupled array of three-state Markovian units. Once again we started with a deterministic mean field noiseless model, with a different form of the interactions, namely, exponential (highly nonlinear). As in the two-state case, there is a coexistence of two stable states. However, in this case the bifurcation is more interesting. Here, the bifurcation involves not only fixed points but a fixed point and a limit cycle. The limit cycle implies that the steady state encompasses a periodic variation of the densities. This is a more striking form of synchronization, that is, one in which the units move together in unison. We moved on to the case of a finite number of globally coupled units in the array, which naturally introduces a noise contribution. Depending on the number of units and other parameter values, the coexistence is still there, but one or the other stable state becomes weaker (i.e. fewer units are in one stable state than in the other) as the number of units increases. Once again the limits $t \to \infty$ and $N \to \infty$ do not commute and the coexistence is washed away.

Finally we considered an entirely different question: is it possible to coarse grain a collection of nonlinearly interacting Kuramoto oscillators to arrive at a three-state (or two-state) model? The Kuramoto system resides in continuous time and continuous phase, the latter ranging continuously from 0 to 2π, while our models have a finite number of states. We showed that reduction by coarse graining is possible, but that our particular three-state model can not be obtained via a simple coarse-graining procedure of the Kuramoto model. We are able to reduce the continuum model to discrete ones in this manner but with more than three states. On the other hand, reduction to a three-state model is possible, but with different critical synchronization behavior than our model.

What remains to be done? Of course in any attempt to model synchronization phenomena there is a very large field that others have populated, and that we have *not yet* approached. There is also a huge literature on synchronization applications. To name just a few directions we may take, we note the problems of the range of interactions in three-state models, the role of the fluctuations due to the finite number of units in systems with growing populations and the establishment of a general framework for the coarse graining of continuous models.

In [13], we showed that for a two-state model the range of interaction produces a transition from a disordered (short range interactions) to an ordered state (long range interactions). Preliminary studies for a three-state model have indicated that increasing the range of interactions interpolates the disordered and ordered (synchronized) states with patterned spatial formations (spirals appear for middle range interactions). A variant of Wood's model for which there is a birth term was proposed in [14]. Such a model also presents coexistence in the mean field approach. The question we pose is whether the fluctuations produced by the finite number of units also destroys the coexistence for this model. Finally, we mention again that we studied the coarse graining of Kuramoto's model. However, the possibility of mapping the phase transition of other continuous models into discrete coarse grained models is still an open question. We hope our work along these lines will continue to shed light on these phenomena.

Acknowledgements A.R. acknowledges the financial support of CPNq (Grant No.308344/2018-9). D.E. thanks funding from Fondecyt-Chile (Grant No.1170669).

References

1. Synchronize (The Merriam-Webster.com Thesaurus), https://www.merriam-webster.com/thesaurus/synchronize. Accessed 27 January 2020
2. Firefly (Wikipedia, The Free Encyclopedia), https://en.wikipedia.org/w/index.php?title=Firefly&oldid=925079214. Accessed 27 January 2020
3. A. Pikovsky, M. Rosenblum, J. Kurths, *Synchronization: A Universal Concept in Nonlinear Sciences* (Cambridge University Press, Cambridge, 2001)
4. G.V. Osipov, J. Kurths, C. Zhou, *Synchronization in Oscillatory Networks* (Springer, Berlin, 2007)
5. Y. Kuramoto, *International Symposium on Mathematical Problems in Theoretical Physics*, ed. by H. Araki. Kyoto University, January 1975. Lecture Notes in Physics, Vol. 39 (Springer, New York, 1975), p. 420
6. I.L.D. Pinto, D. Escaff, U. Harbola, A. Rosas, K. Lindenberg, Phys. Rev. E **89**, 052143 (2014)
7. A. Rosas, D. Escaff, I.L.D. Pinto, K. Lindenberg, J. Phys. A: Math. Theor. **49**, 095001 (2016)
8. S.H. Strogatz, *Nonlinear Dynamics and Chaos: With Applications to Physics, Biology, Chemistry and Engineering*, 2nd edn. (CRC Press, Boca Raton, 2015)
9. B. West, A. Bulsara, K. Lindenberg, V. Seshadri, K. Shuler, Physica A **97**, 211 (1979)
10. D. Escaff, U. Harbola, K. Lindenberg, Phys. Rev. E **86**, 011131 (2012)
11. K. Wood, C. Van den Broeck, R. Kawai, K. Lindenberg, Phys. Rev. E **76**, 041132 (2007)
12. D. Escaff, A. Rosas, R. Toral, K. Lindenberg, Phys. Rev. E **94**, 052219 (2016)
13. A. Rosas, D. Escaff, I.L.D. Pinto, K. Lindenberg, Phys. Rev. E **95**, 032104 (2017)
14. W. Yu, K. Wood, Phys. Rev. E **91**, 062708 (2015)

Chapter 14
Physics-Informed Learning Machines for Partial Differential Equations: Gaussian Processes Versus Neural Networks

Guofei Pang and George Em Karniadakis

Abstract We review and compare physics-informed learning models built upon Gaussian processes and deep neural networks for solving forward and inverse problems governed by linear and nonlinear partial differential equations. We define a unified data model on which Gaussian processes, physics-informed Gaussian processes, neural networks, and physics-informed neural networks are based. We develop continuous-time and discrete-time models to facilitate different application scenarios. We present a connection between a Gaussian process and an infinitely wide neural network, which enables us to obtain a "best" kernel, which is determined directly by the data. We demonstrate the implementation of physics-informed Gaussian processes and physics-informed neural networks using a pedagogical example. Additionally, we compare physics-informed Gaussian processes and physics-informed neural networks for two nonlinear partial differential equations, i.e. the 1D Burgers' equation and the 2D Navier–Stokes, and provide guidance in choosing the proper machine learning model according to the problem type, i.e. forward or inverse problem, and the availability of data. These new methods for solving partial differential equations governing multi-physics problems do not require any grid, and they are simple to implement, and agnostic to specific application. Hence, we expect that variants and proper extensions of these methods will find broad applicability in the near future across different scientific disciplines but also in industrial applications.

14.1 Introduction

Unlike everyday commercial applications which are characterized by big data, in scientific applications the data are scarce, noisy, unstructured, and of multifidelity.

G. Pang · G. E. Karniadakis (✉)
Division of Applied Mathematics, Brown University, 170 Hope Street, Providence, RI 02912, USA
e-mail: george_karniadakis@brown.edu

G. Pang
e-mail: guofei_pang@brown.edu

P. G. Kevrekidis et al. (eds.), *Emerging Frontiers in Nonlinear Science*, Nonlinear Systems and Complexity 32, https://doi.org/10.1007/978-3-030-44992-6_14

In the *small data* regime, supervised learning, where machine learning models only act as black-box function approximators, is not adequate. However, in scientific machine learning we have conservation laws expressed by partial differential equations (PDEs), which can provide additional information to a machine learning model. For example, let us consider a parameter estimation problem, aiming to estimate $\boldsymbol{\gamma}$ in the PDE $\mathcal{L}^{\boldsymbol{\gamma}} u = f$, where $\mathcal{L}$ is a differential operator and $\boldsymbol{\gamma}$ is a collection of unknown parameters of $\mathcal{L}$. We can identify $\boldsymbol{\gamma}$ by solving the PDE N times under N different $\boldsymbol{\gamma}$'s. From the training set $\mathcal{D} = \{\boldsymbol{\gamma}_i, u(\boldsymbol{x}; \boldsymbol{\gamma}_i)\}_{i=1}^N$, we can train a machine learning model $\tilde{\mathcal{M}}$ to approximate the mapping from $\boldsymbol{\gamma}$ to solution u, i.e, $u(\cdot\,; \boldsymbol{\gamma}) = \mathcal{M}(\boldsymbol{\gamma})$, after which we search for the optimal $\boldsymbol{\gamma}$ that minimizes the misfit $\|\tilde{\mathcal{M}}(\boldsymbol{\gamma}) - u_{\text{obs}}\|$, where u_{obs} are observations of u and the norm $\|\cdot\|$ is a norm in the function space where u resides. In this setup, the machine learning model is a black-box function approximator. A large number of training points, namely, large N, is needed to train the model well. In contrast, by incorporating the expression of the PDE into the misfit function explicitly, namely, $\|\tilde{u}(\boldsymbol{\theta}) - u_{\text{obs}}\| + \|\mathcal{L}^{\boldsymbol{\gamma}} \tilde{u}(\boldsymbol{\theta}) - f\|$, where we minimize the misfit with respect to both $\boldsymbol{\gamma}$ and the parameters $\boldsymbol{\theta}$ of the machine learning model $\tilde{u}$, we can reduce the number of training points $\{\boldsymbol{x}_i, u_{\text{obs}}(\boldsymbol{x}_i)\}_{i=1}^N$ substantially. We call this paradigm *physics-informed learning*.

In this chapter, we give a brief introduction to two typical machine learning models, i.e. Gaussian processes (GPs) and deep Neural Networks (NNs), which can be easily incorporated into the physics-informed learning paradigm. Gaussian processes and physics-informed Gaussian processes (PIGPs) [1–4] are among the class of probabilistic machine learning algorithms [5, 6]. Both of the models provide the posterior probabilistic distribution of their predictions in the context of Bayesian inference, and enjoy the analytical tractability in doing Bayesian inference. Despite the mathematical elegance, these models compromise the accuracy of their predictions for strongly nonlinear PDEs by linearizing nonlinear terms in time. Physics-informed neural networks (PINNs) [7–10] can handle nonlinear PDEs without linearization, and can also be easily modified to handle high-order temporal discretization, such as the 500-stage Runge–Kutta method considered in [10]. Although GPs and NNs are both universal approximators for arbitrary continuous functions [11, 12], NNs are preferred in real-world applications, such as image recognition and natural language processing, due to the linear computational cost with respect to the number of model parameters, compared to the cubic cost with respect to training set size in standard GPs. Unlike GPs, the standard NNs are, however, not a probabilistic machine learning model, and thus it cannot estimate the uncertainty associated with its predictions.

The rest of the chapter is organized as follows. In Sect. 14.2, we introduce a unified data model working for all of GPs, PIGPs, NNs, and PINNs. In Sect. 14.3 we first briefly review the basic ideas of GPs and NNs for function approximation, then introduce the methodology of PIGPs and PINNs in solving a linear PDE, and finally discuss the connection between GPs and NNs. A pedagogical example for linear diffusion problem is given in Sect. 14.4 to illustrate how to implement PIGPs and PINNs. The corresponding Python codes are also shared. Section 14.5 compares

PIGPs and PINNs for solving forward and inverse problems of nonlinear PDEs. Summary and outlook are provided in the last section.

14.2 Data Models for Machine Learning

We consider the following data model relating the model input $\boldsymbol{x}$ to m groups of observations $\{y_j\}_{j=1}^m$

$$\begin{aligned}
y_1(\boldsymbol{x}) &= \mathcal{L}_1^{\gamma_1}\{f_1(\boldsymbol{x}), f_2(\boldsymbol{x}), \ldots, f_p(\boldsymbol{x})\} + \epsilon_1, \quad \boldsymbol{x} \in \Omega_1, \\
y_2(\boldsymbol{x}) &= \mathcal{L}_2^{\gamma_2}\{f_1(\boldsymbol{x}), f_2(\boldsymbol{x}), \ldots, f_p(\boldsymbol{x})\} + \epsilon_2, \quad \boldsymbol{x} \in \Omega_2, \\
&\vdots \quad \vdots \\
y_m(\boldsymbol{x}) &= \mathcal{L}_m^{\gamma_m}\{f_1(\boldsymbol{x}), f_2(\boldsymbol{x}), \ldots, f_p(\boldsymbol{x})\} + \epsilon_m, \quad \boldsymbol{x} \in \Omega_m.
\end{aligned} \tag{14.1}$$

There are n quantities of interest (QoIs): $f_1(\boldsymbol{x}), f_2(\boldsymbol{x}), \ldots, f_n(\boldsymbol{x})$. These quantities can be velocity, pressure, concentration, etc. The argument $\boldsymbol{x}$ can be spatial and/or temporal coordinates. The input space $\Omega \subset \mathbb{R}^d$, where $\boldsymbol{x}$ resides, is assumed to be a compact set. We will typically denote a column vector by a boldface letter and a scalar by a standard letter. Note the difference between $\boldsymbol{x} \in \mathbb{R}^d$ and $x \in \mathbb{R}$. It is reasonable to assume that the QoIs $\{f_k(\boldsymbol{x})\}_{k=1}^p$ and their transformation $\mathcal{L}^\gamma\{f_1(\boldsymbol{x}), f_2(\boldsymbol{x}), \ldots, f_p(\boldsymbol{x})\}$ can be observed at certain subsets of the input space. For instance, letting $\boldsymbol{x} = [x, t]^\top$, $p = 1$, $m = 2$, $\mathcal{L}_1 = \mathcal{I}$ (identity operator), and $\mathcal{L}_2 = \frac{\partial^2}{\partial x^2}$, we observe y_1 at N_1 different locations: $y_1(\boldsymbol{x}_{1,1}), y_1(\boldsymbol{x}_{1,2}), \ldots, y_1(\boldsymbol{x}_{1,N_1})$ for $\boldsymbol{x}_{1,i} \in \Omega_1 = (0, 1) \times (0, 0.5]$. We call $y_1(\boldsymbol{x}_{1,i}) = f(\boldsymbol{x}_{1,i}) + \epsilon_1$ the ith observation of the data model in the subset of the input space Ω_1. Also, we observe $y_2(\boldsymbol{x}_{2,1}), y_2(\boldsymbol{x}_{2,2}), \ldots, y_2(\boldsymbol{x}_{2,N_2})$ for $\boldsymbol{x}_{2,i} \in \Omega_2 = [1, 2] \times (0.5, 1.0]$, where $y_2(\boldsymbol{x}_{2,i}) = \frac{\partial^2 f(x_{2,i}, t_{2,i})}{\partial x^2} + \epsilon_2$. The symbol ϵ_j denotes the additive white Gaussian noise for the jth group of observations, and unless otherwise stated, $\epsilon_1, \epsilon_2, \ldots, \epsilon_m$ are assumed to have zero mean and variance σ_n^2. The symbol $\boldsymbol{\gamma}$ denotes a collection of parameters in the operator $\mathcal{L}$, e.g. $\mathcal{L}^\gamma = (c_1^2 + c_2)\frac{\partial^2}{\partial x^2}$, where $\boldsymbol{\gamma} = [c_1, c_2]^\top$. Let us look at three typical data models:

- *Data model for a scalar-valued function* $f(\boldsymbol{x})$. $y(\boldsymbol{x}) = f(\boldsymbol{x}) + \epsilon$, *where* $p = m = 1$, $\mathcal{L}^\gamma = \mathcal{I}$
- *Data model for the forward problem governed by Poisson equation*

$$\begin{aligned}
y_1(\boldsymbol{x}_1) &= c\nabla^2 u(\boldsymbol{x}_1) + \epsilon_1 = s(\boldsymbol{x}_1) + \epsilon_1, \ \boldsymbol{x}_1 \in \Omega_1 \\
y_2(\boldsymbol{x}_2) &= u(\boldsymbol{x}_2) + \epsilon_2 = g(\boldsymbol{x}_2) + \epsilon_2, \quad \boldsymbol{x}_2 \in \Omega_2 = \partial\Omega_1,
\end{aligned} \tag{14.2}$$

where $p = 1, m = 2$, $f_1(\boldsymbol{x}) = u(\boldsymbol{x})$, $\gamma_1 = c$, $\mathcal{L}_1^{\gamma_1} = c\nabla^2$, and $\mathcal{L}_2^{\gamma_2} = \mathcal{I}$. This model corresponds to the Poisson problem

$$\begin{aligned} c\nabla^2 u(\boldsymbol{x}) &= s(\boldsymbol{x}), \quad \boldsymbol{x} \in \Omega_1, \\ u(\boldsymbol{x}) &= g(\boldsymbol{x}), \quad \boldsymbol{x} \in \Omega_2 - \partial\Omega_1. \end{aligned} \tag{14.3}$$

Here y_1 and y_2 are the observations for the source term s and the boundary condition g, respectively. The observations are assumed to be contaminated by noise.

Remark 1 There are no established theories that would inform us a priori how many data points are required for the well-possedness of the problems we address here.

- *Data model for parameter estimation of 2D incompressible Navier–Stokes equations (corresponding to continuous-time machine learning models, see Sects. 14.3.3.1 and 14.3.4.1)*

$$\begin{aligned} y_1 &= u_t + \lambda_1(uu_x + vu_y) + p_x - \lambda_2(u_{xx} + u_{yy}) = 0, \quad \boldsymbol{x}_1 \in \Omega_1, \\ y_2 &= v_t + \lambda_1(uv_x + vv_y) + p_y - \lambda_2(v_{xx} + v_{yy}) = 0, \quad \boldsymbol{x}_2 \in \Omega_2 = \Omega_1, \\ y_3 &= u_x + v_y = 0, \quad \boldsymbol{x}_3 \in \Omega_3 = \Omega_1, \\ y_4 &= u + \epsilon_4, \quad \boldsymbol{x}_4 \in \Omega_4 = \Omega_1, \\ y_5 &= v + \epsilon_5, \quad \boldsymbol{x}_5 \in \Omega_5 = \Omega_1, \end{aligned} \tag{14.4}$$

where $p = 3$, $m = 5$, $\boldsymbol{x} = [x, y, t]^{\mathrm{T}}$, $f_1 = u$, $f_2 = v$, $f_3{=}p$, $\boldsymbol{\gamma}_1{=}\boldsymbol{\gamma}_2 = [\lambda_1, \lambda_2]^{\mathrm{T}}$, and $\mathcal{L}_1^{\boldsymbol{\gamma}_1}$ and $\mathcal{L}_2^{\boldsymbol{\gamma}_2}$ are nonlinear operators acting on the velocity components u and v and the pressure p. $\mathcal{L}^{\boldsymbol{\gamma}_3}\{u, v, p\} = u_x + v_y$, $\mathcal{L}^{\boldsymbol{\gamma}_4}\{u, v, p\} = u$, and $\mathcal{L}^{\boldsymbol{\gamma}_5}\{u, v, p\} = v$.

14.3 Two Machine Learning Models

We define the machine learning task considered in this chapter as follows: given m groups of input-output pairs

$$\begin{aligned} \mathcal{D} &= \{\mathcal{D}_1, \mathcal{D}_2, \ldots, \mathcal{D}_m\} \\ &= \{\{\boldsymbol{x}_{1,i}, y_1(\boldsymbol{x}_{1,i})\}_{i=1}^{N_1}, \{\boldsymbol{x}_{2,i}, y_2(\boldsymbol{x}_{2,i})\}_{i=1}^{N_2}, \ldots, \{\boldsymbol{x}_{m,i}, y_m(\boldsymbol{x}_{m,i})\}_{i=1}^{N_m}\} \end{aligned} \tag{14.5}$$

and employing the structure of the data model (14.1), we are asked to predict QoIs $\{f_k\}_{k=1}^{p}$ at never-before-seen location $\boldsymbol{x}^*$. We also need to identify the parameters of the operator (i.e. $\boldsymbol{\gamma}$) if these parameters are not known beforehand.

To learn $\{f_k\}_{k=1}^{p}$ and $\boldsymbol{\gamma}$, first of all, we need to select a machine learning model $f_{\mathrm{ML}}(\boldsymbol{x}; \boldsymbol{\theta})$, which approximates f_k in the data model. Here, $\boldsymbol{\theta}$ is a collection of model (hyper-)parameters to be learned. GPs $f_{\mathrm{GP}}(\boldsymbol{x}; \boldsymbol{\theta})$ and deep NNs $f_{\mathrm{NN}}(\boldsymbol{x}; \boldsymbol{\theta})$ are two widely used machine learning models since they are both universal approximators for arbitrary continuous function [11, 12]. In other words, the function spaces generated by these two models are both dense in $C^0(\Omega)$. The key step of employing the machine learning models is to obtain the model (hyper-)parameters $\boldsymbol{\theta}$ after minimizing certain cost (or loss) functions.

In Sects. 14.3.1 and 14.3.2, we give a brief introduction to GPs and deep NNs models corresponding to the data model for a function, respectively. This introduction aims at demonstrating how to approximate a function using the two models. In Sects. 14.3.3 and 14.3.4, we discuss the extensions of the two models to the data models for PDEs, illuminating data-driven approaches for solving forward and inverse problems governed by PDEs.

14.3.1 Gaussian Process Regression

Recall the data model for a scalar-valued function: $y(\boldsymbol{x}) = f(\boldsymbol{x}) + \epsilon$. A Gaussian process is generally written as $f_{\mathrm{GP}}(\boldsymbol{x}) \sim \mathcal{GP}(m(\boldsymbol{x}), k(\boldsymbol{x}, \boldsymbol{x}'; \boldsymbol{\theta}))$, where the mean function is $m(\boldsymbol{x}) = \mathbb{E}(f_{\mathrm{GP}}(\boldsymbol{x}))$ and the covariance function is $k(\boldsymbol{x}, \boldsymbol{x}'; \boldsymbol{\theta}) = \mathbb{E}\left[(f_{\mathrm{GP}}(\boldsymbol{x}) - m(\boldsymbol{x}))(f_{\mathrm{GP}}(\boldsymbol{x}') - m(\boldsymbol{x}'))\right]$. The symbol $\boldsymbol{\theta}$ represents *ad hoc* parameters in the kernel, a.k.a. hyper-parameters of GPs. A Gaussian process is a collection of random variables, any finite number of which have joint Gaussian distribution [14]. From the definition, the random vector $f_{\mathrm{GP}}(\mathbf{X}) = [f_{\mathrm{GP}}(\boldsymbol{x}_{,1}), \ldots, f_{\mathrm{GP}}(\boldsymbol{x}_{,N})]^{\mathrm{T}}$ obeys the multivariate Gaussian distribution, where $\mathbf{X} = [\boldsymbol{x}_{,1}, \boldsymbol{x}_{,2}, \ldots, \boldsymbol{x}_{,N}]$. A well-known property of multivariate Gaussian distribution is that the conditional distribution of a Gaussian random vector given another Gaussian random vector is still a multivariate Gaussian distribution. Let $\mathbf{f}$ and $\boldsymbol{y}$ be jointly Gaussian random vectors (column vectors)

$$\begin{bmatrix} \mathbf{f} \\ \boldsymbol{y} \end{bmatrix} \sim \mathcal{N}\left(\begin{bmatrix} \boldsymbol{\mu}_{\mathbf{f}} \\ \boldsymbol{\mu}_{\boldsymbol{y}} \end{bmatrix}, \begin{bmatrix} \mathbf{A} & \mathbf{C} \\ \mathbf{C}^{\mathrm{T}} & \mathbf{B} \end{bmatrix}\right), \tag{14.6}$$

where $\boldsymbol{\mu}$ is the mean (column) vector, and $\mathbf{A}$, $\mathbf{B}$, and $\mathbf{C}$ are the covariance matrices, then the conditional distribution of $\mathbf{f}$ given $\boldsymbol{y}$ is [14]

$$\mathbf{f}|\boldsymbol{y} \sim \mathcal{N}(\boldsymbol{\mu}_{\mathbf{f}} + \mathbf{C}\mathbf{B}^{-1}(\boldsymbol{y} - \boldsymbol{\mu}_{\boldsymbol{y}}), \mathbf{A} - \mathbf{C}\mathbf{B}^{-1}\mathbf{C}^{\mathrm{T}}). \tag{14.7}$$

From this formula, we note that the unknown information $\mathbf{f}$ can be inferred from the known information $\mathbf{y}$ as long as we know the correlation between the unknown and known information, which is embedded in the covariance matrix $\mathbf{C}$. We expect to predict f at q never-before-seen locations $\mathbf{X}^* = [\boldsymbol{x}^*_{,1}, \ldots, \boldsymbol{x}^*_{,q}]$, i.e. $f_{\mathrm{GP}}(\mathbf{X}^*) \approx f(\mathbf{X}^*)$. To make a prediction, we assume $f_{\mathrm{GP}}(\mathbf{X}^*)$ and the observations $\boldsymbol{y} = [y(\boldsymbol{x}_{,1}), y(\boldsymbol{x}_{,2}), \ldots, y(\boldsymbol{x}_{,N})]^{\mathrm{T}}$ have the joint Gaussian distribution

$$\begin{bmatrix} f_{\mathrm{GP}}(\mathbf{X}^*) \\ \boldsymbol{y} \end{bmatrix} \sim \mathcal{N}\left(\begin{bmatrix} \mathbf{0}_{q\times 1} \\ \mathbf{0}_{N\times 1} \end{bmatrix}, \begin{bmatrix} k(\mathbf{X}^*, \mathbf{X}^*)_{q\times q} & k(\mathbf{X}^*, \mathbf{X})_{q\times N} \\ k(\mathbf{X}, \mathbf{X}^*)_{N\times q} & k(\mathbf{X}, \mathbf{X})_{N\times N} + \sigma_n^2 \mathbf{I}_{N\times N} \end{bmatrix}\right), \tag{14.8}$$

where $k(\mathbf{X}^*, \mathbf{X}^*)_{ij} = k(\boldsymbol{x}^*_{,i}, \boldsymbol{x}^*_{,j})$ and $\mathbf{K}_{ij} = k(\mathbf{X}, \mathbf{X})_{ij} = k(\boldsymbol{x}_{,i}, \boldsymbol{x}_{,j})$. Using the formula (14.7) yields the posterior mean and covariance functions [14]

$$\mathbb{E}\{f_{\mathrm{GP}}(\mathbf{X}^*)\} = k(\mathbf{X}^*, \mathbf{X})(\mathbf{K} + \sigma_n^2\mathbf{I})^{-1}\boldsymbol{y}, \tag{14.9}$$

$$\mathrm{Cov}\left(f_{\mathrm{GP}}(\mathbf{X}^*), f_{\mathrm{GP}}(\mathbf{X}^*)\right) = k(\mathbf{X}^*, \mathbf{X}^*) - k(\mathbf{X}^*, \mathbf{X})(\mathbf{K} + \sigma_n^2\mathbf{I})^{-1}k(\mathbf{X}, \mathbf{X}^*). \tag{14.10}$$

For single test point $\boldsymbol{x}^*$, the covariance reduces to variance at $\boldsymbol{x}^*$. To make the above predictive formulas work, we also need to know the hyper-parameters $\boldsymbol{\theta}$ and the noise standard deviation (*std* for short) σ_n. They are generally learned by minimizing the negative log marginal likelihood function [14]

$$-\log p(\boldsymbol{\theta}, \sigma_n; \boldsymbol{y}, \mathbf{X}) = \frac{1}{2}\boldsymbol{y}^{\mathrm{T}}(\mathbf{K} + \sigma_n^2\mathbf{I})^{-1}\boldsymbol{y} + \frac{1}{2}\log|\mathbf{K} + \sigma_n^2\mathbf{I}| + \frac{N}{2}\log 2\pi. \tag{14.11}$$

14.3.2 Deep Neural Network Approximation

We still consider the data model for a scalar-valued function: $y(\boldsymbol{x}) = f(\boldsymbol{x}) + \epsilon$, where $\boldsymbol{x} \in \Omega \subset \mathbb{R}^d$. After being fed the input $\boldsymbol{x}$, the deep feedforward neural network will output

$$\begin{aligned} f_{\mathrm{NN}}(\boldsymbol{x}) &= \boldsymbol{x}^l = \mathbf{W}_l\boldsymbol{x}^{l-1} + \mathbf{b}_l, \\ \boldsymbol{x}^j &= G(\mathbf{W}_j\boldsymbol{x}^{j-1} + \mathbf{b}_j),\ j = 1, 2, \ldots, l-1, \\ \boldsymbol{x}^0 &= \boldsymbol{x} \in \mathbb{R}^d. \end{aligned} \tag{14.12}$$

The input $\boldsymbol{x}$ is first subject to a linear transformation with $\mathbf{W}_1\boldsymbol{x}^0 + \mathbf{b}_1$, where $\mathbf{W}_1 \in \mathbb{R}^{n_1\times d}$ and $\mathbf{b}_1 \in \mathbb{R}^{n_1}$ are called weight matrix and bias vector, respectively. Subsequently, the resulting vector is fed to a nonlinear function, $G(\cdot)$, which acts on the vector element-wise; this function is called *activation function*. The activated state $\boldsymbol{x}^1$ is viewed as the updated input, and it will be subject to another linear transformation with $\mathbf{W}_2 \in \mathbb{R}^{n_2\times n_1}$ and $\mathbf{b}_2 \in \mathbb{R}^{n_2}$ and then the nonlinear activation function $G(\cdot)$. This process is repeated many times until the activated state $\boldsymbol{x}^{l-1}$ is reached. The output of the NN $f_{\mathrm{NN}}(\boldsymbol{x})$ is a linear transformation of $\boldsymbol{x}^{l-1}$ with $\mathbf{W}_l \in \mathbb{R}^{1\times n_{l-1}}$ and $\mathbf{b}_l \in \mathbb{R}$. The NN consists of one input layer, $l-1$ hidden layers, and one output layer. The widths of the hidden layers are $n_1, n_2, \ldots, n_{l-1}$. The depth of the NN is l. Single output is assumed here. For multiple outputs, the width of the output layer will become $n_l > 1$, and $\mathbf{W}_l \in \mathbb{R}^{n_l\times n_{l-1}}$ and $\mathbf{b}_l \in \mathbb{R}^{n_l}$. Note that in practical computation, the widths n_i and depth l, which are hyperparameters of the NN, are usually selected empirically. However, the on-going progress on meta-learning will enable us to automate the selection process in the future [15]. The tunable parameters in the NN are the collection of $\{\mathbf{W}_j\}_{j=1}^l$ and $\{\mathbf{b}_j\}_{j=1}^l$, which are denoted by $\boldsymbol{\theta}$. The NN can be rewritten as $f_{\mathrm{NN}}(\boldsymbol{x}; \boldsymbol{\theta})$. For the regression problem, the mean squared error loss function is usually employed, i.e.

$$\min_{\boldsymbol{\theta}} \frac{1}{N}\sum_{i=1}^{N}[y(\boldsymbol{x}_i) - f_{\text{NN}}(\boldsymbol{x}_i; \boldsymbol{\theta})]^2 + \mathcal{J}\{f_{\text{NN}}\}. \tag{14.13}$$

The regularization functional $\mathcal{J}\{f_{\text{NN}}\}$ can be L_1, L_2, or L_∞ norm of $\mathbf{W}_k$, e.g. $\mathcal{J}\{f_{\text{NN}}\} = \lambda \sum_{k=1}^{l} \|\mathbf{W}_k\|_{L_2}^2$, where λ is a pre-selected parameter indicating the penalization strength. Another regularization approach that prevents overfitting is Dropout [16]. Adding regularization can avoid fitting the noise in the data and thus lower the possibility of overfitting.

14.3.3 Physics-Informed Gaussian Process: PIGP

The operators in the general data model (14.1) can be differential operators in PDEs, as the data models for Poisson and Navier–Stokes equations show. Machine learning models built upon the data models involving differential operators of PDEs are physics-informed; physical laws, such as conservation of mass, momentum and energy, are expressed by PDEs. We start from a simple linear PDE to explain the basic idea of the PIGP model. Nonlinear PDEs will be considered in Sect. 14.5. Depending on whether temporal discretization is performed, we divide the PIGP model into two categories: continuous-time GP and discrete-time GP models. The data models corresponding to the two categories are different.

14.3.3.1 Continuous-Time GP

Consider the 1D diffusion problem

$$\begin{aligned} \frac{\partial u(x,t)}{\partial t} &= c\frac{\partial^2 u(x,t)}{\partial x^2} + s(x,t),\ [x,t] \in (0,1) \times (0,1], \\ u(0,t) &= u(1,t) = g(t),\ t \in (0,1], \\ u(x,0) &= u_0(x). \end{aligned} \tag{14.14}$$

For the forward problem, the goal is to solve for the solution u given the diffusivity c, the source term s, the boundary condition g, and the initial condition u_0. At first sight, we have two groups of observations for u: u on the boundary $x = 0$ and $x = 1$, and u at the initial time $t = 0$. In fact, the equation can be viewed as the third group of observations after we put the terms including the unknown solution u on the left-hand-side and keep the known term s on the right-hand-side of the diffusion equation. The data model takes the form

$$\begin{aligned} y_1(\boldsymbol{x}_1) &= \left(\tfrac{\partial}{\partial t} - c\tfrac{\partial^2}{\partial x^2}\right)\{u(\boldsymbol{x}_1)\} + \epsilon_1 = s(\boldsymbol{x}_1) + \epsilon_1, & \boldsymbol{x}_1 &\in (0,1) \times (0,1], \\ y_2(\boldsymbol{x}_2) &= u(\boldsymbol{x}_2) + \epsilon_2 = g(\boldsymbol{x}_2) + \epsilon_2, & \boldsymbol{x}_2 &\in \{0,1\} \times (0,1], \\ y_3(\boldsymbol{x}_3) &= u(\boldsymbol{x}_3) + \epsilon_3 = u_0(\boldsymbol{x}_3) + \epsilon_3, & \boldsymbol{x}_3 &\in (0,1) \times \{0\}. \end{aligned} \tag{14.15}$$

The symbol $\boldsymbol{x}$ instead of x, t is used for simplicity. To approximate u, we consider a GP $u(\boldsymbol{x};\boldsymbol{\theta}) \sim \mathcal{GP}(0, k_u(\boldsymbol{x}, \boldsymbol{x}';\boldsymbol{\theta}))$. We need input-output pairs to learn the hyper-parameters $\boldsymbol{\theta}$. The datasets $\mathcal{D}_2 = \{\boldsymbol{x}_{2,i}, g(\boldsymbol{x}_{2,i}) + \epsilon_2\}_{i=1}^{N_2}$ and $\mathcal{D}_3 = \{\boldsymbol{x}_{3,i}, u_0(\boldsymbol{x}_{3,i}) + \epsilon_3\}_{i=1}^{N_3}$ can be directly applied to the GP u, but the dataset $\mathcal{D}_1 = \{\boldsymbol{x}_{1,i}, s(\boldsymbol{x}_{1,i}) + \epsilon_1\}_{i=1}^{N_1}$ cannot, since $\mathcal{D}_3$ are the observations of a linear transformation of u. Thanks to the fact that a linear transformation of a GP is still a GP [17], $U(\boldsymbol{x};\boldsymbol{\theta}) := \mathcal{L}_{\boldsymbol{x}}^c u(\boldsymbol{x};\boldsymbol{\theta}) \sim \mathcal{GP}(0, k_U(\boldsymbol{x}, \boldsymbol{x}';\boldsymbol{\theta}))$ is a GP sharing hyper-parameters with u, where $\mathcal{L}_{\boldsymbol{x}}^c = \partial/\partial t - c\partial^2/\partial x^2$. The $\boldsymbol{x}$ in the subscript of $\mathcal{L}$ emphasizes that the partial differentiation acts on $\boldsymbol{x}$. It should be noted that the operator $\mathcal{L}_{\boldsymbol{x}}^c$ must be a linear operator. For nonlinear operators in other equations, they need to be linearized first before GP regression works. For example, for nonlinear dynamical systems, explicit time integration automatically provides this linearization about the state of the previous time step. The covariance function for the GP U is computed as $k_U(\boldsymbol{x}, \boldsymbol{x}';\boldsymbol{\theta}) = \mathcal{L}_{\boldsymbol{x}}^c \mathcal{L}_{\boldsymbol{x}'}^c k_u(\boldsymbol{x}, \boldsymbol{x}';\boldsymbol{\theta})$. We have two GPs u and U now, and they are correlated. Hence, we also need to know the cross-covariance function between them: $k_{uU} = \mathrm{Cov}(u, \mathcal{L}_{\boldsymbol{x}'}^c u) = \mathcal{L}_{\boldsymbol{x}'}^c \mathrm{Cov}(u, u) = \mathcal{L}_{\boldsymbol{x}'}^c k_u$; similarly, $k_{Uu} = \mathrm{Cov}(\mathcal{L}_{\boldsymbol{x}}^c u, u) = \mathcal{L}_{\boldsymbol{x}}^c \mathrm{Cov}(u, u) = \mathcal{L}_{\boldsymbol{x}}^c k_u$. Due to $k_u(\boldsymbol{x}, \boldsymbol{x}';\boldsymbol{\theta}) = k_u(\boldsymbol{x}', \boldsymbol{x};\boldsymbol{\theta})$, we have $k_{uU}(\boldsymbol{x}, \boldsymbol{x}') = k_{Uu}(\boldsymbol{x}', \boldsymbol{x})$. Assuming the QoI $u(\boldsymbol{x}^*;\boldsymbol{\theta})$ and the observation vectors $\boldsymbol{y}_1, \boldsymbol{y}_2, \boldsymbol{y}_3$ to have joint Gaussian distribution, we obtain a GP prior similar to (14.8)

$$\begin{bmatrix} u(\mathbf{x}^*;\boldsymbol{\theta}_u) \\ y_1 \\ y_2 \\ y_3 \end{bmatrix} \sim \mathcal{N}\left(\begin{bmatrix} 0 \\ \mathbf{0} \\ \mathbf{0} \\ \mathbf{0} \end{bmatrix}, \begin{bmatrix} k_u(\boldsymbol{x}^*, \boldsymbol{x}^*) & k_{uU}(\boldsymbol{x}^*, \mathbf{X}_1) & k_u(\boldsymbol{x}^*, \mathbf{X}_2) & k_u(\boldsymbol{x}^*, \mathbf{X}_3) \\ k_{Uu}(\mathbf{X}_1, \boldsymbol{x}^*) & k_{Un}(\mathbf{X}_1, \mathbf{X}_1) & k_{Uu}(\mathbf{X}_1, \mathbf{X}_2) & k_{Uu}(\mathbf{X}_1, \mathbf{X}_3) \\ k_u(\mathbf{X}_2, \boldsymbol{x}^*) & k_{uU}(\mathbf{X}_2, \mathbf{X}_1) & k_{un}(\mathbf{X}_2, \mathbf{X}_2) & k_u(\mathbf{X}_2, \mathbf{X}_3) \\ k_u(\mathbf{X}_3, \boldsymbol{x}^*) & k_{uU}(\mathbf{X}_3, \mathbf{X}_1) & k_u(\mathbf{X}_3, \mathbf{X}_2) & k_{un}(\mathbf{X}_3, \mathbf{X}_3) \end{bmatrix} \right), \tag{14.16}$$

where $\mathbf{X}_k = [\boldsymbol{x}_{k,1}, \boldsymbol{x}_{k,2}, \ldots, \boldsymbol{x}_{k,N_k}]$, $k_{Un}(\mathbf{X}, \mathbf{X}) = k_U(\mathbf{X}, \mathbf{X}) + \sigma_n^2 \mathbf{I}$, and $k_{un}(\mathbf{X}, \mathbf{X}) = k_u(\mathbf{X}, \mathbf{X}) + \sigma_n^2 \mathbf{I}$. The subscript n in k_{Un} and k_{un} emphasizes that these covariance matrices take into account "noise". We omit $\boldsymbol{\theta}$ in covariance function for simplicity. Note that the covariance matrix is symmetric. Next, we can compute the predictive mean and variance according to the formulas (14.9) and (14.10), respectively. The hyper-parameters and noise *std* are learned by minimizing the negative log marginal likelihood (14.11), where $\boldsymbol{y} = [\boldsymbol{y}_1^{\mathrm{T}}, \boldsymbol{y}_2^{\mathrm{T}}, \boldsymbol{y}_2^{\mathrm{T}}]^{\mathrm{T}}$, $\mathbf{K}$ consists of 3×3 sub-matrices in the lower right of the covariance matrix in (14.16), and $N = N_1 + N_2 + N_3$.

In addition to the forward problem, PIGPs can also handle inverse problems, such as parameter estimation. The problem is defined as

$$\begin{aligned} \frac{\partial u(x,t)}{\partial t} &= c\frac{\partial^2 u(x,t)}{\partial x^2} + s(x,t),\ [x,t] \in (0,1) \times (0,1], \\ u(x,t) &= g(x,t),\ [x,t] \in [0,1] \times [0,1], \end{aligned} \tag{14.17}$$

where we aim to estimate the PDE parameter c given the source term s and the observations g. The data model is

$$\begin{aligned} y_1(\boldsymbol{x}_1) &= \left(\tfrac{\partial}{\partial t} - c\tfrac{\partial^2}{\partial x^2}\right)\{u(\boldsymbol{x}_1)\} + \epsilon_1 = s(\boldsymbol{x}_1) + \epsilon_1, && \boldsymbol{x}_1 \in (0,1) \times (0,1], \\ y_2(\boldsymbol{x}_2) &= u(\boldsymbol{x}_2) + \epsilon_2 = g(\boldsymbol{x}_2) + \epsilon_2, && \boldsymbol{x}_2 \in [0,1] \times [0,1]. \end{aligned} \tag{14.18}$$

Compared to the forward problem, the sampling space for $\boldsymbol{x}_2$ (i.e. Ω_2) now is the whole spatio-temporal domain. Note that $\boldsymbol{x}_2$ is not necessarily sampled on the boundary of spatio-temporal domain. The GP prior is similar to (14.16) except that $\boldsymbol{y}_3$ does not exist. The negative log marginal likelihood is the same as that for the forward problem. Other than $\boldsymbol{\theta}$ and σ_n, one extra hyper-parameter, c, can be learned by minimizing

$$-\log p(c, \boldsymbol{\theta}, \sigma_n; \mathbf{y}, \boldsymbol{X}) = \frac{1}{2}\mathbf{y}^{\top}(\mathbf{K} + \sigma_n^2\mathbf{I})^{-1}\mathbf{y} + \frac{1}{2}\log|\mathbf{K} + \sigma_n^2\mathbf{I}| + \frac{N}{2}\log 2\pi. \tag{14.19}$$

The PDE parameter c enters the matrix $\mathbf{K}$ through the derivation of k_U, k_{uU} and k_{Uu}. The optimal value of c is what we expect for parameter estimation.

14.3.3.2 Discrete-Time GPs

An alternative approach is to treat spatial and temporal coordinates differently, by discretizing the temporal domain of the problem (14.14) but leaving the spatial domain approximated by GPs. For temporal discretization scheme, we start from the backward Euler

$$\frac{u^{n+1}(x) - u^n(x)}{\Delta t} = c\frac{d^2 u^{n+1}(x)}{dx^2} + s^{n+1}(x),\ x \in (0, 1), n = 0, 1, \ldots, T-1, \tag{14.20}$$

where $\Delta t = 1/T$, $u^n(x) := u(x, n\Delta t)$ and $s^n(x) := s(x, n\Delta t)$. The initial and boundary conditions are $u^0(x) = u_0(x)$ and $u^n(0) = g^n(0) := g(0, n\Delta t)$, $u^n(1) = g^n(1) := g(1, n\Delta t)$, respectively. The data model for the forward problem is

$$\begin{aligned} y_1(x_1) &= \left(I - c\Delta t\frac{d^2}{dx^2}\right)\{u^{n+1}(x_1)\} + \epsilon_1 = u^n(x_1) + \Delta t s^{n+1}(x_1) + \epsilon_1,\ x_1 \in (0, 1), \\ y_2(x_2) &= u^{n+1}(x_2) + \epsilon_2 = g^{n+1}(x_2) + \epsilon_2,\ x_2 \in \{0, 1\} \end{aligned} \tag{14.21}$$

with $n = 0, 1, \ldots, T-1$. The datasets are $\mathcal{D}_1 = \{x_1, u^n(x_1) + \Delta t s^{n+1}(x_1) + \epsilon_1\}$ and $\mathcal{D}_2 = \{x_2, g^{n+1}(x_2) + \epsilon_2\}$. In the data model, the nth snapshot $u^n(x_1)$ is only known for $n = 0$. We first obtain the estimate of $u^1(x_1)$, denoted by $\tilde{u}^1(x_1)$, using GPs, and then repeat the GP regression to predict $u^2(x^*)$ after replacing $u^1(x_1)$ with $\tilde{u}^1(x_1)$. Every time we move from nth snapshot to $(n+1)$th snapshot, we use the estimate of $u^n(x_1)$ predicted from the previous step to predict $u^{n+1}(x^*)$. It is not hard to see that the numerical error of GP approximation will be accumulated as we move step by step in time.

To approximate u^{n+1}, we assume a GP $u^{n+1}(x; \boldsymbol{\theta}^{n+1}) \sim \mathcal{GP}(0, k_u^{n+1}(x, x'; \boldsymbol{\theta}^{n+1}))$. To utilize the observations y_1, we derive the covariance function for $U^{n+1}(x; \boldsymbol{\theta}^{n+1}) = \mathcal{L}^c_{x,\Delta t} u^{n+1}(x; \boldsymbol{\theta}^{n+1})$ where $\mathcal{L}^c_{x,\Delta t} = I - c\Delta t\frac{d^2}{dx^2}$. Similar to the continuous-time GPs, the covariance function of U^{n+1} is $k_U^{n+1}(x, x'; \boldsymbol{\theta}^{n+1}) = \mathcal{L}^c_{x,\Delta t}\mathcal{L}^c_{x',\Delta t}k_u^{n+1}(x, x'; \boldsymbol{\theta}^{n+1})$; the cross-covariance functions between u^{n+1} and U^{n+1} are $k_{uU}^{n+1} = \mathcal{L}^c_{x',\Delta t}k_u^{n+1}$ and

$k_{Uu}^{n+1} = \mathcal{L}_{x,\Delta t}^{c} k_u^{n+1}$. Note that $k_{uU}^{n+1}(x, x'; \boldsymbol{\theta}^{n+1}) = k_{Uu}^{n+1}(x', x; \boldsymbol{\theta}^{n+1})$. The GP prior for the data model is

$$\begin{bmatrix} u^{n+1}(\mathbf{x}^*; \boldsymbol{\theta}^{n+1}) \\ \mathbf{y}_1 \\ \mathbf{y}_2 \end{bmatrix} \sim \mathcal{N}\left(\begin{bmatrix} 0 \\ \mathbf{0} \\ \mathbf{0} \end{bmatrix}, \begin{bmatrix} k_u^{n+1}(\boldsymbol{x}^*, \boldsymbol{x}^*) & k_{uU}^{n+1}(\boldsymbol{x}^*, \mathbf{X}_1) & k_u^{n+1}(\boldsymbol{x}^*, \mathbf{X}_2) \\ k_{Uu}^{n+1}(\mathbf{X}_1, \boldsymbol{x}^*) & k_{Un}^{n+1}(\mathbf{X}_1, \mathbf{X}_1) & k_{Uu}^{n+1}(\mathbf{X}_1, \mathbf{X}_2) \\ k_u^{n+1}(\mathbf{X}_2, \boldsymbol{x}^*) & k_{uU}^{n+1}(\mathbf{X}_2, \mathbf{X}_1) & k_{un}^{n+1}(\mathbf{X}_2, \mathbf{X}_2) \end{bmatrix} \right), \tag{14.22}$$

where $k_{Un}^{n+1}(\mathbf{X}, \mathbf{X}) = k_U^{n+1}(\mathbf{X}, \mathbf{X}) + \sigma_n^2 \mathbf{I}$, and $k_{un}^{N+1}(\mathbf{X}, \mathbf{X}) = k_u^{n+1}(\mathbf{X}, \mathbf{X}) + \sigma_n^2 \mathbf{I}$. Note that the subscript "n" indicates "noise", whereas the superscript "n" denotes the index of time step. The predictive mean and variance are computed according to (14.9) and (14.10), respectively. The hyper-parameters $\boldsymbol{\theta}^{n+1}$ and the noise *std* σ_n are learned by minimizing the negative log likelihood function.

The data model for the inverse problem is the same as that for the forward problem except the sampling space of x_2. For the inverse problem, $x_2 \in [0, 1]$, whereas for the forward problem, $x_2 \in \{0, 1\}$. Solving inverse problems requires performing GP regression only once based on observations of two successive snapshots, which is more efficient than solving the forward problem. For instance, it suffices to estimate the diffusivity c based on the observations $y_1 = u^{10}(x_1) + \Delta t s^{11}(x_1) + \epsilon_1$ and $y_2 = u^{11}(x_2) + \epsilon_2$. The GP prior has a similar form to that of the forward problem. The diffusivity c, the hyper-parameters $\boldsymbol{\theta}^{n+1}$, and the noise *std* σ_n are jointly learned by minimizing the negative log likelihood function.

14.3.4 Physics-Informed Neural Network: PINN

In the PIGPs, the first step is to approximate the PDE solution with a GP prior $u(\boldsymbol{x}; \boldsymbol{\theta}) \sim \mathcal{GP}(0, k(\boldsymbol{x}, \boldsymbol{x}'; \boldsymbol{\theta}))$, where $\boldsymbol{\theta}$ collects all the hyper-parameters in the kernel k. Similarly, in the PINNs, the first step is to approximate the PDE solution with a NN $u(\boldsymbol{x}; \boldsymbol{\theta})$, where $\boldsymbol{\theta}$ represents the parameters of the NN. The PINN models can also be divided into two categories according to whether temporal discretization is performed: continuous-time NNs and discrete-time NNs.

14.3.4.1 Continuous-Time NNs

We still consider the 1D diffusion problem (14.14). The data model is exactly the same as the data model (14.15) of the continuous-time GPs. To learn $\boldsymbol{\theta}$, we minimize the mean squared error loss function, which is similar to (14.13).

$$\mathrm{Loss}(\boldsymbol{\theta}) = \frac{1}{N_1}\sum_{i=1}^{N_1}\left[y_1(\boldsymbol{x}_{1,i}) - \mathcal{L}_{\boldsymbol{x}}^{c} u(\boldsymbol{x}_{1,i};\boldsymbol{\theta})\right]^2 + \frac{1}{N_2}\sum_{i=1}^{N_2}\left[y_2(\boldsymbol{x}_{2,i}) - u(\boldsymbol{x}_{2,i};\boldsymbol{\theta})\right]^2$$
$$+ \frac{1}{N_3}\sum_{i=1}^{N_3}\left[y_3(\boldsymbol{x}_{3,i}) - u(\boldsymbol{x}_{3,i};\boldsymbol{\theta})\right]^2 + \mathcal{J}\{u\}. \tag{14.23}$$

We can evaluate $\mathcal{L}_{\boldsymbol{x}}^{c} u(\boldsymbol{x};\boldsymbol{\theta})$ analytically using the chain rule of differentiation. For instance, the automatic differentiation routine provided in TensorFlow [18] can be employed to compute the derivatives of u. For integral or integro-differential operators, where the chain rule does not work, one can discretize the operator with traditional numerical approaches [19]. The data model for the inverse problem is the same as (14.18), and the loss function takes the form

$$\mathrm{Loss}(c,\boldsymbol{\theta}) = \frac{1}{N_1}\sum_{i=1}^{N_1}\left[y_1(\boldsymbol{x}_{1,i}) - \mathcal{L}_{\boldsymbol{x}}^{c} u(\boldsymbol{x}_{1,i};\boldsymbol{\theta})\right]^2 + \frac{1}{N_2}\sum_{i=1}^{N_2}\left[y_2(\boldsymbol{x}_{2,i}) - u(\boldsymbol{x}_{2,i};\boldsymbol{\theta})\right]^2$$
$$+ \frac{1}{N_3}\sum_{i=1}^{N_3}\left[y_3(\boldsymbol{x}_{3,i}) - u(\boldsymbol{x}_{3,i};\boldsymbol{\theta})\right]^2 + \mathcal{J}\{u\}. \tag{14.24}$$

The diffusivity c enters the loss function through the differential operator $\mathcal{L}_{\boldsymbol{x}}^{c}$ acting on u. Also, c and $\boldsymbol{\theta}$ are jointly learned to minimize the loss function.

14.3.4.2 Discrete-Time NNs

We consider the backward Euler scheme in temporal discretization again. The data model is the same as (14.21) of the discrete-time GPs. We assume $u^{n+1}(x)$ to be a NN $u^{n+1}(x;\boldsymbol{\theta}^{n+1})$. The loss function takes the form

$$\mathrm{Loss}(\boldsymbol{\theta}^{n+1}) = \frac{1}{N_1}\sum_{i=1}^{N_1}\left[y_1(x_{1,i}) - \mathcal{L}_{x,\Delta t}^{c} u^{n+1}(x_{1,i};\boldsymbol{\theta}^{n+1})\right]^2 +$$
$$\frac{1}{N_2}\sum_{i=1}^{N_2}\left[y_2(\boldsymbol{x}_{2,i}) - u^{n+1}(x_{2,i};\boldsymbol{\theta}^{n+1})\right]^2 + \mathcal{J}\{u^{n+1}\}. \tag{14.25}$$

The loss function for the inverse problem is exactly the same as (14.25) except for one extra parameter c to be learned.

14.3.5 Connection Between GPs and NNs

The kernel of GPs, $k(\cdot,\cdot)$, and the activation function of NNs, $G(\cdot)$, are related when the widths of hidden layers of NNs are infinity, namely, $n_j \to \infty$, $j = 1, \ldots, l-1$, and when the weights and biases are independently and identically distributed (i.i.d.) random variables. Specifically, let us recall the matrix-vector form of feedforward NNs (14.12). Suppose entries of the weight matrix and the bias vector are i.i.d. random variables with zero mean and variance σ^2, $[\mathbf{W}_k]_{ij} \sim \Phi(0, \sigma_w^2)$, $[\mathbf{b}_k]_i \sim \Phi(0, \sigma_b^2)$ for $k = 1, 2, \ldots, l$, where Φ is a certain distribution. Entries of weight matrices and bias vectors are independent both in the same layer and across layers. Then, we know that the component of the output of the jth hidden layer, i.e. the component of $\boldsymbol{x}^j$ for $j = 2, \ldots, l$, is a GP indexed by $\boldsymbol{x}$, which is denoted by $\boldsymbol{x}^j[i](\boldsymbol{x}) \sim \mathcal{GP}(0, k^j(\boldsymbol{x}, \boldsymbol{x}'))$ for $i = 1, 2, \ldots, n_j$ and $n_j \to \infty$. The GPs for different components are the same with the kernel $k^j(\boldsymbol{x}, \boldsymbol{x}')$. We rewrite the notation $\boldsymbol{x}^j[i](\boldsymbol{x})$ as $\boldsymbol{x}^j[\cdot](\boldsymbol{x})$. Finally, we have the formula relating the kernel of GPs to the activation function of this infinitely wide NNs

$$k^j(\boldsymbol{x}, \boldsymbol{x}'; \boldsymbol{\theta} = \{\sigma_w, \sigma_b\}) = \sigma_w^2 \mathbb{E}\left[G(\boldsymbol{x}^{j-1}[\cdot](\boldsymbol{x}))\, G(\boldsymbol{x}^{j-1}[\cdot](\boldsymbol{x}'))\right] + \sigma_b^2 \quad (14.26)$$

with $j = 2, 3, \ldots, l$. We refer readers to [20] for the details of deriving the formula. Using this formula, we can construct a GP with the NN-induced kernel $k^l(\boldsymbol{x}, \boldsymbol{x}; \boldsymbol{\theta})$, which could make GPs more flexible in approximating functions involving discontinuities and in solving PDEs whose solutions have large gradients [20].

Remark 2 We note that both PIGPs and PINNs require the solution of a non-convex optimization problem, in which we are very easy to be stuck in local minima, particularly for high-dimensional optimization problems in PINNs. Therefore, even for well-posed initial-boundary value problems, the current formulations do not guarantee unique solutions. Multi-started gradient-based optimization algorithms are commonly used to minimize the non-convex loss functions. However, we have observed that the presence of more data points can greatly increase the rate of training, which is suggestive of attraction to a "relevant" minimum.

14.4 An Illustrative Example

We demonstrate the implementations of the PIGP and the PINN models using a pedagogical example, i.e. how to solve forward and inverse problems of a 1D linear PDE. The Python codes for the example are available at [13].

Let us consider the diffusion problem with the fabricated solution $u(x,t) = \sin(2\pi x)\exp\{-t\}$.

$$\begin{aligned} \frac{\partial u(x,t)}{\partial t} &= c\frac{\partial^2 u(x,t)}{\partial x^2} + s(x,t),\ [x,t] \in (0,1) \times (0,1], \\ u(0,t) &= u(1,t) = 0,\ t \in (0,1], \\ u(x,0) &= \sin(2\pi x),\ x \in (0,1), \end{aligned} \tag{14.27}$$

with $c = 0.1$ and the source term $s(x,t) = \sin(2\pi x)\exp\{-t\}(4c\pi^2 - 1)$. The goal of the forward problem is to predict u given diffusivity c, source term s, and initial-boundary conditions. Here, 5% white Gaussian noise is added to y_j, and the noise *std* is calculated as $\sigma_n = 5\% \times \text{std}(y_j)$, where $\text{std}(y_j)$ is the population standard deviation of the observation y_j.

For continuous-time GPs, the data model is (14.15); $\boldsymbol{x}_1$, $\boldsymbol{x}_2$, and $\boldsymbol{x}_3$ are the inputs of the data model; $\boldsymbol{x}_1$ are scattered in the spatio-temporal domain and are taken from a quasi-random sequence, called the Sobol sequence, and $\boldsymbol{x}_2$ and $\boldsymbol{x}_3$ are uniformly distributed on the boundary of the spatio-temporal domain. We take the Sobol sequence [21] for selecting $\boldsymbol{x}_1$ since the sequence can cover the domain $\Omega_1 = (0,1) \times (0,1]$ as much as possible. The kernel for u is assumed to be a squared exponential

$$k_u(\boldsymbol{x}, \boldsymbol{x}'; \boldsymbol{\theta}) = \sigma^2 \exp\left(-\frac{(x-x')^2}{2\theta_x^2} - \frac{(t-t')^2}{2\theta_t^2}\right), \tag{14.28}$$

where $\boldsymbol{x} = [x, t]^{\mathrm{T}}$ and $\boldsymbol{\theta} = [\sigma, \theta_x, \theta_t]^{\mathrm{T}}$. The selection of a specific kernel depends on the properties of the data, such as the smoothness and periodicity the data demonstrates. Since the fabricated solution we consider is very smooth, we select a smooth kernel, which is infinitely differentiable. Automatic selection of kernels is possible, e.g. [22], but is beyond the scope of this book chapter. The analytical expression for the induced kernels k_U, k_{uU}, and k_{Uu} can be derived by symbolic computation using Maple or other software. Before making predictions using the mean and covariance formulas (14.9) and (14.10), we learn the hyper-parameters $\boldsymbol{\theta}$ and the noise *std* σ_n by minimizing the negative log marginal likelihood (14.11). A stochastic gradient descent algorithm –Adam, written in TensorFlow– is employed to perform the minimization. We take 30 points for $\boldsymbol{x}_1$ ($N_1 = 30$) and 28 points for $\{\boldsymbol{x}_2, \boldsymbol{x}_3\}$ ($N_2 + N_3 = 28$). The initial guesses for the hyper-parameters and noise *std* are all ones. After 30000 Adam iterations, these parameters are $\sigma = 4.300$, $\sigma_n = 8.743 \times 10^{-8}$, $\theta_x = 0.4075$, and $\theta_t = 1.842$ for the noise-free case, and $\sigma = 1.753$, $\sigma_n = 0.0569$, $\theta_x = 0.4098$, $\theta_t = 1.828$ for 5%-noise case. In the noise-free case, the learned noise variance is almost zero, while in the 5%-noise case, the learned noise *std* has the same order of magnitude as the *std* of ϵ_j, namely, $\text{std}(\epsilon_1) = 0.05 \times \text{std}(y_1) \approx 0.07$ and $\text{std}(\epsilon_{2,3}) = 0.05 \times \text{std}(\{y_2, y_3\}) \approx 0.02$. We see that the noise magnitude can be inferred accurately by the GPs model. We note that the hyper-parameters are also known as characteristic lengths. The learned length, θ_x, in x direction is smaller than θ_t, in t direction, which indicates that the solution varies faster in x direction. The L_2 relative error for predicting u is 1.15×10^{-4} and 0.029 for noise-free and 5%-noise cases, respectively.

For continuous-time NNs, the data model is still (14.15). We specify the width (n_j in (14.12)) and the depth (l in (14.12)) of the NN. These two hyper-parameters are selected empirically here, but they can also be automatically obtained based on meta-learning strategies [15, 23]. We choose $n_j \equiv 20$ and $l = 3$ for all NN examples in this section. The automatic differentiation function in TensorFlow is employed to analytically derive $\mathcal{L}_{\boldsymbol{x}}^{c} u$ in the loss function (14.23). The loss function is also minimized using the Adam optimizer in TensorFlow. We set the iteration number of the Adam optimizer to be 0.1 million. The number of iterations can be pre-set to be a large value, and a threshold condition, for instance, that the value of loss function is lower than 10^{-6}, is employed to terminate the iteration before reaching that pre-set iteration number. We compare the L_2 relative error of the prediction u for the cases with and without regularization. For the 5%-noise case, in the absence of regularization, the error is roughly 10%, but if we consider L_2 regularization $\mathcal{J}\{f_{\mathrm{NN}}\} = 10^{-4} \sum_{k=1}^{l} \|\mathbf{W}_k\|_{L_2}^2$, the error decreases to 3.7%. In addition, for the noise-free case, the error is 6.4×10^{-3}, which is one order of magnitude higher than the error for GPs. The non-parametric feature of GPs accounts for its higher accuracy. GPs is a non-parametric model, whereas NNs is a parametric model. There is an infinite number of parameters in GPs (note that these parameters differ from the hyper-parameters conceptually), but the number of the parameters in NNs model is finite. For instance, for the index set $x \in (0, 1)$, the parameters of a GP $f_{\mathrm{GP}}(x)$ are $\{f_{\mathrm{GP}}(x) | x \in (0, 1)\}$, the cardinality of which is infinity.

In addition to the forward problem, we also test the inverse problem in which we identify the diffusivity c based on the two snapshots of u at $t = 0.5$ and $t = 0.51$. Since the observations are only two snapshots, the continuous-time models are not appropriate. We consider instead the discrete-time GP and the discrete-time NN models, whose data models are the same, namely, (14.21). We set $\Delta t = 0.01$ according to the observations. There are 20 equispaced training points for each snapshot, i.e. $N_1 = N_2 = 20$. The initial guess for c is 1.0, and it converges to the true value $c = 0.1$ after 2000 Adam iterations for GPs and 7500 Adam iterations for NNs, in the noise-free case. The initial guess converges to 0.124 for GPs and 0.136 for NNs in the 5%-noise case. The slower convergence for NN can be explained by the fact that there are many more parameters to be optimized in NNs compared to GPs.

14.5 Comparison of PIGPs and PINNs for Nonlinear PDEs

Continuous-time GPs does not work for nonlinear equations, while discrete-time GPs works for nonlinear equations such as Burgers' and Navier–Stokes equations upon proper linearization. Compared to PIGP models, continuous-time and discrete-time NNs are much more flexible for strongly nonlinear equations, without linearization of PDEs. In this section, we demonstrate the results for 1D Burgers' equation and 2D Navier–Stokes equation in order to show certain features of discrete-time GPs and PINNs in handling nonlinear PDEs.

The Python codes for this section are available at [13].

14.5.1 1D Burgers' Equation

The forward problem is to solve for u [4]

$$u_t + \lambda_1 u u_x - \lambda_2 u_{xx} = 0, \quad x \in (-8, 8), \quad t \in (0, 5], \tag{14.29}$$

with $\lambda_1 = 1$, $\lambda_2 = 0.1$, the Gaussian profile initial conditions [24] and the boundary conditions $u(-8, t) = u(8, t) = 0$. In the inverse problem, the parameter estimation task is to estimate the parameters λ_1 and λ_2 based on two snapshots of u at $t = 1.5$ and $t = 1.6$.

For the discrete-time models, instead of a backward Euler for temporal discretization, which has only first-order accuracy, here we consider a fourth-order implicit Runge–Kutta method [3]. The data model for the discrete-time models and for the forward problem takes the form

$$\begin{aligned}
y_1(x_1) &= \mathcal{L}_1^{\gamma}\{u^{n+1}, u^{n+c_2}, u^{n+c_1}\} = u^{n+1} - b_1 \Delta t \mathcal{N}_x^{\gamma}\{u^{n+c_1}\} - b_2 \Delta t \mathcal{N}_x^{\gamma}\{u^{n+c_2}\} = u^n,\\
y_2(x_2) &= \mathcal{L}_2^{\gamma}\{u^{n+1}, u^{n+c_2}, u^{n+c_1}\} = u^{n+c_2} - a_{21} \Delta t \mathcal{N}_x^{\gamma}\{u^{n+c_1}\} - a_{22} \Delta t \mathcal{N}_x^{\gamma}\{u^{n+c_2}\} = u^n,\\
y_3(x_3) &= \mathcal{L}_3^{\gamma}\{u^{n+1}, u^{n+c_2}, u^{n+c_1}\} = u^{n+c_1} - a_{11} \Delta t \mathcal{N}_x^{\gamma}\{u^{n+c_1}\} - a_{12} \Delta t \mathcal{N}_x^{\gamma}\{u^{n+c_2}\} = u^n,\\
y_4(x_4) &= u^{n+1} = 0, \; x_4 \in \{-8, 8\},\\
y_5(x_5) &= u^{n+c_2} = 0, \; x_5 \in \{-8, 8\},\\
y_6(x_6) &= u^{n+c_1} = 0, \; x_6 \in \{-8, 8\},
\end{aligned} \tag{14.30}$$

where $x_1, x_2, x_3 \in (-8, 8)$, $a_{11} = 0.25$, $a_{12} = 0.25 - \frac{\sqrt{3}}{6}$, $a_{21} = 0.25 + \frac{\sqrt{3}}{6}$, $a_{22} = 0.25$, $b_1 = b_2 = 0.5$, $c_1 = 0.5 - \frac{\sqrt{3}}{6}$, $c_2 = 0.5 + \frac{\sqrt{3}}{6}$, and $u^{n+c_i} := u(x, (n + c_i)\Delta t)$. The differential operator $\mathcal{N}_x^{\gamma}\{u^k\} = -\lambda_1 u^k u_x^k + \lambda_2 u_{xx}^k$. To make the discrete-time GPs work, we linearize the operator as $\tilde{\mathcal{N}}_x^{\gamma}\{u^k\} = -\lambda_1 u^n u_x^k + \lambda_2 u_{xx}^k$, $k = n + 1, n + c_2$ and $n + c_1$, where u^n is known from the initial conditions or estimated in the previous time step. Three GP priors are, respectively, placed on u^{n+1}, u^{n+c_2}, and u^{n+c_1}: $u^{n+1}(x; \boldsymbol{\theta}^{n+1}) \sim \mathcal{GP}(0, k_u^{n+1}(x, x'; \boldsymbol{\theta}^{n+1}))$, $u^{n+c_2}(x; \boldsymbol{\theta}^{n+c_2}) \sim \mathcal{GP}(0, k_u^{n+c_2}(x, x'; \boldsymbol{\theta}^{n+c_2}))$, and $u^{n+1}(x; \boldsymbol{\theta}^{n+c_1}) \sim \mathcal{GP}(0, k_u^{n+c_1}(x, x'; \boldsymbol{\theta}^{n+c_1}))$. The cross-covariance functions between observations $k_{y_i, y_j}(x, x')$ are derived in a similar manner as in Sect. 14.3.3.2. The derived cross-covariance functions can be found in [3]. On the other hand, for the discrete-time NNs, rather than linearized operator $\tilde{\mathcal{N}}_x^{\gamma}$, we still consider the original operator $\mathcal{N}_x^{\gamma}$, and assume u^{n+1}, u^{n+c_2}, and u^{n+c_1} to be the multiple outputs of a NN $f_{\mathrm{NN}}(x, \boldsymbol{\theta})$. In other words, we need to set the width of the output layer of the NN to be three, or $n_l = 3$. Note that the three outputs share the same parameters $\boldsymbol{\theta}$.

We first compare the discrete-time GPs and the continuous-time NNs for the forward problem. For the discrete-time GPs, we select 70 equispaced training points for u^0 ($N_1 = N_2 = N_3 = 70$) and fix the locations of these points for $n > 0$. The hyper-parameters we have learned in the nth time step are fed to the $(n + 1)$th time step as the initial guesses for the hyper-parameters in $(n + 1)$th time step. We train the GP u^1 for 5000 Adam iterations and train the GPs u^{n+1} only for 500 iterations for $n > 0$. For continuous-time NNs, we select 100 ($N_2 = 100$) and 256 ($N_3 = 256$) equispaced training points for enforcing the boundary and initial conditions, respectively. We place 2000 training points, generated by the Sobol sequence, in the spatio-temporal domain to enforce the PDE ($N_1 = 2000$), and train the model by 0.2 million Adam iterations. We consider the noise-free case here and refer readers to [4] for noisy cases. The width for the NN is 10, and the depth is 9. A hyperbolic tangent activation function is employed. The fabricated solution is taken from [24]. The L_2 relative error of u for the discrete-time GPs is 0.0158 for $\Delta t = 0.05$, 0.0032 for $\Delta t = 0.01$, and 0.0016 for $\Delta t = 0.005$. The error for the continuous-time NN is 0.0014. Decreasing the temporal step size will increase the solution accuracy for GPs, and GPs achieve comparable accuracy to NNs with $\Delta t = 0.005$. Since NNs have many more parameters to be learned than GPs, NNs requires more training data-points than GPs do to achieve comparable accuracy. Figure 14.1 compares the patterns in the absolute error fields of GPs and NNs. We see from Fig. 14.1(b) that GPs approximation and linearization errors are accumulated when we move step by step in time. Also, the absolute error is basically proportional to the solution magnitude. In contrast, for NNs, there is no specific trend for the absolute error distribution because the training points $\boldsymbol{x}_1$ are scattered in the spatio-temporal domain.

The estimated parameters for the inverse problem are $\lambda_1 = 1.019$ and $\lambda_2 = 0.1012$ for the discrete-time GPs, and $\lambda_1 = 0.9997$ and $\lambda_2 = 0.09942$ for discrete-time NNs, whereas the true parameters are $\lambda_1 = 1.0$ and $\lambda_2 = 0.1$. We consider the discrete-time NNs instead of the continuous-time NNs because of the availability of small dataset. The same training points are taken as in the discrete-time GPs case, and we take $\Delta t = 0.1$ as the observed data requires. The width of the NN is 10, and the depth is 5. The hyperbolic tangent activation function is employed. In terms of the accuracy of the estimated parameters, NNs outperform GPs slightly, which could be explained by the fact that NNs are free of the linearization error.

14.5.2 2D Navier–Stokes Equations

Here we consider the inverse problem formulated as a parameter estimation problem, aiming at estimating the parameters λ_1 and λ_2 of the equations [4]

$$\begin{aligned} u_t + \lambda_1(uu_x + vu_y) &= -p_x + \lambda_2(u_{xx} + u_{yy}), \\ v_t + \lambda_1(uv_x + vv_y) &= -p_y + \lambda_2(v_{xx} + v_{yy}), \end{aligned} \tag{14.31}$$

based on two snapshots of u, v at $t = 0.18$ and $t = 0.20$. The equations describe a fluid flow past a cylinder, and here $(x, t) \in (-1.7, 1.7) \times (0, 10]$. The flow is incompressible $u_x + v_y = 0$. For convenience of implementation, we consider the backward Euler scheme, instead of the higher-order schemes such as the aforementioned Runge–Kutta scheme, for the temporal discretization. The data model for the discrete-time models is

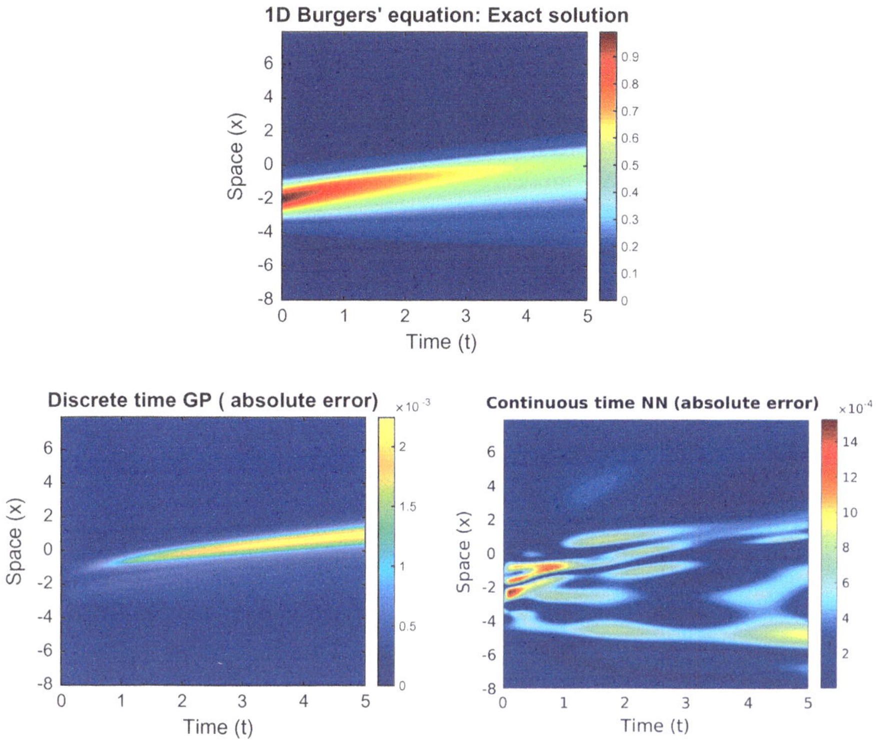

Fig. 14.1 1D Burgers' equation (Forward problem): Comparison of the discrete-time GP and the continuous-time NN. **a** Fabricated solution in spatio-temporal domain with data taken from [24], **b** absolute error of the discrete-time GP in spatio-temporal domain, and **c** absolute error of the continuous-time NN in spatio-temporal domain. Time step size $\Delta t = 0.005$ for the discrete-time GP

$$
\begin{aligned}
y_1(x_1) &= \mathcal{L}_1^{\gamma}\{u^{n+1}, v^{n+1}, p^{n+1}\} = u^{n+1} - \Delta t \mathcal{N}_{1,x,y}^{\gamma}\{u^{n+1}, v^{n+1}, p^{n+1}\} = u^n, \\
y_2(x_2) &= \mathcal{L}_2^{\gamma}\{u^{n+1}, v^{n+1}, p^{n+1}\} = v^{n+1} - \Delta t \mathcal{N}_{2,x,y}^{\gamma}\{u^{n+1}, v^{n+1}, p^{n+1}\} = v^n, \\
y_3(x_3) &= u_x^{n+1} + v_y^{n+1} = 0, \\
y_4(x_4) &= u^{n+1}, \\
y_5(x_5) &= v^{n+1},
\end{aligned}
\tag{14.32}
$$

where $x_j \in (-1.7, 1.7) \times (0, 10]$, $j = 1, 2, 3, 4, 5$ and $\mathcal{N}_{1,x,y}^{\gamma}\{u^{n+1}, v^{n+1}, p^{n+1}\} = -\lambda_1(u^{n+1}u_x^{n+1} + v^{n+1}u_y^{n+1}) - p_x^{n+1} + \lambda_2(u_{xx}^{n+1} + u_{yy}^{n+1})$ and $\mathcal{N}_{2,x,y}^{\gamma}\{u^{n+1}, v^{n+1}, p^{n+1}\} = -\lambda_1(u^{n+1}v_x^{n+1} + v^{n+1}v_y^{n+1}) - p_y^{n+1} + \lambda_2(v_{xx}^{n+1} + v_{yy}^{n+1})$. The linearized operators are $\tilde{\mathcal{N}}_{1,x,y}^{\gamma}\{u^{n+1}, v^{n+1}, p^{n+1}\} = -\lambda_1(u^n u_x^{n+1} + v^n u_y^{n+1}) - p_x^{n+1} + \lambda_2(u_{xx}^{n+1} + u_{yy}^{n+1})$ and $\tilde{\mathcal{N}}_{2,x,y}^{\gamma}\{u^{n+1}, v^{n+1}, p^{n+1}\} = -\lambda_1(u^n v_x^{n+1} + v^n v_y^{n+1}) - p_y^{n+1} + \lambda_2(v_{xx}^{n+1} + v_{yy}^{n+1})$. Three GP priors are placed on u^{n+1}, v^{n+1}, and p^{n+1}, and the multiple outputs of the NNs are also u^{n+1}, v^{n+1}, and p^{n+1}. Alternatively, to reduce the number of unknown functions, it is convenient to define an auxiliary function ψ, which relates to the velocity by $u^{n+1} = \psi_y^{n+1}$ and $v^{n+1} = -\psi_x^{n+1}$. We thus leave out the third group of observations y_3, and the GP priors and NN outputs both change to ψ^{n+1} and p^{n+1}. We refer readers to [4, 10] for more implementation details of the discrete-time GP and NN models for this problem. We take 250 scattered points for each snapshot and consider the noise-free case. We take $\Delta t = 0.02$ according to the observed snapshots. The Adam iterations for GPs and NNs are 3×10^4 and 2×10^6, respectively. The width of the NNs is 10, and the depth is 9; a hyperbolic tangent activation function is used. In the noise-free case, the estimated parameters are $\lambda_1 = 1.001$ and $\lambda_2 = 0.0092$ for the discrete-time GPs and $\lambda_1 = 0.9955$ and $\lambda_2 = 0.0094$ for the discrete-time NNs, whereas the true parameters are $\lambda_1 = 1.0$ and $\lambda_2 = 0.01$. GPs and NNs produce comparable accuracy for the estimated parameters. It should be noted that the NNs are easily extended to a very high order Runge–Kutta method, such as the 500-stage Runge–Kutta method considered in [10], in which we only need to let the NNs output u^{n+1} and u^{n+c_i}, $i = 1, 2, \ldots, 500$. For the discrete-time GPs, however, the resulting covariance matrix will become too large to make the model computationally efficient.

14.6 Summary and Outlook

We review and compare recent physics-informed learning models built upon GPs and deep NNs. In Sect. 14.2 we start with a unified data model on which GPs, PIGPs, NNs, and PINNs are based. We then illustrate the concepts of GPs and NNs for function approximation in Sects. 14.3.1 and 14.3.2. Thanks to the analytical tractability of differentiation of the GP kernel and of the deep NNs, it is straightforward to extend GPs and NNs to solve PDEs. The differential operators in PDEs enter the likelihood function of GPs and the loss function of NNs by acting on the GP kernel and deep NNs, respectively, as discussed in Sects. 14.3.3 and 14.3.4. Continuous-time and discrete-time models are distinguished, depending on whether it is necessary to perform temporal discretization. Moreover, a connection between the GP kernel and the activation function of NNs is briefly discussed in Sect. 14.3.5. In Sect. 14.3.4, a pedagogical example is given to demonstrate the implementation of PIGPs and PINNs, and we include the Python codes to demonstrate the simplicity of the implementation. In Sect. 14.3.5, we compare PIGPs and PINNs for two nonlinear equations: the Burgers' equation and the Navier–Stokes equations. Continuous-time GPs are not possible for nonlinear PDEs, and hence we compare the discrete-time GPs, the continuous-time NNs, and the discrete-time NNs. For a forward problem, the discrete-time GPs and the discrete-time NNs both suffer from the accumulated temporal discretization error, while the discrete-time GPs also suffer from the linearization error. High-order temporal discretization schemes may mitigate the accumulated error, but it is not easy to extend GPs to handle high-order schemes since forming the covariance matrix by symbolic computation becomes very tedious. In contrast, it is much easier to modify

NNs to handle high-order schemes by only changing the width of the output layer. The continuous-time NN is trained only once, whereas the discrete-time NNs is successively trained from one time instant to the next one. Nevertheless, the continuous-time NN requires many more data-points than the discrete-time NNs, since it places training points in the entire spatio-temporal domain. For parameter estimation, the discrete-time models require fewer training data-points than the continuous-time models. However, if the goal is to simultaneously identify the PDE parameters and reconstruct the solution field, the continuous-time models may perform better [25].

The PIGPs could be optimized to assimilate heterogeneous datasets by employing the kernels induced by NNs [20, 26] or kernels automatically selected from the kernel space [22]; moreover, PIGPs could handle *big data* by leveraging scalable GPs, e.g, [27]. On the other hand, PINNs may be more efficient in tackling long-time integration of time-dependent PDEs by leveraging *parareal* methods [28]. Apart from PDEs, PINNs can be easily extended to solve ODEs and integral/integro-differential equations [19, 29]. Another attractive feature of PIGPs and PINNs is that both of them can be modified to be trained based on multifidelity framework [30–32], which could utilize experimental data and synthetic datasets to construct a surrogate for the highest fidelity dataset.

The progress that has been made so far for physics-informed learning is encouraging, but there are still several open issues related to the accuracy and efficiency of both PIGPs and PINNs. In both cases, the underlying difficulty is related to the solution of a non-convex optimization problem whose solution is not guaranteed. This makes the solution of forward problems inefficient compared to traditional numerical methods. At this juncture, our experience suggests that PIGPs and PINNs are more effective for inverse problems for which the given data helps the minimizer to reach a minimum at a faster rate. Moreover, the implicit regularization of machine learning models allows them to solve ill-posed problems, e.g. due to lack of boundary conditions, which is not possible with traditional methods. A fundamental question still remains open as to the possibility of simulating nonlinear systems with chaotic dynamics without the need of observers as suggested in the work of [33].

Despite the aforementioned current technical issues, we expect the methods presented herein and appropriate extensions and variants to form the algorithmic backbone of "digital twins" across scientific domains and applications. This is based on the already clear advantages of PIGPs and PINNs over traditional numerical methods, e.g. in integrating seamlessly multi-fidelity data and mathematical models, their simplicity of implementation and maintenance of the computer codes, and the lack of need for laborious mesh generation. Moreover, they can be readily extended to nonlinear multi-physics problems of industrial complexity and can be employed in modern design as in the recent example where NVIDIA engineers have used PINNs running on their GPUs to improve the design of the next generation of GPUs [34]. On the fundamental side, physics-informed learning machines are likely to tackle grand challenges, e.g. the problems of turbulence, which is the last frontier in classical physics. In addition, they can be used to discover new principles in the form of conservation laws, e.g. in social dynamics, where there are no governing principles but there is an enormous amount of daily data thanks to social media.

Acknowledgements This work was partially supported by AFOSR (FA9550-17-1-0013), Army Research Laboratory (W911NF-12-2-0023), and PHILMS (DE-SC0019453).

References

1. M. Raissi, P. Perdikaris, G.E. Karniadakis, J. Comput. Phys. **348**, 683 (2017)
2. M. Raissi, P. Perdikaris, G.E. Karniadakis, J. Comput. Phys. **335**, 736 (2017)
3. M. Raissi, P. Perdikaris, G.E. Karniadakis, SIAM, J. Sci. Comput. **40**, A172 (2018)
4. M. Raissi, G.E. Karniadakis, J. Comput. Phys. **357**, 125 (2018)
5. Z. Ghahramani, Nature **521**(7553), 452 (2015)
6. M.I. Jordan, T.M. Mitchell, Science **349**(6245), 255 (2015)
7. M. Dissanayake, N. Phan-Thien, Commun. Numer. Meth. Eng. **10**, 195 (1994)
8. B.P. van Milligen, V. Tribaldos, J. Jiménez, Phys. Rev. Lett. **75**, 3594 (1995)
9. I.E. Lagaris, A. Likas, D.I. Fotiadis, I.E.E.E. Trans, Neural Netw. **9**, 987 (1998)
10. M. Raissi, P. Perdikaris, G.E. Karniadakis, J. Comput. Phys. **378**, 686 (2019)
11. C.A. Micchelli, Y. Xu, H. Zhang, J. Mach. Learn. Res. **7**, 2651 (2006)
12. K. Hornik, M. Stinchcombe, H. White, Neural Netw. **2**, 359 (1989)
13. G. Pang, Python codes for the numerical examples in this chapter (2019), https://github.com/Pang1987/pedagogical-example-PIGP-PINN. Accessed 27 Jan 2020
14. C.E. Rasmussen, in *Advanced Lectures on Machine Learning*, ed. by O. Bousquet, U. von Luxburg, G. Rätsch (Springer, Berlin, 2003), p. 63
15. C. Finn, P. Abbeel, S. Levine, in *34th International Conference on Machine Learning*. Proceedings of Machine Learning Research, vol. 70, ed. by D. Precup, Y.W. Teh (Sidney, 2017), p. 1126
16. N. Srivastava, G. Hinton, A. Krizhevsky, I. Sutskever, R. Salakhutdinov, J. Mach, Learn. Res. **15**, 1929 (2014)
17. S. Särkkä, in *Artificial Neural Networks and Machine Learning – ICANN 2011, 21st International Conference on Artificial Neural Networks, Espoo, Finland, June 2011*. Lecture Notes in Computer Science, vol. 6791, ed. by T. Honkela, W. Duch, M. Girolami, S. Kaski (Springer, Berlin, 2011), p. 151
18. M. Abadi et al., in *Proceedings of OSDI '16. 12th USENIX Symposium on Operating Systems Design and Implementation, Savannah, USA* (2016), p. 265
19. G. Pang, L. Lu, G.E. Karniadakis, SIAM, J. Sci. Comput. **41**, A2603 (2019)
20. G. Pang, L. Yang, G.E. Karniadakis, J. Comput. Phys. **384**, 270 (2019)
21. I.M. Sobol', Zh Vychisl, Mat. Mat. Fiz. **7**, 784 (1967)
22. D. Duvenaud, Ph.D. thesis (University of Cambridge, 2014)
23. B. Zoph, Q.V. Le, Neural architecture search with reinforcement learning, arXiv:1611.01578
24. S.H. Rudy, S.L. Brunton, J.L. Proctor, J.N. Kutz, Sci. Adv. **3**, e1602614 (2017)
25. M. Raissi, A. Yazdani, G.E. Karniadakis, Hidden fluid mechanics: a Navier-Stokes informed deep learning framework for assimilating flow visualization data, arXiv:1808.04327
26. A.G. Wilson, Z. Hu, R. Salakhutdinov, E.P. Xing, in *Proceedings of the 19th International Conference on Artificial Intelligence and Statistics, AISTATS 2016*. JMLR Workshop and Conference Proceedings, vol. 51, ed. by A. Gretton, C.C. Robert. Cádiz (Spain, 2016), p. 370
27. A. Wilson, H. Nickisch, in *Proceedings of the 32nd International Conference on Machine Learning*. Proceedings of Machine Learning Research, vol. 37, ed. by F. Bach, D. Blei (Lille, France, 2015), p. 1775
28. X. Meng, Z. Li, D. Zhang, G.E. Karniadakis, PPINN: parareal physics-informed neural network for time-dependent PDEs, arXiv:1909.10145
29. L. Lu, X. Meng, Z. Mao, G.E. Karniadakis, DeepXDE: a deep learning library for solving differential equations, arXiv:1907.04502
30. L. Le Gratiet, Ph.D. thesis (Université Paris Diderot, 2013)

31. G. Pang, P. Perdikaris, W. Cai, G.E. Karniadakis, J. Comput. Phys. **348**, 694 (2017)
32. X. Meng, G.E. Karniadakis, J. Comput. Phys. **401**, 109020 (2020)
33. J. Pathak, B. Hunt, M. Girvan, Z. Lu, E. Ott, Phys. Rev. Lett. **120**, 024102 (2018)
34. J. Huang, SC19: NVIDIA CEO Jensen Huang on the expanding universe of HPC, https://www.youtube.com/watch?v=69nEEpdEJzU. Accessed 27 Jan 2020

Chapter 15
Nonlinear Systems for Unconventional Computing

Kirill P. Kalinin and Natalia G. Berloff

Abstract The search for new computational machines beyond the traditional von Neumann architecture has given rise to a modern area of nonlinear science—development of unconventional computing—requiring the efforts of mathematicians, physicists and engineers. Many analogue physical systems including nonlinear oscillator networks, lasers, and condensates were proposed and realised to address hard computational problems from various areas of social and physical sciences and technology. The analogue systems emulate spin Hamiltonians with continuous or discrete degrees of freedom to which actual optimisation problems can be mapped. Understanding of the underlying physical process by which the system finds the ground state often leads to new classes of system-inspired or quantum-inspired algorithms for hard optimisation. Together physical platforms and related algorithms can be combined to form a hybrid architecture that may one day compete with conventional computing. In this chapter, we review some of the systems and physically-inspired algorithms that show such promise.

15.1 Introduction

We live in a world dominated by information. Systems that enable faster information processing and decision making are becoming more integrated into our daily lives. This data-intensive science relies on continual improvements in hardware for solving ever growing, e.g. in number of variables and constraints, optimisation problems. Digital electronics can no longer satisfy this trend, as exponential hardware scaling (Moore's law) and the von Neumann architecture are reaching their limits [1, 2].

K. P. Kalinin · N. G. Berloff (✉)
University of Cambridge, Cambridge CB3 0WA, United Kingdom
e-mail: N.G.Berloff@damtp.cam.ac.uk

K. P. Kalinin
e-mail: kpk26@cam.ac.uk

N. G. Berloff
Skolkovo Institute of Science and Technology Russian Federation,
Bolshoy Boulevard 30, bld. 1, 121205 Moscow, Russia

P. G. Kevrekidis et al. (eds.), *Emerging Frontiers in Nonlinear Science*, Nonlinear Systems and Complexity 32, https://doi.org/10.1007/978-3-030-44992-6_15

Looking beyond the traditional computing one turns to physical platforms that with their superior speed and reconfigurability and internal parallel processing can provide faster alternatives to solving a specialised class of nonlinear problems. Despite a number of physical systems that were proposed as quantum or analogue simulators and further elucidated in active applied research, significant challenges still remain before scalable analogue processors can be realised and show the superior performance in comparison with the von Neumann computing architecture. Over the years, various unconventional computing techniques were proposed that enable simultaneous communication, computation, and memory access throughout their architecture with the purpose to alleviate the device and system architectural challenges faced by conventional computing platforms.

Neuromorphic computing based on neural networks promises to make processors that use low energies while integrating massive amounts of information. Quantum annealer devices promise to find the global minimum of a combinatorial optimisation problem faster than classical computers. Physical (natural) systems aim to become analogue machines by bridging the physics of a particular system with hardware platforms to enhance the performance of machine learning algorithms.

A central challenge is in the development of mathematical models—system-inspired computing—linking physical platforms to models of complex analogue information processing. Among such models, those based on principles of neural networks and quantum annealing are perhaps the most widely studied.

A large class of problems that can be solved on physical platforms includes nonlinear programming problems. They seek to minimise some nonlinear objective function $E(\mathbf{x})$ of real or complex variables $\mathbf{x}$ subject to a series of constraints represented by equalities or inequalities, i.e. $g(\mathbf{x}) = 0$ and $h(\mathbf{x}) \leq 0$. Numerous applications in social sciences and telecommunications, finance and aerospace, biological and chemical industries can be described in this basic framework [3–5].

Nonlinear optimisation problems are notoriously difficult to solve, and often involve specialised techniques such as genetic algorithms, particle swarm optimisation, simulation and population annealing. Around the vicinity of the optimal solution nonlinear optimisation problems are quadratic to second order, and therefore, quadratic programming for minimising quadratic functions subject to linear constraints is a usual simplification to such problems that can be used with a wide array of applications. Quadratic programming occurs in various machine learning problems, such as the support vector machine training and least squares regression. At the same time, quadratic programming and other nonlinear optimisation problems can be mapped to spin Hamiltonians which can be emulated by real physical systems: the degrees of freedom $\mathbf{x}$ become "spins," the cost function $E(\mathbf{x})$ is a "Hamiltonian" that specifies the interaction pattern between spins. In this chapter we discuss two possible ways by which the system can find the optimal solution—the ground state of the corresponding spin Hamiltonian—depending on the nature of the system. The system in thermodynamic equilibrium may find the optimal solution by quantum annealing which is executed with the time-dependent Hamiltonian

$$H(t) = \left(1 - \frac{t}{\tau}\right) H_0 + \frac{t}{\tau} H_{\text{objective}}, \tag{15.1}$$

where H_0 is the initial trivial Hamiltonian whose ground state is known, and $H_{\text{objective}}$ is the final Hamiltonian at $t = \tau$ which encodes an original objective function $E(\mathbf{x})$. If the system is in thermal equilibrium at all times then it stays close to the ground state, as Hamiltonian parameters are adiabatically varied. A linear time dependence in (15.1) is assumed for simplicity but more complex annealing schedules can be used. The time τ for obtaining the result of optimisation is much larger than that defined by the inverse of the spectral gap (the distance between the ground state and the lowest excited state) of $H(t)$ [6]. When spectral gap is large, the coupling to the environment helps the annealer by cooling the system towards its ground state, however, as the system becomes larger and the spectral gap shrinks (typically exponentially fast with the system size) the excited states lead to large errors at the same time slowing down the annealing procedure.

Non-equilibrium systems rely on a different principle of approaching the ground state from below rather than via quantum tunnelling during the adiabatic annealing. The principle of the gain-dissipative simulator is based on a two-stage process: gain increase below the threshold and the coherence of operations at the threshold. Ramping up the gain allows system to overcome its linear losses and to stabilise by the nonlinearity of the gain saturation. The emergent coherent state minimises the losses and, therefore, maximises the total number of particles as it will be explained further below, which leads to minimising a particular functional that can be written as the objective spin Hamiltonian. Close to the threshold, the resulting evolution of the system elements resembles the dynamics of Hopfield networks which were shown to be able to solve quadratic optimisation problems more than thirty years ago [7, 8] by using a system of differential equations that describe the evolution of individual neurons:

$$\frac{\mathrm{d}x_i}{\mathrm{d}t} = -\sum_{j=1}^{N} J_{ij} S_j(x_j), \tag{15.2}$$

where x_i is an input, $S_j(x_j)$ is the activation function (e.g. sigmoid), and J_{ij} is the connectivity matrix among the neurons. Various modifications of Hopfield networks were extensively proposed and studied [9], however, the optimisers based on Hopfield networks were surpassed by other computational methods. This is largely due to the high connectivity between neurons that neural networks require and the concomitant time it takes to evolve large networks on classical hardware. The recent interest in Hopfield networks re-emerged as it became possible to create them in analogue physical systems such as electronic circuits or photonic neural networks. Photonic systems have an advantage over their electronic counterparts due to the picosecond to femtosecond time scale of their operation and as hundreds of high bandwidth signals can flow through a single optical waveguide. This means that a photonic implementation of Hopfield networks as optimisers can have a large dimensionality and dense connectivity as well as a fast convergence time. However, the evolution of Hopfield networks does not necessarily lead to the optimal solution.

In this chapter we will review recent progress in building the analogue devices that implement either quantum annealing or gain-dissipative principle in their architecture while illustrating the idea of devising system-inspired algorithms.

15.1.1 Spin Hamiltonians

The majority of optimisation problems are computationally impractical for conventional classical computers with classic examples of a so-called "hard optimisation task" being the travelling salesman problem, the dynamic analysis of financial markets, the prediction of new chemical materials, and machine learning hardware [10]. Mathematically, it is possible to reformulate many of these optimisation problems from vastly different areas as a problem of finding the ground state of a particular spin New Hamiltonian with discrete or continuous degrees of freedom. Throughout this chapter, we will refer to this spin Hamiltonian optimisation simply as solving spin model. The spin Hamiltonian can be emulated with a given simulator, e.g. solid-state system, that would need to have an easy mapping of the variables of the desired Hamiltonian into the elements (spins, currents etc.) of the simulator, independently tunable short and long range interactions between them, and would allow one to perform measurements to obtain the answer with the required precision. Such spin model Hamiltonians are experimentally challenging to implement and control but their possible advantageous performance over classical computers, which struggle solving sufficiently large problem sizes, leads to an intensive search for a superior simulator. Such simulators have been proposed and realised to a various extent in disparate physical systems. Among these systems, two classes of spin Hamiltonians are more common: Ising and XY Hamiltonians. For instance, the Ising Hamiltonian is widely used for a vast variety of hard discrete combinatorial optimisation problems, so that travelling salesman, graph colouring, graph partitioning, and others can be mapped into it with a polynomial overhead [11]. This model is formulated for N classical "spins" s_j that take discrete values $\{-1, 1\}$ to minimise the quadratic unconstrained binary optimisation (QUBO) problem:

$$\min \quad -\sum_{i=1}^{N} \sum_{j=1, j<i}^{N} J_{ij} s_i s_j + \sum_{i=1}^{N} h_i s_i \quad \text{subject to} \quad s_i \in \{-1, 1\} \tag{15.3}$$

where h_i represents external (magnetic) field. This term can be incorporated in $\mathbf{J}$ matrix by considering $N + 1$ spins and thus will be omitted for the rest of the chapter. Experimental realisation of the nonlinear terms beyond quadratic in the Ising Hamiltonian would lead to a k-local spin Hamiltonian with $k > 2$ and would allow for a direct mapping of higher order binary optimisation (HOBO) problems including Max-SAT [12] or number factorisation [13]:

$$\min \quad - \sum_{i_1,i_2,\ldots i_k}^{N} Q_{i_1,i_2,\ldots i_l,\ldots,i_k} s_{i_1} s_{i_2} \ldots s_{i_l} \ldots s_{i_k} \quad \text{subject to} \quad s_{i_l} \in \{-1, 1\}. \quad (15.4)$$

In the XY model "spins" are continuous $s_j = \cos\theta_j + i \sin\theta_j$ and the corresponding quadratic continuous optimisation (QCO) problem can be formulated as

$$\min \quad - \sum_{i<j} J_{ij} \mathbf{s}_i \cdot \mathbf{s}_j = \min - \sum_{i<j} J_{ij} \cos(\theta_i - \theta_j) \quad \text{subject to} \quad \theta_i \in [0, 2\pi). \quad (15.5)$$

When phases θ_j are limited to discrete values $2\pi/n$ with an integer $n > 2$ the model (15.5) recovers the n-state Potts model (Clock model) with applications in protein folding [14].

QCO, QUBO, and HOBO problems are all examples of **NP**-hard problems. The corresponding spin models are universal. The connection between these notions are detailed in the next section.

15.1.2 P, NP, NP-Complete Problems

The computational complexity of a problem can be revealed by looking at the dependence of the problem's size on time or the number of operations required to solve it. In a simple case of such polynomial dependence, i.e. when a polynomial time algorithm exists, a problem belongs to a **P** class. If a polynomial time algorithm of finding a solution is not known but there exists a polynomial algorithm for verifying a solution when presented, then a problem belongs to non-deterministic polynomial-time (**NP**) class that clearly includes the **P** class. Whether **P** = **NP** is true or not is a major unsolved problem in computer science although it is widely believed to be untrue [15]. A problem is **NP**-hard when every problem in **NP** can be reduced in polynomial time to it. The problems that are both **NP**-hard and **NP** are called **NP**-complete. All **NP**-complete problems are equivalent in a sense that either all of them or none of them admit a polynomial-time algorithm. Examples include the travelling salesman problem, spin glass models, and integer linear programming. The computational complexity of finding the ground state of the Ising Hamiltonian (Ising model) on finite lattices has been studied before [16] where the two-dimensional Ising model with a magnetic field (15.3) and equal antiferromagnetic couplings has been shown to be **NP**-complete for planar graphs. In addition, **NP**-completeness was demonstrated for the three-dimensional Ising model with nearest neighbour interactions and coupling strengths from $\{-1, 0, 1\}$ [16]. Consequently, the above mentioned hierarchy of complexity classes allows one to conclude the impossibility of existence of a polynomial algorithm for computing the ground state energy of the Ising model without the existence of a polynomial algorithm for all **NP**-complete problems.

The existence of universal spin Hamiltonians has been established. Universality means that all classical spin models with any range of interactions can be reproduced within such a model, and certain simple Hamiltonians such as the 2D Ising model on a square lattice with transverse fields and nearest neighbour interactions are universal [17]. Thus, due to **NP**-hardness of the Ising model, there should exist a polynomial time mapping of many practically relevant **NP**-complete problems to the Ising Hamiltonian, whose decision version solves the **NP**-complete problem of interest. The mapping of various **NP** problems, including Karp's 21 **NP**-complete problems [18], to Ising models with a polynomial overhead was demonstrated [11]. For example, the travelling salesman problem for N cities, that are connected with weighted edges $w_{uv} \geq 0$ from the set E (distances between cities), can be formulated as the following Ising problem of size N^2:

$$H_{\mathrm{TSP}} = A\sum_{i=1}^{N}\left(1-\sum_{v=1}^{N}x_{v,i}\right)^2 + A\sum_{v=1}^{N}\left(1-\sum_{i=1}^{N}x_{v,i}\right)^2 + A\sum_{(uv)\notin E}\sum_{i=1}^{N}x_{u,i}x_{v,i+1} + B\sum_{(uv)\in E}w_{u,v}\sum_{i=1}^{N}x_{u,i}x_{v,i+1}. \tag{15.6}$$

Each spin $x_{v,i} \in \{0, 1\}$ in (15.6) represents the vertex v and its order i in a path. All valid routes in this representation are regulated by the first three terms: each city should be in the route (first term) and should appear in it only once (second term), any adjacent cities in the route should be connected (third term), while the search for the optimal route is realised by minimising the sum of weights of all cities in a route (forth term). The reasonable choice of constants A and B (e.g. A should be big enough with respect to $B > 0$) guarantees that only the space of valid routes is explored. Reshaping this two-dimensional spin matrix with elements $x_{v,i}$ to a spin vector of size N^2 allows one to recover the coupling matrix $\mathbf{J}$ and magnetic field $\mathbf{h}$ to formulate the corresponding Ising Hamiltonian. The size of the Ising problem can be reduced to $(N-1)^2$ by fixing a particular city to be the first in the route. Note, that the Hamiltonian H_{TSP} can represent both directed and undirected graphs, and the generalisation for the cycles optimisation problem is straightforward. We also note that a polynomial overhead does not always apply and some combinatorial optimisation problems can be mapped to the Ising model of the same size N. For example, the maximum cut problem

$$\max_{S^+,S^-}\sum_{i\in S^+,j\in S^-} w_{ij} \tag{15.7}$$

seeks for the cut of a graph into two subsets with a largest sum of their connecting weighted edges. By assigning $+1$ and -1 spins to all vertices in subsets S^+ and S^-, respectively, this optimisation problem can be formulated as

$$\max_{s_i}\frac{1}{2}\sum_{i<j}w_{ij}(1-s_is_j) = \frac{1}{2}\sum_{i<j}w_{ij} + \min_{s_i}\frac{1}{2}\sum_{i<j}w_{ij}s_is_j \tag{15.8}$$

and thus a maximum cut of any graph can be converted to minimisation of the corresponding Ising Hamiltonian with the coupling matrix $J_{ij} = -w_{ij}$ with an addition of an offset. A well-known standardised set of maximum cut type of problems often serve as a metric for comparison of newly proposed simulators and algorithms [19–21].

Another example of a universal spin model is the XY model which is directly related to the notoriously hard to solve phase retrieval problem. The problem's objective is to recover a general signal (or image) from the magnitude of its Fourier transform [22–24]. This problem arises from the fact that the signal detectors can usually record only modulus of the diffraction pattern, therefore, losing the information about the phase of the optical wave. Mathematically, one needs to recover a signal $\mathbf{x} \in \mathbb{C}^m$ from the amplitude $\mathbf{b} = |A\mathbf{x}|$, where $A \in \mathbb{C}^{n\times m}$, $\mathbf{b} \in \mathbb{R}^n$. Then the phase recovery problem [25] can be formulated as:

$$\min_{x_j,u_i} \sum_i \left(\sum_j A_{ij} x_j - b_i u_i \right)^2 \tag{15.9}$$

where $\mathbf{u} \in \mathbb{C}^n$ is a phase vector that satisfies $A\mathbf{x} = diag(\mathbf{b})\mathbf{u}$, $|u_i| = 1$ for $i = \overline{1, n}$. This optimisation problem can be further rewritten as

$$\min \sum_{ij} M_{ij} u_i u_j \quad \text{subject to} \quad |u_i| = 1, i = \overline{1, n}, \tag{15.10}$$

where $M = diag(\mathbf{b})(I - AA^{\dagger})diag(\mathbf{b})$ is the Hermitian matrix, I is the identity matrix, and $A^{\dagger}$ is the Moore–Penrose pseudoinverse of a matrix A (see [25] for details).

It is important to note that when we refer to a spin problem as **NP**-complete we understand that for some specific coupling matrix $\mathbf{J}$ ("problem instances") finding the solution can be easy (belong to **P** class). The term **NP**-completeness reflects worst case behaviour and may allow a polynomial time to solution for most instances on average. This leads to the cornerstone question of how to distinguish hard instances from simple ones. The answer is especially important for the rivalry between classical hardware and unconventional computing machines which have to compete on problems of known complexity. It is believed that the way to create "hard" instances for spin Hamiltonians resides at the intersection of computational complexity and physics, e.g. the hardness of problems can be connected to the existence of a first-order phase transition in a system (see [26] and references therein). If an instance is indeed hard then it would be difficult to solve even for a medium size on a classical computer since the number of operations grows as an exponential function with the matrix size. Thus, the time required to find reliably the ground state energy should highly depend on the coupling matrix structure $\mathbf{J}$ and the way it was constructed. For instance, finding the global minimum of the XY model for positive definite matrices remains **NP**-hard due to the non-convex constraints but can be effectively approximated using a semidefinite programming relaxation with some performance guarantee [27, 28]. Sparsity also plays an important role and for sufficiently sparse matrices fast methods exist [29]. For spin models, the generation of matrix instances **J** with tunable algorithmic hardness and, preferably, with a specifiable ground state, is an ongoing problem studied by many research teams. An elegant way of creating such problems has been recently proposed by a Microsoft research team [26] who suggested to use the Wishart planted ensemble technique. Having a unified set of optimisation problems with a tunable hardness and known solutions would allow for an objective benchmark of quantum simulators on various physical platforms as well as for classical algorithms. Otherwise, announcements of state-of-the-art platforms and methods, which demonstrated their performance on some random and not necessarily hard instances, would continue to happen.

15.2 Physical Platforms for Large-Scale Optimisation

Rather than trying to model nature one can consider a reverse idea of exploiting physical phenomena for solving **NP**-complete problems. Such problems can be tackled by quantum computers or simulators to produce solutions in reasonable time. In the last five years we have seen a competition of different physical platforms in solving classical optimisation problems faster than it can be achieved on a classical hardware for a given problem size. This rivalry resulted in the rapid emergence of a new field at the intersection of laser and condensed matter physics, engineering and complexity theories, which aims to develop quantum or classical analogue devices to simulate spin Hamiltonians. Next we discuss the achieved success in such simulations for a range of physical systems.

15.2.1 Cold Atoms in Optical Lattices

Ultracold atoms in optical lattices constitute a well-controlled experimental setting to realise various spin Hamiltonians [30, 31]. Optical lattices are formed by directing several laser beams to interfere and to create standing wave configurations. Such waves provide practically loss-free external potentials in which ultracold atoms may condense, move and interact with one another [32, 33]. The unprecedented control and precision with which one can engineer such lattices and load the atoms there led to many suggestions to consider such systems as possible candidates for unconventional computing in quantum information processing and quantum simulations.

Here we will only discuss a weakly interacting Bose gas in an optical lattice. The description of particles in the strongly-correlated regime is possible with Bose– and Fermi–Hubbard models as well as with extended Hubbard models with nearest-neighbour, next nearest-neighbour interactions, etc. [34]. If the bosonic gas is dilute, the time evolution of the condensate wave function ψ is governed by the Gross–Pitaevskii equation [35–37]

$$i\hbar\frac{\mathrm{d}}{\mathrm{d}t}\psi(\mathbf{r},t) = -\frac{\hbar^2}{2m}\nabla^2\psi(\mathbf{r},t) + V_{\mathrm{ext}}(\mathbf{r})\psi(\mathbf{r},t) + g|\psi(\mathbf{r},t)|^2\psi(\mathbf{r},t), \quad (15.11)$$

where g is the strength of the delta-function interactions and the external potential V_{ext} describes an optical lattice—periodic potential—usually combined with a weak harmonic trapping potential.

The condensate evolution and particles' interactions at different local minima of the optical lattice in superfluid regime can be described with the tight-binding approximation, which is valid when the barrier between the neighbouring sites is much higher than the chemical potential. In this approximation the condensate wavefunction ψ is written as a sum of normalized wave functions $\phi_i = \phi(\mathbf{r} - \mathbf{r}_i)$ localized in each minimum of the periodic potential, i.e. $\mathbf{r} = \mathbf{r}_i$:

$$\psi(\mathbf{r},t) = \sum_i \Psi_i(t)\phi(\mathbf{r} - \mathbf{r}_i), \quad (15.12)$$

where $\Psi_i(t) = \sqrt{\rho_i(t)}e^{i\theta_i(t)}$ is the complex amplitude of the ith lattice site, ρ_i and θ_i are the number of particles and the phase in the ith site, respectively. The amplitude Ψ_i describes the state of the so-called "coherent center" located at $\mathbf{r}_i$. By inserting this anzats into (15.11) and integrating the spatial degrees of freedom out one obtains the discrete nonlinear Schrödinger (DNLS) equation (see e.g. [38–40])

$$i\hbar\frac{\partial\Psi_i}{\partial t} = -J(\Psi_{i+1} + \Psi_{i-1}) + \epsilon_i\Psi_i + U|\Psi_i|^2\Psi_i, \quad (15.13)$$

where J is the nearest-neighbour tunnelling rate,

$$J = -\int \mathrm{d}\mathbf{r}\left[\frac{\hbar^2}{2m}\nabla\phi_i \cdot \nabla\phi_{i+1} + \phi_i V_{\mathrm{ext}}\phi_{i+1}\right], \quad (15.14)$$

ϵ_i is the on-site energy given by

$$\epsilon_i = \int \mathrm{d}\mathbf{r}\left[\frac{\hbar^2}{2m}(\nabla\phi_i)^2 + V_{\mathrm{ext}}\phi_i^2\right], \quad (15.15)$$

and U is the nonlinear coefficient given by

$$U = g\int d\mathbf{r}\,\phi_i^4. \quad (15.16)$$

Such classical lattice models described by DNLS equations represent the mean-field limit of Bose–Hubbard models [41]. The mean-field limit of the non-standard Bose–Hubbard models includes the interactions beyond the nearest neighbours which leads to a generalised DNLS

$$i\hbar\frac{\partial \Psi_i}{\partial t} = -\frac{1}{2}\sum_{\langle i,j\rangle} J_{ij}\Psi_j + (\epsilon_i + U|\Psi_i|^2)\Psi_i, \tag{15.17}$$

where J_{ij} is the coupling strength between the ith and jth coherent centers. If one loads an equal number of particles in each site of the lattice, the ground state of (15.17) realises the minimum of the XY Hamiltonian $-\sum_{\langle i,j\rangle} J_{ij}\cos(\theta_i - \theta_j)$. This has been experimentally demonstrated in triangular lattices using the atoms motional degrees of freedom and tunable artificial gauge fields [42, 43].

The quantum annealing protocol can in principle be implemented in such a system by using (15.1) with $H_0 = \sum_{\langle i,j\rangle}\cos(\theta_i - \theta_j)$ and $H_{\text{objective}} = -\sum_{\langle i,j\rangle} J_{ij}\cos(\theta_i - \theta_j)$. A similar principle of adiabatic quantum annealing has been realised in the famous D-Wave machine that we discuss below.

15.2.2 D-Wave Quantum Annealer

D-Wave is a first commercially available quantum annealer that is built on superconducting qubits with programmable couplings and specifically designed to solve QUBO problems (15.3) [44]. By specifying the interactions J_{ij} between qubits, a desired QUBO problem is solved [45] via a quantum annealing process as in (15.1). Adiabatic (slow) transition in time from an initial state of a specially prepared "easy" Hamiltonian to the objective Ising Hamiltonian guarantees that the system remains in the low energy state, which gives the final energy that corresponds to the optimal solution of the QUBO problem.

Many benchmarks on different QUBO problems were performed on a D-Wave One and D-Wave Two machines without a solid demonstration of quantum speedup of annealer over classical algorithms [46–48]. A better performance was shown for the last 2000-qubit D-Wave machine released in 2017 on a newly proposed synthetic problem class in which the computational hardness is created through frustrated global interactions. The major limitations of D-wave simulators is that each qubit can be connected to maximum of six other qubits which is the consequence of creating chips with Chimera structure. The next generation of D-Wave quantum computer is expected to be announced in 2020 with a different architecture which would allow for 15 connections per each node. Together with reverse annealing and virtual graphs features a significant performance improvement could be possibly demonstrated.

15.2.3 Complex Laser Networks

A new generation of complex lasers such as degenerate cavity lasers, multimode fibre amplifiers, large-aperture VCSEL or random lasers have many advantages in comparison with the relatively simple traditional laser resonators in terms of their computing properties [49]. They have a large number of spatial degrees of freedom, their nonlinear interactions within the gain material can be controlled by adjusting the spatial structures of lasing modes, the spatial coherence of emission can be tuned over a wide range, and the output beams may have arbitrary profiles. These properties allow the complex lasers to be used for reservoir computing [50] or for solving hard computational problems.

In laser networks the coupling can be engineered by mutual light injection from one laser to another. This introduces losses that depend on the relative phases between the lasers. Such dissipative coupling drives the system to a phase locking and therefore to a steady state solution of QCO (15.5), i.e. to the minimum of the XY Hamiltonian [51–53]. Degenerate cavity lasers are particularly useful as solvers as all their transverse modes have nearly identical quality factor. This implies that a large number of transverse modes lase simultaneously since they all have similar lasing thresholds [49].

The evolution of the N single transverse and longitudinal modes class-B lasers can be described by the rate equations [54, 55] on the amplitude A_i, phase θ_i, and gain G_i of the ith laser

$$\frac{\mathrm{d}A_i}{\mathrm{d}t} = (G_i - \alpha_i)\frac{A_i}{\tau_p} + \sum_j J_{ij}\frac{A_j}{\tau_p}\cos(\theta_i - \theta_j), \tag{15.18}$$

$$\frac{\mathrm{d}\theta_i}{\mathrm{d}t} = \Omega_i - \sum_j J_{ij}\frac{A_j}{\tau_p A_i}\sin(\theta_i - \theta_j), \tag{15.19}$$

$$\frac{\mathrm{d}G_i}{\mathrm{d}t} = \frac{1}{\tau_c}[P_i - G_i(1 + |A_i|^2)], \tag{15.20}$$

where P_i, α_i, Ω_i represent the pump strength, loss, frequency detuning of laser i, respectively, whereas τ_p and τ_c denote the cavity round trip time and the carrier lifetime, respectively. The coupling strengths between ith and jth lasers are represented by J_{ij}. If the amplitudes of all lasers are equal, (15.19) reduces to the system of coupled phase oscillations, $\Omega_i = \Omega$,

$$\frac{\mathrm{d}\theta_i}{\mathrm{d}t} = \Omega - \frac{1}{\tau_p}\sum_j J_{ij}\sin(\theta_i - \theta_j). \tag{15.21}$$

Equation (15.21) is a celebrated Kuramoto model of identical oscillators which is widely used to describe the emergence of coherent behaviour in complex systems [56, 57]. By LaSalle Invariance Principle [58] every trajectory of the Kuramoto model converges to a minimum of the XY Hamiltonian.

It was shown that the probability of finding the global minimum of the XY Hamiltonian agrees between experimental realisations of the laser array and numerical simulations of (15.18)–(15.20). However, simulating the Kuramoto model of (15.21) on the same matrix of coupling strengths gives a much lower probability of finding the global minimum. The conclusion was made that the amplitude dynamics described by (15.18) provides a mechanism to reach the global minimum [55] by pumping from below. This suggested that the cavity lasers can be used as an efficient physical simulator for finding the global minimum of the XY Hamiltonian, and therefore, for solving phase retrieval problems.

A digital degenerate cavity laser has recently been shown to solve phase retrieval problems rapidly [59]. It is an all-optical system that uses nonlinear lasing process to find a solution that best satisfies the constraint on the Fourier magnitudes of the light scattered from an object. To make sure that the solution to the phase retrieval problem is found the compact support aperture is introduced inside the cavity that ensures that different configurations of laser phases compete to find the one with the minimal losses. The system combines the advantages of short round-trip times of the order of 20 ns and high parallelism in selecting the winning mode.

15.2.4 Coherent Ising Machine

Network of coupled optical parametric oscillators (OPOs) is an alternative physical system for solving the Ising problem (cf. [60] and references therein). Each OPO is a nonlinear oscillator with two possible phase states above the threshold that can be interpreted as binary spin states $\{-1, 1\}$ with respect to the reference beam. The OPO is stimulated with pulses of light which are then loaded into a loop of optical fiber. Below threshold, pulses of low intensity have random phase fluctuations. Depending on the enforced pulse interactions, the intensities are continuously modulated so that after multiple runs around the loop the final binary phases are formed for all OPOs at about the same time. Driving the system close to this near-threshold regime, the lowest loss configuration state can be found. This state corresponds to the optimal solution of the Ising Hamiltonian and, therefore, the OPO-based simulator is known as the coherent Ising Machine (CIM).

The currently most successful implementations of CIMs have been realised using a fiber-based degenerate optical parametric oscillators (DOPOs) and a measurement based feedback coupling, in which a matrix-vector multiplication is performed on a field-programmable gate array (FPGA) embedded in the feedback loop. The computational performance of such scalable optical processor, that is bounded by the electronic feedback, was demonstrated for various large-scale Ising problems [60–62], while a speedup over classical algorithms is an ongoing study [63, 64]. The ability to implement arbitrary coupling connections between any two spins [60] was apparently the main reason to claim a better scalability of the CIM than the quantum annealer, i.e. D-Wave machine [61].

In a Coherent Ising Machine each Ising spin corresponds to a DOPO that is described by a rate equation for the complex amplitude of the signal field a_i:

$$\frac{\mathrm{d}a_i}{\mathrm{d}t} = pa_i^* - a_i - |a_i|^2 a_i + \sum_j J_{ij} a_j, \tag{15.22}$$

where the dynamics is defined by a linear pump term p, normalised linear and nonlinear losses, and mutual couplings J_{ij}. To experimentally realise these couplings, a portion of light is extracted from the cavity after each round trip. That light is then homodyned against a reference pulse to produce a_i that is next supplied to FPGA where a feedback signal is computed for each pulse. Lastly, an optical modulator is applied to convert the signal back to light that can be used for the next round trip. The equations (15.22) are often reformulated in terms of the in-phase and quadrature components $a_i = c_i + is_i$ giving the equations in real terms:

$$\frac{\mathrm{d}c_i}{\mathrm{d}t} = \left(p - 1 - (c_i^2 + s_i^2)\right)c_i + \sum_j J_{ij}c_j \tag{15.23}$$

$$\frac{\mathrm{d}s_i}{\mathrm{d}t} = \left(-p - 1 - (c_i^2 + s_i^2)\right)s_i + \sum_j J_{ij}s_j. \tag{15.24}$$

The computational effectiveness of these equations has been demonstrated [65] by tackling small size Ising type problems of order up to 20. In a part devoted to polariton condensates we will show that for achieving the global minimum the realisation of an individual pump variation p_i for equalising all signal amplitudes $|a_i|$ is crucial.

Phase-stability for the whole length of the cavity is required which makes the DOPOs system highly susceptible to external perturbations that can affect performance [61]. Furthermore, the nonlinear DOPO generation process demands powerful laser systems and temperature-controlled nonlinear materials, which result in large and complex optical setups. These issues lead to recent proposals of other physical platforms for implementing a CIM-like machine. A CIM based on opto-electronic oscillators with self-feedback was suggested to be more stable and less expensive for solving Ising optimisation problems on regular and frustrated graphs up to 100 spins with similar or better performance compared to the original DOPO-based CIM [66]. An analogue all-optical implementation of a CIM based on a network of injection-locked multicore fiber lasers [67] demonstrated a possibility to solve Ising Hamiltonians for up to thirteen nodes. The dynamics of a network of injection-locked lasers was based on nonlinear coupled photon rate equations and the couplings were implemented using spatial light modulators. The couplings were reported to be dependent on the photon numbers that are not known beforehand, which can be a major obstacle on the way of solving a given Ising Hamiltonian with the proposed photonic CIM. To solve this issue, approaches similar to gain variation [20, 68] may be considered in the future. Another large-scale optical Ising machine based on the use of a spatial light modulators was experimentally demonstrated by using the binary phases in separated spatial points of the optical wave front of an amplitude-modulated laser beam and realising configurations with thousands of spins with tunable all-to-all pairwise interactions [69].

15.2.5 Photon and Polariton Networks

Microcavity exciton-polaritons, or simply polaritons, are quasi-particles that result from the hybridisation of light confined inside semiconductor microcavities and bound electron hole pairs (excitons). The steady states in these nonequilibrium systems are set by the balance between the pumping intensity, coming from the interconversion rate of the exciton's reservoir into polaritons, and losses, happening due to the leakage of photons. As polaritons are bosons and obey Bose–Einstein statistics, they can form a condensed (coherent) state above a critical density [70]. Thus, polaritons offer a unique playground to explore nonequilibrium condensation and related effects in solids. The advantage for such explorations comes from the polariton's small effective mass that is 4–5 orders of magnitude smaller than the electron's mass. The design and choice of material allows one to control the polariton mass and to realise such solid state nonequilibrium condensates not only at cryogenic temperatures but even at room temperature in organic structures. The weak coupling at high temperatures and high pumping intensities transits continuously to strong coupling at lower temperatures and lower pumping intensities. In the limit of small gain and small losses, solid state condensates resemble equilibrium Bose–Einstein condensates (BECs) and in the regime of high gain and high losses, they approach lasers. This transition from the equilibrium BECs to normal lasers was described with a unified approach via polariton condensates [71].

In another system, closely resembling the physics of polariton condensates, macroscopic occupation of the lowest mode for a gas of photons confined in a dye-filled optical microcavity was recently shown [72–75]. The rapid thermalization of rovibrational modes of the dye molecules by their collisions with the solvent and phonon dressing of the absorption and emission by the dye molecules leads to the thermal equilibrium distribution of photons and concomitant accumulation of low-energy photons. Such systems resemble microlasers [76], but unlike microlasers exhibit a sharp threshold which occurs far below inversion.

To realise the lattices of polariton or photon condensates many techniques have been proposed and realised in experiments. Polariton lattices can be optically engineered by injecting polaritons in specific areas of the sample using a spatial light modulator [77–81]. A variety of potential landscapes to confine polariton or photons have also been engineered [82–84]. The rate equations describing the evolution of gain-dissipative condensates in a lattice were derived from the space and time resolved mean-field equations [68, 85] and take a form of the Stuart-Landau equations

$$\dot{\Psi}_i = -iU|\Psi_i|^2\Psi_i + (\gamma_i - |\Psi_i|^2)\Psi_i + \sum_{j\neq i} C_{ij}\Psi_j, \tag{15.25}$$

where $\Psi_i = \sqrt{\rho_i}\exp[i\theta_i]$ is the complex amplitude of the ith condensate, U is the strength of self-interactions between the quasi-particles, γ_i is the effective injection rate (the difference between the pumping of the quasi-particles into the system and linear losses). The coupling strength $C_{ij} = J_{ij} + iG_{ij}$ is generally a complex number and consists of the Heisenberg coupling J_{ij} mediated by the injection reservoir and the Josephson part G_{ij} that comes from exchange interactions between the condensates. The system described by (15.25) reaches the fixed point when $J_{ij} \gg G_{ij}$ and the pumping feedback is introduced in the system [68]. The feedback on the pumping intensity ensures that all the occupations are the same at the fixed point, by adjusting the pumping if the occupation exceeds the set threshold value $|\Psi_i|^2 = \rho_{\rm th}$. The total injection of the particles in the system of N condensates at the fixed point is given by

$$\sum_{i=1}^{N} \gamma_i = N\rho_{\rm th} - \sum_{i=1}^{N}\sum_{j<i}^{N} J_{ij}\cos(\theta_i - \theta_j). \tag{15.26}$$

Choosing the lowest possible total particle injection $\sum \gamma_i$ that leads to the occupation $\rho_{\rm th}$ for each condensate guarantees that the minimum of the XY Hamiltonian is reached. In order to find the true global minimum the system has to slowly be brought to the condensation threshold while spending enough time in its neighbourhood to span various phase configurations driven by the system noise (classical and quantum fluctuations). When the system reaches a phase configuration in the vicinity of the minimum of the XY Hamiltonian it quickly converges to it by the gradient descent given by the imaginary part of (15.25):

$$\dot{\theta}_i = -U\rho_{\rm th} - \sum_{j\neq i}^{N} J_{ij}\sin(\theta_i - \theta_j). \tag{15.27}$$

This idea has been theoretically justified [68] and experimentally realised for simple polariton graphs [81]. It was also proposed how to extend the scheme to discrete optimisation problems such as QUBO (minimising the Ising Hamiltonian) or n-states Potts Hamiltonians [86]. When the resonant excitation is combined with a non-resonant one, the spins are forced to take the discrete values aligning with the directions set by the resonant excitation. If $n:1$ resonant drive is added to the system, the dynamics of the coherent centres obeys

$$\dot{\Psi}_i = -iU|\Psi_i|^2\Psi_i + (\gamma_i - |\Psi_i|^2)\Psi_i + \sum_{j\neq i} J_{ij}\Psi_j + h(t)\Psi_i^{*(n-1)}, \tag{15.28}$$

where $h(t)$ is an increasing function that reaches some value $H > \max_i \sum_j |J_{ij}|$ at the threshold density $\rho_{\rm th}$. The adjustment of injection rates leads to the equal-density fixed point and (15.26) becomes

$$\sum_{i=1}^{N} \gamma_i = N\rho_{\rm th} - \sum_{i=1}^{N}\sum_{j<i}^{N} J_{ij}\cos(\theta_i - \theta_j) - H\rho_{\rm th}^{n/2-1}\cos(n\theta_i). \tag{15.29}$$

At $n = 2$, the last term on the right-hand side provides the penalty to phases deviating from 0 or π reducing the optimisation problem to QUBO. For $n > 2$, the n-state Potts Hamiltonian is minimised. The minimisation of HOBO may be achieved when the system operates much above the threshold and higher order terms can not be neglected [87].

If the time evolution of the reservoir of noncondensed particles is slow, the system of N interacting coherent centres is better described by the following equations [85]:

$$\dot{\Psi}_i = -iU|\Psi_i|^2\Psi_i + (R_i - \gamma_c)\Psi_i + \sum_{j\neq i} J_{ij}\Psi_j, \tag{15.30}$$

$$\dot{R}_i = \Gamma_i - \gamma_R R_i - R_i|\Psi_i|^2, \tag{15.31}$$

where R_i is the occupation of the ith reservoir, Γ_i, γ_R and γ_c characterize the rate of particle injection into the reservoir and the linear losses of the reservoir and condensate, respectively. If one replaces Ψ_i by the electric field and R_i by the population inversion of the ith laser, the result is a form of the Lang–Kobayashi equations normally derived to describe the dynamical behavior of coupled lasers from Lamb's semiclassical laser theory [88, 89]. The total injection of the particles in the system of N condensates at the fixed point is given by

$$\sum_{i=1}^{N} \Gamma_i = (\gamma_R + \rho_{\text{th}}) \left[N\gamma_c - \sum_{i=1}^{N}\sum_{j<i}^{N} J_{ij}\cos(\theta_i - \theta_j) \right]. \tag{15.32}$$

Similar to (15.26), if the total injection into the system is minimal, the phases of coherent centres minimise the XY Hamiltonian. Next we discuss the current and future role of physical systems mentioned above in a rapidly developed machine learning applications.

15.3 Analogue Physical Systems for Recurrent Neural Networks and Reservoir Computing

The artificial neuron (perceptron) is a weighted decision-making procedure that takes binary input vector and produces a single binary output [90]. For a stable learning process, a small change in a weight (or bias) should cause only a small change in the output so that a few updates of parameters with a technique such as backpropagation [91] can produce a better output. Such condition doesn't hold for a perceptron network that is sensitive to small changes in parameters of any single neuron. Introducing a nonlinear activation function helps to overcome this problem with common choices for the function being the sigmoid, tanh, and the rectified linear unit. With an addition of intermediate layers of neurons, hidden layers, the architecture and the training of the network can become nontrivial. The final design usually is a trade off between the number of hidden layers against the time required to train the network. In fact, even optimising the design for single hidden-layer networks is an **NP**-hard problem [10]. The outputs in each layer can be calculated given the outputs from the lower layers, so the information is always transmitted forward.

Such feedforward neural networks are powerful learning models that have shown state-of-the-art results on a variety of machine learning applications [92] though they were still limited to the tasks with independent training and test data points. Addressing the data with time and space relationships (e.g. video frames and audio snippets) requires backward connection between layers of neurons, and so better modelled with recurrent neural networks (RNNs).

The capability of RNNs with nonlinear activations to perform nearly arbitrary computation was demonstrated with a simulation of a universal Turing machine [93]. The two well-known RNN architectures for sequence learning were introduced later: long short-term memory [94] and bidirectional recurrent neural networks [95]. The former introduced the memory cell, a unit of computation that replaces traditional nodes in the hidden layer of a network, while the latter proposed that the output at any point in the sequence should be dependent on information from both the future and the past (comprehensive reviews of RNNs can be found in [96, 97]). Following [97], the input (target) sequence to a simple RNN with one hidden layer can be denoted as a sequence of vectors $\mathbf{x}^{(t)}$ ($\mathbf{y}^{(t)}$) for $t = \overline{1, T}$. The RNN will then produce predicted vectors $\mathbf{y}^{(t)}$ at each time step:

$$\mathbf{h}(t) = f^{(1)}_{\text{activation}}\left(W^{hx}\mathbf{x}(t) + W^{hh}\mathbf{h}(t-1) + \mathbf{b}_h\right) \tag{15.33}$$

$$\widehat{\mathbf{y}}(t) = f^{(2)}_{\text{activation}}\left(W^{yh}\mathbf{h}(t) + \mathbf{b}_y\right) \tag{15.34}$$

where W^{hx} (W^{yh}) is the weight matrix between the input (output) and the hidden layer, W^{hh} is the matrix of recurrent weights connecting the hidden layer with itself at adjacent time steps, $\mathbf{b}_h$ and $\mathbf{b}_y$ are bias vectors, and $f^{(i)}_{\text{activation}}$ are nonlinear activation functions. In such RNN, nodes with recurrent edges, i.e. edges that connect adjacent time steps, receive input from both the current state $\mathbf{x}(t)$ and from the previous state via hidden node values $\mathbf{h}(t-1)$. Given the hidden node values $\mathbf{h}(t)$, the output values $\widehat{\mathbf{y}}(t)$ are affected by the input $\mathbf{x}(t-1)$ and $\mathbf{x}(t)$. Such cyclic network can still be trained across many time steps using backpropagation through time [98] since RNNs can be interpreted as a deep network with one layer per time step and shared weights across time steps.

RNNs constitute a natural approach to numerous problems ranging from handwriting generation [99, 100] to character prediction [101] to machine translation [102, 103]. Such successful results and computational hardness of learning process make recurrent networks a great candidate for simulations with a physically based platform. Such physical systems can be much more efficient than any of known classical hardware implementations due to the inherent physical properties such as quantum or classical parallellism and neural network-type architecture. A recent mapping of dynamics of acoustic and optical waves to RNNs was suggested [104].

Other computational frameworks for data processing such as reservoir computing (RC) (originally referred to as echo state networks [105] or liquid state machines [106]) can benefit from an implementation with analogue physical platforms. A particular useful property is that RC systems require less accuracy of individual control of all forward and backward neuron connections. Various nonlinear dynamical systems, including electronic [107–109], photonic [110, 111], spintronic [112–114], mechanical [115], and biological [116] systems, have been recently employed as potential reservoirs for RC (see [117] and references therein).

RC methods have been successfully applied to many practical problems involving real data, with focus on machine learning applications. The role of the reservoir (physical system) in RC is to nonlinearly map sequential inputs into a higher-dimensional space so that features can then be extracted from its output with a simple learning algorithm. Therefore, such reservoirs become attractive for an experimental implementation in many physical systems with a motivation of realising fast information processing devices with low learning cost. Networks of polaritons or lattices of atomic condensates, discussed above, can serve as interacting nonlinear elements for an efficient network-type RC system with a possible approach suggested recently [118].

15.4 System-Inspired Algorithms

The discovered principles of operation of the analogue physical systems for finding optimal solutions open opportunities for new optimisation algorithms that can be realised on specialised but classical computing architectures: FPGAs, GPUs, etc.

The principle of operation of the Coherent Ising Machine was implemented as the network of nonlinear oscillators described by simplified equations [119]:

$$\frac{\mathrm{d}x_j}{\mathrm{d}t} = -\frac{\partial V}{\partial x_j} \quad \text{with} \quad V = \sum_j V_b(x_j) + \epsilon V_H(x_j), \tag{15.35}$$

where x_j are N analogue variables, $V_b(x_j) = -0.5\alpha x_j^2 + 0.25x_j^4$ is the paradigmatic bistable potential, $\alpha = -1 + p$ is the bifurcation parameter given by the normalized decay rate and linear gain p for the signal field, $\epsilon \ll 1$ is a positive coefficient, $V_H(x_j) = -\sum_i J_{ij}x_i x_j$ is the analogue of the Hopfield network. The addition of the amplitude variation to these equations [21], i.e. $\epsilon \to \epsilon e_j$, resulted in

$$\frac{\mathrm{d}e_j}{\mathrm{d}t} = \beta(\rho_{\mathrm{th}} - x_j^2)e_j, \tag{15.36}$$

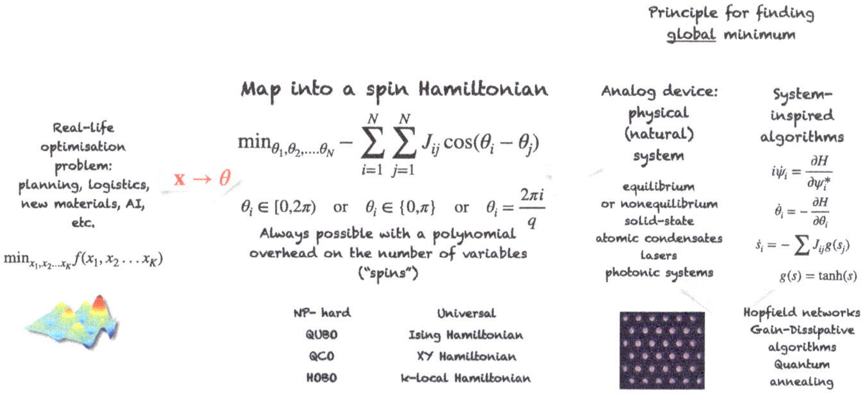

Fig. 15.1 Schematics of Hybrid Computing. A computationally hard real life problem can be mapped into a spin Hamiltonian, where spins represent the degrees of freedom (discrete or continuous) of the problem and coupling strengths represent the structure and the constraints of the objective function. Such mapping is possible if the original problem is nondeterministic polynomial, and the resulting spin Hamiltonian is universal. The ground state of the spin Hamiltonian can be found by an analogue physical system using quantum annealing or the gain-dissipative principle of its operation coupled to a digital device (CMOS, FPGA, GPU, etc.) compatible with and based on the system-inspired algorithms

where e_i is a positive error variable and β is a positive rate of change of error variables. The initialisation of such additional control of the target amplitude $\rho_{\rm th}$ of all analogue variables x_j allowed to numerically optimise the medium-scale Ising type problems ($\sim$1000 spins). A similar approach has been recently realised using FPGAs for a network of Duffing oscillators [120].

Another example of a gain-dissipative algorithm was inspired by the operation of polariton networks [20]. By gradually increasing the pumping strength γ_i for ith oscillator while inducing enough noise to span large volume of high dimensional space of the problem one can explore the low energy part of the spin Hamiltonian. This can be achieved by numerical integration of complex fields Ψ_i

$$\dot{\Psi}_i = (\gamma_i - |\Psi_i|^2)\Psi_i + \sum_{j\neq i} J_{ij}\Psi_j + h(t)\Psi_i^{*(n-1)}, \tag{15.37}$$

$$\dot{\gamma}_i = \epsilon(\rho_{\rm th} - \rho_i), \tag{15.38}$$

with the meaning of parameters explained in Sect. 15.2.5. The performance of this algorithm was demonstrated on the medium scale Ising and XY models [20].

15.5 Conclusions and Future Challenges

What promise does unconventional computing platform hold? Would it be able to find a better solution in a fixed time? Or find a solution for a fixed precision faster? Or solve more complex problems at fixed and limited cost? Would it appear as an accelerator in neural networks, machine learning and artificial intelligence platforms? Would it be able to solve a full range of different problems or perform a single computationally intensive task or operation as a part of a hybrid platform? The answer "yes" to any of these challenges could imply high technological gains from predicting and developing new materials and designs to creating fully automated AI-controlled systems.

In the short term one may anticipate appearance of a plethora of disparate unconventional computing systems designed to execute specific algorithms or solving a particular practically relevant task. These will be more energy-efficient and will demonstrate a superior performance over classical hardware architectures. These systems could be further orchestrated together to perform larger tasks. The role of orchestra conductor could be devoted to traditional computing resources including CPUs, GPUs, FPGAs, and TPUs. Such symbiosis of classical and unconventional hardware together with the development of system-inspired optimisation algorithms will form an ultimate hybrid computing platform that may allow for continued scaling beyond the physical limits of Moore's law (Fig. 15.1). Many recent proposals exploit this idea, including, for instance, a photonic accelerator for neural networks with weights and inputs encoded in optical signals allowing for the neural network to be reprogrammed and trained on the fly at high speed [121] and the recurrent Ising machine in a photonic integrated circuit [122].

To establish the advantages of unconventional computing platforms we can apply the "quantum supremacy" test formulated by John Preskill in 2012 to characterise superior performance of a quantum simulator over any existing classical computing machines [123]. A particular milestone of quantum supremacy has been recently demonstrated with the Google's Sycamore processor [124] based on 53 programmable superconducting qubits for a certain application, namely random circuit sampling. Their claimed speedup of $200s$ against 10000 years was immediately scaled down to $200s$ compared to 2 days by using the absolute state-of-the-art classical supercomputer "Summit" at Oak Ridge National Lab [125]. Nevertheless, this is still a remarkable, at least three orders of magnitude acceleration provided by quantum simulations in comparison with classical conventional computing. This achievement is bound to ignite a quantum supremacy race, as already shown by the subsequent boson sampling experiment with tens of photons covering the Hilbert space similar to that of 48 qubits [126]. In upcoming years, we envision many more systems to actively participate in this race and demonstrate their superior suitability for solving particular classes of problems.

The unconventional computing systems we described in this chapter are not quantum computers in the traditional meaning of this term. Although quantum effects contribute to the system operation (e.g. Bose–Einstein condensation is a quantum process that obeys quantum statistics), it is not clear if these system offer any quantum speed-up during the search for a solution, e.g. via entanglement and superposition of states. Although, unlike quantum computers these systems have a crucial ingredient that drives their operation: nonlinearity! Nonlinearity leads to the emergence of coherence within each "bit" of the unconventional computers we discussed: a laser, a condensate, an optical parametric oscillator. Nonlinearity drives the gain saturation, mode locking, and coupling.

The physical systems that we described aim at finding the global minimum of hard optimisation problems. All these systems have advantages and limitations. They vary in nonlinearity of the underlying modes of operation, scalability, ability to engineer the required couplings, flexibility of turning the interactions, precision of read-out, factors facilitating the approach the global rather than local minimum. These issues have to be addressed from the experimental point of view in all newly proposed physical platforms.

All of the considered systems in this chapter have some parts of their operation that promise increased performance over the classical computations. Combined with system-inspired computational algorithms these systems may indeed one day revolutionise our computing.

References

1. G.E. Moore, IEEE solid state circuits society newsletter **11**, 33 (2006). [Reprinted from Electronics 38, 114 (1965)]
2. M.M. Waldrop, Nat. News **530**, 144 (2016)
3. G.E. Santoro, R. Martoňák, E. Tosatti, R. Car, Science **295**(5564), 2427 (2002)
4. R. Babbush, P.J. Love, A. Aspuru-Guzik, Sci. Rep. **4**, 6603 (2014)
5. A. Perdomo-Ortiz, N. Dickson, M. Drew-Brook, G. Rose, A. Aspuru-Guzik, Sci. Rep. **2**, 571 (2012)
6. E. Farhi, J. Goldstone, S. Gutmann, M. Sipser, Quantum Computation by Adiabatic Evolution, arXiv:quant-ph/0001106
7. J.J. Hopfield, D.W. Tank, Biol. Cybern. **52**, 141 (1985)
8. G. Wilson, G. Pawley, Biol. Cybern. **58**, 63 (1988)
9. S.V. Aiyer, M. Niranjan, F. Fallside, I.E.E.E. Trans, Neural Netw. **1**, 204 (1990)
10. A. Blum, R.L. Rivest, in *Advances in Neural Information Processing Systems*, ed. by D.S. Touretzky (Morgan Kaufmann, San Francisco, 1989), p. 494
11. A. Lucas, Front. Phys. **2**, 5 (2014)
12. B. Molnár, F. Molnár, M. Varga, Z. Toroczkai, M. Ercsey-Ravasz, Nat. Commun. **9**, 4864 (2018)
13. C.J.C. Burges, Microsoft Research Technical Report MSR-TR-2002-83 (2002)
14. M. Ekeberg, C. Lövkvist, Y. Lan, M. Weigt, E. Aurell, Phys. Rev. E **87**, 012707 (2013)
15. M. Sipser, in *Stoc'92 – Proceedings of the Twenty-Fourth Annual ACM Symposium on Theory of Computing*. Victoria, May 1992 (ACM, New York, 1992), p. 603
16. F. Barahona, J. Phys. A: Math. Gen. **15**, 3241 (1982)
17. G. de las Cuevas, T.S. Cubitt, Science **351**, 1180 (2016)
18. R.M. Karp, in *Complexity of Computer Computations*, ed. by R. Miller (Springer, New York, 1972), p. 85
19. U. Benlic, J.K. Hao, Eng. Appl. Artif. Intel. **26**, 1162 (2013)

20. K.P. Kalinin, N.G. Berloff, Sci. Repos. **8**, 17791 (2018)
21. T. Leleu, Y. Yamamoto, P.L. McMahon, K. Aihara, Phys. Rev. Lett. **122**, 040607 (2019)
22. R.W. Harrison, J. Opt. Soc. Amer. A **10**, 1046 (1993)
23. O. Bunk, A. Diaz, F. Pfeiffer, C. David, B. Schmitt, D.K. Satapathy, J.F. van der Veen, Acta Crystallogr. A **63**, 306 (2007)
24. J. Miao, T. Ishikawa, Q. Shen, T. Earnest, Annu. Rev. Phys. Chem. **59**, 387 (2008)
25. I. Waldspurger, A. d'Aspremont, S. Mallat, Math. Program. **149**, 47 (2015)
26. F. Hamze, J. Raymond, C.A. Pattison, K. Biswas, H.G. Katzgraber, The Wishart planted ensemble: a tunably-rugged pairwise Ising model with a first-order phase transition, arXiv:1906.00275
27. M.X. Goemans, D.P. Williamson, J. Comput. Syst. Sci. **68**, 442 (2004)
28. S. Zhang, Y. Huang, SIAM J. Optimiz. **16**, 871 (2006)
29. N. Krislock, J. Malick, F. Roupin, A.C.M.T. Math, Software **43**, 32 (2017)
30. D. Jaksch, P. Zoller, Ann. Phys. **315**, 52 (2005)
31. I. Bloch, J. Dalibard, W. Zwerger, Rev. Mod. Phys. **80**, 885 (2008)
32. R. Grimm, M. Weidemüller, Y.B. Ovchinnikov, Adv. Atom. Mol. Opt. Phys. **42**, 95 (2000)
33. P. Windpassinger, K. Sengstock, Rep. Prog. Phys. **76**, 086401 (2013)
34. O. Dutta, M. Gajda, P. Hauke, M. Lewenstein, D.S. Lühmann, B.A. Malomed, T. Sowiński, J. Zakrzewski, Rep. Prog. Phys. **78**, 066001 (2015)
35. E.P. Gross, Il Nuovo Cimento **20**, 454 (1961)
36. L. Pitaevskii, Sov. Phys. JETP **13**, 451 (1961)
37. L. Pitaevskii, S. Stringari, *Bose-Einstein Condensation and Superfluidity* (Oxford University Press, Oxford, 2016)
38. J.C. Eilbeck, P. Lomdahl, A.C. Scott, Phys. D **16**, 318 (1985)
39. G.L. Alfimov, P.G. Kevrekidis, V.V. Konotop, M. Salerno, Phys. Rev. E **66**, 046608 (2002)
40. A. Trombettoni, A. Smerzi, Phys. Rev. Lett. **86**, 2353 (2001)
41. R. Mishmash, L. Carr, Math. Comput. Simul. **80**, 732 (2009)
42. J. Struck, C. Ölschläger, R. Le Targat, P. Soltan-Panahi, A. Eckardt, M. Lewenstein, P. Windpassinger, K. Sengstock, Science **333**, 996 (2011)
43. J. Struck, M. Weinberg, C. Ölschläger, P. Windpassinger, J. Simonet, K. Sengstock, R. Höppner, P. Hauke, A. Eckardt, M. Lewenstein et al., Nat. Phys. **9**, 738 (2013)
44. M.W. Johnson, M.H. Amin, S. Gildert, T. Lanting, F. Hamze, N. Dickson, R. Harris, A.J. Berkley, J. Johansson, P. Bunyk et al., Nature **473**, 194 (2011)
45. T. Kadowaki, H. Nishimori, Phys. Rev. E **58**, 5355 (1998)
46. D.S. Steiger, B. Heim, T.F. Rønnow, M. Troyer, Proc. SPIE **9468**, 964816 (2015)
47. V.S. Denchev, S. Boixo, S.V. Isakov, N. Ding, R. Babbush, V. Smelyanskiy, J. Martinis, H. Neven, Phys. Rev. X **6**, 031015 (2016)
48. I. Zintchenko, M.B. Hastings, M. Troyer, Phys. Rev. B **91**, 024201 (2015)
49. H. Cao, R. Chriki, S. Bittner, A.A. Friesem, N. Davidson, Nat. Rev. Phys. **1**, 156 (2019)
50. D. Brunner, M.C. Soriano, C.R. Mirasso, I. Fischer, Nat. Commun. **4**, 1364 (2013)
51. L. Bao, N.H. Kim, L.J. Mawst, N.N. Elkin, V.N. Troshchieva, D.V. Vysotsky, A.P. Napartovich, Appl. Phys. Lett. **84**, 320 (2004)
52. V. Eckhouse, M. Fridman, N. Davidson, A.A. Friesem, Phys. Rev. Lett. **100**, 024102 (2008)
53. M. Nixon, M. Friedman, E. Ronen, A.A. Friesem, N. Davidson, I. Kanter, Phys. Rev. Lett. **106**, 223901 (2011)
54. F. Rogister, K.S. Thornburg Jr., L. Fabiny, M. Möller, R. Roy, Phys. Rev. Lett. **92**, 093905 (2004)
55. V. Pal, C. Tradonsky, R. Chriki, A.A. Friesem, N. Davidson, Phys. Rev. Lett. **119**, 013902 (2017)
56. Y. Kuramoto, *Chemical Oscillations, Waves, and Turbulence* (Springer, Berlin, 1984)
57. Y. Kuramoto, Lect. Notes Phys. **30**, 420 (1975)
58. H.K. Khalil, *Nonlinear Systems* (Prentice-Hall, Upper Saddle River, 2002)
59. C. Tradonsky, O. Raz, V. Pal, R. Chriki, A.A. Friesem, N. Davidson, Sci. Adv. **5**, eaax4530 (2019)

60. P.L. McMahon, A. Marandi, Y. Haribara, R. Hamerly, C. Langrock, S. Tamate, T. Inagaki, H. Takesue, S. Utsunomiya, K. Aihara et al., Science **354**, 614 (2016)
61. R. Hamerly, T. Inagaki, P.L. McMahon, D. Venturelli, A. Marandi, T. Onodera, E. Ng, C. Langrock, K. Inaba, T. Honjo, et al., Sci. Adv. **5**, eaau0823 (2019)
62. F. Böhm, T. Inagaki, K. Inaba, T. Honjo, K. Enbutsu, T. Umeki, R. Kasahara, H. Takesue, Nat. Commun. **9**, 5020 (2018)
63. Y. Haribara, H. Ishikawa, S. Utsunomiya, K. Aihara, Y. Yamamoto, Quantum Sci. Technol. **2**, 044002 (2017)
64. A.D. King, W. Bernoudy, J. King, A.J. Berkley, T. Lanting, Emulating the coherent Ising machine with a mean-field algorithm, arXiv:1806.08422
65. K. Takata, A. Marandi, Y. Yamamoto, Phys. Rev. A **92**, 043821 (2015)
66. F. Böhm, G. Verschaffelt, G. van der Sande, Nat. Commun. **10**, 1 (2019)
67. M. Babaeian, D.T. Nguyen, V. Demir, M. Akbulut, P.A. Blanche, Y. Kaneda, S. Guha, M.A. Neifeld, N. Peyghambarian, Nat. Commun. **10**, 1 (2019)
68. K.P. Kalinin, N.G. Berloff, New J. Phys. **20**, 113023 (2018)
69. D. Pierangeli, G. Marcucci, C. Conti, Phys. Rev. Lett. **122**, 213902 (2019)
70. J. Kasprzak, M. Richard, S. Kundermann, A. Baas, P. Jeambrun, J. Keeling, F. Marchetti, M. Szymańska, R. André, J. Staehli et al., Nature **443**, 409 (2006)
71. N. Berloff, J. Keeling, in *Physics of Quantum Fluids*, ed. by A. Bramati, M. Modugno (Springer, Berlin, 2013), p. 19
72. J. Klaers, J. Schmitt, F. Vewinger, M. Weitz, Nature **468**, 545 (2010)
73. J. Klaers, F. Vewinger, M. Weitz, Nat. Phys. **6**, 512 (2010)
74. J. Klaers, J. Schmitt, T. Damm, F. Vewinger, M. Weitz, Appl. Phys. B **105**, 17 (2011)
75. J. Schmitt, T. Damm, F. Vewinger, M. Weitz, J. Klaers, New J. Phys. **14**, 075019 (2012)
76. H. Walther, B.T. Varcoe, B.G. Englert, T. Becker, Rep. Prog. Phys. **69**, 1325 (2006)
77. E. Wertz, L. Ferrier, D. Solnyshkov, R. Johne, D. Sanvitto, A. Lemaître, I. Sagnes, R. Grousson, A.V. Kavokin, P. Senellart et al., Nat. Phys. **6**, 860 (2010)
78. F. Manni, K.G. Lagoudakis, T.C.H. Liew, R. André, B. Deveaud-Plédran, Phys. Rev. Lett. **107**, 106401 (2011)
79. G. Tosi, G. Christmann, N. Berloff, P. Tsotsis, T. Gao, Z. Hatzopoulos, P. Savvidis, J. Baumberg, Nat. Phys. **8**, 190 (2012)
80. G. Tosi, G. Christmann, N. Berloff, P. Tsotsis, T. Gao, Z. Hatzopoulos, P. Savvidis, J. Baumberg, Nat. Commun. **3**, 1243 (2012)
81. N.G. Berloff, M. Silva, K. Kalinin, A. Askitopoulos, J.D. Töpfer, P. Cilibrizzi, W. Langbein, P.G. Lagoudakis, Nat. Mat. **16**, 1120 (2017)
82. C. Schneider, K. Winkler, M. Fraser, M. Kamp, Y. Yamamoto, E. Ostrovskaya, S. Höfling, Rep. Prog. Phys. **80**, 016503 (2016)
83. A. Amo, J. Bloch, C.R. Phys. **17**, 934 (2016)
84. D. Dung, C. Kurtscheid, T. Damm, J. Schmitt, F. Vewinger, M. Weitz, J. Klaers, Nat. Photonics **11**, 565 (2017)
85. K.P. Kalinin, N.G. Berloff, Phys. Rev. B **100**, 245306 (2019)
86. K.P. Kalinin, N.G. Berloff, Phys. Rev. Lett. **121**, 235302 (2018)
87. N. Stroev, N.G. Berloff, Discrete polynomial optimization with coherent networks of condensates and complex coupling switching, arXiv:1910.00842
88. R. Lang, K. Kobayashi, IEEE J. Quantum Elect. **16**, 347 (1980)
89. J.A. Acebrón, L.L. Bonilla, C.J.P. Vicente, F. Ritort, R. Spigler, Rev. Mod. Phys. **77**, 137 (2005)
90. F. Rosenblatt, *Principles of Neurodynamics: Perceptrons and the Theory of Brain Mechanisms* (Spartan Book, Washington D.C., 1962)
91. D.E. Rumelhart, G.E. Hinton, R.J. Williams, Nature **323**, 533 (1986)
92. O. Russakovsky, J. Deng, H. Su, J. Krause, S. Satheesh, S. Ma, Z. Huang, A. Karpathy, A. Khosla, M. Bernstein et al., Int. J. Comput. Vision **115**, 211 (2015)
93. H.T. Siegelmann, E.D. Sontag, Appl. Math. Lett. **4**, 77 (1991)
94. S. Hochreiter, J. Schmidhuber, Neural Comput. **9**, 1735 (1997)

95. M. Schuster, K.K. Paliwal, IEEE T. Signal Proces. **45**, 2673 (1997)
96. H. Zhang, Z. Wang, D. Liu, I.E.E.E.T. Neural Netw. Lear. **25**, 1229 (2014)
97. Z.C. Lipton, J. Berkowitz, C. Elkan, A critical review of recurrent neural networks for sequence learning, arXiv:1506.00019
98. P.J. Werbos, Proc. IEEE **78**, 1550 (1990)
99. A. Graves, M. Liwicki, S. Fernández, R. Bertolami, H. Bunke, J. Schmidhuber, IEEE T. Pattern Anal. **31**, 855 (2008)
100. A. Graves, Generating sequences with recurrent neural networks, arXiv:1308.0850
101. I. Sutskever, J. Martens, G.E. Hinton, in *ICML'11 – Proceedings of the 28th International Conference on Machine Learning*, ed. by L. Getoor, T. Scheffer. Bellevue, Washington, June 28—July 2, 2011 (Omnipress, Madison, 2011), p. 1017–1024
102. N. Kalchbrenner, P. Blunsom, in *EMNLP'13 – Proceedings of the 2013 Conference on Empirical Methods in Natural Language Processing*, ed. by D. Yarowsky, T. Baldwin, A. Korhonen, K. Livescu, S. Bethard. Seattle, October 2013 (Association for Computational Linguistics, Stroudsburg, 2013), p. 1700
103. I. Sutskever, O. Vinyals, Q.V. Le, in *Advances in Neural Information Processing Systems 27*, ed. by Z. Ghahramani, M. Welling, C. Cortes, N.D. Lawrence, K.Q. Weinberger. Conference on Neural Information Processing Systems 2014, Montreal, December 2014 (Curran Associates, New York, 2014), p. 3104
104. T.W. Hughes, I.A. Williamson, M. Minkov, S. Fan, Sci. Adv. **5**, eaay6946 (2019)
105. H. Jaeger, GMD Report 148, German National Research Center for Information Technology (2001)
106. W. Maass, T. Natschläger, H. Markram, Neural Comput. **14**, 2531 (2002)
107. M.C. Soriano, S. Ortín, L. Keuninckx, L. Appeltant, J. Danckaert, L. Pesquera, G. Van der Sande, I.E.E.E.T. Neural Netw. Lear. **26**, 388 (2014)
108. P. Antonik, *Application of FPGA to Real-Time Machine Learning: Hardware Reservoir Computers and Software Image Processing* (Springer, Cham, 2018)
109. A. Polepalli, N. Soures, D. Kudithipudi, in *2016 IEEE International Conference on Rebooting Computing (ICRC)*. San Diego, October 2016 (IEEE, 2016) https://doi.org/10.1109/ICRC.2016.7738687
110. A. Katumba, J. Heyvaert, B. Schneider, S. Uvin, J. Dambre, P. Bienstman, Sci. Rep. **8**, 2653 (2018)
111. L. Larger, A. Baylón-Fuentes, R. Martinenghi, V.S. Udaltsov, Y.K. Chembo, M. Jacquot, Phys. Rev. X **7**, 011015 (2017)
112. J. Torrejon, M. Riou, F.A. Araujo, S. Tsunegi, G. Khalsa, D. Querlioz, P. Bortolotti, V. Cros, K. Yakushiji, A. Fukushima et al., Nature **547**, 428 (2017)
113. R. Nakane, G. Tanaka, A. Hirose, IEEE Access **6**, 4462 (2018)
114. D. Prychynenko, M. Sitte, K. Litzius, B. Krüger, G. Bourianoff, M. Kläui, J. Sinova, K. Everschor-Sitte, Phys. Rev. Appl. **9**, 014034 (2018)
115. G. Urbain, J. Degrave, B. Carette, J. Dambre, F. Wyffels, Front. Neurorobotics **11**, 16 (2017)
116. P. Vincent-Lamarre, G. Lajoie, J.P. Thivierge, J. Comput. Neurosci. **41**, 305 (2016)
117. G. Tanaka, T. Yamane, J.B. Héroux, R. Nakane, N. Kanazawa, S. Takeda, H. Numata, D. Nakano, A. Hirose, Neural Netw. **115**, 100 (2019)
118. A. Opala, S. Ghosh, T.C. Liew, M. Matuszewski, Phys. Rev. Appl. **11**, 064029 (2019)
119. T. Leleu, Y. Yamamoto, S. Utsunomiya, K. Aihara, Phys. Rev. E **95**, 022118 (2017)
120. H. Goto, K. Tatsumura, A.R. Dixon, Sci. Adv. **5**, eaav2372 (2019)
121. R. Hamerly, L. Bernstein, A. Sludds, M. Soljačić, D. Englund, Phys. Rev. X **9**, 021032 (2019)
122. M. Prabhu, C. Roques-Carmes, Y. Shen, N. Harris, L. Jing, J. Carolan, R. Hamerly, T. Baehr-Jones, M. Hochberg, V. Čeperić, et al., A recurrent Ising machine in a photonic integrated circuit, arXiv:1909.13877
123. J. Preskill, Quantum computing and the entanglement frontier. Rapporteur talk at the 25th Solvay Conference on Physics, Brussels, October 2011. arXiv:1203.5813
124. F. Arute, K. Arya, R. Babbush, D. Bacon, J.C. Bardin, R. Barends, R. Biswas, S. Boixo, F.G. Brandao, D.A. Buell et al., Nature **574**, 505 (2019)

125. E. Pednault, J.A. Gunnels, G. Nannicini, L. Horesh, R. Wisnieff, Leveraging secondary storage to simulate deep 54-qubit sycamore circuits, arXiv:1910.09534
126. H. Wang, J. Qin, X. Ding, M.C. Chen, S. Chen, X. You, Y.M. He, X. Jiang, Z. Wang, L. You, Z. Wang, C. Schneider, J.J. Renema, S. Höfling, C.Y. Lu, J.W. Pan, Phys. Rev. Lett. **123**, 250503 (2019)

Index

P. G. Kevrekidis et al. (eds.), *Emerging Frontiers in Nonlinear Science*, Nonlinear Systems and Complexity 32, https://doi.org/10.1007/978-3-030-44992-6

The manufacturer's authorised representative in the EU is Springer Nature Customer Service Centre GmbH, Europaplatz 3, 69115 Heidelberg, Germany. If you have any concerns regarding our products, please contact ProductSafety@springernature.com

Printed and bound by CPI Group (UK) Ltd, Croydon, CR0 4YY
30/09/2026
02225646-0015